Optimum Decision Making in Asset Management

María Carmen Carnero
University of Castilla - La Mancha, Spain

Vicente González-Prida
University of Seville, Spain

A volume in the Advances in Logistics, Operations, and Management Science (ALOMS) Book Series

IGI GLOBAL
DISSEMINATOR OF KNOWLEDGE

www.igi-global.com

Published in the United States of America by
IGI Global
Business Science Reference (an imprint of IGI Global)
701 E. Chocolate Avenue
Hershey PA, USA 17033
Tel: 717-533-8845
Fax: 717-533-8661
E-mail: cust@igi-global.com
Web site: http://www.igi-global.com

Library of Congress Cataloging-in-Publication Data

Names: Carnero, Maria Carmen, 1970- editor. | Gonzalez-Prida, Vicente, 1975- editor.
Title: Optimum decision making in asset management / Maria Carmen Carnero and Vicente Gonzalez-Prida, editors.
Description: Hershey, PA : Business Science Reference, [2017] | Series: Advances in logistics, operations, and management science | Includes bibliographical references and index.
Identifiers: LCCN 2016023216| ISBN 9781522506515 (hardcover) | ISBN 9781522506522 (ebook)
Subjects: LCSH: Maintenance--Decision making. | Industrial equipment--Maintenance and repair--Management. | Public works--Management. | Capital--Management. | Operations research.
Classification: LCC TS192 .O65 2017 | DDC 620/.0046--dc23 LC record available at https://lccn.loc.gov/2016023216

This book is published in the IGI Global book series Advances in Logistics, Operations, and Management Science (ALOMS) (ISSN: 2327-350X; eISSN: 2327-3518)

British Cataloguing in Publication Data
A Cataloguing in Publication record for this book is available from the British Library.

Advances in Logistics, Operations, and Management Science (ALOMS) Book Series

John Wang
Montclair State University, USA

ISSN: 2327-350X
EISSN: 2327-3518

Mission

Operations research and management science continue to influence business processes, administration, and management information systems, particularly in covering the application methods for decision-making processes. New case studies and applications on management science, operations management, social sciences, and other behavioral sciences have been incorporated into business and organizations real-world objectives.

The **Advances in Logistics, Operations, and Management Science** (ALOMS) Book Series provides a collection of reference publications on the current trends, applications, theories, and practices in the management science field. Providing relevant and current research, this series and its individual publications would be useful for academics, researchers, scholars, and practitioners interested in improving decision making models and business functions.

Coverage

- Marketing engineering
- Political Science
- Information Management
- Production management
- Decision analysis and decision support
- Networks
- Organizational Behavior
- Services management
- Operations Management
- Risk Management

IGI Global is currently accepting manuscripts for publication within this series. To submit a proposal for a volume in this series, please contact our Acquisition Editors at Acquisitions@igi-global.com or visit: http://www.igi-global.com/publish/.

The Advances in Logistics, Operations, and Management Science (ALOMS) Book Series (ISSN 2327-350X) is published by IGI Global, 701 E. Chocolate Avenue, Hershey, PA 17033-1240, USA, www.igi-global.com. This series is composed of titles available for purchase individually; each title is edited to be contextually exclusive from any other title within the series. For pricing and ordering information please visit http://www.igi-global.com/book-series/advances-logistics-operations-management-science/37170. Postmaster: Send all address changes to above address.

Titles in this Series

For a list of additional titles in this series, please visit: www.igi-global.com

Green Supply Chain Management for Sustainable Business Practice
Mehmood Khan (Abu Dhabi University, UAE) Matloub Hussain (Abu Dhabi University, UAE) and Mian M. Ajmal (Abu Dhabi University, UAE)
Business Science Reference • copyright 2017 • 330pp • H/C (ISBN: 9781522506355) • US $185.00 (our price)

Multinational Enterprise Management Strategies in Developing Countries
Mohammad Ayub Khan (Tecnológico de Monterrey, Mexico)
Business Science Reference • copyright 2016 • 490pp • H/C (ISBN: 9781522502760) • US $200.00 (our price)

Strategic Management of Sustainable Manufacturing Operations
Rameshwar Dubey (Symbiosis International University, India) and Angappa Gunasekaran (University of Massachuesetts, Dartmouth, USA)
Business Science Reference • copyright 2016 • 419pp • H/C (ISBN: 9781522503507) • US $215.00 (our price)

Handbook of Research on Global Fashion Management and Merchandising
Alessandra Vecchi (London College of Fashion, U.K.) and Chitra Buckley (London College of Fashion, U.K.)
Business Science Reference • copyright 2016 • 862pp • H/C (ISBN: 9781522501107) • US $335.00 (our price)

Handbook of Research on Managerial Strategies for Achieving Optimal Performance in Industrial Processes
Giner Alor-Hernández (Instituto Tecnológico de Orizaba, Mexico) Cuauhtémoc Sánchez-Ramírez (Instituto Tecnológico de Orizaba, Mexico) and Jorge Luis García-Alcaraz (Universidad Autónoma de Ciudad Juárez, Mexico)
Business Science Reference • copyright 2016 • 674pp • H/C (ISBN: 9781522501305) • US $365.00 (our price)

Supply Chain Strategies and the Engineer-to-Order Approach
Richard Addo-Tenkorang (University of Vaasa, Finland) Jussi Kantola (University of Vaasa, Finland) Petri Helo (University of Vaasa, Finland) and Ahm Shamsuzzoha (Sultan Qaboos University, Oman)
Business Science Reference • copyright 2016 • 283pp • H/C (ISBN: 9781522500216) • US $200.00 (our price)

Optimal Inventory Control and Management Techniques
Mandeep Mittal (Amity School of Engineering and Technology, India) and Nita H. Shah (Gujarat University, India)
Business Science Reference • copyright 2016 • 406pp • H/C (ISBN: 9781466698888) • US $170.00 (our price)

Stochastic Processes and Models in Operations Research
Neelamegam Anbazhagan (Alagappa University, India)
Business Science Reference • copyright 2016 • 338pp • H/C (ISBN: 9781522500445) • US $185.00 (our price)

To my parents Manuel and Carmen and my husband José Luis.
María Carmen Carnero

To my wife Margarita and my daughters Pilar and Paloma.
Vicente González-Prida

Yuliia Parfeneko, *Sumy State University, Ukraine*
Carlos Parra, *University of Seville, Spain*
Manuel Rodríguez Mendez, *ESeyPro S.L., Spain*
Richard Ruitenburg, *University of Twente, The Netherlands*
Jose Maria Sarriegi, *Tecnun, University of Navarra, Spain*
Vira Shendryk, *Sumy State University, Ukraine*
Wieger Tiddens, *University of Twente, The Netherlands*
Tiedo Tinga, *University of Twente, The Netherlands*
Jasper Veldman, *University of Groningen, The Netherlands*
Pablo Viveros Gunckel, *University Federico Santa Maria, Chile*
Patzi Xabier Zubizarreta, *IKERLAN, Spain*

Table of Contents

Detailed Table of Contents

considering multiple decision centres: The Hospital's technical services, environment and occupational risk prevention managers, healthcare managers (operating theatres and health activity programming), healthcare staff, technicians, purchasing service managers and Hospital executives. The model uses functional, safety and technical-economical criteria, amongst which is availability. Mean availability for repairable systems has been measured to assess this criterion, using Markov chains from the data obtained over three years from the subsystems of a Hospital operating theatre. All this is aimed at increasing the operating theatre's availability and, consequently, increasing physical safety in patient operations and reducing the number of delayed operations due to technical malfunctions.

The increase in the frequency of disastrous events and society's dependence on Critical Infrastructures (CIs) has led to greater concern about the need to increase resilience in order to improve Critical Infrastructure Protection. CIs are basic service providers for society and they need to be effectively protected against hazards. Nowadays, CIs can be owned by private entities. However, although they can be privately owned or managed, they provide a public service that directly affects the whole society. Consequently, those activities that increase the overall resilience level of CIs need to be under the supervision of public entities. Increasing resilience requires special attention be paid to correct infrastructure and crisis response equipment maintenance. This chapter explains why effective Public-Private Partnerships (PPP) are valuable for correctly maintaining CIs and illustrates examples of real situations that demonstrate the need for effective PPPs in maintenance activities.

GAMM (Graphical Analysis for Maintenance Management) is a method that supports decision-making in maintenance management through the visualization and graphical analysis of data. GAMM also allows the identification of anomalous behavior in equipment, derived from its own operations, maintenance activities, improper use of equipment or even as a result of design errors. As a basis for analysis, the GAMM method uses a nonparametric estimator of the reliability function using historical data, sometimes in very limited amounts. However, for successful results, experience and advanced knowledge in maintenance management are strictly necessary. In order to ease the interpretations of the GAMM method results, with the intention that the method becomes really amicable for managers, a set of basic rules have been developed. This set of rules leads to a proper and objective interpretation of GAMM results, improving the decision making.

Knowledge management has been analyzed in numerous areas of the industrial enterprise, especially in the areas of strategic management, innovation, trade, or administration. However in operational areas with technicians working mainly on the basis of its experience gained over the years, such as the departments of industrial maintenance, there are no deep analysis of the incidence of the knowledge management in these areas. The peculiarities in this type of activity on the inside of the company, knowledge of these people is strongly based on your experience (strong tacit component), difficult to measure and articulate, and however, on numerous occasions, this knowledge not transmitted, can be a high cost for the company (many times assumed as inevitable) due to the increase of production and services downtime, loss of efficiency, or time of coupling of new personnel to these areas.

Buildings lifecycle management is an area of great interest. Due this, the R&D is being promoting worldwide looking for new effective maintenance tools and methodologies. In this scenario there are two development lines whose convergence can bring great advances in this area: Asset Management (AM) and Building Information Modeling (BIM). BIM models are transforming the way buildings are conceived, designed, constructed and managed. But current use of BIM concentrates on preplanning, design, construction and integrated project delivery of buildings and facilities, rather than maintenance and building operation management. Asset Management tools, including Facilities Management (FM), and application frameworks provide the approach and required elements to get more efficiency and efficacy in the building lifecycle management. This chapter introduces the application of AM for building and how the development of BIM models is the key element to allow its effective implementation.

In recent years, the concept of Industry 4.0 has been significantly advanced in industrial circles as an aspect that provides a competitive differential. Through the technologies involved, machines can now monitor and relay information on their operating conditions for analysis and decision-making, as well as for prompting action. These new functions generally involve the development of technological projects and significant investments. This renders it expedient to explain why certain systems should be monitored, but not others, as well as the use to be given to the data gathered as a way of generating income for a firm. This approach is especially important in certain corporate operations, such as after-sales maintenance. This article introduces a reference framework that permits the effective and efficient management of after-sales maintenance services. This framework relates after-sales service technologies with product technologies (Industry 4.0), and therefore covers the reasons and purposes of Industry 4.0 within the ambit of after-sales service.

Many investigations led to show that no crack begins to propagate to an increase of stress intensity factor. The life of the components of a structure containing premature cracks, can be governed by the degree of subcritical crack propagation. Thus, knowledge of crack propagation to determine the fatigue of the structure is necessary. One problem of steels of high resilience is their low toughness in the HAZ, when they are welded with a high heat input. In this work we have studied nine specimens that have been welded under a submerged arc welding process controlling the welding parameters and checking in the HAZ of such specimens, critical tensions at the ends of the cracks, the critical cracks lengths and stress intensity factors. It is intended to check that the parameters that indicate the values of fracture mechanics in the HAZ, after heat cycle to which the steel has undergone, under a process with a maximum heat input of 2.327kJ /mm, they are still valid, with the welding parameters applied. It is checked a correlation between the theoretical and experimental values.

Advanced maintenance techniques (AMTs) are practices that can help practitioners to make better-informed maintenance decisions, such as ensuring just-in-time maintenance, corporate business planning or lifetime extension of physical assets. These techniques take the current, but preferably also the future, state of assets into account. Although there is much literature on the development of specific techniques, reports on their adoption and use show that only few companies have successfully applied these methods. Guidance is needed on their selection and application. In this chapter, a typology of companies–based on an ongoing multiple-case study–that apply these maintenance techniques is proposed. This typology identifies typical difficulties practitioners experience in applying AMTs. Finally, a four-step procedure is offered with the aim of helping practitioners to overcome the discussed difficulties in the application of AMTs.

The chapter focuses on implementation of information technologies in energy management of district heating system. It captures the current state of energy management in district heating. The main standards of energy management are described. The specific features of district heating system and the existing tools of energy management in district heating are explored. The chapter studies the existing models, which can be applied in the decision support system for heat energy management. It describes the decision support system "HeatCAM" as a tool of energy management in district heating.

Megaprojects are large and complex projects that entail multi-actor management, non-standard technology and processes. This chapter aims to explore offshore wind projects (OWPs) as megaprojects, particular in the planning phase. Based on interviews with 26 experts from a variety of backgrounds in the offshore wind industry in The Netherlands, the risks and uncertainties in the planning phase of OWPs and key factors in the decision making process are explored. A framework is presented that depicts the planning phase of an OWP, as well as ten risks and seven uncertainties that are most common in an OWP. The role of the government and the project structure are further highlighted. The findings of this research allow practitioners to gain a better overview of the planning process of an OWP and can help to improve asset management decision making.

Getting wood for cooking, heating, processing into charcoal and/or selling is a fundamental part of most household survival strategies in Developing Countries. Entangling in complex and dynamic ways local and global ecosystems, poverty, technology and business Wood Fuel Energy Systems (WES) are fundamental and require simple to use design tools to support the strategic and optimised used of available socio-ecological resources/assets. However, there are very few tools able to support relevant actors (e.g. charcoal makers, experts, policy makers) in that task. To bridge that gap the 2MBio, a participatory conceptual design tool to support the strategic design of WES, is introduced and its practical results in Mozambique presented. The 2MBio explicit in a simple and intuitive layout the set of necessary and sufficient resources/assets required to produce comprehensive and meaningful WES designs/strategies, which represent in themselves a strategic asset, while further stimulates knowledge and creativity as a tacit asset.

Within aerospace and defence sectors, maintaining asset availability during operational service has become more important than quality of service throughout the system life cycle. This requires organisations to establish cost effective strategies to manage uncertainties within their value led services e.g. maintenance activities. In large organizations, it is not always apparent whose decision affects the outcome the most. Often, accountability moves away from the designated organization personnel in unforeseen ways, and depending on the decisions of individual decision makers, the structure of the organization, or unregulated operating procedures may change. This can have far more effect on the overall reliability leading to inadequate troubleshooting, repeated down-time and reduced availability. This chapter focuses on discussing the No Fault Found problems in aviation and highlights the drivers that influence its decision making process. It further articulates the contents of tacit knowledge and discusses the knowledge gaps with current management policies.

Foreword

I am pleased and honoured at the opportunity to write a short Foreword to Optimum Decision Making in Asset Management. The book's editors, Vicente González-Prida Díaz and María Carmen Carnero Moya, bring a wealth of experience to bear on their topic, they are good friends and well recognized authors of research papers and manuscripts within the field of assets management, and more precisely, assets' warranty and predictive maintenance technologies management.

Vicente González-Prida Díaz is a member of the SIM Research Group (*Sistemas Inteligentes de Mantenimiento*) of the University of Seville. He has managed warranty and technical services departments in various multinational companies in the military and in the utilities sectors. María Carmen Carnero Moya is Associate Professor (with tenure) at the Higher Technical School of Industrial Engineering of the University of Castilla-La Mancha in Ciudad Real, in the Department of Business Administration. He has published extensively in the area of predictive maintenance management and has been involved in many maintenance projects, especially within the Health sector.

Many of the contributors of this book, included the two Editors, are members of the Network of Excellence on Assets Management (DPI2014-56547-REDT), Promoted in Spain by the *Plan Estatal de Investigación Científica y Técnica y de Innovación 2013-2016*, released by the Economy and Competitiveness Ministry of the Government of Spain. Because of my responsibility as the coordinator of this network, I am glad to see this work as another interesting outcome of our academic and research community (Section 1 of the book contains papers in which at least one of the authors is a member of our network), dealing with the topic of assets management.

The field of Physical Assets Management is an emerging field of action in industry and in other sectors, which has given renewed attention to the maintenance of industrial equipment, buildings and infrastructure in general. The main concern of assets management is to ensure the integrity and sustainability of the assets, paying at the same time special attention to their eco-efficiency over their life cycle.

This book is a good example of how the industrial maintenance concept has evolved over time with the advent of new technical and organizational capabilities to the industry. Concepts like risk and uncertainty management, or life cycle cost analysis and management are becoming a must for modern assets and maintenance managers. Good examples of this reality are also presented in Section 2 of the book, from very diverse and enriching perspectives.

It is also importance to notice the emergence of standards such as the family ISO 55000, introducing new challenges especially for industrial organizations. The open nature of the standard, besides the number of different areas it deals with, makes essential the search for collaboration between companies and research institution interested in this field.

I think this book will contribute to connect different research centres together with practitioners and other interested parties, in order to facilitate the identification and to face the challenges of emerging needs of modern industry, within the context of asset management.

Adolfo Crespo Márquez
University of Seville, Spain

Adolfo Crespo Márquez *is currently Full Professor at the School of Engineering of the University of Seville, and Head of the Department of Industrial Management. He holds a PhD in Industrial Engineering from this same University. His research works have been published in journals such as the International Journal of Production Research, International Journal of Production Economics, European Journal of Operations Research, Journal of Purchasing and Supply Management, International Journal of Agile Manufacturing, Omega, Journal of Quality in Maintenance Engineering, Decision Support Systems, Computers in Industry, Reliability Engineering and System Safety and International Journal of Simulation and Process Modeling, among others. Prof. Crespo is the author of 7 books, the last four with Springer-Verlag in 2007, 2010, 2012 and 2014 about maintenance, warranty and supply chain management. Prof. Crespo leads the Spanish Research Network on Dependability Management and the Spanish Committee for Maintenance Standardization (1995-2003). He also leads a research team related to maintenance and dependability management currently with 5 PhD students and 4 researchers. He has extensively participated in many engineering and consulting projects for different companies, for the Spanish Departments of Defense, Science and Education as well as for the European Commission (IPTS). He is the President of INGEMAN (a National Association for the Development of Maintenance Engineering in Spain) since 2002.*

Preface

OVERVIEW

Asset management, once considered a tactical area, is now a matter of strategy, given the implications it has for availability of facilities and equipment, delivery time, product quality, costs, safety and the environment. In addition, the introduction of advanced manufacturing techniques and new production management systems, which lead to increased automation and reduced delivery times, has given great importance to asset management.

In manufacturing, production, finance, etc., decisions are increasingly taken based on models or techniques which provide satisfactory, objective decision making, which guarantees improved competitiveness, reducing risk and uncertainty, and that can be justified to management. However, maintenance managers have taken decisions based only on their experience or supported by the advice of system sales staff or consultants. This lack of models and techniques in the area of asset management leads to underperforming maintenance departments characterized by a reactive approach, underutilized maintenance information systems, inaccurately managed costs, no scheduled maintenance hours, feedback on work quality not being provided, etc.

This book looks to promote and address the application of objective and effective decision making in asset management based on mathematical models and practical techniques that can be easily implemented in organizations. This comprehensive and timely publication aims to be an essential reference source, building on the available literature in the field of asset management while providing for further research breakthroughs in this field. This text provides the necessary resources for managers, technology developers, scientists and engineers to adopt and implement optimum decision making based on models and techniques that contribute to recognizing risks and uncertainties and, in general terms, to the important role of asset management to increase competitiveness in organizations.

SUMMARY OF TOPICS

The relevance of maintenance in organizations has increased considerably over the last two decades; this importance is linked to the introduction of a growing number of factors with an influence on efficient asset management. The existence of increasingly complex equipment and processes, the increase in the number of assets, the speed of technological change, the need to reduce costs in the modern world, together with increases in the level of excellence of commercial goals such as quality and delivery time, and concern for the safety of workers and the environment, make asset management an important source of benefits and competitive advantages for present and future world class enterprises. This book analyses these factors, which are divided into, although not limited to, the following categories:

- Maintenance policy selection.
- After-sales management.
- Knowledge management.
- Critical asset and infrastructure management.
- Asset life-cycle management.
- After-sales maintenance.
- Performance measurement system.
- Sensors and health monitoring systems.
- Reliability centred maintenance.
- Building information modelling.
- Advanced maintenance techniques.
- The set-up process.

Nevertheless, the areas of application of these categories should undoubtedly be underlined as having attracted few contributions, as having special characteristics, and as being of great present and future importance; these areas include:

- Operating theatre.
- District heating.
- Offshore wind farms.
- Panama Canal.
- Buildings.
- Wood-fuel energy systems.
- Aerospace industry.
- Travelling crane.

TARGET AUDIENCE

Industrial and manufacturing engineers, managers and plant supervisor, academicians, researchers, advanced-level students (both postgraduate and doctoral), technology developers, and managers who take decisions in this field will find in this book a source of ideas, models and techniques which mark out a path for future research in this field, and may also serve to encourage original ideas and in many cases practical application in business. *Optimum Decision Making in Asset Management* is aimed at the above-mentioned target audience worldwide and because of the number of chapters it contains and the variety of the subjects analysed, it provides an in-depth look at current global concerns.

IMPORTANCE OF EACH CHAPTER

The book is structured in two parts. The first part consists of 11 chapters which include contributions by researchers from the Spanish Network of Excellence in the management of Physical Assets, which brings together most of the best Spanish researchers in the field of asset management. The second part consists of 9 chapters with contributions by authors from different countries including: The UK, France, the Netherlands, Portugal, the Ukraine, Saudi Arabia, Chile, Venezuela, Panama, the United States and India. This gives a more complete view of the state of asset management around the world.

Of the 57 members of the editorial board and authors who have participated in this book, at least 20 have carried out activities related to asset management in different commercial organizations, bringing practical vision, but also motivating improvements, advances and new techniques to be applied to companies. This favours orientation of research towards real application in organizations, which is part of the value added by this book. There is also a significant flow of ideas and experience through collaboration as reviewers with a number of authors who have contributed to the book. The suggestions, comments and ideas they provide their colleagues have been enriching for all of us, seeding meaningful developments in knowledge. For all these reasons, we wish to thank each of them.

A brief description of the twenty chapters, with regard to their research material and the conclusions reached, are collected and summarized as follows:

Chapter 1 describes a maintenance and reliability management model for the project: Design and Construction of the Third Set of Locks in the Panama Canal. This model consists of eight sequential management building blocks. The first three building blocks are concerned with condition maintenance effectiveness, the fourth and fifth ensure maintenance efficiency; blocks six and seven deal with maintenance and asset life-cycle cost assessment, and finally block number eight ensures continuous maintenance management improvement. This chapter could be used to assist different plant teams in designing the optimal strategies for maintenance and inspection of the assets; additionally, recommendations for optimizing the processes in the areas of maintenance and reliability management are also included.

Chapter 2 presents a multicriteria model constructed by means of Measuring Attractiveness by a Categorical Based Evaluation Technique (MACBETH) to select the most suitable combination of maintenance policies in the different subsystems that make up an operating theatre. A decision-making group including the Hospital's technical services, environmental and occupational risk prevention managers, healthcare managers (operating theatres and health activity programming), healthcare staff, technicians, purchasing services managers and Hospital executives was used to determine the relevant decision criteria and their weightings. Markov chains were used to calculate the mean availability for repairable systems. This is aimed at increasing the availability of the operating theatre, thereby increasing physical safety during patient operations and reducing the number of delayed operations due to technical malfunctions.

Chapter 3 focuses on critical infrastructures which are basic service providers for society and so need to be effectively protected against hazards. Current economic and political trends mean that private companies are nowadays the owners or operators of most critical infrastructure. The prevention activities required to anticipate or reduce the impact of critical events affecting critical infrastructures are analysed. Additionally, this chapter explains the main reasons why public-private partnerships are sometimes valuable for ensuring that critical infrastructures are correctly maintained. The aim is to promote the sharing of resources and information between partners involved to enhance the overall resilience level of critical infrastructures and achieve better decisions by critical infrastructures managers.

Chapter 4 describes the Graphical Analysis for Maintenance Management (GAMM) method, which uses a non-parametric estimator of the reliability function using historical data. GAMM provides a visual representation and graphical analysis of data from a complete or partial historical record of the maintenance work performed, allowing anomalous behaviour in equipment, whether caused by its own operations, maintenance activities, improper use of equipment or even as a result of design errors, to be identified. Different patterns can be observed, providing useful information for decision making and problem solving. A set of rules has been developed to provide a proper and objective interpretation of the GAMM method results.

Chapter 5 analyses how maintenance departments of companies generate, transfer and use knowledge, and the impact this can have on the organization. Maintenance workers essentially work based on their experience or tacit knowledge, which is difficult to measure or express, but nevertheless this untransmitted knowledge can lead to high costs for the company because of increased stoppage times for production and services, loss of energy efficiency, or an increase in adaptation time of new staff. This chapter describes the factors which influence knowledge management in maintenance engineering, and their effect on industrial production, as well as identifying how knowledge management in asset management works, and how it may be impeded or made more efficient.

Chapter 6 shows how maintenance management for buildings has become a research area of great interest because they are required to operate efficiently. However, while traditionally building information models have mainly addressed the construction sector, recently the research focus has shifted from early life-cycle stages to maintenance, refurbishment, deconstruction and end-of-life considerations, especially of complex structures. Asset management tools such as facilities management provide the approach and required elements to achieve greater efficiency and effectiveness in building life-cycle management. This chapter introduces the application of asset management in buildings and how the development of building information modelling is the key element in allowing effective implementation.

Chapter 7 describes the concept of Industry 4.0 in which machines can monitor and transmit information about their operating conditions, both to internal control mechanisms and to other machines or external systems for analysis, decision making or maintenance activity. This chapter shows a reference model developed from a number of case studies (sheet-metal processing machines, tool machines, special machinery, aerogenerators, compressors, etc.), which allows efficient management of the after-sales service. This model relates after-sales service technology with product technologies (Industry 4.0) in the area of after-sales maintenance. Three levels have been identified at which companies in the machine goods sector may be classified, depending on the level of excellence of the after-sales maintenance service.

Chapter 8 performs experimental analysis of crack propagation by fatigue in high resilience steels. The life of the components of a structure containing premature cracks can be determined by the degree of subcritical crack propagation. One problem of high resilience steels is their low toughness in the HAZ, when they are welded with a high heat input. This chapter analyses nine specimens that have been welded by a submerged arc welding process to check that the parameters that indicate the values of fracture mechanics in the HAZ, after the heat cycle steel has undergone, in a process with a maximum heat input of 2,327kJ /mm, are still valid. A correlation between the theoretical and experimental values is confirmed. The experimental work carried out ensures that the parameters regulating the mechanism of fracture remain valid under the rules compatible with the design of the base material and that as far as possible a correlation is established between welding parameters and those obtained in fracture tests, so that if the results of the tests are not satisfactory, the appropriate solution may be applied to the welding for the parameters governing fracture tests to be acceptable.

Chapter 9 analyses different indicator systems such as the balanced scorecard, the Alsyouf Model, the maintenance scorecard model, the metrics of the Maintenance of the Society of Maintenance and Reliability Professionals and European Standard EN-15341, through which maintenance management can be evaluated. Additionally, the chapter proposes a maintenance-management indicator system based on the definition of indicators for the various hierarchical levels and different functions and processes taking place in a maintenance department. This system provides an assessment of these aspects at all levels of the organization and shows the relationship between the various indicators to understand the overall performance of maintenance management and thus align departmental objectives with the strategic objectives. The chains of key performance indicators for maintenance are carried out by means of the C-KPI-M model which defines the sequence of key indicators in maintenance management directly contribute to the achievement of maximum economic value added.

The development and the implementation of advanced actuation systems such as electromechanical actuators has increased because they intensify the ease of control of the system, provide options for re-configuration, maintain functionality during faults, and make it possible to carry out advanced diagnostics and prognostics for a more intelligent maintenance, leading to an increase of aircraft availability with long-term planning for maintenance activities. Chapter 10 shows how to develop health-monitoring algorithms based on AI and statistical technologies to detect and predict early stages of failure in a gearbox, taking into account vibration signals obtained from the electromechanical actuators by means of tri-axial accelerometers in on-ground testing. The testing is experimental, involving data collection of healthy and faulty gears and extraction of a set of features with different pre-processing techniques.

Chapter 11 presents four complementary approaches to quantify the interpretation of signals exchanged in a network of sensors in the presence of uncertainty. The aim is to be able to assess measurement error and the corresponding risk in order to reduce oversizing of the monitoring architectures and better define the level of confidence placed in the information received from the network. A comparative analysis is presented, according to different criteria of quality, quantity and/or type of data collected by the sensor network between Bayesian networks, Transferable Belief Model (TBM), Directed Evidential Networks (DEN) and Deep Belief Networks (DBN). This includes the resources necessary to implement each method in order to understand measurement or to reduce errors related to uncertainty in a sensor network. This analysis will assist in identifying the most suitable method for each problem detected by a sensor network.

Chapter 12 sets out a number of implications for practitioners of asset management, such as that is a multidisciplinary and strategic practice which should look at the complete lifetime of assets, and where best practice is to close the loop of objectives, performance, interventions, expected performance, and new performance figures, taking account of the knowledge of experts. It presents a practical twelve-step approach and develops an asset life-cycle plan in which sessions with experts are used to identify the main lifetime impacts that may be useful in guaranteeing or increasing the value of the asset to the company.

Chapter 13 provides insights into real-life applications of advanced maintenance techniques, in order to support companies and practitioners in moving towards better-informed maintenance decision making. Three case studies are introduced, corresponding to companies with different maturity levels in the usage of advanced maintenance techniques, to highlight typical problems businesses experience in the application of prognostic techniques. A four-step procedure is described that guides practitioners in the application of advanced maintenance techniques, thereby overcoming the difficulties of application.

Chapter 14 focuses on improving energy efficiency in district heating by implementation of information technologies. The models, features and existing tools of energy management, which represent the state of the art of district heating management, are set out here. This includes the decision support system HeatCAM for making decisions on energy management for district heating system operation in the demand-driven model. This system can provide real-time monitoring and decision making on the regulation of heat consumption modes of the particular building or group of buildings as required by the consumer.

Chapter 15 explores offshore wind projects considered to be megaprojects, that is, with a high level of complexity, especially at the beginning, which accumulate high risks resulting in a higher likelihood of failure. The current literature regarding offshore wind projects fails to address the planning aspect, in particular the risks and uncertainties occurring in this phase. In this chapter, a Delphi study is described, involving interviews with 26 experts from a variety of backgrounds in the offshore wind industry in The Netherlands. A framework is presented that depicts the planning phase of an offshore wind project, along with ten risks and seven uncertainties that are commonly found in offshore wind projects. The structure of the project and the role of government is also analysed.

Chapter 16 analyses the process by which the machinery is prepared, so that the product is available in the quantity and at the moment required by the customer. These set-up activities have traditionally been considered a wasteful operation, as are other logistical operations in companies; however, it is in fact of great importance in the modern competitive world. This chapter analyses the variables that can affect the development of set-up process, the typology of operations in the set-up process, the times involved, a methodology for analysis and improvement of the set-up process and a control system for the improvements gained.

Chapter 17 describes an amenable management of risk, considering all the types of risk that can be applicable to asset management through a holistic approach, which is a key element in taking optimal decisions. It analyses different factors that determine whether risk assessment has been done effectively, such as process information availability, understanding of risks, internal control, change management and personnel management. It presents, in addition, a practical approach using a Sarbanes - Oxley (SOX) framework in order to implement a proper internal control framework, which is sustainable and adaptable to each company or case.

Chapter 18 introduces the reliability-based maintenance philosophy with the goal of enhancing the reliability of the asset under consideration at optimal cost. This mostly qualitative methodology can determine maintenance intervals based on past experience and age-exploration methods. The method provides a quantitative way of determining maintenance intervals using the principles of reliability engineering for repairable systems. This is data-based, relying on failure and other data generated by assets. It complements the reliability-centred maintenance philosophy while helping to improve the efficiency and effectiveness of industrial asset maintenance.

Chapter 19 analyses the wood-fuel energy systems which represent around 70% of the energy consumed in Mozambique. These systems require design tools to support the strategic and optimised used of available socio-ecological resources/assets, and which also include the various agents involved in the decision. This chapter therefore develops a novel tool for the strategic design of wood-fuel energy systems, called 2MBio. 2MBio is a participatory conceptual design tool that provides: a formalized common space for dialogue, allowing participants to express their knowledge and experience on the question of "why" and "how" to address problems and solutions, and facilitates participatory conceptual design of comprehensive and integrated strategies, policies, projects and solutions for wood-fuel energy systems.

Chapter 20 focuses on discussing the occurrence of faults where the root cause cannot be determined, usually called no-fault-found problems. In the aerospace industry this is an important topic. When faced with no fault found issues, how organisations make decisions to control them (regulations and control of the amount of time, resources and budget to be spent), and how impact information flows within organisations, (organisational issues and gaps that will require extensive research efforts, etc.) are explained.

CONCLUSION

At the current time making better-informed maintenance decisions is of key importance for organizations. Thus, the process of data collection, treatment and management is increasingly important over the life-cycle of assets, in accordance with asset management standards such as PAS 55 and ISO 55000. This can be seen in the contributions in this book, which analyse areas ranging from the interpretation of signals provided by a network of sensors, through algorithms to determine the fatigue of the structure and to detect and predict early stages of failure, to different approaches to strategic decision making in asset life cycle management. This book analyses from specific operational decisions relating failure mechanisms and detection, to strategic decisions involving the whole life-cycle of industrial plants. This is all done without losing sight of the risks and uncertainties involved in any real-life decision-making process.

This book, therefore, will stimulate the idea that decisions related to the management of physical assets should cease to be based purely on empirical evidence from experience and understanding of maintenance engineers and managers, and should begin to incorporate techniques, tools and models using integrated mathematical techniques, together with better data and analytical and clarification techniques, the judgements of experts, and common sense. This should lead not only to satisfactory or good decisions, as optimal decisions do not generally exist, due to frequently conflicting criteria, but it will also lead to efficient and effective maintenance systems which separate world class companies from the rest.

The interplay of techniques required to achieve this, the application to ever more complex and exclusive areas, and the inclusion of risk and uncertainty in decision making are questions whose analysis this book attempts to address, and they represent a challenge for future research in asset management.

María Carmen Carnero
University of Castilla – La Mancha, Spain

Vicente González-Prida
University of Seville, Spain

Section 1

Chapter 1
Case Study on a Maintenance and Reliability Management Model Proposal:
A Third Set of Locks Project in the Panama Canal

Carlos Parra
University of Seville, Spain

Adolfo Crespo Márquez
University of Seville, Spain

Vicente González-Prida
University of Seville, Spain

Fredy Kristjanpoller
Universidad Técnica Federico Santa María, Chile

Pablo Viveros
Universidad Técnica Federico Santa María, Chile

Gabriel Llort
MWH Global, USA

Alfredo R. Aguilar
MWH Global, Panama

ABSTRACT

The purpose of this chapter, is to provide a Maintenance and Reliability Management Model for the project: Design and Construction of the Third Set of Locks in the Panama Canal, with the approach of the process of asset management optimization. A practical vision of the maintenance and reliability management process and framework is presented with the idea of: Structuring the maintenance management process by grouping management activities within a series of so-called management building blocks; Structuring the framework grouping techniques that can be used to support decisions to be taken within each of these building block. This chapter presents not only a process but also the framework and techniques to manage and improve maintenance and reliability effectiveness and efficiency. This report will be used to assist different plant teams to elaborate the optimal strategies for maintenance and inspection for the assets, specified for the project: Third Set of Locks in the Panama Canal.

DOI: 10.4018/978-1-5225-0651-5.ch001

1. MAINTENANCE MANAGEMENT MODEL PROPOSED FOR THE PROJECT: THIRD SET OF LOCKS IN THE ACP (AUTORIDAD DEL CANAL DE PANAMÁ)

1.1. Introduction to Maintenance Management Model

The Maintenance Management Models are frequently associated with a wide range of difficulties. Why is this function, at least in appearance, so difficult to manage? We have carried out a review of literature to find out some of the reasons:

- Lack of maintenance management models (Parra and Crespo, 2012). There is a lack of models that could improve the understanding of the underlying dimensions of maintenance. Maintenance is somewhat "under-developed" with a lack of effective prevention methodologies and the integration of said methods in manufacturing companies in most continents;
- Wide diversification in the maintenance problems. Maintenance is composed of a set of activities for which it is very difficult to find procedures and information support systems in one place to ease the improvement process. Normally, there is a very wide diversification in the problems that maintenance encounters, sometimes a very high level of variety in the technology used to manufacture the product, even in businesses within the same productive sector; therefore, it has been difficult to design an operative methodology of general applicability;
- Lack of plant/process knowledge and data. Managers, supervisors and operators typically find that the lack of plant and process knowledge is the main constraint, followed by the lack of historical data, to implement suitable maintenance policies;
- Lack of time to complete the analysis required. Many managers indicate how they do not have the required time to carry out suitable maintenance problems analysis. Day to day actions and decision making activities distract them from these fundamental activities to improve maintenance;
- Lack of top management support. Lack of leadership to foster maintenance improvement programs, fear of an increase in production disruptions, etc., are other common causes of maintenance underdevelopment in organizations;
- Exigent safety and environmental factors. In addition to process and technology related issues mentioned above, new and more exigent safety and environmental factors such as emerging regulations put pressure on a maintenance manager and add complexity to this function.

Some authors (Parra & Crespo, 2012) have worked on the characterization of the complexity found in managing the maintenance function in a production environment, creating tools where we are able to value each one of previously reviewed factors for a certain organization (with a degree of fulfilment – DFi), and evaluate them according to environmental aspects (with a relevance factor – RFi). The maintenance management complexity index can be helpful as one way of comparing across different production environments to help decide the relative effort and resources required to maintain them.

1.2. Proposal for a Generic Model of Maintenance Management for the Project: Third Set of Locks in the ACP

The generic model proposed for maintenance management that will now be proposed and defined integrates other models found in the literature for built and in-use assets, and consists of eight sequential management building blocks (Parra and Crespo, 2012). Each block is, in fact, a key decision area for asset maintenance and life cycle management. Within each of these decision areas we can find methods and models that may be used to order and facilitate the decision-making processes *(this model is being used in the ACP since 2012)*.

The maintenance management process can be divided into two parts: the definition of the strategy, and the strategy implementation. The first part, definition of the maintenance strategy, requires the definition of the maintenance objectives as an input, which will be derived directly from the business plan. This initial part of the maintenance management process conditions the success of maintenance in an organization, and determines the effectiveness of the subsequent implementation of the maintenance plans, schedules, controls and improvements. Effectiveness shows how well a department or function meets its goals or company needs, and is often discussed in terms of the quality of the service provided, viewed from the customer's perspective. Effectiveness concentrates then on the correctness of the process and whether the process produces the required result (Vagliasindi, 1989; Wireman, 1998; Palmer, 1999).

The second part of the process, the implementation of the selected strategy has a different significance level. Our ability to deal with the maintenance management implementation problem (for instance, our ability to ensure proper skill levels, proper work preparation, suitable tools and schedule fulfilment), will allow us to minimize the maintenance direct cost (labour and other maintenance required resources). In this part of the process, we deal with the efficiency of our management, which should be less important. Efficiency is acting or producing with minimum waste, expense, or unnecessary effort. Efficiency is then understood as providing the same or better maintenance for the same cost.

In this report, we present a generic model proposed for maintenance management integrates other models found in the literature (Pintelon & Gelders, 1992; Vanneste & van Wassenhove, 1995) for built and in-use assets, and consists of eight sequential management building blocks, as shown in Figure 1 (Parra & Crespo, 2012). The first three building blocks condition maintenance effectiveness, the fourth and fifth ensure maintenance efficiency, blocks six and seven are devoted to maintenance and assets life cycle cost assessment, finally block number eight ensures continuous maintenance management improvement. The maintenance management model proposed to the project: Third Set of Locks in the ACP, is presented below *(model is based on the Asset Management Standards ISO 55000, 55001 and 55002)*.

In this section, we will briefly introduce each block and discuss methods that may be used to improve each building block decision-making process (Figure 2) (Parra & Crespo 2012).

1.2.1 Definition of Maintenance Objectives and Strategy

Regarding the definition of maintenance objectives and key performance indicators – KPI's (Phase 1), it is common the operational objectives and strategy, as well as the performance measures, are inconsistent with the declared overall business strategy (Gelders et al., 1994). This unsatisfactory situation can indeed be avoided by introducing the balanced scorecard – BSC (Kaplan & Norton, 1992). The BSC is specific for the organization for which it is developed and allows the creation of KPIs for measuring maintenance management performance which are aligned to the organization's strategic objectives.

1.2.2 Asset Priority and Maintenance Strategy Definition

Once the maintenance objectives and strategy are defined, there are a large number of quantitative and qualitative techniques which attempt to provide a systematic basis for deciding what assets should have

Figure 1. Maintenance Management Model
Source: Crespo Marquez, A. (2007), The Maintenance Management Framework, Models and Methods for Complex Systems Maintenance, Springer, London

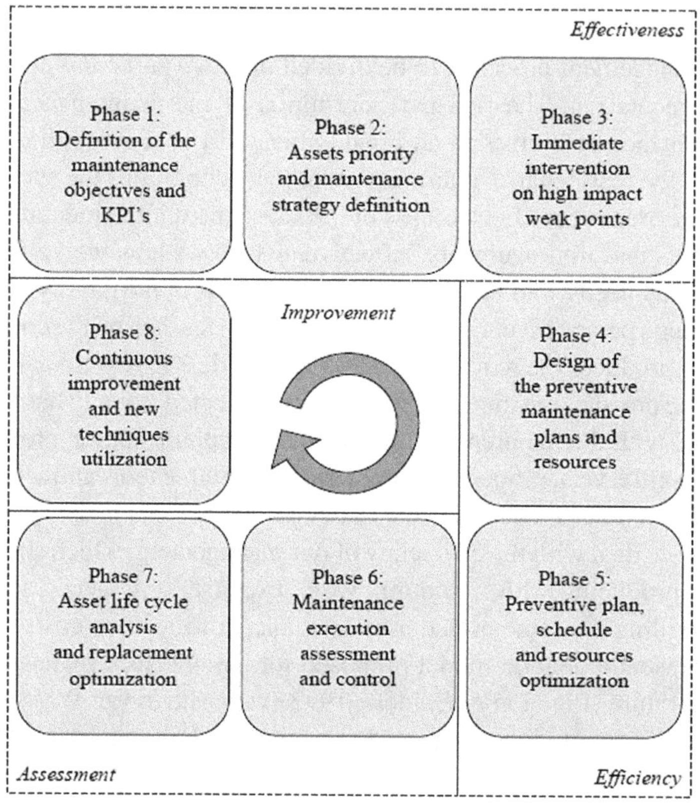

Figure 2. Sample of techniques within the maintenance management framework
Source: Crespo Marquez, A. (2007), The Maintenance Management Framework, Models and Methods for Complex Systems Maintenance, Springer, London

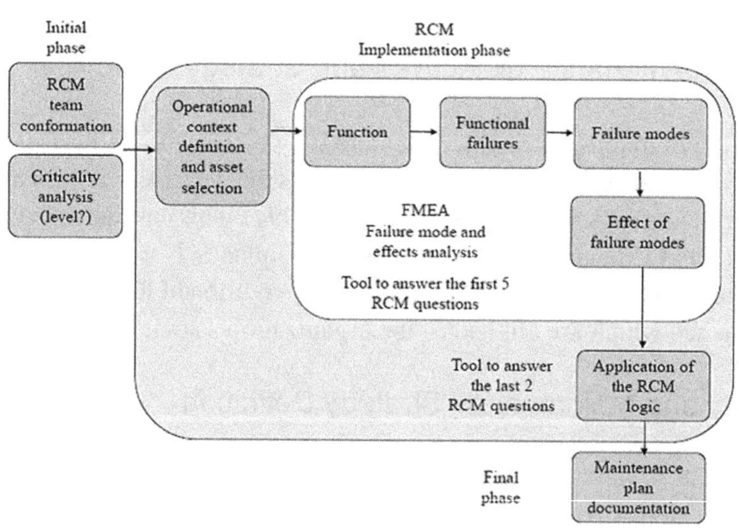

priority within a maintenance management process (Phase 2), a decision that should be taken in accordance with the existing maintenance strategy. Most of the quantitative techniques use a variation of a concept known as the "probability/risk number" – PRN (Moubray, 1997).

Assets with the higher PRN will be analysed first. Often, the number of assets potentially at risk outweighs the resources available to manage them. It is therefore extremely important to know where to apply available resources to mitigate risk in a cost-effective and efficient manner. Risk assessment is the part of the ongoing risk management process that assigns relative priorities for mitigation plans and implementation. In professional risk assessments, risk combines the probability of an event occurring with the impact that event would cause. The usual measure of risk for a class of events is then $R = P \times C$, where P is probability and C is consequence. The total risk is therefore the sum of the individual class-risks (Parra and Crespo, 2012). The procedure to follow in order to carry out an assets criticality analysis following risk assessment techniques could be then depicted as follows:

1. Define the purpose and scope of the analysis;
2. Establish the risk factors to take into account and their relative importance;
3. Decide on the number of asset risk criticality levels to establish; and
4. Establish the overall procedure for the identification and priorization of the critical assets.

Notice that assessing criticality will be specific to each individual system, plant or business unit. For instance, criticality of two similar plants in the same industry may be different since risk factors for both plants may vary or have different relative importance.

1.2.3 Immediate Intervention on High Impact Weak Points

Once the assets have been prioritized and the maintenance strategy to follow defined, the next step would be to develop the corresponding maintenance actions associated with each category of assets. Before doing so, we may focus on certain repetitive – or chronic – failures that take place in high-priority items (Phase 3).

Finding and eliminating, if possible, the causes of those failures could be an immediate intervention providing a fast and important initial payback of our maintenance management strategy. The entire and detailed equipment maintenance analysis and design could be accomplished, reaping the benefits of this intervention if successful.

There are different methods developed to carry out this weak point analysis, one of the most well known being RCFA (Root Cause Failure Analysis). This method consists of a series of actions taken to find out why a particular failure or problem exists and to correct those causes. Causes can be classified as physical, human or latent. The physical cause is the reason why the asset failed, the technical explanation on why things broke or failed. The human cause includes the human errors (omission or commission) resulting in physical roots. Finally, the latent cause includes the deficiencies in the management systems that allow the human errors to continue unchecked (flaws in the systems and procedures). Latent failure causes will be our main concern at this point of the process. Note that although informal RCFA techniques are usually used by individual or groups to determine corrective actions for a problem, they have limitations that can make the development of long-term solutions difficult

1.2.4 Design of the Preventive Maintenance Plans and Resources

Designing the preventive maintenance (PM) plan for a certain system (Phase 4) requires identifying its functions, the way these functions may fail and then establish a set of applicable and effective PM tasks, based on considerations of system safety and economy. A formal method to do this is the RCM (Parra and Crespo, 2012). RCM methodology allows the identification of real maintenance needs starting from the analysis of the 7 questions:

- What are the functions and associated performance standards of the asset in its present operating context?
- In what ways does it fail to fulfil its functions?
- What causes each functional failure?
- What happens when each failure occurs?
- In what way does each failure matter?
- What can be done to prevent each failure?
- What should be done if a suitable preventive task cannot be found?

1.2.5 Preventive Plan, Schedule, and Resources Optimization

Optimization of maintenance planning and scheduling (Phase 5) can be carried out to enhance the effectiveness and efficiency of the maintenance policies resulting from an initial PM plan and program design.

Models to optimize maintenance plan and schedules will vary depending on the time horizon of the analysis. Long-term models address maintenance capacity planning, spare parts provisioning and the maintenance/replacement interval determination problems, mid-term models may address, for instance, the scheduling of the maintenance activities in a long plant shut down, while short-term models focus on resources allocation and control (Duffuaa, 2000). Modelling approaches, analytical and empirical, are very diverse. The complexity of the problem is often very high and forces the consideration of certain assumptions in order to simplify the analytical resolution of the models, or sometimes to reduce the computational needs.

For example, the use of Monte-Carlo simulation modelling can improve PM scheduling, allowing the assessment of alternative scheduling policies that could be implemented dynamically on the plant/shop floor (Parra and Crespo, 2012).

1.2.6 Maintenance Execution Assessment and Control

The execution of the maintenance activities, once designed planned and scheduled using techniques described for previous building blocks have to be evaluated and deviations controlled to continuously pursue business targets and approach stretch values for key maintenance performance indicators as selected by the organization (Phase 6). Many of the high-level maintenance KPIs, are built or composed using other basic level technical and economical indicators. Therefore, it is very important to make sure that the organization captures suitable data and that data are properly aggregated/disaggregated according to the required level of maintenance performance analysis.

1.2.7 Asset Life Cycle Analysis and Replacement Optimization

A life cycle cost analysis (Phase 7) calculates the cost of an asset for its entire life span (Figure 4) (Parra and Crespo, 2012). The analysis of a typical asset could include costs for planning, research and development (R&D), production, operation, maintenance and disposal. Costs such as up-front acquisition (research, design, test, production and construction) are usually obvious, but life cycle cost analysis crucially depends on values calculated from reliability analyses such us failure rate, cost of spares, repair times, and component costs. A life cycle cost analysis is important when making decisions about capital equipment (replacement or new acquisition) (Campbell and Jardine, 2001), it reinforces the importance of locked in costs, such as R&D, and it offers three important benefits:

1. All costs associated with an asset become visible. Especially: upstream; R&D, downstream; maintenance.
2. Allows an analysis of business function interrelationships. Low R&D costs may lead to high-maintenance costs in the future.
3. Differences in early stage expenditure are highlighted, enabling managers to develop accurate revenue predictions.

1.2.8 Continuous Improvement and New Techniques Utilization

Continuous improvement of maintenance management (Phase 8) will be possible due to the utilization of emerging techniques and technologies in areas that are considered to be of higher impact as a result of the previous steps of our management process. Regarding the application of new technologies to maintenance, the "e-maintenance" concept (Parra and Crespo, 2012) is put forward as a component of the e-manufacturing concept (Lee, 2003), which profits from the emerging information and communication technologies (ICTs) to implement a cooperative and distributed multi-user environment. e-Maintenance can be defined (Tsang et al., 1999) as a maintenance support which includes the resources, services and management necessary to enable proactive decision process execution.

Figure 3. Life cycle cost analysis
Source: Parra, C. and Crespo, A, (2012), Ingeniería de Mantenimiento y Fiabilidad Aplicada en la Gestión de Activos. Desarrollo y aplicación práctica de un Modelo de Gestión del Mantenimiento (MGM), Primera Edición. Editado por INGEMAN, Escuela Superior de Ingenieros Industriales, Universidad de Sevilla, España

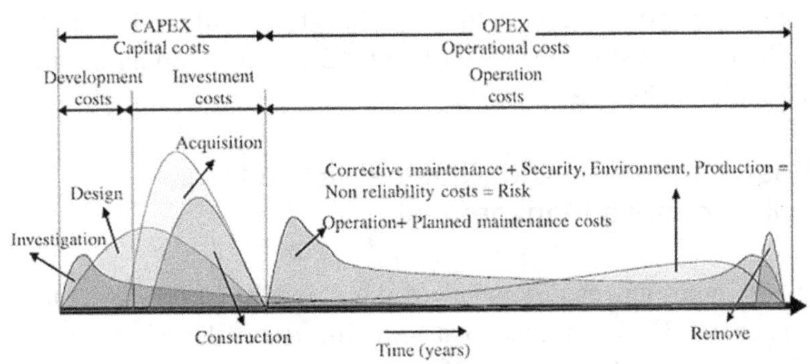

Figure 4. RCM implementation process
Source: Crespo Marquez, A. (2007), The Maintenance Management Framework, Models and Methods for Complex Systems Maintenance, Springer, London

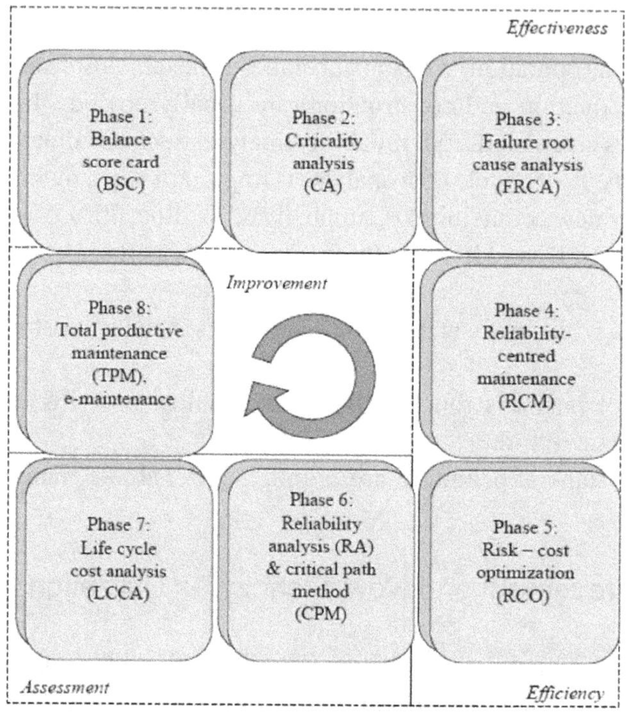

This section summarizes the process (the course of action and the series of stages or steps to follow) and the framework (the essential supporting structure and the basic system) needed to manage and optimize the maintenance strategies in theThird Set of Locks of the ACP.

1.3. Proposal of a Reliability Team to Implement the Activities within the Maintenance Management Model

With the aim to cover different activities to be developed within each block of the maintenance management model proposed for the Third Set of Locks of the ACP, it is necessary to create a support group in Maintenance and Reliability Engineering, to promote and run a set of activities at different stages of the maintenance model presented in Figure 1. Initially, this group will implement the MAXIMO system and subsequently, shall control and manage the MAXIMO and develop an optimization process from Maintenance and Reliability techniques.

1.3.1 Minimum Requirements of Knowledge for Expert in Reliability and Maintenance Management

This section contains the minimum requirements of the theoretical knowledge for a maintenance and reliability manager in general. However, this document aims to fulfil the intention to be comprehensive enough and include the essential and fundamental knowledge that any expert in maintenance management needs to have, regardless of which company or in which country he or she is working. The requirements cover the following areas *(EFNMS publication of June 19th 2006. The requirements of Competencies and Responsibilities for an European Expert in Maintenance and Realibility)*:

1.3.1.1 Management and Organization

Within this area, it is essential to have a very good knowledge about the importance of maintenance for the economy in the company, for achieving production goals and for the quality of the product, and so on. It is important to have good knowledge of how maintenance activities are organized. Therefore, the following knowledge is necessary:

- *How to set up a company management policy to be able to participate in its definition as far as maintenance is concerned to:*
    - *Describe why a policy has to be set up and what the requirements are for that policy;*
    - *Give examples on in which way the maintenance aspects are in a company management policy.*
- *How to formulate the maintenance policy within a company to:*
    - *Give an example of a maintenance policy;*
    - *Describe the requirements for a maintenance policy;*
    - *Describe the process of the development of a maintenance policy.*
- *How to formulate the maintenance goals to:*
    - *Describe the general requirements for maintenance goals;*
    - *Describe the process of the development of maintenance goals;*
    - *Give examples of maintenance goals;*
    - *Describe the relationship between goals and policy.*
- *Different maintenance strategies and how to choose the right strategy to:*
    - *Formulate different maintenance strategies;*
    - *Describe the reasons behind the choice of a certain strategy.*
- *How to specify the requirements for the maintenance activities to:*
    - *Describe the different maintenance activities;*
    - *Describe different requirements for the maintenance activities;*
    - *Describe the process of the identification, the formulation and the communication of the requirements.*
- *How to organize the maintenance activities, how to choose a suitable organization and assure the right competence within the organization to:*
    - *Describe different types of maintenance organizations (e.g. centralized, decentralized, co-operation with the equipment supplier and/or servicing companies and integration with the production);*
    - *Describe the advantages and the disadvantages with the different types of organizations and the combination of them;*
    - *Describe how to develop the competence in all the different types of organizations.*
- *How to determine the human and material resources in order to implement the organization to:*
    - *State the different types of maintenance resources (e.g. tools, material, personnel, transportation, documentation, shops);*

  - *Describe how to develop and optimize the maintenance resources (personnel and material), their location, quality and quantity.*
- *How to assure (by maintenance activities) the health and safety and the right environment conditions (inside and outside the company) to:*
  - *Describe different conditions in the production equipment that may cause risks for health, safety and the environment (inside and outside the company);*
  - *Describe the possibility to prevent such incidents by maintenance activities, including co-operation with other departments in the company and external parties;*
- *How to guide, control and analyze the maintenance activities to:*
  - *Describe different methods and techniques to achieve an optimized result for the company by the maintenance activities, including the economical and safety aspects for these methods and techniques;*
  - *Describe different general aspects that are to be taken into account for analysis;*
  - *Describe the methods and techniques for analyzing and the betterment process;*
- *How to develop and use key-figures for the economical control to:*
  - *Describe how to use the key-figures in the control and development of the maintenance activities;*
  - *Describe what the fundamental requirements are for key-figures;*
  - *Describe the most useful key figures for different maintenance organizations.*
- *LCC (Life Cycle Cost) techniques/methods to:*
  - *Describe the methods of LCC, and when they can be used;*
  - *Be able to make some fundamental calculations of LCC;*
  - *Describe how to organize the work when using the concepts of LCC;*
  - *Describe how the concepts of LCC can be used in different situations;*
  - *Describe how to specify the LCC requirements in a procurement process;*
  - *Describe different methods to control the maintenance activities;*
  - *Understand the different maintenance concepts (e.g. TPM, RCM).*
  - *Describe how to verify the LCC values and the consequences if the verified result is not in accordance with the specified requirements [Good Knowledge].*
- *Logistics support, material and store handling, methods for spare part calculations to:*
  - *Describe the different factors that will have an influence on an optimized organization of the spare part consumption (e.g. cost for lack of spare parts, cost for storage, cost for interest);*
  - *Describe routines and organization for an optimized logistic support of spare parts (e.g. purchasing, quality control, delivery systems inside the maintenance organization);*
  - *Describe different ways of organizing the spare part store (e.g. centralized, decentralized, at the supplier);*
  - *Describe how to calculate the total amount of spare parts and how many of each type, inclusive the typical mathematical formulas for this purpose.*
- *How to measure and analyze the results of the maintenance activities, e.g. efficiency and economy to:*
  - *Describe different methods to measure the result of the maintenance activities, the advantages and disadvantages with the methods and their handling of the economical aspects;*
  - *Describe what is not covered by these methods;*

    - *Understand different economical models regarding maintenance and understand the fundamental principles regarding the economical results for a company;*
    - *Be able to develop a model for measurement and analysis of the maintenance activities.*
- *The maintenance activities in the development and procurement of new production equipment to:*
    - *Be able to transfer production requirements into functional requirements (e.g. equipment*
    - *Dependability) and into quantitative and qualitative maintenance requirements (e.g. reliability and maintainability) and optimize the resources;*
    - *Understand the importance for maintenance of taking part in the development phase;*
    - *Describe how the maintenance experience can be used during the design phase.*
- *How to define the future maintenance needs of a company to:*
    - *Understand which factors that are important for the need of maintenance activities and how they might be changed in the future (e.g. new requirements regarding goals, strategies and results);*
    - *Understand the future needs of maintenance and its influence on the actual activities in the long run (e.g. work load, type of work, quality and quantity);*
    - *Be able to describe different future scenarios. (EFNMS, 2006)*

2. IMPLEMENTATION PROCESS OF THE RELIABILITY CENTERED MAINTENANCE (RCM) IN THE PROJECT: THIRD SET OF LOCKS IN THE ACP (AUTORIDAD DEL CANAL DE PANAMÁ)

2.1. Introduction to RCM

RCM serves as a guide to identify maintenance activities with their respective frequencies in the most important elements of an operative context. This is not a mathematical formula; its success is based on the functional analysis of a certain operational context undertaken by a review team. The effort developed by the review team allows the generation of a flexible maintenance management system, adapted to the needs of real maintenance in the organization. Keeping in mind, personal security, environment, operations and benefit/cost reason (Crespo, 2007, Parra and Crespo, 2012). RCM specifically allows: a) detection of failures early enough to ensure minimum interruptions to system's operation, b) elimination of causes of some failures before they appear, c) elimination of the causes of some failures through changes in design, and d) identification of those failures that may happen without any decrease in the system's safety.

2.2. RCM Implementation

2.2.1 The Process and the Review Team

Hereafter we present the proposed scheme for implementation of RCM. The success of this process will depend basically on the selection of the proper *RCM review team*. This team will have the responsibility of answering the seven basic questions of the RCM, according to the scheme in Figure 4.

People with different functions in the organization will form the RCM review team. This team will work jointly for a certain period of time in a positive atmosphere, to analyses common problems with the different departments and with a common goal.

The facilitator plays an important role in this team; his/her basic function consists in guiding and leading the implementation of the RCM process. The activities that the instructor should undertake are:

- Guide the review team during the failure modes and effects analysis (FMEA) and the selection of maintenance activities;
- Help to select the decision level to be used in the failure mode and effects analysis;
- Help to identify the critical assets that should be analyzed under this methodology;
- Ensure that the work meetings are led in a professional manner;
- Ensure real consensus;
- Motivate the team;
- Ensure that all required information is in place when needed;
- Ensure that results are correctly recorded.

2.2.2 System Selection and Definition of the Operational Context

The selection of the assets, systems or equipment where RCM will be applied can be carried out using techniques explained in the previous chapters. The correct definition of the operational context will be a requirement to do this phase properly. Operational context definition will require certain information to be gathered. A graphical tool that eases the visualization of the overall operational context is the IPO (Input—Process—Output) diagram, which can be synthesized and represented as in Figure 5. In Figure 5, insumes are raw materials or resources to be transformed or converted. The process will be divided into systems that will have a certain function (or group of functions). Maintenance efforts will be then concentrated on each one of the systems functions.

Figure 5. Input Process Output diagram (IPO)
Source: Parra, C. and Crespo, A, (2012), Ingeniería de Mantenimiento y Fiabilidad Aplicada en la Gestión de Activos. Desarrollo y aplicación práctica de un Modelo de Gestión del Mantenimiento (MGM), Primera Edición. Editado por INGEMAN, Escuela Superior de Ingenieros Industriales, Universidad de Sevilla, España

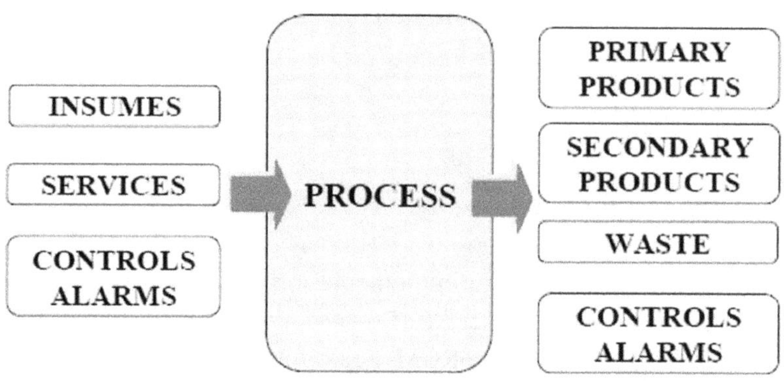

2.2.3 Failure Mode and Effect Criticality Analysis (FMECA)

FMECA is recognized as the most fundamental tool that is used in RCM. Due to its practical and qualitative approach, it is also the most widely understood and applied form of reliability and risk analysis found throughout industry. Given a specific process, FMECA deals with the identification of its failure modes, failure causes and frequencies (reliability), and the effects that might result if any specific failure occurs during the process operation (risk). Based on the information provided by FMECA, design and management personnel are better informed about the way to determine what can be done in order to avoid or mitigate failure modes.

The FMECA process is divided into the following four steps:

1. Description of functions;
2. Description of functional failures;
3. Failure modes (failure rate data) definition;
4. Description of failure mode effects and criticality (RPN: Risk priority number).

2.2.3.1 Describing Functions

Each item or equipment usually has more than one function. They can be divided into five categories:

- **Primary Functions:** These are functions required to fulfil the intended purpose of the item;
- **Auxiliary Functions:** These are functions that support the primary function;
- **Protective and Control Functions:** These are functions intended to control a process and protect people, equipment, or the environment;
- **Information Functions:** These are related to alarms, and the monitoring of several conditions;
- **Interface Functions:** These are functions that apply to the interface between two items. The interface may be active or passive.

An example of definition of functions for heat exchangers is shown in Table 1.

Table 1. FMECA example: Function description

FMECA	
Item: Heat Exchanger without Change of Phase	
Class of Function	**Function Description**
Primary	*-Provide correct heat exchange at a desired rate*
Auxiliary	*-Contain cooling and heating fluids* *-Prevent mixing of cooling and heating fluid*
Protective and Control	*-Prevent damage to heat exchanger and downstream equipment* *-Control the process*
Information	*-Condition monitoring*
Interface	*- Heat exchanger supports (oil cooler)*

2.2.3.2 Defining Functional Failures

A functional failure is defined as the inability of the equipment to keep a desired standard of performance (function). Functional failures vary in degree of magnitude; for example, a pump may have no output or may have its output restricted. An example of definition of functional failures for heat exchangers is shown in Table 2.

2.2.3.3 Definition of the Failure Modes

A failure mode is defined, in RCM as the physical cause of the functional failure. Only failure modes with a high occurrence possibility are recorded. It is not recommended to list every single failure possibility. Reasonably likely failure modes include the following:

- Failures which have occurred before on the same or similar assets;
- Failures modes which are already the subject of preventive maintenance routines, and which would occur if no preventive maintenance is done;
- Any other failure modes that have not yet occurred, but nevertheless have a real possibility of occurrence.

Additionally, we need to estimate the mean time to failure (MTTF) for each failure mode. The MTTF is expressed as the expected mean time to failure expected of a given failure mode. In this case, the ideal situation is to have valid historical data for the equipment in the operational context. In most cases, plant-specific data is unavailable or may have a low reliability level to allow its use without corroborating it. The uncertainties of data selection can be reduced by learning as much possible about data sets, taxonomy, equipment boundaries, used equipment type, equipment design and construction, process medium, plant operation, maintenance programs, and failure modes. OREDA, IEEC, PERD - Std 500-1994, Reliability Data Book for components in Chemical Process (CCPS), etc., are examples of data

Table 2. FMECA example: Functional failures

FMECA		
Item: Heat Exchanger without Change of Phase		
Class of Function	**Function Description**	**Functional Failure**
Primary	*Provide correct heat exchange at a desired rate*	*Unable to provide any heat /catastrophic*
		Provide reduced/excessive heat exchange /degraded
Auxiliary	*-Contain cooling and heating fluids*	*Unable to contain cooling and heating fluids / catastrophic*
		Contain partially cooling and heating fluid /degraded
	-Prevent mixing of cooling and heating fluids	*Unable to prevent mixing of fluids / catastrophic*
		Partial mixing of fluids (incipient)
Protective and control	*- Control the process*	*Unable to control the process (catastrophic)*
	-Prevent damage to heat exchanger and downstream equipment	*Unable to prevent damage to heat exchanger equipment (catastrophic)*
Information	*-Condition monitoring*	
Interface	*-None*	

sets that provide details of taxonomy, data origin, treatment, and limitations. An example of definition of failure modes and MTTF data for heat exchangers is shown in Table 3.

2.2.3.4 The Description of the Failure Modes Effects and Criticality Analysis

The failure effects describe what would happen if the failure mode occurs, and are related to issues such as downtime, effects on product quality, evidence that the failure has occurred, and threats to safety and environment. The description of these effects should include all the information needed to support the evaluation of the consequences of the failure (criticality analysis). When describing the effects of a failure, the following issues should be recorded:

- Evidence (if any) that the failure has occurred;
- Hidden failures (not evident)
- Threats (if any) to safety or environment;
- Effects (if any) in production or operations;
- Physical damages (if any) caused by the failure;
- Repairs needed to correct the effects of the failure.

Table 3. FMECA example: Failure mode

FMECA				
Item: Heat Exchanger without Change of Phase				
Class of Function	**Function Description**	**Functional Failure**	**Failure Mode**	**MTTF**
Primary	*Provide correct heat exchange*	*Unable to provide any heat /catastrophic*	*Complete stoppage of fluid*	--
			External Rupture	--
		Provide reduce/excessive heat exchange / degraded	*Partial reduction in fluid flow*	--
			Partial external rupture	*48 months*
			Plugged	--
Auxiliary	*-Contain cooling and heating fluids*	*Unable to contain cooling and heating fluids / catastrophic*	*External rupture*	*48 months*
		Contain partially cooling and heating fluid / degraded	*Partial external rupture*	--
	-Prevent mixing of cooling and heating fluids	*Unable to prevent mixing of fluids / catastrophic*	*Internal rupture*	*28 months*
		Mixing partial of fluids (incipient)	*Partial internal rupture*	*18 months*
Protective and Control	*- Control the process*	*Unable to control the process (catastrophic)*	*Control system fail*	*6 months*
	-Prevent damage to heat exchanger	*Unable to prevent damage to heat exchanger equipment (catastrophic)*	*Structure of support fail*	*72 months*
Information	*Condition monitoring*			
Interface	*None*			

The impact that a failure mode has on the organization depends on the operating context of the asset, the performance standards, which apply to each function and the physical effects of each failure mode. This combination of context, standards and effects means that every failure has a specific set of consequences connected to it.

An example of failure mode effects definition for heat exchangers is shown in Table 4.

2.2.3.4.1 RCM and the Hidden Failures

Equipment, in most cases, has more than one function. When one of these functions fails and someone may notice the failure or the fault state of the equipment, failures are said to be evident. However, in some occasions, no one knows that the equipment is in fault state unless another failure takes place. The first failure, the one that remained unnoticed until another took place, was not evident on its own. These are known as hidden failures.

To be able to understand this, suppose you have two pumps in a given operational context. Pump C (reserve pump) is not available (fault state), this fact shall not be evident under normal circumstances, since pump B is working under normal operation. In other words, the fault/failure of pump C shall not have any direct impact on its own unless or until pump B also fails. Pump C failure will not be evident under normal operating conditions unless other failures do occur. Notice that failures in pump C will only have some consequence if another failure – in this case in pump B - also takes place. When pump C is in fault state, the failure in pump B is known as a multiple failure. Regarding this point, the review team must know that all sole hidden failures will not have any direct consequences; however, they shall have an indirect consequence increasing the risk level of multiple faults/failures. "The only consequence of a hidden failure is the consequent increase of a multiple failure".

Table 4. FMECA example: Failure effects

FMECA. Item: Heat exchanger without change of phase				
Function Type	**Functional Failure**	**Failure Mode**	**MTTF months**	**Effects of Failures Modes**
Primary: Provide correct heat exchange at a desired rate	Unable to provide any heat/ catastrophic	Pump stoppage of fluid	-	Total loss of heat exchange/this failure has operational consequences
		External rupture	-	Total loss of heat exchange/this failure could have safety and environmental consequences
	Provide reduce/excessive heat exchange/degraded	Pump partial reduction in fluid flow	-	Partial loss of heat/operational consequences
		Partial external rupture	48	Partial loss of heat/operational consequences
		Plugged	-	Partial loss of heat/operational consequences
Auxiliary: Contain cooling and heating fluids	Unable to contain cooling and hating fluids/ catastrophic	External rupture	48	Major loss of process fluid to atmosphere/ this failure could have safety and environmental consequences
Auxiliary: Prevent mixing of cooling and heating fluids	Unable to prevent mixing of fluids/catastrophic	Internal Rupture	28	Major leakage between media/operational consequences
	Mixing partial of fluids/ incipient	Partial internal rupture	18	Leakage between media/operational consequences

The appearance of hidden failures, on their own, are not evident, in the normal operation process; therefore, in order to identify or recognize hidden faults, the RCM review team should answer the following question:

If the functional failure is caused by a failure mode on its own, is it evident under normal operation conditions?

If the answer to the question is NO, the failure mode is a hidden failure (not evident), and if the answer is YES, the mode is evident.

2.2.3.4.2 Risk Priority Number (RPN). Criticality Analysis for Failures Modes

Risk is the potential impact (positive or negative) to an asset or characteristic of value that may arise from some present process or from some future event. In everyday usage, "risk" is often used synonymously with "probability" and restricted to negative risk or threat. Risk management is the ongoing process of identifying these risks and implementing plans to address them. Some industries manage risk in a highly-quantified and numerate way. These include, for instance, the nuclear power and aircraft industries, where the possible failure of a complex series of engineered systems could result in highly undesirable outcomes (Crespo, 2007; Parra & Crespo, 2012).

Often, the number of assets potentially at risk outweighs the resources available to manage them. It is therefore extremely important to know where to apply available resources to mitigate risk in a cost-effective and efficient manner. Risk assessment is the part of the ongoing risk management process that assigns relative priorities for mitigation plans and implementation. In professional risk assessments, risk combines the probability of an event occurring with the impact that event would cause. The usual measure of RPN for a class of events is then R = P x C, where P is probability and C is consequence. The total risk is therefore the sum of the individual class-risks.

Risk assessment techniques can be used to prioritize assets and to align maintenance actions to business targets at any time. By doing so we ensure that maintenance actions are effective, that we reduce the indirect maintenance cost, the most important maintenance costs, those associated to safety, environmental risk, production losses, and ultimately, to customer dissatisfaction.

The procedure to follow in order to carry out an assets criticality analysis following risk assessment techniques could be then depicted as follows:

1. Define the purpose and scope of the analysis;
2. Establish the risk factors to take into account and their relative importance;
3. Decide on the number of asset risk criticality levels to establish;
4. Establish the overall procedure for the identification and priorization of the critical assets.

Risk factors considered in the RPN analysis were: safety, environment affection, operation downtime, maintenance and direct and indirect cost of operations, failure frequency and mean time to repair (Crespo, 2007, Parra and Crespo, 2012).

The assessment of RPN for each failure mode considered was

$$RPN = F \times C \qquad (1)$$

where F is the frequency factor or number of failures in a certain time period (year) and C is consequence of the failure measured as follows:

$$C = E \times Cr \times MTTR \quad (2)$$

With:

E: Effect on Lane Availability
Cr: Failure Mode Criticality
MTTR: Mean Time to Repair

Concerning the frequency of failures (F), the team decided to establish the classification and scale in Table 5, to rank the different failures modes.

Regarding the different consequence factors (defining C), they were classified and scaled as in Tables 6, 7 and 8.

Table 5. Failure frequency, classification and scale

Failure Frequency (F)	Failures per Hours	Model Value
Poor	> 1x10^-5	4
Average	< 1x10^-5	3
Good	< 1x10^-7	2
Excellent	< 1x10^-8	1

Table 6. Effect on lane availability, classification and scale

Effect on Lane Availability (E)	Consequence	Model Scale
Extremely high	Single point of failure (no backup): lane unavailable	5
Very high	2nd order failure: If backup also fails: lane unavailable	4
High	Single point of failure (no backup): lane available, but lockage times increase	3
Low	2nd order failure: If backup also fails: lane available, but lockage times increase	2
None	No effect	1

Table 7. Failure mode criticality, classification and scale

Failure Mode Criticality (Cr)	Criticality	Model Scale
High	Yes	4
Low	No	1

Table 8. Mean time to repair, classification and scale

Mean Time to Repair	MTTR Hours	Model Value
Poor	> 8	4
Average	< 8	3
Good	< 4	2
Excellent	< 2	1

As a result of the above mentioned classification, maximum value of RPN for an failure mode was set to 320 risk dimensionless units (notice that 320=4x5x4x4 when substituting in Equations 2 and 1). The team established three levels of assets criticality as in Table 9.

In Table 10, presents an example for calculating the RPN. A total of 16 failures modes of the sub-system are presented already sorted by their priority resulting from their estimated RPN. A number of 2 failures modes out of 16 were found to be extremely critical, 2 very critical, 6 critical, 2 low criticality and 3 non-critical.

2.2.3.5. Selection of Maintenance Activities within RCM

Once the FMECA has been done, the natural RCM equipment should select the maintenance activity that aides in the appearance of each previously identified failure mode, starting at the logical decision tree (tool designed by RCM, that permits selection of the most adequate maintenance activity to prevent the appearance of a fault mode or decease its possible effects). After the selection of the type of maintained activity from the logical decision tree, one has to specify the maintenance action to undertake, associated to the selected maintenance activity, with the respective frequency of execution, keeping in mind that one of the main objectives of RCM, is preventing or at least reducing possible consequences to human, environmental and operative security, that may arise from the appearance of different fault modes (Crespo, 2007). The first step to the selection of maintenance activities, consists on the identification of the consequences that generate failure modes (see Figure 6).

2.2.3.5.1 Preventive Activities: Condition Maintenance Tasks

Many preventive maintenance activities can be based on the equipment condition. This is due to the fact that the equipment conditions do not change instantaneously when the failure takes place (function loss), but they normally follow a certain continuous deterioration process during a period of time. A

Table 9. RPN criticality level for failure mode

Failure Mode Criticality Level	Adimensional Risk Value	Value
Extremely Critical	$256 \leq RPN \leq 320$	5
Very Critical	$192 \leq RPN < 256$	4
Critical	$128 \leq RPN < 192$	3
Low Criticality	$64 \leq RPN < 128$	2
Non-Critical	$RPN < 64$	1

Table 10. Failures modes priority according to their RPN

Failure Mode	F	E	Cr	MTTR	C	RPN	PRIORITY
-PUMP 101A	4	5	4	4	80	320	EXTREMELY CRITICAL
-MOTOR 105B	4	4	4	4	64	256	EXTREMELY CRITICAL
-PRES. VALVES 101V	4	3	4	4	48	192	VERY CRITICAL
-SEAL GAS	3	4	4	4	64	192	VERY CRITICAL
COMPRES. 214C							
-PRES. VALVES	3	3	4	4	48	144	CRÍTICAL
109V							
-MAIN COL. M123	3	4	4	3	48	144	CRÍTICAL
-HEAT	2	4	4	4	64	128	CRÍTICAL
EXCHANGER E23							
-SECONDARY	3	3	4	4	48	144	CRÍTICAL
ABSORBER S45							
-H2S INHIBITOR I32	3	3	4	4	48	144	CRÍTICAL
-PRECIPITATOR P12	4	3	4	3	36	144	CRÍTICAL
-3 STAGE	3	3	4	3	36	108	LOW CRÍTICALITY
SEPARATOR S12							
-PUMP BOILER 23	2	3	4	4	48	96	LOW CRÍTICALITY
-PRE-WARMING	2	3	4	4	48	96	LOW CRÍTICALITY
TRAIN WT124							
-NAFTA DESPOILER VALVE 141V	2	3	4	1	12	24	NON-CRITICAL
-ALC. DESPOILER MOTOR 138V	1	2	1	2	4	6	NON-CRITICAL
-APC DESPOILER VALVE 176V	1	1	1	2	2	2	NON-CRITICAL

potential failure can then be defined as an identifiable physical equipment condition that indicates that a functional failure is about to happen, or is happening, during the process. The moment in time when it is possible to detect the occurrence of a functional failure, or that a failure is about to occur, is known as the potential failure time. Common examples of potential failures are:

- Vibration readings indicating imminent bearings failure;
- Existing cracks in metals indicating imminent failure due to fatigue;
- Oil particles in any gearbox, indicating imminent faults due to excessive teeth wearout;
- Hot spots indicating deterioration/wearing of the isolating material in a boiler, etc.

The behavior over time of the equipment condition is illustrated in Figure 7. This figure shows how a certain equipment condition that starts to deteriorate (beginning point "I"; often this point may not be

Figure 6. Consequences of failure modes
Source: Parra, C. and Crespo, A, (2012), Ingeniería de Mantenimiento y Fiabilidad Aplicada en la Gestión de Activos. Desarrollo y aplicación práctica de un Modelo de Gestión del Mantenimiento (MGM), Primera Edición. Editado por INGEMAN, Escuela Superior de Ingenieros Industriales, Universidad de Sevilla, España

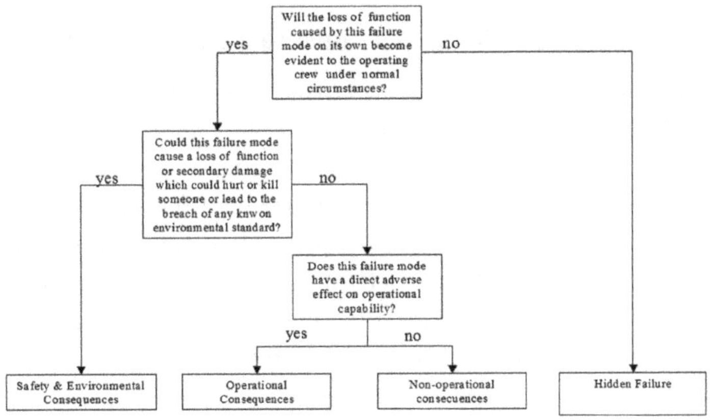

detected); then this condition reaches a point when the failure may be detected (potential failure point "P"); finally, if the failure is not detected nor corrected the equipment condition gets to a point where the functional failure takes place (point "F", where the equipment does not fulfil the function any longer).

2.2.3.5.2 Preventive Activities: Scheduled Restoration Tasks (Fixed Time)

These are periodical activities carried out to restore a part of an item (system, equipment, part) to its original condition. Of course the time interval between two consecutive scheduled actions will be shorter than the operative life limit of the item part to be restored. During this type of preventive maintenance action, items are taken out of service, unarmed, put aside and inspected in a general manner, corrected and replaced if necessary, in order to prevent the appearance of possible failure modes. In case of large equipment or systems, these scheduled restoration tasks are generally known as "overhauls" and they

Figure 7. Behavior Curve of potential faults
Source: Parra, C. and Crespo, A, (2012), Ingeniería de Mantenimiento y Fiabilidad Aplicada en la Gestión de Activos. Desarrollo y aplicación práctica de un Modelo de Gestión del Mantenimiento (MGM), Primera Edición. Editado por INGEMAN, Escuela Superior de Ingenieros Industriales, Universidad de Sevilla, España

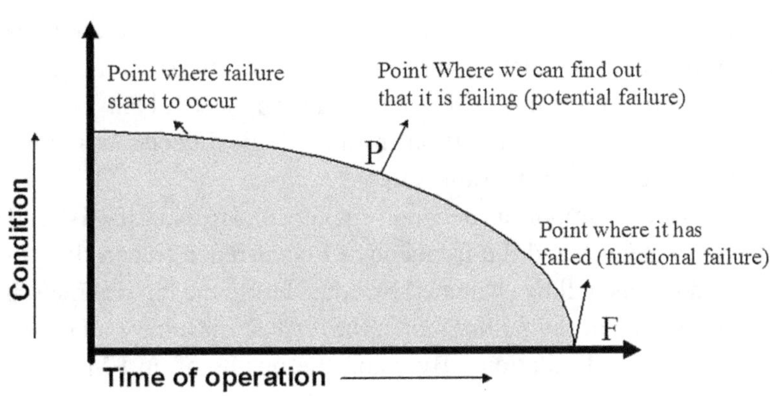

are common in equipment such as compressors, turbines, broiler, furnaces, engines, *etc*. Restoration includes different actions such as: adjustment, inspection, improvement, cleaning, restoration and even replacement.

2.2.3.5.3 Preventive Activities: Scheduled Discards Tasks (Fixed Time)

This type of preventive activity is oriented towards replacement of components or used parts of an active, for new ones, at intervals of time shorter than that of their useful life (before they fail). The programmed discarding activities shall return the component to its original condition, since the old component shall be replaced with a new one. The difference between the discarding and the restoration tasks is that the first are applied to components and or parts of an active and not to complex actives (actives with many components), and then the undertaking of the programmed discarding is specifically to the replacement of an old component for a new one. In the case for programmed restoration tasks the actions to be undertaken may be: adjusting, inspection, improvement, cleaning, restoration and even exchanging old parts or pieces for new ones.

2.2.3.5.4 Preventive Activities: Failure Finding Tasks (for Hidden Fault)

As was defined earlier, the hidden fault modes are no evident under normal operation conditions, therefore this type of fault does not have direct consequences, but these do originate appearance of multiple faults in a determined operational context. One of the ways to reduce the occurrence possibility of hidden faults is to periodically check if a hidden function is working correctly. These checks are known as revision tasks for hidden faults. In conclusion, the revision task consist on the checking action of the actives with hidden functions at regular interval of time, in order to detect if the hidden functions are at normal operation or at failure stage.

When preventive activities are not technically feasible, or they are ineffective, for a certain failure mode, corrective activities shall be those that apply. Possible corrective actions can be as below.

2.2.3.5.5 Corrective Activities: Redesign

In the case, that one may not finding preventive actions that aide in the reduction of fault modes that affect security or the environment to an acceptable level, a re-designing is necessary to minimize or eliminate the consequences of failure modes.

2.2.3.5.6. Corrective Activities: Non Scheduled Maintenance (Run to Failure)

In the that there are no preventive activities that may result least costly than the possible effects that may arise from the fault modes with safety, environment or operational consequences (minimal risk), the decision to wait for a failure or act in a corrective manner may be taken.

An example of maintenance activity definition from the logical decision tree is shown in Table 11 (failures modes and RPN were taken from Table 10).

RCM methodology provides an important decision-making tool to quantify risk and reliability in terms of the severity of the consequences and the frequency of occurrence (mean time to failure – MTTF). The severity of the consequences will be evaluated by considering the environment in which the failure mode occurs. To evaluate MTTF of the failure mode, the analyst must have an understanding of failure mode rates data, their origin and limitations. By using the results of an FMECA, analysts are better

Table 11. Maintenance activity definition from the RCM (logical decision tree)

Failure Mode	Effects of Failures Modes	RPN	RCM Decision Tree	Maintenance Activities / Scheduling Frequency
Pump 101	- Not affect safety or environment - Total loss of pumping, affecting operations (unavailability)	320 (extremely critical)	- Condition Task	1. Vibration Analysis / 15 days 2. Lubrication Analysis / 3 months
Seal Gas 214C	- Not affect safety or environment - Partial loss of compression, affecting operations (unavailability)	192 (very critical)	- Condition Task - Discard Task	1. Pressure Analysis / Daily 2. Replacement / 3 years
Alc. Despoiler Motor 138v	- Not affect safety or environment - Not affect the operations	6 (non-critical)	-Non-scheduled Task	1. Run to failure

equipped to answer questions such as: "Which of several candidate systems poses the least risk?" "Are risk reduction modifications necessary?" and, "Which modifications would be most effective in reducing risk or increasing reliability?"

RCMs can be used effectively for several purposes, besides identifying safety or reliability failure modes and effects. The organizations can use RCM for:

- Preparation of diagnostic routines such as flowcharts or fault-finding tables. FMEA provides a convenient listing of failure modes, which produce particular failure effects, or symptoms, and their relative likelihood of occurrence;
- Preparation of preventive maintenance requirements. The effects and mean time to failures can be considered in relation to the need for scheduled inspection, servicing or replacement;
- Retention as formal records of the safety and reliability analysis, to be used as evidence if required in reports to customers or in product safety litigation.

Finally, it is important to coordinate these activities, so that the most effective use can be made of RMC in all of them, and to ensure that RCM are available at the right time and to the right people.

3. KPIs (KEY PERFOMANCE INDICATORS) FOR MAINTENANCE MANAGEMENT

KPIs selection is an important decision making process that may have many potential implications (Parra and Crespo, 2012, Crespo, 2007, Kaplan and Norton, 1992). For this report, the KPIs proposed, are related with the three basic technical indicators: Availability, Reliability and Maintainability (Jardine, 1999). We hereafter shall present the parameters to be used in the calculation of these indexes, see Figure 8 (it is recommended that these indicators can be calculated within the software MAXIMO).

Where:

1 = operational condition of the equipment.

Figure 8. Distribution of Equipment Failures

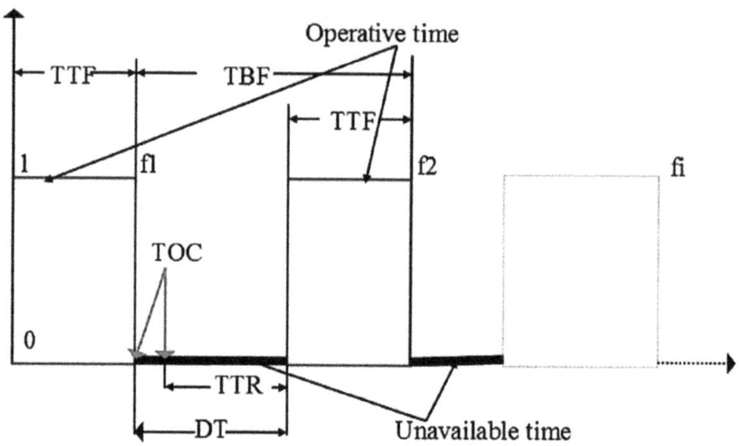

0 = non- operational condition of the equipment.
Fi = nth fault/failure.
TTF = time to failure.
TBF = time between failure.
DT = down time between failures.
TTR = time to repair.
TOC = time out of control (difficult estimation time, related to maintenance logistics: supplier, transportation, delays, lead-time, etc.).

3.1 Reliability Indicator

Reliability may be defined as:

The probability that a team fulfils a specific mission (no failure) under specific operation conditions during a specific period.

Reliability is related to the failure rate (mount of failures) and with the mean time to failure (MTTF, operative time to failure). Whilst the number of faults of a specific asset is increasing or while the MTTF decreases, its reliability will be less.

Basic indicator of Reliability: MTTF = mean time to failure.

$$MTTF = \sum TTF / n \quad (3)$$

n = number of failures.

3.2 Maintainability Indicator

Maintainability may be defined as:

The possibility for an equipment to be returned to the stage to be able to fulfill the mission in a given time period, after the appearance of a fault, using pre-established maintenance procedures.

The maintainability relates to the design and complexity of the equipment, with the qualifies personal to undergo the maintenance, with the available tools and the maintanance procedures. The fundamental parameter to calculate maintainability is constituted by the mean time to repair (MTTR). When the MTTR for specific equipment is high, the equipment has a low maintainability (its time to repair shall be reduced). In the contrary, if the mean time to repair for specific equipment is low, it is considered that the equipment has a high maintainability.

Basic indicator of Maintainability: MTTR = mean time to repair.

$$MTTR = \sum TTR / n \quad (4)$$

n = number of failures.

3.3 Availability Indicator

This term may be expressed in first approximation, as proportion to the time for the equipment to be ready to be put into action again, in the given conditions, with respect to the time to fulfill its mission and did not do so. This last period is denominated unavailability and is obviously, unproductive. The concept of availability may be defined as:

The possibility for an equipment to be able to fulfill its mission at any given time

The availability relates the mean downtime (MDT, represents maintainability) and the mean time to failure (MTTF, represents, reliability).

From the three mentioned indicators, availability constitutes information from the parameter most represented and useful of the management. Upon calculating availability is easier in comparison to the calculus of the other two parameters and interrelated to reliability and maintainability.

We shall now present the way to calculate: Operational Availability (Ao)

- **Availability (Ao):** operational availability takes into account the non-operative time of the team in a general manner (from the time comes out of service until it is put into action again), i.e. includes the delay (but does not estimate nor quantifies) which bring about the logistic of the maintenance (purchase of parts, transportation, unspecified idleness, etc.). The equation to calculate operational availability (Ao) is:

$$A_o = \frac{MTTF}{MTTF + MDT} \times 100\% \quad (5)$$

where,

n: number of failures
MTTF: mean time to failure
MDT: mean downtime
MTOC: mean time out control

$$MDT = MTTR + MTOC \quad (6)$$

$$MDT = \sum DT / n \quad (7)$$

$$MTTR = \sum TTR / n \quad (4)$$

$$MTOC = \sum TOC / n \quad (8)$$

$$MTTF = \sum TTF / n \quad (3)$$

3.4 Example of Calculation: Reliability, Maintainability, and Availability

- Given the following failure distribution (Figure 9) of the equipment X, for a period of 53 hours, calculate:
  - Reliability: MTTF
  - Maintainability: MTTTR and MDT
  - Availability: Ao

1. To calculate MTTF (reliability), apply equation 3:

Figure 9. Failure distribution of equipment X in an operation time of 53 hours

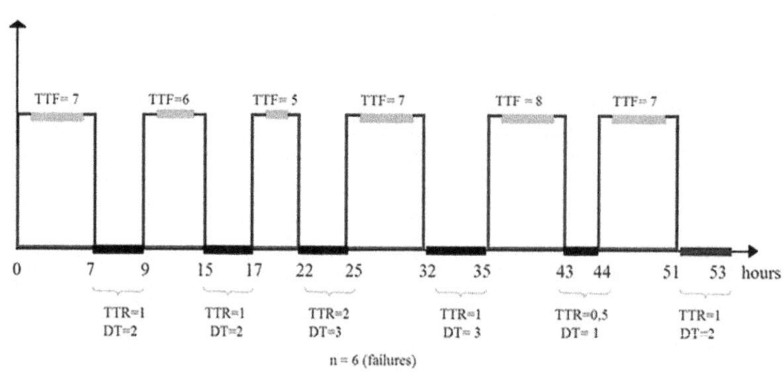

MTTF = (7 + 6 + 5 + 7+ 8+ 7) / 6 = 40/6 = 6.66 hours

2. To calculate MTTR y MDT (maintainability), apply equations 4 and 8:

MTTR = (1 + 1 + 2 + 1+ 0.5+ 1) / 6 = 40/6 = 1.08 hours

MDT = (2 + 2 + 3 +3 +1+2) / 6 = 13/ 6 = 2.16 hours

3. To calculate availability, apply equation 5:

MTTF = (7 + 6 + 5 + 7+ 8+ 7) / 6 = 40/6 = 6.66 hours

MDT = (2 + 2 + 3 +3 +1+2) / 6 = 13/ 6 = 2.16 hours

Ao = (40/6) / ((40/6) + (13/6)) =

Ao = 40 / (40 + 13) = 40 / 53 =

Ao = 0.754 = 75.4% availability of the equipment X in a period of 53 hours.

The KPIs plays an important role in the efficiency of the maintenance organization (Jardine, 1999). Generally, equipment records are classified under four categories: inventory, maintenance cost, files, and maintenance work performed. KPIs are used in various areas including troubleshooting breakdowns, investigating incidents, procuring new equipment to determine operating performance trends, performing life cycle cost and design studies, conducting replacement and modification studies, and conducting reliability and maintainability studies. Progressive maintenance organizations measure their performance on a regular basis through various means. KPIs analyses play an important role in maintenance organization efficiency and are useful in revealing equipment downtime, peculiarities in operational behavior of the organization, and so on. Finally, maintenance management must ensure calculate the three KPIs shown in this section, in order to minimize uncertainty in the process of decision making in the areas of: reliability, maintainability and availability.

4. GENERAL RECOMMENDATIONS FOR MAINTENANCE AND RELIABILITY MANAGEMENT

This section presents some general recommendations to MWH Global for optimizing the processes in the areas of Maintenance and Reliability Management within the project: Design and Construction of the Third Set of Locks in the ACP (AUTORIDAD DEL CANAL DE PANAMÁ). The following key aspects to consider in this project are shown below (Peters, 2006):

- View maintenance as a priority and as an internal business opportunity. The process of performing maintenance and managing physical assets must be recognized as a top priority. It must also be viewed as an internal business and not as a necessary evil. It will be viewed as an area that contributes directly to the bottom line when a profit- and customer-centered strategy and continuous maintenance improvement are adopted.

- Develop leadership and technical understanding. Maintenance leaders must understand the challenges of maintenance and provide effective maintenance leadership to operate maintenance as an internal business. Maintenance leadership must continually develop the skills, abilities, and attitudes to lead maintenance into the future.
- Develop pride in maintenance. Maintenance operations will experience fundamental improvements in work ethic, attitude, values, job performance, and customer service to achieve real pride in maintenance excellence. Tangible savings and improvements will occur as a result of continuous maintenance improvement.
- Recognize importance of the maintenance and reliability profession. The profession of maintenance and reliability will gain greater importance as a key profession for success within all types of organizations as the role of the chief maintenance officer (CMO) becomes well established. Maintenance and Reliability leaders will be recognized as critical resources that are absolutely necessary for the success of the total operation.
- Increase core competencies of your maintenance personnel. A significant upgrade in the level of personnel involved with maintenance will take place to keep pace with new technologies and responsibilities. Maintenance operations will achieve a significant upgrade in the skill level of maintenance craftspeople in order to keep pace with new technology and responsibilities.
- Establish effective maintenance planning, estimating and scheduling. Maintaining customer satisfaction and the utilization of available craft time will improve through more effective planning and scheduling systems. The development of more effective planning and scheduling systems will be a top priority for a profit- and customer centered strategy. As reductions in breakdown repairs occur through more effective preventive/predictive maintenance, the opportunity to increase planned maintenance work will result.
- Develop pride in ownership. Equipment operators and maintenance will develop a partnership for maintenance service and prevention and take greater pride in ownership through operator-based maintenance. Equipment operators will assume greater responsibility for cleaning, lubricating, inspecting, monitoring, and making minor repairs to equipment. Maintenance will provide training support to operators to achieve this transfer of responsibility and to help operators with early detection and prevention of maintenance problems. Operators will develop greater pride in ownership of their equipment with their expanded responsibilities.
- Improve equipment effectiveness. A leadership-driven, team-based approach will be used by maintenance and manufacturing operations to totally evaluate and subsequently improve all factors related to equipment effectiveness. The goal is maximum availability of the asset for performing its primary function.
- Maintenance and engineering: a partnership for profitable technology application. Maintenance and engineering will work closely during systems specification, installation, start-up, and operation to provide maintenance with the technical depth required for maintaining all assets and systems. Engineering will provide technical resources and support to ensure that maintenance has the total technical capability to maintain all equipment and systems.

- Continuously improve reliability and maintainability. Machines and systems will be specified, designed, retrofitted, and installed with greater reliability and ease of maintainability. Equipment design will focus on maintainability and reliability and not primarily on performance. Design for maintainability will become an accepted philosophy that fully recognizes the high cost of maintenance in the life-cycle cost of equipment. The causes for high life-cycle costs will be reduced through the application of good maintainability and reliability principles during design.
- Manage life-cycle cost and obsolescence. The life-cycle costs of physical assets and systems will be closely monitored, evaluated, and managed to reduce total costs. A profit- and customer-centered strategy will achieve significant reductions in total life-cycle costs through an effective design process prior to purchase and installation. During the equipment's operating life, systems will be developed to continually monitor equipment costs. Information to identify trends will be available to highlight equipment with high maintenance costs.
- Minimize uncertainty and eliminate root causes. Uncertainty will be minimized through effective preventive/predictive maintenance programs and through continuous application of reliability-centered maintenance techniques and continuous monitoring systems. Effective preventive/predictive maintenance programs will be used to anticipate and predict maintenance problems in order to eliminate the uncertainty of expected breakdowns and high repair costs. Preventive/predictive maintenance tasks will be adequately planned based upon criticality of failure and will cover all major assets within the operation.
- Maximize use of computerized maintenance management and enterprise asset management. Systems that support the total maintenance operation will improve the quality of maintenance and physical asset management and be integrated with the overall business system of the organization. Computerized maintenance management systems (CMMS) will provide greater levels of manageability to maintenance operations. CMMS will cover the total scope of the maintenance operation providing the means to improve the overall quality of maintenance management. Enterprise asset management (EAM) will provide a broader scope of integrated software to manage physical assets, human resources, and parts inventory in an integrated system for maintenance management, maintenance, procurement, inventory management, human resources, work management, asset performance, and process monitoring.
- Use maintenance information to manage the business of maintenance. The maintenance information system and database will encompass the total maintenance function and provide real-time information to improve maintenance management. The implementation of CMMS and EAM provides the opportunity for improved maintenance information systems. With CMMS and EAM, the maintenance information system can be developed and tailored to support maintenance as a true "business operation."
- Ensure an effective maintenance storeroom operation. The maintenance storeroom will be orderly, space efficient, labor efficient, responsive, and provide the effective cornerstone for maintenance excellence. The maintenance storeroom for maintenance repair operations (MRO) items will be recognized as an integral part of a successful maintenance operation. Initial storeroom design or modernization will include effective planning for space, equipment, and personnel needs while providing a layout that ensures efficient inventory control and includes maximum loss control measures. It will be professionally managed and maintained in a clean, orderly, and efficient manner.
- Aggressive support compliance to environmental, health, and safety requirements. Maintenance must provide proactive leadership and support to regulatory compliance actions. Maintenance

leaders must maintain the technical knowledge and experience to support compliance with all state and federal regulations. The issue of indoor air quality must receive constant attention to eliminate potential problems. Maintenance must work closely with other staff groups in the organization such as quality and safety to provide a totally integrated and mutually supportive approach to regulatory compliance.
- Continuously evaluate, measure, and improve maintenance performance and service. Broad-based measures of maintenance performance and customer service will provide a continuous evaluation of the value of maintenance. CMMS will allow for a broad range of measurement for maintenance performance and service. Investment in best maintenance practices will require valid return on investment. Projected savings will be established and results will be validated. Measures will be developed in areas such as labor performance/utilization, compliance to planned repair and preventive/predictive maintenance schedules, current backlog levels, emergency repair hours, storeroom performance, asset uptime, and availability.

REFERENCES

BSI PAS 55. (2008). Publicly Available Specification, London.

Campbell, J. D., & Jardine, A. K. S. (2001). *Maintenance Excellence*. New York, NY: Marcel Dekker.

CEN. (2001). *Maintenance Terminology. European Standard, EN 13306:2001*. Brussels: European Committee for Standardization.

Crespo Marquez, A. (2007). *The Maintenance Management Framework, Models and Methods for Complex Systems Maintenance*. London: Springer.

Duffuaa, S. O. (2000). Mathematical models in maintenance planning and scheduling. In M. Ben-Daya, S. O. Duffuaa, & A. Raouf (Eds.), *Maintenance, Modelling and Optimization*. Boston, MA: Kluwer Academic Publishers. doi:10.1007/978-1-4615-4329-9_2

EFNMS. (2006). The requirements of Competencies and Responsibilities for an European Specialist in Maintenance Supervision.

European Federation of National Maintenance Societies (2006). *The Requirements of Competencies and Responsibilities for a European Specialist in Maintenance Supervision.*

Gelders, L., Mannaerts, P., & Maes, J. (1994). Manufacturing strategy, performance indicators and improvement programmes. *International Journal of Production Research*, *32*(4), 797–805. doi:10.1080/00207549408956971

ISO 55000. (2014). Asset management — Overview, principles and terminology.

ISO 55001. (2014). Asset management — Management systems — Requirements.

ISO 55002. (2014). Asset management — Management systems — Guidelines on the application of ISO 55001.

Jardine, A. (1999). Measuring maintenance performance: A holistic approach. *International Journal of Operations & Production Management*, *19*(7), 691–715. doi:10.1108/01443579910271674

Kaplan, R. S., & Norton, D. P. (1992). The balanced scorecard – measures that drive performance. *Harvard Business Review*, *70*(1), 71–79. PMID:10119714

Kaplan, R. S., & Norton, D. P. (1992). The Balanced Scorecard - measures that drive performance. *Harvard Business Review*, *70*(1), 71–79. PMID:10119714

Lee, J. (2003). E-manufacturing: Fundamental, tools, and transformation. *Robotics and Computer-integrated Manufacturing*, *19*(6), 501–507. doi:10.1016/S0736-5845(03)00060-7

Moubray, J. (1997). *Reliability-centred Maintenance* (2nd ed.). Oxford: Butterworth-Heinemann.

Palmer, R. D. (1999). *Maintenance Planning and Scheduling*. New York, NY: McGraw-Hill.

Parra, C., & Crespo, A. (2012). *Ingeniería de Mantenimiento y Fiabilidad Aplicada en la Gestión de Activos. Desarrollo y aplicación práctica de un Modelo de Gestión del Mantenimiento (MGM)*. Primera Edición. Editado por INGEMAN, Escuela Superior de Ingenieros Industriales, Universidad de Sevilla, España.

Peters, R. A. (2006). *Maintenance Benchmarking and Best Practices* (1st ed.). New York, NY: McGraw-Hill.

Pintelon, L. M., & Gelders, L. F. (1992). Maintenance management decision making. *European Journal of Operational Research*, *58*(3), 301–317. doi:10.1016/0377-2217(92)90062-E

Prasad Mishra, R., Anand, D., & Kodali, R. (2006). Development of a framework for world-class maintenance systems. *Journal of Advanced Manufacturing Systems*, *5*(2), 141–165. doi:10.1142/S0219686706000807

Tsang, A., Jardine, A., & Kolodny, H. (1999). Measuring maintenance performance: A holistic approach. *International Journal of Operations & Production Management*, *19*(7), 691–715. doi:10.1108/01443579910271674

Vagliasindi, F. (1989). *Gestire la manutenzione. Perche e come*. Milan: Franco Angeli.

Vanneste, S. G., & van Wassenhove, L. N. (1995). An integrated and structured approach to improve maintenance. *European Journal of Operational Research*, *82*(2), 241–257. doi:10.1016/0377-2217(94)00261-A

Wireman, T. (1998). *Developing Performance Indicators For Managing Maintenance*. New York, NY: Industrial Press.

Chapter 2
Multicriteria Model for the Selection of Maintenance Policies in Subsystems of an Operating Theatre

María Carmen Carnero
University of Castilla – La Mancha, Spain & University of Lisbon, Portugal

Andrés Gómez
University General Hospital of Ciudad Real, Spain

ABSTRACT

The aim of this article is to select the most suitable combination of maintenance policies in the different subsystems that make up an operating theatre. To do so a multicriteria model will be developed using the Measuring Attractiveness by a Categorical Based Evaluation Technique (MACBETH) technique considering multiple decision centres: The Hospital's technical services, environment and occupational risk prevention managers, healthcare managers (operating theatres and health activity programming), healthcare staff, technicians, purchasing service managers and Hospital executives. The model uses functional, safety and technical-economical criteria, amongst which is availability. Mean availability for repairable systems has been measured to assess this criterion, using Markov chains from the data obtained over three years from the subsystems of a Hospital operating theatre. All this is aimed at increasing the operating theatre's availability and, consequently, increasing physical safety in patient operations and reducing the number of delayed operations due to technical malfunctions.

INTRODUCTION

The most up-to-date hospitals are experiencing increasing automation and monitoring of installation and equipment and, in general, greater complexity in the assets to be maintained (Wolf, 2008). To adjust to this new environment, a hospital's technical services (including maintenance engineering, safety and environmental services) should adapt to modern strategies and management tools that have, in fact,

DOI: 10.4018/978-1-5225-0651-5.ch002

already been operating for some time in the manufacturing, energy, etc. industries. Currently, due to the lack of structured criteria in hospital activity, it is normal to find different strategies and approaches to achieve similar ends with a great variety of costs and results. However, the activities carried out are essentially corrective and technical-legal (Fernández & Canals, 2006).

It is essential to analyse, from the qualities perceived by patients and healthcare professionals, the aspects that have a bearing on a hospital as a building answering to customer needs and increased safety. It is in these areas that the technical services have a predominant role, although there are deficiencies in recognizing the importance of maintenance management in hospitals (Rani, Baharum, Akber, & Nawawi, 2015). In general, users of the Spanish health system believe that the scientific-technical quality of its health professionals is excellent (SESCAM, 2007). They do not think the same about the other services, and particularly those regarding buildings, facilities, cleaning and those necessary for minimum comfort standards.

The application of a methodology such as that described in this research should therefore lead to a very significant change in the appreciation that people in general have of the technical services of the hospital responsible for carrying out maintenance

Furthermore, health care professionals should be made aware of the situation of the facilities and equipment they use as internal users, as they need greater safety in their working environment. To do so, procedures should be established so that the condition of the equipment, facilities and working environment is precisely determined, and the staff need to be included in the assessment process.

The aim of this article is to select the most suitable maintenance policies in the different subsystems that make up an operating theatre. To do so, a multi-criteria model is designed, using the Measuring Attractiveness by a Categorical Based Evaluation Technique (MACBETH) (Bana e Costa & Vansnick, 1994), considering multiple decision centres: managers from the Hospital's technical services, environmental and occupational risk prevention sections, healthcare managers (operating theatre and health activity programming), healthcare staff, technicians, purchasing managers and Hospital executives. The model uses safety, functional and technical-economic criteria, among which is availability. Mean availability for repairable systems has been measured to assess this criterion, using Markov chains from the data obtained over three years from the subsystems of a Hospital operating theatre. All this is aimed at increasing the operating theatre's availability and, consequently, increasing physical safety in patient operations and reducing the amount of delayed operations due to technical malfunctions.

Thus, the application of this research could lead to a significant increase in care quality, when carried out successfully in a state general hospital.

To choose the type of Multi-Criteria Decision Making (MCDM) to apply in this case study, it was felt that, given the multi-faceted profile of the decision-making groups, it should be easy to handle, intuitive, and respond to qualitative stimuli, without the need for the decision-making group to have to establish a numerical scale for subsystems. The choice of MACBETH is justified latter.

The originality of this chapter rests on the following points:

- The inclusion of the multicriteria technique MACBETH in group decision-making and its use with Markov chains in repairable systems to assess the availability of subsystems in the Hospital.
- It is the first research to analyse the selection of the best combination of maintenance policies, assessing all the subsystems which make up the operating theatre system. To this end the interrelations between the different subsystems of the Hospital were analysed.

- Although practical research on choice of maintenance policy in manufacturing companies has been done, this is one of the first studies that analyses this matter in a public service company such as a Hospital, and specifically an operating theatre. The environment, the facilities and the need for availability and quality of service in those facilities or systems may be very different from other industrial settings, due to the fact that its efficiency has a very important impact on the way in which health services are provided and on the life of patients. This is expressed in the decision criteria used, which are adapted to the context; for example, Risk to users (patients).

The rest of this chapter is structured as follows. First, there is a review of the literature about maintenance policy selection; next it describes the characteristics of the MACBETH multi-criteria technique and the Markov theory for stable repairable systems; then the multi-criteria model for the selection of maintenance policies developed for an operating theatre's subsystems is presented, followed by a sensitivity analysis. The paper ends with future lines of research and conclusions.

BACKGROUND

Different types of maintenance policy can be identified:

- **Corrective Maintenance:** Carried out after a failure with the aim of returning the machine to a state in which it can perform its desired function (Kelly, 2006).
- **Preventive Maintenance:** Carried out at set intervals throughout the working life of the machine (Knezevic, 1996), with the aim of reducing the probability of problems or wear in the machine (Kelly, 2006).
- **Condition-Based Maintenance:** Based on the control of physical parameters of a working machine that can be registered, periodically or continuously, by a set of sensors, to analyse and compare data which allow decisions to be made about the working and maintenance of the machine (Rao, 1996).
- **Opportunity-Based Maintenance:** Involves using the stoppage of a system due to a fault in a component to carry out preventive maintenance in other components which are working normally (Pintelon & Gleders, 1992).
- **Proactive Maintenance:** Detects and removes problems before they come about, hence avoiding their negative effects (Gómez & Crespo, 2009).
- **Perfective Maintenance:** Based on continuous improvement. The aim is to improve the behaviour, ease of maintenance and the service which the system can provide (Gómez & Crespo, 2009).
- **Total Productive Maintenance (TPM):** Defined by Nakajima (1988) as productive maintenance involving total participation in addition to maximizing equipment effectiveness and establishing a thorough system of planned preventive maintenance.
- **Reliability Centred Maintenance (RCM):** A systematic methodology for the allocation of efficient predictive and preventive maintenance aimed at preventing the dominant causes of failure of critical equipment, and, in turn, towards achieving acceptable levels of equipment availability and costs by reducing corrective maintenance (Martorell et al., 1999).

There is little research that analyses the choice of maintenance policy despite its importance in industry (Bertolini & Bevilacqua, 2006; Wang, Chu, & Wu, 2007); those worth mentioning include Al-Najjar (1999), Azadivar and Shu (1999), Sharma et al. (2005)Bashiri et al. (2011), Zaim et al., (2012), Chan and Prakash (2012), Shahin, Shirouyehzad and Pourjavad (2012). A review of contributions which apply MCDM in the choice of maintenance policies may be seen in Shafiee (2015).

In most of these previous papers, cost is the key determinant. However, the maintenance of medical equipment or installations should bear in mind other factors in addition to cost, such as the safety of the patient and professionals, the impact on healthcare activity and the environment. MCDM constitutes a rational and objective tool to improve understanding of underlying decision processes in the systemic processes and, to help the decision-making groups to compare the alternatives (Romero, 1996). The application of an MCDM for optimising maintenance policies provides the advantages of analysing each criterion from the point of view of the different actors that have decision-making capacity on the equipment analysed. Thus, Almeida and Bohoris (1995) show that many maintenance decisions need to include criteria such as ease of maintenance, reliability, availability and safety in industrial settings, such as a Hospital, due to the high technology of the facilities and devices and the need for availability (Martorell et al., 2005).

The use of multicriteria techniques in choosing optimal maintenance policies can be found in the following work. Vatn (1997) uses utility functions and influence diagrams for the choice of maintenance actions (scheduled failure-finding task, scheduled on-condition task, scheduled restoration task and scheduled discard task) within an RCM programme. Labib et al. (1998) develop a maintenance decision model using the Analytic Hierarchy Process (AHP) together with the available data for equipment failure. Ramadhan et al. (1999) use AHP to decide priority factor weights which in order to develop a pavement maintenance priority ranking procedure for a road network. Bevilacqua and Braglia (2000) use AHP to select the best maintenance strategy in an integrated gasification and combined cycle factory. Kodali and Chandra (2001) use AHP to justify the application of TPM in Indian industries; the alternatives considered in this model are the application of a TPM system or a traditional maintenance system. Emblemsvag and Tonning (2003) determine appropriate maintenance policy for a weapon of the Norwegian Army by means of AHP; in this case, four maintenance echelons are considered and Monte Carlo simulations are used to evaluate the robustness of the decision. Al-Najjar and Alsyouf (2003) use fuzzy multicriteria decision making taking as criteria the causes of failure and as alternative corrective, preventive, and condition monitoring maintenance based on vibration analysis with continuous and periodic data collection, RCM, TPM and Total Quality Maintenance. Labib (2004) analyses the application of maintenance policy to equipment with different level of criticality, using fuzzy logic to obtain criteria in the form of a set of functions. In Carnero (2005), AHP and factorial analysis are used to choose the most suitable diagnostic techniques and instrumentation for conditional maintenance programmes based on analysis of lubricants and vibrations, taking into account different levels of technology, categorized by cost and diagnostic capacity. Bertolini and Bevilacqua (2006) set out a model combining goal programming and AHP to define the best maintenance policy to be applied to critical centrifugal pumps in an oil refinery, giving a priority vector of the possible policies: corrective, preventive and predictive for each type of failure analysed. By the use of multicriteria classification for critical equipment, Gómez de León and Ruiz (2006) give indices of criticality for each subsystem and the contribution of each device or piece of equipment to the criticality value obtained. Thus, it is possible to quantify the degree of importance of each subsystem and apply the most suitable maintenance policy depending on the resources available. In Carnero (2006), AHP is used to obtain a strategy for introducing predictive maintenance (internal, subcontracting, carrying out improvements in advance and not introducing any strategy) in a petrochemical plant, in a mixed form with Bayesian techniques to minimize uncertainty, where there is insufficient information. Wang et al. (2007) select an optimal combination of maintenance policies which increase availability and reliability of productive facilities without significantly increasing investment. This is achieved via a fuzzy modification of AHP so that the uncertain and imprecise judgements of decision makers are translated into fuzzy numbers. Carnero (2007) describes a model using AHP to choose the most suitable predictive maintenance techniques by vibration analysis, lubricant analysis and the integration of both techniques depending on the different technological levels. Ghosh and Roy (2009) apply fuzzy AHP to select the optimal mix of maintenance approaches, including corrective maintenance, time-based preventive maintenance and condition-based predictive maintenance in a safety process plant. Also, a goal-programming model is included to optimise risk reduction and cost minimisation that are subject to resource constraints. Ilangkumaran and Kumanan (2009) combine fuzzy AHP and TOPSIS to select an optimum maintenance strategy for a textile plant. Ishizaka and Nemery (2014) apply ELECTRE-SORT, to consider an unlimited number of criteria in order to assign machines to non-comparable strategies, because maintenance strategies cannot always be completely ordered. Cavalcante and Lopes (2015) use a multi-attribute value function to select the maintenance policy for a cogeneration system in an ethanol and sugar plant. Goossens and Basten (2015) apply AHP to select the suitable maintenance policy in naval ships.

However, there is a significant lack of literature related to the application of multi-criteria techniques in the selection of the most suitable maintenance policy for a Hospital. The only precedent Taghipour et al. (2011) which prioritizes medical devices, but according to their criticity it assigns higher or lower priority in the planning of maintenance and sets guidelines to select appropriate maintenance strategies.

This research may serve as an example to other hospitals that can develop their own decision groups and make assessments depending on the specific conditions, the combination of maintenance policies best applied based on objective mathematical tools, instead of taking decisions based solely on experience. This is the only research which integrates MCDM techniques and Markov chains. Furthermore, it should be noted that it has been successfully applied to a state-run general hospital.

Among the decision criteria used in the multi-criteria model presented in this paper is availability. To assess them in the different subsystems that exist in an operating theatre, it was necessary to assess the mean availability of repairable systems with Markov chains. Markov chains in stable repairable systems are frequently used to predict the behaviour of complex repairable elements in the pipework of nuclear power plants with constant failure and repair rates (Fleming, 2004; Gosselin & Fleming, 1997) so that periodical revisions can be programmed for the cost of the shutdowns to be minimum, according to Crespo and Sánchez (2002), and decide on when to intervene depending on the condition of the system (Castanier et al. 2005).

MACBETH INTRODUCTION

MACBETH (Measuring Attractiveness by a Categorical Based Evaluation Technique) (Bana e Costa & Vasnick, 1999) builds a cardinal scale of the alternatives, based on the information provided by the decision centre, for each pair of criteria submitted for judgement, so that this information is a group of attractiveness differences between the two stimuli. For a decision-making group without a great deal of technical knowledge on the information to be assessed, this method is very useful, as it builds a scale of preferences from the indirect information obtained from the answers from simple and direct questions, without this group needing mentally to make a numerical scale, for which extensive knowledge of the material is required.

MACBETH is applied in numerous real cases (see Bana e Costa, Nunes da Silva, & Vansnick, 2001; Bana e Costa, Da Costa-Lobo, Ramos, & Vansnick, 2002; Bana e Costa, Oliveira, & Vieira, 2008; Carvalho & Lourenço, 2002; Dhouib, 2014; Rodrigues, 2014).

The procedure by which MACBETH builds a scale of quantitative values is the following. A facilitator asks the decision-making group to judge the attractiveness difference between the two stimuli $(x, y) \in S$ in 6 semantic attractiveness categories, C_k, $(k=1, 2, \ldots, 6)$; these categories are: very weak, weak, moderate, strong, very strong and extreme. Furthermore, the decision-making group may consider x and y equally attractive, or consider that they are not comparable in attractiveness terms.

The application of the method is carried out in three stages:

1. Obtaining the cardinal scale for each criterion. In each criterion the impact levels that constitute the overall judgements are established by J so that between the elements $S = \{s_1, \ldots s_n\}$ there is a strict weak order, so that s_1 is more attractive than s_2, this than s_3 and successively, regarding the considered criterion.

Two reference levels "good" and "neutral" are chosen from the S group.

M-MACBETH assesses the degree of inconsistency of the judgement matrix and tells the assessor the corrections to make to solve this. M-MACBETH determines, from all the possible scales, a special numerical scale μ in S, with linear programming.

The numerical scale presented by M-MACBETH, called the basic scale, is represented in a standard way or with the thermometer tool, so that the new scale is obtained making the good and neutral levels coincide with 100 and 0 respectively.

In Figure 1, the numerical scale and thermometer are represented as well as the margin of values offered by M-MACBETH, due to the scale obtained not being rigid, but rather there being an infinite number of scales inside the preference margin given by the decision-making group and, therefore, that do not alter the consistency of the judgement matrix. Thus, the decision-making group can choose the values to adopt within these margins, according to their judgement after observing the thermometer scale and comparing margins between the different attractivenesses.

2. Obtaining the MACBETH scale for the criteria. To measure the attractiveness of each stimulus, the weight of each criterion has to be determined. To do so, the previous process is applied, comparing the attractiveness difference between pairs of criteria. M-MACBETH analyses the consistency of the judgement matrix, as it is filled in and allows the decision-making group to correct them at the same time as creating the weight scale in ascending order and from greater to lesser weighting.

The scale offered by M-MACBETH is, therefore, linear. To obtain the standardised MACBETH scale, $\sum_{i=1..n} K_i = 1$; K_i is applied as the specific standardised weight of each criterion and n the number of criteria.

As in the previous case, M-MACBETH offers the possibility of modifying the values obtained within a range.

Figure 1. MACBETH, thermometer and variation margin in the M-MACBETH thermometer

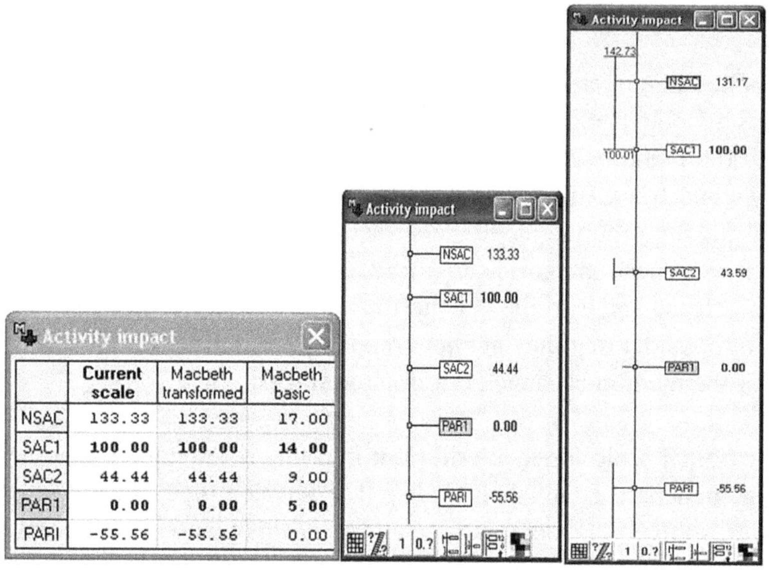

3. Impact of the alternatives on the criteria. The third judgement expression level by the decision-making group is considering what impact value each alternative has on each criterion. To do so, the decision-making group considers a criterion and assigns it a given impact level.

For a detailed explanation of the MACBETH technique, see Bana e Costa and Vansnick (1994), Bana e Costa and Vasnick (1999), Bana e Costa et al. (2006) and Bana e Costa et al. (2011).

MARKOV CHAINS FOR REPAIRABLE SYSTEMS

In a Markov process, "the occurrence of a future state depends on the immediately previous state and only on it" (Taha, 2004). If the system's different states are considered, so that they are exhaustive and mutually exclusive, a special case is obtained for a Markov process, called a Markov Chain. To describe a Markov chain, we have to know the current state and the transition possibilities. If transition probabilities between states are maintained constant throughout time, they are called Homogenous Markov Chains. The Homogeneous Markov chains' most important properties are (Prawda, 1999):

- For a high n number of transitions and considering infinite time, the rows of the transition matrix tend to become equal to the value of the system in each state and at a constant value $\prod$.

$$\lim_{t\to\infty} P_i(t) = \Pi_i \tag{1}$$

- The sum of probable values of the system in each state is equal to 1.

$$\sum_{i=0}^{m} \Pi_i = 1 \tag{2}$$

- The transition between states is considered a Markov chain, and the change in state occurs in a sufficiently small Δt so that only a fault or a repair occurs.

According to international standard IEC 61165 "Application of Markov techniques", the probability of a system being in state *i*, after applying a Δt is called P_i (t + Δt). The system passing from any state to *i* may occur by means of a fault or a repair, not simultaneously. Therefore, two transit situations are considered: states *j (0* to *i-l)* for states in which the transit is carried out by means of a fault and states *h (i+1* to *n)* for those which the transit is carried out by means of a repair.

$$P_i(t+\Delta t) = \sum_{j=0}^{i-1} P_j(t)P_{ji}(\Delta t) + P_i(t)(1-(\sum_{j=i+1}^{n} P_{ij}(\Delta t) + \sum_{g=0}^{i-1} P_{ig}(\Delta t))) + \sum_{h=i+1}^{n} P_h(t)P_{hi}(\Delta t) \tag{3}$$

where:

$P_i(t+\Delta t)$ is the probability of the system being in state *i*, after applying a Δt.
$P_i(t)$, $P_j(t)$, P_h are the probabilities that the system is in the *i, j, h* states in *t*.
P_{ij}, P_{ji}, P_{ig} are the probabilities that the system changes state in a Δt.

The probabilities of a change of state to another imply the non-reliability or the maintainability of the system in this state. They are called:

- $R_0(\Delta t)$, $R_1(\Delta t)$, ..., $R_n(\Delta t)$: Reliability of the system in each state between *t* and $t+\Delta t$ (probabilities of the system being in state *i* in $t+\Delta t$).
- $M_1(\Delta t)$, $M_2(\Delta t)$, ..., $M_n(\Delta t)$ **:** maintainability function of the system in each state between *t* and $t+\Delta t$ (probabilities of the system passing from state *i+1* to state *i* in $t+\Delta t$).
- $F_1(\Delta t)$, $F_2(\Delta t)$, ..., $F_n(\Delta t)$: unreliability function of the system in each state between *t* and $t+\Delta t$ (probabilities of the system passing from state *i* to state *i+1* in $t+\Delta t$). $F_i(\Delta t) = 1 - R_0(\Delta t)$.

Replacing in Equation (3) and considering that fault and repair density functions are taken to be exponential, the following is obtained for each state:

$$P_0(t+\Delta t) = P_0(t)(1-\sum_{j=1}^{n} \lambda_{oj}\Delta t + \sum_{h=1}^{n} P_h(t)\mu_{h0}(\Delta t)$$

$$P_1(t+\Delta t) = P_0(t)\lambda_{01}\Delta t + P_1(t)(1-(\sum_{j=2}^{n} \lambda_{1j}\Delta t + \mu_{10}\Delta t)) + \sum_{h=2}^{n} P_h(t)\mu_{h1}(\Delta t)$$

$$P_{n-1}(t+\Delta t) = \sum_{j=0}^{n-2} P_j(t)\lambda_{j,\,n-1}\Delta t + P_{n-1}(t)(1-(\lambda_{n-1}\Delta t + \sum_{g=0}^{n-2} \mu_{n-1,\,g}\Delta t)) + P_n(t)\mu_{n,\,n-1}(\Delta t)$$

$$P_n(t+\Delta t) = \sum_{j=0}^{n-1} P_j(t)\lambda_{j,\,n}\Delta t + P_n(t)(1-\sum_{g=0}^{n-2} \mu_{n,\,g}\Delta t) \tag{4}$$

Operating on Equation (4) and letting $\Delta t \to 0$, the derivatives in each state can be represented in vector form; the system of equations to be considered is that equivalent to:

$$P' = A * P \tag{5}$$

applying transposition, the system ends up as shown in Equation (6).

$$\begin{pmatrix} P'_o \\ P''_1 \\ \cdot \\ \cdot \\ P'_{n-1} \\ P'_n \end{pmatrix} = \begin{pmatrix} -\sum_{j=1}^{n} \lambda_{oj} & \mu_{10} & \cdot & \cdot & \mu_{n-1,\,0} & \mu_{no} \\ \lambda_{01} & -(\sum_{j=2}^{n} \lambda_1 j + \mu_{10}) & \mu_{21} & \cdot & \cdot & \mu_{n1} \\ \cdot & \cdot & \cdot & \cdot & \cdot & \cdot \\ \cdot & \cdot & \cdot & \cdot & \cdot & \cdot \\ \lambda_{0,\,n-1} & \lambda_{1,\,n-1} & \cdot & \cdot & -(\lambda_{n-1,\,n} + \sum_{j=0}^{n-2} \mu_{n-1,\,g}) & \mu_{n,\,n-1} \\ \lambda_{0,\,n} & \lambda_{1,\,n} & \cdot & \cdot & \lambda_{n-1,\,n} & -\sum_{g=0}^{n-1} \mu_{n,\,g} \end{pmatrix} \begin{pmatrix} P_o \\ P_1 \\ \cdot \\ \cdot \\ P_{n-1} \\ P_n \end{pmatrix} \tag{6}$$

In the matrix in Equation (6), Equation (2) must be satisfied.

For stable Markov processes, the probability at each instant P_i tends to be a constant for infinite t, where, according to Equation (1):

$$\lim_{t \to \infty} P'_i(t) = 0 \tag{7}$$

and in algebraic form, the equation system is:

$$A\vec{\Pi} = \vec{0} \tag{8}$$

where $\vec{\Pi} = (\Pi_o, \ldots, \Pi_n)^T$. The sum of all probabilities is, according to (3), equal to 1. Replacing Equation (6) a new, non-homogeneous equation system is obtained with a unique solution:

$$A_1\vec{\Pi} = \vec{0} \tag{9}$$

where:

$$A_1 = \begin{pmatrix} -\sum_{j=1}^{n}\lambda_{oj} & \mu_{10} & . & . & \mu_{n-1,0} & \mu_{no} \\ \lambda_{01} & -(\sum_{j=2}^{n}\lambda_{1}j + \mu_{10}) & \mu_{21} & . & . & \mu_{n1} \\ . & . & . & . & . & . \\ . & . & . & . & . & . \\ \lambda_{0,n-1} & \lambda_{1,n-1} & . & . & -(\lambda_{n-1,n} + \sum_{j=0}^{n-2}\mu_{n-1,g}) & \mu_{n,n-1} \\ 1 & 1 & . & . & 1 & 1 \end{pmatrix} \tag{10}$$

Thus, the mean availability D_M for *k* minimum operating states is:

$$D_M = \pi_0 + \pi_1 + \ldots + \pi_k \tag{11}$$

MULTICRITERIA MODEL FOR THE SELECTION OF MAINTENANCE POLICIES IN OPERATING THEATRE SUBSYSTEMS

The choice of MACBETH for studying maintenance policies in medical installations and equipment has been made by the technical services manager of the Hospital, acting as decision analyst. He has extensive knowledge of multicriteria techniques and has previously used ELECTRE, Analytic Hierarchy Process (AHP) and MACBETH. The choice of MACBETH as opposed to other techniques was due to the possibility of obtaining a quantitative scale value from the qualitative judgements based on good and neutral reference levels and of making a comparison between criteria based on those references rather than subjectively. The analytical ability of the M-MACBETH software was another important factor since it allowed an assessment of the quantitative ranges associated with each level on the scale of a criterion or possible ranges in the weighting of criteria, according to the judgements given. The existence of detailed procedures for decision making; for example, how to construct the judgement matrix between levels of a descriptor associated to a criterion, how to fill out the judgement matrix over the criteria (Bana e Costa & Chagas, 2002) has also been considered. Techniques like the multi-attribute utility theory could have introduced probabilistic information into the decision problem; however, in the case of the availability criterion, the calculation of mean availability for each subsystem has been made by the analysis of Markov chains. There therefore exist data for the mean availabilities associated with each subsystem. For this reason, the decision makers have preferred to work with deterministic mean availability values for each subsystem, which are also those typically used in a maintenance department, and also in the Hospital, rather than using utility functions, which were, in any case, not understood by some decision makers.

In the case of the criterion Risk to employees, although the risk has a certain associated probability, the maintenance activities to be carried out for each action have been performed countless times and have been deeply analysed, and so the association of a probability to these actions is close to the value 1. Thus, it was decided to define criteria and descriptors with deterministic scale levels.

For the application of the MACBETH method in this paper, a decision-making group has been established made up of the managers of the following areas of the General Hospital of Ciudad Real: installations, safety, environment and medical equipment, and technical services. The decision-making group has been coordinated by the technical services manager, who acted as facilitator, helping the decision makers in the decision process.

The group decision-making process took into account, among others, the following factors, described in León (1993), Munda (2005) and Zahir (1999): the solution obtained is more easily put into practice if the members who will be affected by the decision have participated in making it; by Pareto's principle, the decision-making process has not imposed the preference of any individual on the group and if, after the group has given a classification of alternatives, the alternative of lowest preference is eliminated, the classification should subsist as made. To collate the judgements given by the members of the group the models set out in Forman and Peniwati (1998) have been considered: joining together individual judgements and individual preferences. The first of these has been used, since the individuals form a homogeneous group.

The structuring phase includes the following activities:

1. Identifying the criteria and organizing them into a hierarchical structure;
2. Constructing an impact descriptor for each criterion, to make the evaluation of the alternatives operational;

Carrying out an impact estimate of the different alternatives on each criterion.

Markov Chain Analysis

In this case study, the subsystems included in the operating theatre system are: power supply, air conditioning, operating theatre lighting, sterile water and medicinal gases. All of them are critical. These subsystems include connections with the IT subsystem of the Hospital. This subsystem, though very important in a hospital, is not considered an independent subsystem of the operating theatre; that is, the connections or devices are considered parts of the power supply, air conditioning, operating theatre lighting, sterile water and medicinal gases subsystems and therefore their failures are analysed as failures of the subsystem they form part of.

To be able to study the behaviour of each subsystem, they must be modelled according to the operating states, bearing in mind different eventualities that might appear. This section shows an example of application to the power supply subsystem. Similar developments have been carried out on the rest of the subsystems in an operating theatre in the University General Hospital of Ciudad Real.

The power supply subsystem includes the elements that provide uninterrupted power supply to the vital attendance and operating theatre lighting, when there is a general power failure. The elements that make it up are as follows: insulation transformers, automatic circuit breakers, uninterrupted power supply system, electric leakage detection system and operating theatre lamp.

The electricity supply comes from an Uninterrupted Power Supply (UPS), powered by an insulation transformer. That of the subsystem has the following states:

1. In this state, the power supply is provided via the corresponding transformer centre by the electricity supplier company. This supply passes through the operating theatre's protection panel and insulation and through the UPS, before powering the operating theatre equipment.
2. If the insulation fails, an audible alarm goes off in the operating theatre, indicating the existence of an electrical leakage in a phase. If the electrical connection has a malfunction, all the charge is powered by the UPS batteries. Ten minutes later, all the power socket power circuits are disconnected, except for the vital attendance and operating theatre lamp. The latter, respirator and monitor are disconnected one hundred and twenty minutes later.
3. This state occurs when a second derivation occurs in another phase, with the risk of electrocution of patients or staff, if an electrical leakage occurs. The UPS may automatically go into bypass, if any internal anomaly is detected, disconnecting the accumulators of the charge, and these will not have any backup supply.
4. Fault in the main supply. All the charge is supplied from the UPS batteries.
5. UPS fault. The UPS goes into by-pass and, therefore, the equipment is not powered if a fault occurs in the main supply.

The subsystem is considered to fail when it is in states 2 and 4. Figure 2 represents the Markov state diagram. The system is considered operative when it is in states 0, 1 or 2. The mean availability D_M is,

$$D_M = D_0 + D_1 + D_2$$

Matrix A_1 is that shown in Equation (12). The row of ones allows the constraint of a unique solution to be included.

Figure 2. Markov availability diagram.

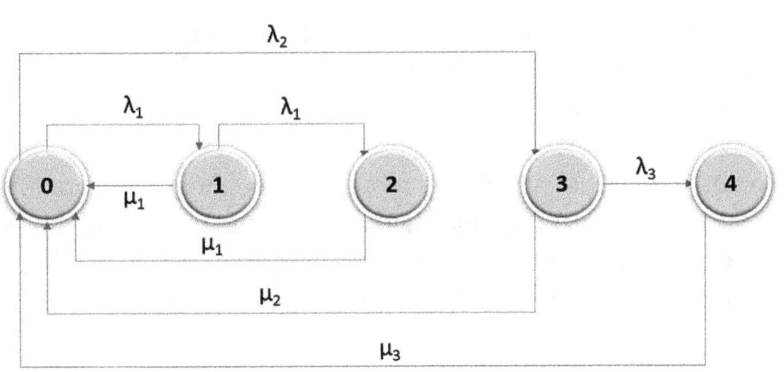

$$\begin{pmatrix} -(\lambda_1+\lambda_2+\lambda_3) & \mu_1 & \mu_2 & 0 & \mu_4 \\ \lambda_1 & -(\lambda_2+\mu_1) & 0 & 0 & 0 \\ \lambda_3 & 0 & -(\lambda_4+\mu_2) & 0 & 0 \\ \lambda_2 & 0 & 0 & -\lambda_5 & 0 \\ 1 & 1 & 1 & 1 & 1 \end{pmatrix} \tag{12}$$

Description of Criteria and Impact Levels

In the first stage of the structuring phase, the decision criteria were chosen by consensus. The technical services manager acted as facilitator, helping the decision makers in this decision process. If a disagreement occurred among the decision makers, the facilitator asked each member to state and discuss his opinion until a final consensus was reached. Nevertheless, the initially established criteria do not differ greatly from those chosen. A hierarchical structure was obtained made up of the following three areas and six decision criteria:

1. **Technical-Economic:** Availability, failure detection time and cost.
2. **Safety:** Risk to employees and risk to users.
3. **Functional:** Activity impact.

Any criterion must be associated to a descriptor in order to make it operational. A descriptor is a set of impact levels that can measure, either quantitatively or qualitatively, a criterion's level of compliance (Bana e Costa & Vasnick, 1999).

There exist the following types of descriptor: natural, when the impact levels reflect the effects directly; proxy, when the impact levels mainly reflect the causes rather than the effects; and constructed, when the impact levels are defined as a finite combination of reference levels.

In order to define each descriptor a meeting of the group was arranged in which each member set out, anonymously, five scale levels per descriptor, in order of importance, and the normal and good reference levels. During this stage, changes were made to the definition of the criterion to make it intelligible to all members of the group. The final scale levels were obtained by consensus of all members of the group, following the procedure of the Delphi method. Fulfilment of the properties of completeness, mutual independence preference, non-redundancy, concision, degree of consensus, measurability, operationality and non-ambiguity was checked during the process.

The different decision criteria used and impact levels of the associated descriptors are defined below, with an indication of the neutral (N) and good (G) levels.

The considered decision criteria are:

- **Availability (Avail):** The mean availability foreseen throughout its operating life is considered, bearing in mind the historical data obtained and applying the Markovian model, according to section 3. Therefore, the descriptor is the mean value obtained for the availability using the repairable system Markovian model. The scale levels considered in the mean availability indicator are:

- 0: $99 \leq D_M < 100$
- 1: $95 \leq D_M < 99$ (G)
- 2: $85 \leq D_M < 95$
- 3: $60 \leq D_M < 85$ (N)
- 4: $D_M < 60$

The mean availability descriptor and its scale levels are defined based on the results of a Markov chain analysis of the more than 700 subsystems that exist in the Hospital. Therefore, the mean availabilities associated with each subsystem are known, which also happens to be the information most commonly used in a maintenance department, and so in the Hospital.

The normal and good values have been set by the decision-making group such that the range corresponding to the normal value is that most often obtained among the values calculated in the Markov chain analysis, and the good value corresponds to the range of the best availabilities of the 700 subsystems analysed, until an availability of 100% is reached. The other values, which correspond to levels 2, 3 and 4, have been distributed in proportion up to a value 0.85 below the value at which the least favourable level is reached. This is established as the quality standard of the Hospital.

- **Activity Impact (Activ):** Defined as how much it affects the last operating function, which in this example is surgery. The descriptor is the foreseen suspension percentage of the operating theatre activity as a result of a fault in the subsystem, considering as scale levels:
  - 0: Never suspends activity.
  - 1: Suspends activity in less than 10% of the stoppages (G).
  - 2: Suspends activity from 10 to 25% of the stoppages.
  - 3: Suspends activity from 25 to 50% of the stoppages (N).
  - 4: Always suspends activity.
- **Failure Detection Time (Tfailur):** It is the opportunity of the moment when the fault is detected, due to the implications it may have. The descriptor is the estimate of the time when the fault is detected progressively, where the scale margin is between before the start of the activity and after, with its intermediate values varying in percentage. The following scale levels have been established, according to the accumulated experience:
  - 0: Always before affecting the service.
  - 1: Before affecting the service at least in 90% of cases (G).
  - 2: Before affecting the service at least in 70% of cases.
  - 3: At the time of the fault (N).
  - 4: Indirectly, after the fault.
- **Cost (Cost):** The necessary resources for applying any of the maintenance policies considered, are considered generally as in practice, in-house and subcontracted resources intervene. Therefore, the decision-making group has estimated that the associated indicator is the type of complexity of the required resources:
  - 0: Specific non-programmed resources.
  - 1: Standard programmed resources (G).
  - 2: Specific non-programmed resources and specialised programmed resources.
  - 3: Non-programmed, programmed and specialised services (N).
  - 4: A change to the existing technology.

- **Risk to Employees (Risk1):** This is defined as the risk to which the workers located in the operating theatre area are subjected. The descriptor measures the electrical hazard in the operating theatre. The associated scale levels are:
    - Does not involve a risk for the professional in the event of a fault (G).
    - The professional always detects the risk before the fault occurs.
    - There may be a risk of electric shock, without the professional having time to detect it (N).
    - There is a risk of an electric shock before the professional detects it.
    - High probability of electric shock.
- **Risk to Users (Risk2):** This is defined as the risk the user is subjected to, where the user is understood to be the medical service requester. The associated descriptor is the level of proximity and severity of the possible damage suffered by the patient. The established scale levels are:
    - 0: Does not have implications for the patient (G).
    - 1: Has indirect implications for the patient, without causing damage: deprogrammings.
    - 2: The patient is subjected to direct action without damage: temporary suspension of the operation (N).
    - 3: The patient is subjected to direct action with mild damage: mild burns.
    - 4: Has direct, very serious effects on the patient with the possibility of death: electrocution, respiratory failure, etc.

All the criteria included in the model are independent. This should be especially noted for the Availability and Activity impact criteria due to the fact that in the case of a failure of one of the subsystems, it will have a different level of influence in the surgery, depending on the type of subsystem and independently of the time that it is available. For example, a failure of the electrical installation subsystem will have different levels of severity or impact on an activity, depending on the nature of the failure and independently of the time that it is not operational (availability). In the same way an electrical fault, above the permitted values, will mean the cessation of activity, and it may be traced and repaired immediately or it may require a time-consuming, thorough review of the electrical system.

To determine the impact level in the availability criterion, stable system Markov chains have been used, with a 3-year breakdown and repair log for each operating theatre subsystem.

Criteria Weighting Process and Degree of Attractiveness between Levels in Each Criterion

To perform the weighting process between levels of a single criterion and to establish the degree of attractiveness between criteria, a decision-making group was created made up of the maintenance manager of the Hospital installations area, the manager of the medical equipment, and the person in charge of the management of the operation of the different equipment and installations, from a healthcare point of view (operating theatre supervisor). The person in charge of the technical services of the Hospital acted as facilitator, without making any judgements, performing the following actions: explaining the procedure to all members of the group, implementing the judgements given by each member of the group in M-MACBETH and analysing the results.

Each member of the decision-making group has filled out, anonymously, the opinion forms shown in Table 1 in the Appendix. In most of the judgements given, the results were consensual, even though no one was aware of the judgements made by the other decision makers. Only when comparing the availability criterion with the criteria failure detection time, risk to employees and risk to users, were there discrepancies in the judgements (see Appendix); in these cases, the final judgements have been obtained by applying the most voted judgement method. Thus, if inconsistencies appear, the following most voted judgement is chosen following a descending order. The results obtained from the decision-making group are shown in Appendix.

The third stage of the structuring phase considers the impact of each criterion, with the managers of each area involved intervening. Thus, the experts were the technical managers for the technical-economic criteria, the environmental and occupational risk prevention managers for safety criteria and finally, the healthcare managers (operating theatres and activity programming) for functional criteria.

The results of the judgements expressed regarding the weighting of the availability criterion levels corresponding to the electrical installation subsystem are shown in Figure 3. The same process has been applied to the different criteria of each subsystem. Figure 4 shows the results of the judgements made for comparing criteria, as well as the weightings of the criteria obtained for all subsystems. In this case, the same result is obtained for all subsystems, as the study is carried out in an operating theatre environment, where the importance of each criterion is the same regardless of the subsystem considered.

Figure 3. Judgement matrix between levels of availability criteria of the power supply subsystem

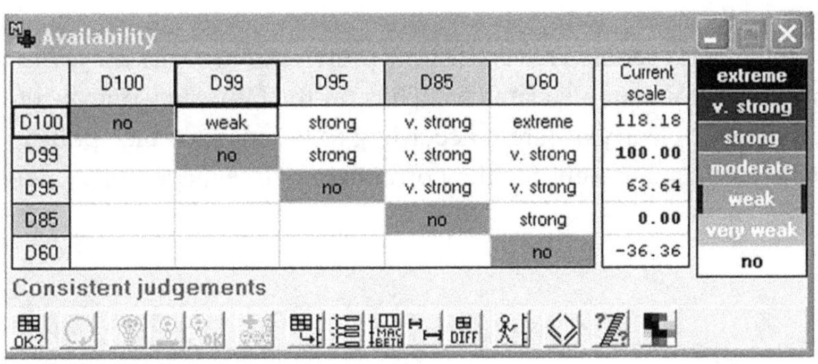

Figure 4. Judgement matrix between criteria via M-MACBETH and bar chart of the weights obtained for each criterion

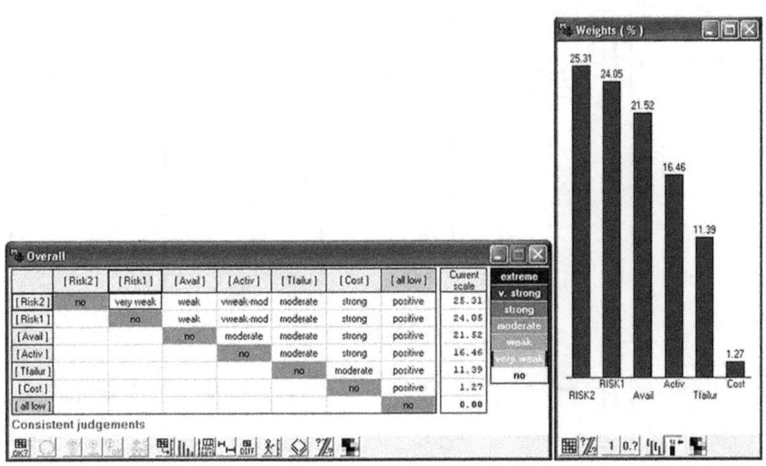

Definition of Actions

The actions or maintenance policies considered are:

- Corrective maintenance (Corr).
- Preventive maintenance together with corrective maintenance by means of daily, weekly, monthly and yearly plans (Prev+Corr).
- Periodical predictive maintenance together with corrective maintenance (Pre+Corr).
- Corrective together with preventive and predictive maintenance (C+Prev+Pred).

The possibility of applying proactive, perfective and total productive maintenance among the actions was considered, however, after assessing the Hospital resources they were not held to be viable in the short term, although the possibility of including them in future analyses was not completely rejected.

RESULTS

Figure 5 shows the ranking obtained from the different maintenance policies in the different subsystems analysed in the operating theatre system.

The maintenance policies considered have a distinct preference in each subsystem, as different impact levels have been considered for a single criterion. Thus, in the Climate Control subsystem, the impact on the availability criterion is clearly different according to the maintenance policy, while for the other criteria it is practically the same, except for the corrective maintenance policy, where the acceptance level is very low.

Figure 5. Maintenance policy classification in the Hospital's different subsystems.
a) Air conditioning b) Sterile water c) Power supply d) Medicinal gases e) Operating theatre lighting

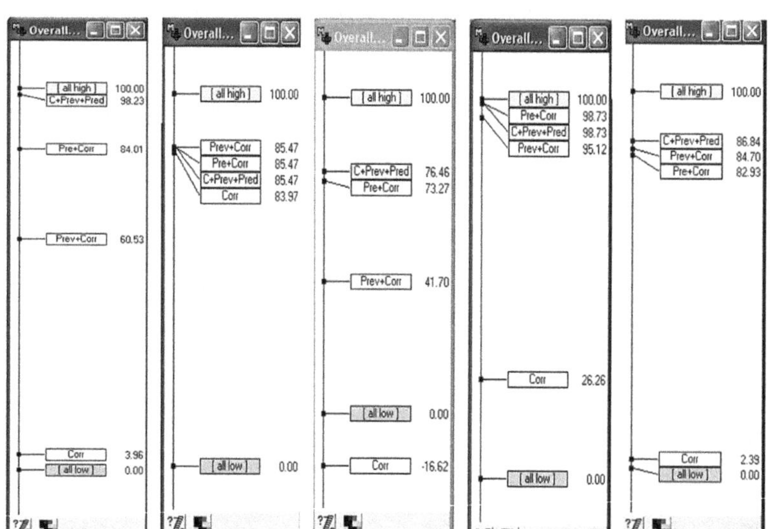

The availability criterion, therefore, has a higher relative weight than the others, which means that the variations within it are highly significant in defining the maintenance policy preference graph. A 99% mean availability value corresponds to a C+Prev+Pred combination of maintenance policies, giving an overall valuation index of 98.23%; for the Pre+Corr action, with a mean availability of 97%, this index is 84.01%. For the Prev+Corr, and Corr actions, with a mean availability of 95% and 75%, the overall valuations are 60.52% and 3.96% respectively.

In the Electrical Supply subsystem the situation is almost the same, with identical mean availability values to the Climate Control subsystem, although the C+Prev+Pred and Pre+Corr actions give practically the same preferences, 76.46% and 73.27% respectively. This is because the Risk to users and Risk to employees criteria assume the same impact in both policies.

In the Supply of Medicinal Gases and Lighting of the Operating Area subsystems the C+Prev+Pred, Pre+Corr and Prev+Corr combinations of maintenance policies give practically the same preferences; since the impact of each action and the mean availability on the criteria varies little they may be taken as more or less the same for each one. This is because predictive maintenance is based on a study of the evolution of the parameters of gas leaks and the electrical consumption of the surgical lamps respectively. This study is carried out using manually acquired data from preventive reviews, and so the difference in the results obtained by using Pre+Corr and Prev+Corr combinations of maintenance policies is virtually non-existent.

In the Supply of Sterilized Water subsystem, although the mean availability is slightly lower for the corrective maintenance policy, 85% as opposed to 90% for the other combinations of policies, the remaining criteria give almost the same impact for each policy combination; therefore, the degree of preference is slightly lower, 83.97% as against 85.47% for the other combinations of maintenance policies.

Currently, maintenance policies applied to the subsystems considered in this paper are daily corrective and preventive maintenance. In the study carried out, a considerable increase in the preferences on applying corrective plus preventive plus predictive maintenance policies is observed in the power supply and air conditioning subsystems, compared to the currently applied policy. In the rest of the subsystems, the relevance of predictive maintenance is practically insignificant. Furthermore, it can be observed that in the sterile water subsystem, including corrective maintenance has a preference practically the same as the rest of the policies.

In the analysis of the results obtained, technical services have opted for assessing the introduction of predictive maintenance in the power supply and air conditioning subsystems and only assigning corrective maintenance to the sterile water subsystem. In the first case, the introduction of a control system and charge measuring of the UPS batteries is considered, with an indication on the centralised control panel, that allows the operator to monitor them at all times, and be able to take the necessary actions sufficiently in advance: battery replacement, revision of environmental work conditions, etc. The cost of introduction per operating theatre is estimated at 6,000 euros.

In the air conditioning subsystem, the centralisation of alarms was studied due to filter dirt and wear of the fan transmission system and air extractor in the centralised control panel. The cost of introduction per operating theatre is estimated at 10,000 euros.

In the sterile water subsystem preventive maintenance has been eliminated with a saving of 0.5 euros/day.

Investment in predictive maintenance represents a single initial investment of 16,000 euros, with a break-even period of 10 years, including maintenance costs. Furthermore, with predictive maintenance, an increase in availability of 95 to 99% is foreseen in the power supply and air conditioning subsystems.

This implies an increase in productivity per operating theatre, bearing in mind that the usage rate under normal conditions is 75%, of 0.4*0.75 = 3%, only in these subsystems, also obtaining an increase in the safety of operation.

The results obtained for the operating theatre system are considered satisfactory by the Hospital Management. The model described has given results which allow decisions to be taken and quantified and which provide a rational approach to the choice of combination of maintenance policies, changing the starting position and the decision-making process considerably from now on.

SENSITIVITY ANALYSIS

Sensitivity analysis allows us to examine how the ranking of actions might change under different weighting.

M-MACBETH allows us to analyse the effect of the variations in the weights of the criteria in the considered combination of maintenance policies or actions. The technical services manager considered assessing possible modifications in the weightings of the cost and availability criteria of the power supply

Figure 6. a) (top) Sensitivity analysis in the cost criterion of the operating theatre lighting subsystem. b) (bottom). Sensitivity analysis in the cost criterion of the power supply subsystem.

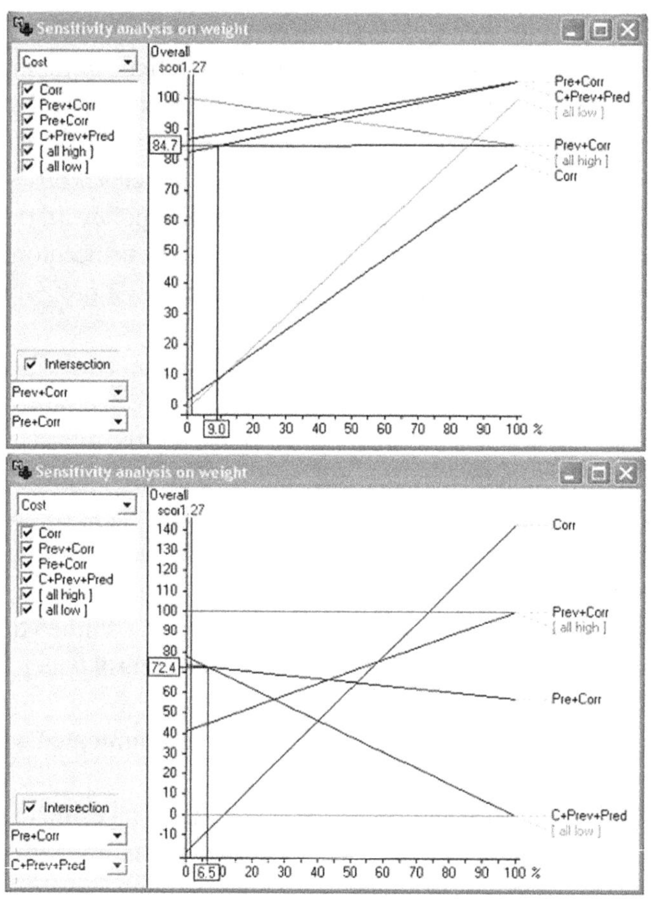

and operating theatre lighting subsystems. These subsystems were chosen because they were the subject of longer discussions during the meeting of the decision-making group.

Figure 6 shows the evolution of the actions when the weighting of the cost criterion in the operating theatre lighting and power supply subsystems is adjusted. The original weighting assigned by the decision makers to the cost criterion are shown on the vertical axis. The value of this weighting is 1.27. It can be seen that a very high increase in the weighting of the cost criteria will be necessary, to change the classification of actions in both subsystems. In the case of lighting subsystem, the cost should reach 9, while for the power supply subsystem it should reach 6.5. The weighting change would be so high that it would not be considered a possible value, and there is therefore no modification in the classification of alternatives in the two considered subsystems.

Figure 7 shows the changes the classification of actions would undergo when the weighting of the criterion availability in the operating theatre lighting and power supply subsystems is adjusted. For the

Figure 7. a) (top) Sensitivity analysis in the availability criterion of the operating theatre lighting subsystem. b) (bottom). Sensitivity analysis in the availability criterion of the power supply subsystem.

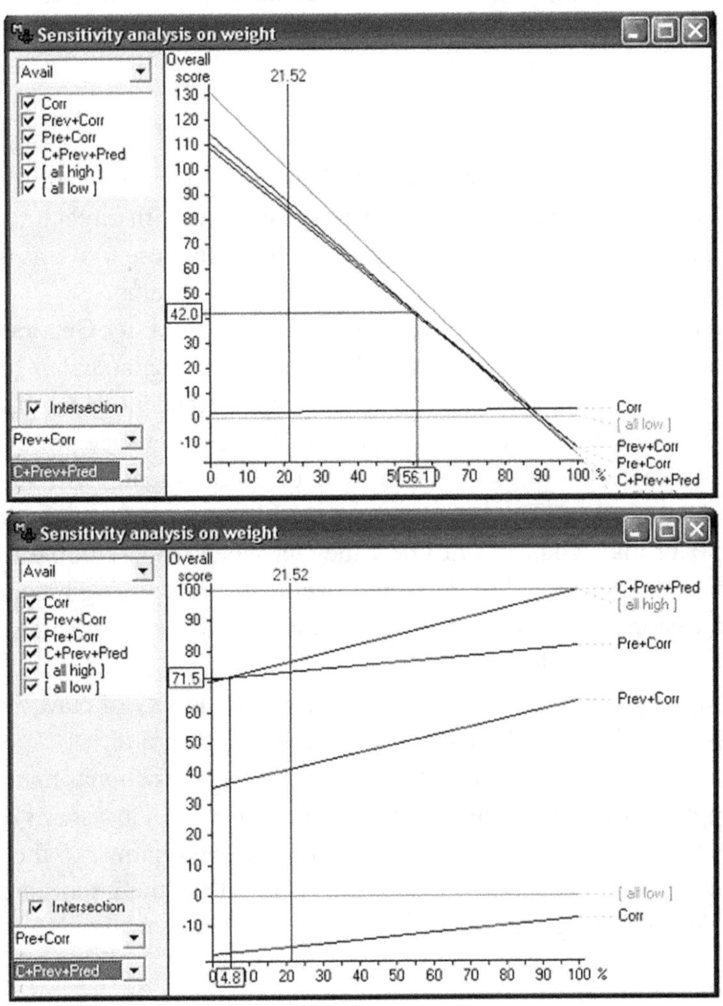

availability criterion, the original weightings set by the decision makers 21.52 (identified by a vertical line). It can be seen that the weighting should be changed significantly, over 100%, in both subsystems to change the classification of alternatives.

The results obtained from the sensitivity analysis have been considered satisfactory, since, as can be seen, there are no modifications in the action classification in any subsystem when the weightings are modified, considering logical modification ranges in the weights in the different criteria.

FUTURE RESEARCH DIRECTIONS

New maintenance policies could be applied to some or all of the subsystems of the operating theatre, for example TPM. In this case, new alternatives, in isolation or in combination with others should be proposed.

Additionally, judgements made by the decision centres could be brought together by geometric means, rather than considering the consensus. The results obtained by each procedure could be compared.

The application of other multicriteria techniques, other than MACBETH, and the comparison of the results achieved by both techniques, would confirm the reliability of the results.

CONCLUSION

This paper sets out a case study of integrating Markov chain techniques and multi-criteria techniques for the selection of the most suitable maintenance policy in subsystems with direct interaction with hospital patients. Therefore, this is an innovative application on integrating these tools, as well as being applied to a health environment, in this case to an operating theatre's subsystems.

The analyses carried out are currently being applied in the University General Hospital of Ciudad Real, and it is notable as one of the first cases of applying advanced maintenance tools, such as MCDM techniques and Markov chains in hospitals. There is only one precedent in the literature where multi-criteria decision techniques for the choice of the most suitable maintenance policy in a health environment, in spite of the large repercussions subsystems with direct interaction with the patient can have on their treatment and/or on their lives. Thus, the research could lead to a change in the mindset of technical services responsible for maintenance. Until now they have been focused on traditional maintenance systems, however, this research shows that small improvements in designing maintenance policies may have significant repercussions not only on the programming and cost of maintenance, but also on the satisfaction of care staff and especially on patients.

Finally, this methodology can mean a significant increase in quality of care. An important point to be noted is the successful application of the research to a general hospital.

The creation of a decision-making group made up of different functional managers in the hospital has allowed the development of an awareness of the importance these subsystems can have on the treatment given to the patient. This aspect which has traditionally been known to the Hospital's technical sector has now also been acknowledged by the medical sector. Thus the importance of the judgements expressed by the decision-making group.

ACKNOWLEDGMENT

This research was supported by the Junta de Comunidades de Castilla-La Mancha and the European Regional Development Fund under Grant number PPII-2014-013-P.

REFERENCES

Al-Najjar, B., & Alsyouf, I. (2003). Selecting the most efficient maintenance approach using fuzzy multiple criteria decision making. *International Journal of Production Economics*, *84*(1), 85–100. doi:10.1016/S0925-5273(02)00380-8

Al-Najjar, G. (1999). Economic criteria to select a cost-effective maintenance policy. *Journal of Quality in Maintenance Engineering*, *5*(3), 236–247. doi:10.1108/13552519910282692

Almeida, A. T., & Bohoris, G. A. (1995). Decision theory in maintenance decision making. *Journal of Quality in Maintenance Engineering*, *1*(1), 39–45. doi:10.1108/13552519510083138

Azadivar, F., & Shu, V. (1999). Maintenance policy selection for JIT production systems. *International Journal of Production Research*, *37*(16), 3725–3738. doi:10.1080/002075499190013

Bana e Costa, C. A. De Corte, J. M., & Vansnick, J. C. (2006). On the Mathematical Foundation of MACBETH. In J. Figueira, S. Greco & M. Ehrgott (Eds.), Multicriteria Decision analysis: State of the art surveys (pp. 409-444). New York, NY: Springer.

Bana e Costa, C. A., & Chagas, M. P. (2002). A career choice problem: An example of how to use MACBETH to build a quantitative value model based on qualitative value judgments (Working Paper LSEOR 02.53). The London School of Economics and Political Science.

Bana e Costa, C. A., Da Costa-Lobo, M. L., Ramos, I. A., & Vansnick, J. C. (2002). Multicriteria Approach for Strategic Town Planning: The Case of Barcelos. *International Journal of Management Sciences*, *30*(3), 227–242.

Bana e Costa, C. A., De Corte, J. M., & Vansnick, J. C. (2011). MACBETH (Measuring Attractiveness by a Categorical Based Evaluation Technique). In J. J. Cochran, L. A. Cox Jr, P. Keskinocak, J. P. Kharoufeh, & J. C. Smith (Eds.), *Encyclopedia of Operations Research and Management Science*. New York, NY: John Wiley & Sons. doi:10.1002/9780470400531.eorms0970

Bana e Costa, C. A., Nunes da Silva, F., & Vansnick, J. C. (2001). Conflict dissolution in the public sector: A case-study. *European Journal of Operational Research*, *130*(2), 388–401. doi:10.1016/S0377-2217(00)00037-0

Bana e Costa, C. A., Oliveira, C. S., & Vieira, V. (2008). Priorization of bridges and tunnels in earthquake risk mitigation using multicriteria decision analysis: Application to Lisbon. *Omega*, *36*(3), 442–450. doi:10.1016/j.omega.2006.05.008

Bana e Costa, C. A., & Vansnick, J. C. (1994). MACBETH - An interactive path towards the construction of cardinal value functions. *International Transactions in Operational Research*, *1*(4), 489–500. doi:10.1016/0969-6016(94)90010-8

Bana e Costa, C. A., & Vansnick, J. C. (1999). The MACBETH approach: basic ideas, software and an application. In N. Meskens & M. Roubens (Eds.), *Advances in Decision Analysis* (pp. 131–157). Dordrecht: Kluwer Academic Publishers. doi:10.1007/978-94-017-0647-6_9

Bashiri, M., Badri, H., & Hejazi, T. H. (2011). Selecting optimum maintenance strategy by fuzzy interactive linear assignment method. *Applied Mathematical Modelling*, *35*(1), 152–164. doi:10.1016/j.apm.2010.05.014

Bertolini, M., & Bevilacqua, M. (2006). A combined goal programming-AHP approach to maintenance selection problem. *Reliability Engineering & System Safety*, *91*(7), 839–848. doi:10.1016/j.ress.2005.08.006

Bevilacqua, M., & Braglia, M. (2000). The analytic hierarchy process applied to maintenance strategy selection. *Reliability Engineering & System Safety*, *70*(1), 71–83. doi:10.1016/S0951-8320(00)00047-8

Carnero, M. C. (2005). Selection of diagnostic techniques and instrumentation in a predictive maintenance program. A case study. *Decision Support Systems*, *38*(4), 539–555. doi:10.1016/j.dss.2003.09.003

Carnero, M. C. (2006). An evaluation system of the setting up of predictive maintenance programmes. *Reliability Engineering & System Safety*, *91*(8), 945–963. doi:10.1016/j.ress.2005.09.003

Carnero, M. C. (2007). Model for the Selection of Predictive Maintenance Techniques. *INFOR*, *45*(2), 83–94.

Carvalho, R., & Lourenço, J. C. (2002). A multicriteria model for assigning new orders to service suppliers. *European Journal of Operational Research*, *139*(2), 390–399. doi:10.1016/S0377-2217(01)00367-8

Castanier, B., Grall, A., & Bérenguer, C. (2005). A condition-based maintenance policy with non-periodic inspections for a two-unit series system. *Reliability Engineering & System Safety*, *87*(1), 109–120. doi:10.1016/j.ress.2004.04.013

Cavalcante, C. A., & Lopes. (2015). V. R. S. Multi-criteria model to support the definition of opportunistic maintenance policy: A study in a cogeneration system. *Energy*, *80*, 32–40.

Chan, F. T. S., & Prakash, A. (2012). Maintenance policy selection in manufacturing firms using the fuzzy MCDM approach. *International Journal of Production Research*, *50*(23), 7044–7056. doi:10.1080/00207543.2011.653451

Crespo, A., & Sánchez, A. (2002). Models for maintenance optimization: A study for repairable system and finite time periods. *Reliability Engineering & System Safety*, *75*(3), 367–377. doi:10.1016/S0951-8320(01)00131-4

Dhouib, D. (2014). An extension of MACBETH method for a fuzzy environment to analyze alternatives in reverse logistics for automobile tire wastes. *Omega*, *42*(1), 25–32. doi:10.1016/j.omega.2013.02.003

Emblemsvag, J., & Tonning, L. (2003). Decision support in selecting maintenance organization. *Journal of Quality in Maintenance Engineering*, *9*(1), 1–24. doi:10.1108/13552510310466765

Fernández, J. C., & Canals, X. (2006). Quality standards for medical devices. In Proccedings of the XV Winter Course Standards in Telemedicine CATAI (pp. 133-140). Tenerife, Spain.

Fleming, K. N. (2004). Markov models for evaluating risk-informed in-service inspection strategies for nuclear power plant piping systems. *Reliability Engineering & System Safety*, *83*(1), 27–45. doi:10.1016/j.ress.2003.08.009

Forman, E., & Peniwati, K. (1998). Aggregating individual judgments and priorities with the Analytic Hierarchy Process. *European Journal of Operational Research*, *108*(1), 165–169. doi:10.1016/S0377-2217(97)00244-0

Ghosh, D., & Roy, S. (2009). A decision-making framework for process plant maintenance. *European Journal of Industrial Engineering*, *4*(1), 78–98. doi:10.1504/EJIE.2010.029571

Gómez, J. F., & Crespo, A. (2009). Framework for implementation of maintenance management in distribution network service providers. *Reliability Engineering & System Safety*, *94*(10), 1639–1649. doi:10.1016/j.ress.2009.04.003

Gómez de León, F. C., & Ruiz, J. J. (2006). Maintenance strategy based on a multicriterion classification of equipments. *Reliability Engineering & System Safety*, *91*(4), 444–451. doi:10.1016/j.ress.2005.03.001

Goossens, A. J. M., & Basten, R. J. I. (2015). Exploring maintenance policy selection using the Analytic Hierarchy Process; An application for naval ships. *Reliability Engineering & System Safety*, *142*, 31–41. doi:10.1016/j.ress.2015.04.014

Gosselin, S. R., & Fleming, K. N. (1997). Evaluation of pipe failure potential via degradation mechanism assessment.*Proceedings of the 5th International Conference on Nuclear Engineering (INCONE)*. Nice, France.

Ilangkumaran, M., & Kumanan, S. (2009). Selection of maintenance policy for textile industry using hybrid multi-criteria decision making approach. *Journal of Manufacturing Technology Management*, *20*(7), 1009–1022. doi:10.1108/17410380910984258

Ishizaka, A., & Nemery, P. (2014). Assigning machines to incomparable maintenance strategies with ELECTRE-SORT. *Omega*, *47*, 45–59. doi:10.1016/j.omega.2014.03.006

Kelly, A. (2006). *Strategic Maintenance Planning*. Oxford, UK: Butterworth-Heinemann.

Knezevic, J. (1996). *Mantenimiento*. Madrid: Isdefe.

Kodali, R., & Chandra, S. (2001). Analytical Hierarchy Process for justification of total productive maintenance. *Production Planning and Control*, *12*(7), 695–705. doi:10.1080/09537280010024045

Labib, W. A. (2004). Towards An Intelligent Holonic Maintenance System. *Maintenance Journal*, *17*(2).

Labib, W. A., O'Connor, F. R., & Glyn, B. W. (1998). An effective maintenance system using the analytic hierarchy process. *Integrated Manufacturing Systems*, *9*(2), 87–98. doi:10.1108/09576069810202005

León, O. (1993). *Análisis de decisiones: técnicas y situaciones aplicables a directivos y profesionales*. Madrid: McGraw-Hill.

Martorell, S., Sánchez, A., Muñoz, A., Pitarch, J. L., Serradell, V., & Roldan, J. (1999). The use of maintenance indicators to evaluate the effects of maintenance programs on NPP performance and safety. *Reliability Engineering & System Safety*, *65*(2), 85–94. doi:10.1016/S0951-8320(98)00088-X

Martorell, S., Villanueva, J. F., Carlos, S., Nebot, Y., Sanchez, A., Pitarch, J. L., & Serradell, V. (2005). RAMS+C informed decision-making with application to multiobjective optimization of technical specifications and maintenance using genetic algorithms. *Reliability Engineering & System Safety*, *87*(1), 65–75. doi:10.1016/j.ress.2004.04.009

Munda, G. (2005). Multiple Criteria Decision analysis and sustainable development. In J. Figueira, S. Greco, & M. Ehrgott (Eds.), Multicriteria Decision analysis. State of the Art Surveys (pp. 953-988). Boston: Springer Verlag. doi:10.1007/0-387-23081-5_23

Nakajima, S. (1988). *TPM: Introduction to total productive maintenance*. Cambridge, MA: Productivity Press Inc.

Pintelon, L., & Gelders, L. (1992). Maintenance management decision-making. *European Journal of Operational Research*, *58*(3), 301–317. doi:10.1016/0377-2217(92)90062-E

Prawda, J. (1999). *Métodos y modelos de investigación de operaciones. Modelos estocásticos*. México: Limusa S.A.

Ramadhan, R. H., Hamad, I., Wahhab, A., & Duffuaa, S. O. (1999). The use of an analytical hierarchy process in pavement maintenance priority ranking. *Journal of Quality in Maintenance Engineering*, *5*(1), 25–39. doi:10.1108/13552519910257041

Rani, N. A. A., Baharum, M. R., Akbar, A. R. N., & Nawawi, A. H. (2015). Perception of Maintenance Management Strategy on Healthcare Facilities. *Procedia: Social and Behavioral Sciences*, *170*(27), 272–281. doi:10.1016/j.sbspro.2015.01.037

Rao, B. K. N. (1996). *Handbook of Condition Monitoring*. Oxford, UK: Elsevier.

Rodrigues, T. C. (2014). The MACBETH Approach to Health Value Measurement: Building a Population Health Index in Group Processes. *Procedia Technology*, *16*, 1361–1366. doi:10.1016/j.protcy.2014.10.153

Romero, C. (1996). *Análisis de Decisiones Multicriterio*. Madrid: Isdefe.

SESCAM. (2007). *Valoración de la calidad de la atención sanitaria pública en hospitales*. Memory of activities of SESCAM.

Shafiee, M. (2015). Maintenance strategy selection problem: An MCDM overview. *Journal of Quality in Maintenance Engineering*, *21*(4), 378–402. doi:10.1108/JQME-09-2013-0063

Shahin, A., Shirouyehzad, H., & Pourjavad, E. (2012). Optimum maintenance strategy: A case study in the mining industry. *International Journal of Services and Operations Management*, *12*(3), 368–386. doi:10.1504/IJSOM.2012.047626

Sharma, R. K., Kumar, D., & Kumar, P. (2005). FLM to select suitable maintenance strategy in process industries using MISO model. *Journal of Quality in Maintenance Engineering*, *11*(4), 359–374. doi:10.1108/13552510510626981

Taghipour, S., Banjevic, D., & Jardine, A. K. S. (2011). Prioritization of medical equipment for maintenance decisions. *The Journal of the Operational Research Society*, *62*(9), 1666–1687. doi:10.1057/jors.2010.106

Taha, H. A. (2011). *Operations research. An introduction*. New York, NY: Prentice Hall.

Vatn, J. (1997). Maintenance optimisation from a decision theoretical point of view. *Reliability Engineering & System Safety*, *58*(2), 119–126. doi:10.1016/S0951-8320(97)00025-2

Wang, L., Chu, J., & Wu, J. (2007). Selection of optimum maintenance strategies based on a fuzzy analytic hierarchy process. *International Journal of Production Economics*, *107*(1), 151–163. doi:10.1016/j.ijpe.2006.08.005

Wolf, M. (2008). Strategic management of medical equipment. Starting and operating a TSC. *Proceedings of the 20th Congress of the International Federation of Hospital Engineering*, Barcelona, Spain.

Zahir, S. (1999). Geometry of decision making and the vector space formulation of the analytic hierarchy process. *European Journal of Operational Research*, *112*(2), 373–396. doi:10.1016/S0377-2217(98)00020-4

Zaim, S., Turkyílmaz, A., Acar, M. F., Al-Turki, U., & Demirel, O. F. (2012). Maintenance strategy selection using AHP and ANP algorithms: A case study. *Journal of Quality in Maintenance Engineering*, *18*(1), 16–29. doi:10.1108/13552511211226166

ADDITIONAL READING

Al-Najjar, B., & Alsyouf, I. (2000). Improving effectiveness of manufacturing systems using total quality maintenance. *Integrated Manufacturing Systems*, *11*(4), 267–276. doi:10.1108/09576060010326393

Burton, J., & Hubaeck, K. (2007). Is small beautiful? A multicriteria assessment of small-scale energy technology applications in local governments. *Energy Policy*, *35*(12), 6402–6412. doi:10.1016/j.enpol.2007.08.002

Carnero, M. C. (2014). Multicriteria model for maintenance benchmarking. *Journal of Manufacturing Systems*, *33*(2), 303–321. doi:10.1016/j.jmsy.2013.12.006

Chen, F. L., & Chen, Y. C. (2010). Evaluating the maintenance performance of the semiconductor factories based on the analytical hierarchy process and grey relational analysis. *American Journal of Applied Sciences*, *7*(4), 567–574. doi:10.3844/ajassp.2010.568.574

Crespo, A. (2007). *The maintenance management framework. Models and methods for complex systems maintenance*. London: Springer.

De Jonge, B., Klingenberg, W., Teunter, R., & Tinga, T. (2015). Optimum maintenance strategy under uncertainty in the lifetime distribution. *Reliability Engineering & System Safety*, *133*, 59–67. doi:10.1016/j.ress.2014.09.013

Iung, B., Véron, M., Suhner, M. C., & Muller A. (2005). Integration of maintenance strategies into prognosis process to decision-making aid on system operation. *CIRP Annals - Manufacturing Technology*, *54*(1), 5-8. doi:.10.1016/S0007-8506(07)60037-7

Macchi, M., & Garetti, M. (2006). Information requirements for e-maintenance strategic planning: A benchmark study in complex production systems. *Computers in Industry*, *57*(6), 581–594. doi:10.1016/j.compind.2006.02.015

Mobley, R. K. (2004). *Maintenance fundamentals*. Burlington, MA: Elsevier Butterworth-Heinemann.

Moore, W. J., & Starr, A. G. (2006). An intelligent maintenance system for continuous cost-based prioritisation of maintenance activities. *Computers in Industry*, *57*(6), 595–606. doi:10.1016/j.compind.2006.02.008

Pintelon, L., Gelders, L., & Van Puyvelde, F. (1997). *Maintenance Management*. Acco Leuven Amersfoort.

Regattieri, A. A., Gamberi, G. M., & Gamberini, R. (2015). An innovative method to optimize the maintenance policies in an aircraft: General framework and case study. *Journal of Air Transport Management*, *44-45*, 8–20. doi:10.1016/j.jairtraman.2015.02.001

Taghipour, S., Banjevic, D., & Jardine, A. K. S. (2011). Prioritization of medical equipment for maintenance decisions. *The Journal of the Operational Research Society*, *62*(9), 1666–1687. doi:10.1057/jors.2010.106

Teixeira de Almeida, A. (2001). Multicriteria decision making on maintenance: Spares and contracts planning. *European Journal of Operational Research*, *129*(2), 235–241. doi:10.1016/S0377-2217(00)00220-4

Vineyard, M., Amoako-Gyampah, K., & Meredith, J. R. (2000). An evaluation of maintenance policies for flexible manufacturing system. A case study. *International Journal of Operations & Production Management*, *20*(4), 409–426. doi:10.1108/01443570010319156

Wang, H. (2002). A survey of maintenance policies of deteriorating systems. *European Journal of Operational Research*, *139*(3), 469–489. doi:10.1016/S0377-2217(01)00197-7

KEY TERMS AND DEFINITIONS

Descriptor: Is an ordered set of impact levels which allows the degree of fulfilment of a criterion by an alternative to be assessed quantitatively and qualitatively.

MACBETH: Measuring Attractiveness by a Categorical Based Evaluation Techniques is a multicriteria decision-making technique based on the additive value model. The methodological background to the technique includes the structuring of value trees, the building of descriptors associated with each criterion, the weighting of criteria, the building of value functions from the judgements of the decision-maker by linear programming, a sensitivity and robustness analysis and finally the considering of alternatives.

Maintenance Policy: Organization of the human and material resources to address faults, wear or breakdowns in machines, devices or facilities, to guide maintenance management decision making, with the aim of having facilities which work continuously in conditions of optimal availability, safety and quality, minimizing the cost of maintenance.

Multicriteria Decision Making: Consists of structuring a complex problem and considering the multiple criteria by which it is affected, which are frequently in conflict, to determine a preferred alternative, choose from a small group of the best alternatives or group alternatives into different preference sets. This is because there is no optimal alternative and so the solution to the problem will be found from a set of non-dominated solutions. In the solution process, the judgements and preferences of the decision maker or decision-making group are taken into account.

Preventive Maintenance: This is a maintenance policy consisting in periodic, systematic revision and repair of machines and facilities, to prevent faults and breakdowns before they happen, so as to guarantee optimal working and to increase the life cycle of assets.

Subsystem: Systems from the decomposition of a group of elements made up of equipment and installations and their interconnections, intended to provide a product or service, so that they can be analysed and assessed more easily.

APPENDIX

This table shows how each decision maker, when comparing one criterion with another, gave a judgement from the semantic values provided by MACBETH: very weak, weak, moderate, strong, very strong and extreme. These judgements are shown in the table with a square symbol. Where the criterion in the left-hand column is more attractive than that in the right-hand column, the values are included in the left side of the table.

On the other hand, if the criterion from the column on the right is more attractive than the one on the left (see Risk to employees against Availability), the judgement given by the decision makers is shown on the right of the table. In some cases, the group as a whole reached a consensus on the judgement given, for example in the case of Cost against Risk to employees; All the members of the decision-making group considered that Risk to employees was more strongly attractive than cost; thus, it only takes up a single square of the table.

Nonetheless, when Availability is compared to Failure detection time, it is seen that there are squares in the very weak, weak and moderate columns. This means that there was no consensus and that the members of the decision group considered that the three vales were appropriate. Finally, the model created with M-MACBETH includes the most voted judgement, identified in the table by *.

Table 1. Judgement attractiveness criteria for intermediate subsystems

	EXPRESSION OF JUDGEMENTS OF ATTRACTION BETWEEN CRITERIA											
	Very weak	**Weak**	**Moderate**	**Strong**	**Very Strong**	**Extreme**	**Very Strong**	**Strong**	**Moderate**	**Weak**	**Very weak**	
Availability									□ *			Activity impact
Availability	□	□ *	□									Failure detection time
Availability			□ *									Cost
Availability								□	□ *			Risk to employees
Availability								□	□ *			Risk to users
Activity impact			□ *									Failure detection time
Activity impact				□ *								Cost
Activity impact											□ *	Risk to employees
Activity impact											□ *	Risk to users
Failure detection time			□ *									Cost
Failure detection time									□ *			Risk to employees
Failure detection time									□ *			Risk to users
Cost								□ *				Risk to employees
Cost								□ *				Risk to users
Risk to employees											□ *	Risk to users

□ Outcome of the judgement expressed by the decision-making group made up of each manager of the following Hospital areas: installations, safety, environment and medical equipment, coordinated by the technical services manager.

* Semantic value included in M-MACBETH.

Chapter 3
Maintenance in Critical Infrastructures:
The Need for Public-Private Partnerships

Patricia Maraña
University of Navarra, Spain

Leire Labaka
University of Navarra, Spain

Jose Mari Sarriegi
University of Navarra, Spain

ABSTRACT

The increase in the frequency of disastrous events and society's dependence on Critical Infrastructures (CIs) has led to greater concern about the need to increase resilience in order to improve Critical Infrastructure Protection. CIs are basic service providers for society and they need to be effectively protected against hazards. Nowadays, CIs can be owned by private entities. However, although they can be privately owned or managed, they provide a public service that directly affects the whole society. Consequently, those activities that increase the overall resilience level of CIs need to be under the supervision of public entities. Increasing resilience requires special attention be paid to correct infrastructure and crisis response equipment maintenance. This chapter explains why effective Public-Private Partnerships (PPP) are valuable for correctly maintaining CIs and illustrates examples of real situations that demonstrate the need for effective PPPs in maintenance activities.

INTRODUCTION

Since September 11th, 2001, the threat of terrorism has raised concerns in many countries about the urgent need to improve the level of protection for their critical infrastructures. Subsequent events—such as the Asian tsunami in 2004, Hurricane Katrina in 2005, and the Fukushima Earthquake and subsequent tsunami in 2011—and their inestimable scope only emphasized this concern. The increase in the fre-

DOI: 10.4018/978-1-5225-0651-5.ch003

quency of crisis occurrence, the difficulty of foreseeing when the next critical event is going to happen, and society's dependence on CIs represent relevant challenges for public decision makers. However, they are not independent challenges and in fact, they are closely related.

When an unexpected critical event occurs, recovery to the normal state becomes more difficult if CIs have been affected, as the impacts of crisis, affecting CIs are significantly bigger. To make this challenge even more critical, the complex interrelationships between CIs and cascading effects (Lorenz, 2009) make it difficult to foresee how a crisis will evolve. Consequently, anticipating the occurrence of these events is critical. This research focuses on the prevention activities that are required to anticipate or reduce the impact of critical events affecting CIs.

Resilience refers to the capacity of a system to prevent a crisis occurrence, to mitigate the impact generated in the case that a triggering event occurs, and to reduce the recovery time needed to bounce back to the initial situation. In fact, it is in CI disruptions when the overall resilience level of a company could make a difference. The difference will be positive when the resilience level is high and negative when the resilience level is low (Maraña, Labaka, Hernantes, & Sarriegi, 2015).

Although we may believe that it is the government's responsibility to protect and maintain the welfare of people living in an area, when we consider how complex society is, we see how it is virtually impossible for just one party to undertake such a difficult task. Given that CIs are required for society's welfare and they are usually managed by private entities (Alexander, 2005; J. Chen, T. H. Y. Chen, Vertinsky, Yumagulova, & Park 2013; Hodge & Greve, 2007), cooperation between public and private companies is valuable for improving the resilience level of CIs and eventually increasing Critical Infrastructure Protection (CIP).

Increasing resilience requires paying special attention to the appropriate maintenance of infrastructures and crisis response equipment. Thus, the main objective of this chapter is to explain the main reasons why Public-Private Partnerships (PPP) are sometimes valuable for ensuring that CIs are correctly maintained. The aim of PPPs is to promote the sharing of resources and information between partners involved. Sharing resources and information among public and private entities enhances the overall resilience level of CIs because the more resources and information are available the better is the quality of the decisions taken by CI managers. To that end, examples of real situations that illustrate the need for an effective PPP in CI maintenance activities will be presented, followed by an explanation of three different types of PPPs that can help to make sure maintenance is performed correctly.

METHODOLOGY

The methodology followed during the development of this book chapter starts with a literature review about Critical Infrastructures (CIs), crisis impacts and resilience. Afterwards, a framework that includes a set of resilience building policies and sub-policies is presented. Some of the policies identified in that framework are directly related to maintenance what justifies the importance of conducting maintenance activities properly in order to increase the overall resilience level of any CI.

Next, the information gathered from a set of semi-structured interviews conducted with experts is explained. The main goals of these semi-structured interviews were the following ones:

- Validate the literature review previously done
- Understand the role of all the partners involved in CIP

- Understand the role of all the agents involved in maintenance activities
- Identify real cases in which a proper maintenance could have reduced the impacts of a crisis
- Identify existing challenges for CIs regarding maintenance
- Obtain information about real PPP examples in CIs

After conducting the interviews, the results obtained were the following ones:

- A list of relevant cases that can illustrate how maintenance is important to provide good CIP
- A better understanding of which are the partners that should be involved in CIP activities
- A better understanding of the different possible roles of the agents involved in maintenance activities of CIs

Considering the information gathered from the semi-structured interviews with experts, in the next section, the most representative six real cases were chosen and are explained in more detail in order to illustrate how effective maintenance activities can help reducing crisis impacts. Examples of maintenance activities that could have helped to reduce crisis impacts are also suggested for each of the presented cases.

Afterwards, an explanation of the PPP concept is given and the reasons why PPPs can increase the efficiency of maintenance activities. Moreover, a framework is proposed to create and develop efficient PPPs to improve CIP. After that and as a result of the research, three different types of PPPs regarding maintenance were identified. In this section, each of the PPP types will be explained in more detail.

Finally, we offer some future research lines and conclusions for further exploration in the hope that many CI owners or operators and crisis managers become better prepared for practical engagement in real-world problem solving.

CRITICAL INFRASTRUCTURE (CI)

Critical Infrastructures are not only important assets for society; the well being of society is also increasingly dependent on CIs. This dependency increases the society's vulnerability to CI disruptions (Boin & McConnell, 2007; Sarriegi, Sveen, Torres, & Gonzalez, 2008). Thus, CIP becomes a critical task for every public manager wanting to preserve the welfare of society.

In Council Directive 2008/114/EC, the European Union defines a CI as "an asset, system or part thereof located in Member States which is essential for the maintenance of vital societal functions, health, safety, security, economic or social well-being of people, and the disruption or destruction of which would have a significant impact on a Member State as a result of failure to maintain those functions" (The Council of the European Union, 2008). The European Union also has a list of the sectors considered to be critical for society (Table 1).

Identifying a CI is the first step in being able to protect it. However, CIs constitute connected and interdependent systems (Chang, McDaniels, Mikawoz, & Peterson, 2007), as each CI needs the outcome of others in order to continue its normal activity. The complex interdependencies that exist between CIs make the task of protecting CIs from disruptions even more difficult, since when a crisis affects one of these infrastructures others may be subsequently affected.

Figure 1. Summary of the methodology followed during the development of this chapter

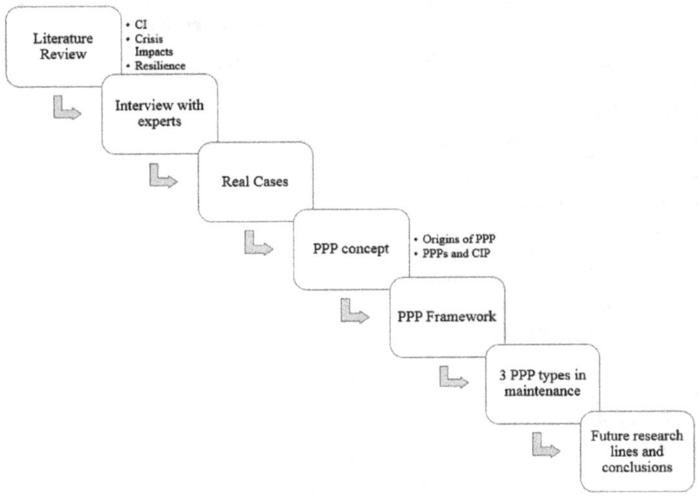

This phenomenon is known as "cascading effects". Researchers in CI protection use the term "cascading effects" to refer to the fact that a triggering event that affects one CI can subsequently impact others (Deshmukh, Ho Oh, & Hastak, 2011; Rinaldi, 2004).

The result is that CI systems and their dependencies make crisis management more difficult— the dependencies make it even more difficult to foresee how a crisis will evolve, and they also complicate the execution of the crisis response and recovery tasks (Lorenz, 2009).

CRISIS IMPACTS AND RESILIENCE

There is a close relationship between crises and resilience. Increasing the resilience level of a system reduces the number of crisis impacts and the amount of recovery time needed to bounce back to the initial situation. In what follows, both terms, crisis impact and resilience, will be explained.

Crisis Impact

Crisis impacts are known as the consequences a crisis generates. Impacts always differ from crisis to crisis for two main reasons: firstly, because of the triggering event that has caused a crisis (Eusgeld, Nan, & Dietz, 2011), and secondly, because of the way the crisis is managed. This will make consequences last for longer or shorter periods.

Crisis impacts can be classified based on two different types of classification. The first type classifies impacts according to their nature (tangible versus intangible and direct versus indirect), and the second type depends on the affected sector (CI, economy, society or environment).

The first classification differentiates between tangible versus intangible and direct versus indirect impact (Laugé, Hernantes, Labaka, & Sarriegi, 2012). Direct impacts represent the physical destruction caused by a triggering event to human beings, buildings, infrastructures, vehicles, capital and stocks, while indirect impacts are a consequence of the direct impact caused by the hazard. Indirect impacts

Table 1. Critical infrastructures list

Sector	Product or Service
Energy	• Oil and gas production, refining, treatment and storage, including pipelines • Electricity generation • Transmission of electricity, gas and oil • Distribution of electricity, gas and oil
Information and Communication Technologies	• Information system and network protection • Instrumentation automation and control systems (SCADA...) • Internet • Provision of fixed telecommunications • Provision of mobile telecommunications • Radio communication and navigation • Satellite communication • Broadcasting
Water	• Provision of drinking water • Control of water quality • Stemming and control of water quantity
Food	• Provision of food and safeguarding food safety and security
Health	• Medical and hospital care • Medicines, serums, vaccines and pharmaceuticals • Bio-laboratories and bioagents
Financial	• Payment services/payment structures (private) • Government financial assignment
Public & Legal Order and Safety	• Maintaining public and legal order, safety and security • Administration of justice and detention
Civil Administration	• Government functions • Armed forces • Civil administration services • Emergency services • Postal and courier services
Transport	• Road transport • Rail transport • Air traffic • Inland waterways transport • Ocean and short-sea shipping
Chemical and Nuclear Industry	• Production and storage/processing of chemical and nuclear substances • Pipelines for dangerous goods (chemical substances)
Space and Research	• Space • Research

(EU Commission, 2005)

are more difficult to quantify than direct impacts, as months or even years may be needed to properly evaluate them (Lequeux & Ciavola, 2011; Pfurtscheller, Lochner, & Thieken, 2011).

Tangible impacts are the impacts that can be measured in monetary terms, such as the destruction of vehicles, buildings or infrastructures, the changes in income or the increase in costs. Estimating these types of impacts, for example in euros, is relatively easy. On the other hand, intangible impacts cannot be translated into monetary terms, as there is no systematic or agreed upon method available to measure them (Bureau of Transport Economics, 2001; Hallegatte & Przyluski, 2010; Middelmann, 2007). This means, then, that intangible impacts are much more difficult to estimate. Examples of intangible impacts are deaths, injured people, human suffering, cultural artefacts, losses of memorabilia, and psychological effects.

The second classification type distinguishes impacts based on the sector affected, for example Critical Infrastructures (CI), economy, society and the environment (Hallegatte & Przyluski, 2010; Middelmann, 2007).

The impact caused by a crisis can evolve differently depending on the previously deployed prevention and preparation policies. Moreover, the consequences of a crisis can be aggravated and prolonged if response and recovery activities are not effective. The concept of resilience embraces prevention, preparation, response and recovery activities. Therefore, improving resilience could help reduce crisis impacts and the recovery time needed to return to the original situation.

Resilience

Resilience has been defined in several different ways in different contexts. Originally, it was a concept used in materials science. In that context resilience refers to the ability of a material to return to its original state (Stewart, Kolluru, & Smith, 2009). It has also been used in behavioural sciences, where it is defined as 'the dynamic process indicating the adaptive functioning of individuals at risk' by Sapountzaki (2007, p.277). In the context of this chapter, we will define resilience as the capacity of a system to prevent a crisis from occurring, to mitigate the impact of a crisis in the event a triggering event does occur, and to return to a normal situation in an effective way in terms of time and resources (Chen et al., 2013; Norris, Stevens, B. Pfefferbaum, Wyche, & R. L. Pfefferbaum, 2008; Sheffi & Rice, 2005). As the definition of resilience clearly implies, there is a close connection between crisis impacts and resilience. In fact, when a crisis occurs in a resilient system, the magnitude of the impact is significantly lower than in cases where no resilient system is in place.

Increasing the resilience level of a system thus helps improve crisis management activities. As with crisis impacts, there are two main classifications for resilience. In one system, resilience is classified as internal or external, depending on the type of stakeholders involved. Internal resilience involves stakeholders from inside the system, while external resilience involves stakeholders from outside the system. Resilience can also be classified in terms of dimensions: technical resilience, organizational resilience, economic resilience and social resilience (Labaka, 2013).

A CI's resilience level is dependent on resilience-building policies. Implementing these policies enhances the overall resilience level of a system. For instance, resilience can be increased through the proper maintenance of the technical equipment that can be used to prevent future crises or to facilitate a more efficient response. Moreover, a CI's resilience level can also be increased by developing organizational skills that can improve the capacity to detect a crisis occurrence early or to cooperate more effectively during the crisis response.

RESILIENCE-BUILDING POLICIES: THE IMPORTANCE OF MAINTENANCE

In order to increase the overall resilience of a system, resilience-building policies and sub-policies need to be implemented for the resilience classifications defined above. Since in this chapter we are only focusing on the resilience level of a CI itself, we only refer to internal resilience. Considering the resilience framework developed by Labaka (2013, p. 199), internal resilience can be broken down into three dimensions: technical resilience, organizational resilience and economic resilience. The policies and sub-policies identified for each of the dimensions of internal resilience are listed in Table 2. The principal aim of this resilience framework for CIs is to help crisis managers to improve the resilience level of their CI.

PREVENTIVE AND CORRECTIVE MAINTENANCE

CI Maintenance is one of the policies included within the technical dimension of resilience. Implementing and periodically carrying out CI maintenance activities will help to guarantee a high level of reliability for the infrastructure. Having a good level of maintenance helps the CI to withstand incidents and reduces the magnitude of the impact and the recovery time needed (Labaka, Hernantes, Comes, & Sarriegi, 2014). In other words, by performing these activities, we ensure the adequacy and reliability of the CI's physical components and their proper functioning. The CI maintenance policy has been disaggregated into two sub-policies: preventive maintenance and corrective maintenance.

Table 2. Resilience policies and sub-policies related to internal resilience

Resilience Dimensions	Resilience Policies	Resilience Sub-policies
Technical Resilience	CI Safety Design and Construction	(Required and Voluntary) Safety systems
		Redundancy
		Simplicity and loose coupling
		(External and internal) Audits
	CI Maintenance	Preventive maintenance
		Corrective maintenance
	CI Data Acquisition and Monitoring System	Data acquisition equipment
		Information monitoring equipment
	CI Crisis Response Equipment	
Organizational Resilience	CI Organizational Procedures for Crisis Management	Coordination procedures with external stakeholders
		Crisis Management procedures
		Incidents management and evaluation
	CI Top Management Commitment	Top Manager commitment and situation awareness
		Activities to promote resilience based in culture
	CI Crisis Manager Preparation	Crisis Manager training
		Crisis Manager situation awareness and commitment
	CI Operator Preparation	Operator training
		Operator situation awareness and commitment
Economic Resilience	CI Crisis Response Budget	

(Labaka, 2013)

Preventive Maintenance

Preventive maintenance covers all the maintenance activities that are performed before an incident occurs. A CI's components need to be periodically inspected in order to verify that they are in their proper condition; old parts need to be replaced or upgraded in order to ensure new regulations are complied with. Moreover, these activities help to prevent future failures by assisting with the identification of early warning signals and dealing with them before a crisis event unfolds.

Corrective Maintenance

Corrective maintenance refers to the unplanned maintenance activities performed after an incident occurs in order to repair damage or strengthen infrastructures. Corrective maintenance helps minor incidents to be detected and solved before the accumulation of accidents has the chance to result in a major incident. For this reason, it is important to manage minor incidents as soon as possible (Turner, 1976). Moreover, it is important to bear in mind that once a failure has occurred it is essential to detect the causes of the incident. This will make it possible to propose new corrective activities that will eventually help prevent the incident from occurring again.

Therefore, following all the maintenance activities required before a disruption occurs could help to prevent a crisis occurrence or to reduce the impacts generated and the recovery time needed after a crisis occurrence.

CI Crisis Response Equipment

Maintenance activities are also implicitly included in the *CI Crisis Response Equipment* policy of the framework. *CI Crisis Response Equipment* refers to the emergency equipment that needs to be available when a crisis occurs. This emergency equipment may be vital in some cases in absorbing the impact of the crisis and increase the safety of workers at the CI. In order to ensure the availability and reliability of the emergency equipment, proper maintenance should be carried out on the equipment in accordance with the specifications and requirements set by manufacturers.

REAL EXAMPLES OF IMPROPER CRITICAL INFRASTRUCTURE MAINTENANCE

In this section, real examples of crises that could have been prevented or at least could have been more effectively managed with better maintenance will be explained in detail, followed by a summary table. In this section, we will pay special attention to the type of CI affected and also to the type of maintenance the example refers to: preventive maintenance, corrective maintenance or emergency equipment maintenance.

Smoke Incident

In January of 2015, a "smoke incident" took place in the metro in Washington D.C. There was no fire. Instead, despite the fact that electricity is supposed to be safely contained within insulated wires, an electrical arcing event caused the problem. This electrical malfunction generated the smoke that eventually escaped into the tunnel (Aratani, 2015).

The first incident filled a tunnel with noxious fumes, which sickened a considerable amount of passengers. Everyone on board was evacuated; 84 people were taken to hospitals and one person died from smoke inhalation. The incident also led to the closure of five metro lines of the subway system due to safety reasons, which caused several delays for passengers.

Inappropriate maintenance of the electrical infrastructure and the dust that had accumulated in the walls of one tunnel were identified as the main causes of the failure in the electrical system. Public authorities assumed the blame and ordered that the washing down of the tunnels be resumed and the electrical equipment be replaced. A dedicated maintenance crew will be in charge of continuously clearing the tunnel passageways of debris, equipment and other potential obstructions across Metro's 100 miles of tunnel segments. Metro also announced an extensive cable-replacement program as well as undertaking a multiyear maintenance program to replace or rehabilitate all 88,044 lights in Metro tunnels.

They also assumed the responsibility of reviewing the protocols for the alarms in the Rail Operation Control Center, with the goal of separating critical alarms from non-critical notifications. The computer system generated so many needless alarms for smoke and other problems that the warnings sometimes went unheeded by controllers.

Due to the economic recession, the budget dedicated to the proper maintenance of public infrastructures in the US had been reduced (Holdeman, 2015). Because of the budget reduction, the number of occurrences of failures in the metro system has increased, including several "smoke incidents" in recent years. We can deduce, then, that corrective maintenance of the infrastructure could have helped to mitigate the impact generated by this event.

Snow Chaos at Heathrow Airport

In December 2010 cold weather and snow caused a collapse at Heathrow airport. The flights of thousands of passengers were delayed or cancelled (Calder, 2010). The collapse of air travel at Heathrow—a major hub serving Europe and other international destinations—meant that the whole air transport system experienced serious disruptions.

Passengers argue that the airports seemed to be unable to cope with snow. They said that there was no more than two inches of snow and yet the runway was totally covered. They asked public authorities why they were not ploughing or salting the snow (Lewis, 2010).

An operator from Heathrow BAA, the airport's owner, said that in the previous year more than £6 million had been invested in technology to move snow and de-ice runways and that staff had worked through the night. He apologised for the "miserable" situation but said problems had been caused by an extreme amount of snow in a very short space of time and safety had to be prioritised. However, other airports more used to this type of freezing weather were able to provide service to passengers. Consequently, a debate emerged about the preparedness level of Heathrow with regard to this type of crisis.

A couple of weeks after the 2010 snow chaos, the aviation minister admitted that the government knew about Heathrow's inadequate preparation for winter situations. The transport secretary had argued with BAA about Heathrow's plans for severe weather before the crisis occurrence. BAA failed to invest in sufficient manpower or equipment to cope with heavy snowfall (Swinford, 2010). For instance, the airport's contingency plan, which was commissioned by the Department of Transport, reveals that the airport had just 10 snowploughs available to clear the runway. Gatwick, which is less than half the size of Heathrow, had 14. Despite the Government having ordered Heathrow airport authorities to update its snow plan in the summer of 2010, BAA made few changes from the previous year's version. Crucially, the airport did not add the necessary runway clearing equipment.

This example illustrates how the availability of the proper amount of emergency equipment necessary would have absorbed part of the impact generated by this crisis. Additionally, regular maintenance of the emergency equipment could have led to the detection of the small amount of available snowploughs.

Hurricane Katrina

Early in the morning of August 29th, 2005, Hurricane Katrina struck the Gulf Coast of the United States. The hurricane caused nearly 2,000 deaths and an estimated economic loss of $125 billion. It also caused the displacement of hundreds of thousands of citizens. It was the costliest natural disaster, as well as one of the five deadliest hurricanes, in the history of the United States (History.com, 2009).

However, the destruction and number of lives lost cannot be attributed entirely to the storm itself. What made this event catastrophic was its aftermath. Katrina brought to light shortfalls in the preparations made in advance of the storm and in the reactions after the storm (Abkowitz, 2008). In fact, the prevention, preparation, response and recovery activities were not as effective as expected. For instance, due to poor design and construction, deferred maintenance and the lack of funding, numerous failures occurred in the city's flood-protection system. In the end, these problems led to nearly 80% of the city being flooded.

Insufficient emergency planning and preparedness, and the inability of responders to communicate compounded the hurricane's effects. This aggravated the difficulty of the situation and left New Orleans vulnerable. Proper preventive maintenance of the New Orleans's levees and drainage canals could have reduced the amount of damage caused by Hurricane Katrina in 2005.

Energy Supply for Hospitals

Energy is a critical for the correct functioning of a hospital, and people's health is dependent on the correct treatment prescribed by doctors. Usually these treatments involve the use of tools that require electric power to function, and thus energy is key for ensuring the well-being of hospital patients. Energy is not only important for the proper functioning of health equipment; it is also key for the correct management of this complex CI. For that reason, it is imperative that hospitals take the necessary measures to preserve electrical power at all times. Surprisingly, power outages happen in big hospitals with alarming frequency, and there are usually backup systems like generators in order to ensure the supply of energy to hospitals.

It is also essential that hospitals have emergency kits with batteries, flashlights, two-way radios, and a non-electronic emergency system for patient identification. Rapid adaptation to unexpected adverse conditions is critical to the successful implementation of any disaster plan (Nates, 2004).

On April 16th, 2002, Rhode Island Hospital and Women & Infants Hospital lost power during the afternoon, leaving various parts of the complexes in the dark for different periods (Harrington, 2002). Although no parts of the hospitals went completely black, the outage affected both operating rooms. This fact disrupted the schedules, postponing procedures and jeopardizing patient health.

At Women & Infants, an emergency generator immediately turned on, emergency lights came on, and no essential services were disrupted. However, at Rhode Island Hospital, the backup electric system did not work. Fortunately, this did not seriously affect the health of any patients, and power was restored an hour and 20 minutes after the failure.

In another event at Rhode Island Hospital in September 1999, a blackout plunged the entire hospital into darkness. The backup system failed and in this instance a patient died after his respirator failed.

Since power outages are quite common in hospitals, proper maintenance of the alternative methods for supplying electricity, such as generators, is essential in order to efficiently respond to power outages in hospitals and reduce the impact of the event.

Road Maintenance

The global economic recession that began in 2008 considerably has decreased budgets dedicated to road maintenance and traffic infrastructure. Although citizens do not perceive the need for investment as long as everything works well (Blanco, 2013), the lack of proper road maintenance is one of the main causes for the increase in traffic accidents over the last years. Reducing expenditures has its negative effects; for example, asphalt grips less.

Spain is a clear example of how reducing the road maintenance budget leads to an increase in the number road fatalities. A study conducted by the Asociacion Española de Carretera (the Spanish Highway Association) deemed the state of the roads to be "the worst since 1985" (Jimenez-Gálvez, 2015). It also claimed that an investment of about 6.2 billion euros is needed to return the roads to their optimal state. Repairing about 330,000 signs, repainting 52,000 kilometres of road markings and inspecting 82% of the lighting are some of the tasks that are expected to ensure a basic level of safety. The government is not investing the amount needed to maintain an acceptable level of road safety, which means that deterioration of the road network is accumulating.

Preventive maintenance of the road infrastructure not only could have decreased the amount of accidents that occur on the roads every year, but also would have eventually decreased the cost of future repairs. Moreover, preserving a road is less risky than building a new one since the demand for its use is far more certain (Jaffe, 2015).

Energy Crisis in South Africa

Eskom is a South African electricity public utility, established in 1923 as the Electricity Supply Commission (ESKOM) by the government of South Africa. The main aim of Eskom is to supply enough energy to meet citizen demand. However, they find this difficult for two main reasons: the continuous growth in the number of customers requiring electricity services and peak periods.

Currently, Eskom faces the challenge of a constrained power system. There is not sufficient capacity to supply all their customers with their available infrastructure. Consequently, until there is new capacity, old power stations are working to full capacity and maintenance activities are carefully scheduled.

In 2007 South Africa started experiencing widespread rolling blackouts as supply fell behind demand, which threatened to destabilize the national electricity grid and forced Eskom to engage in load shedding. Load shedding, or load reduction, is done across the country as a controlled option to be able to respond to unplanned events and protect the power system from a total blackout (Eskom, 2015). It involves taking units offline for maintenance, repairs or re-fuelling (in the case of nuclear units).

In January 2015, Eskom's CEO Tshediso Matona highlighted the lack of maintenance as a major contributor to the country's electricity problems. Matona said although the company had always had a philosophy about properly maintaining the power system, he admitted it has not been faithful to this notion, and this has resulted in unreliable equipment which keeps breaking down (Nicolaides, 2015). This fact highlights the importance of engaging in proper corrective maintenance of the energy infrastructure in order to prevent the occurrence of a national blackout.

Table 3 summarizes all the information for the six real incidents just discussed. For each crisis, the affected CI sector and the related maintenance activities are provided.

PUBLIC-PRIVATE PARTNERSHIP FOR CRITICAL INFRASTRUCTURE PROTECTION

CIs provide basic services that ensure the welfare of society, and therefore protecting CIs is a task that falls under the purview of the government. However, profit-oriented private companies are usually owners of CIs. As basic service providers, private CI owners or managers also have a responsibility towards society and thus are obliged to collaborate with public entities.

Table 3. Summary of the six real incidents. ("X" means there is no relationship to the incident)

Real Incident	Affected CI Sector	Preventive Maintenance	Corrective Maintenance	Emergency equipment Maintenance
Smoke incident	Transport	X	Smoke incidents occurred several times. Corrective maintenance could have prevented the crisis occurrence.	X
Snow chaos at Heathrow airport	Transport	X	X	Periodic maintenance reviews of the emergency equipment could have helped in the absorption of impact.
Flood prevention system	Water	Proper preventive maintenance on the levees and drainage canals could have reduced the consequences of the hurricane.	X	
Energy supply for hospitals	Health	X	X	Proper maintenance of alternative energy supply equipment is highly relevant.
Improper maintenance of traffic infrastructure	Transport	Preventive maintenance of traffic infrastructure.	X	X
Eskom crisis in South Africa	Energy	X	Small blackouts happened. Corrective maintenance is needed to reduce the frequency of load shedding.	X

ORIGINS OF PPP

The concept of partnerships and more particularly of Public-Private Partnerships (PPP) is one of the most widespread ideas of the 21st century. Although there have been several attempts to define the concept, there is still not a formal definition.

Some authors believe that public-private partnership is a "language game" (Teisman & Klijn 2002) or a "hurrah word" that people use without knowing its real meaning. They believe that it is a term that was introduced to replace other controversial terms, for instance privatisation (Hodge & Greve, 2007). During the late 80s some governments, for instance Margaret Thatcher's government in the UK, started to develop and implement a new political strategy whose aim was to restructure the private sector. She called this process privatisation, a term that included all the possible ways in which private parties could be involved in public duties, starting from the sale of national industries to the outsourcing of manufacturing activities. As in some countries, privatisation had a negative connotation; new terms with a more collaborative connotation like Public-Private Partnership (PPP) were introduced into the political vocabulary. Therefore some define PPP as "a long-term contract between a private party and a government entity, for providing a public asset or service, in which the private party bears significant risk and management responsibility, and remuneration is linked to performance" (World Bank, Asian Development Bank & Inter-American Development Bank, 2014).

Others see PPPs as a new governance tool that will replace the traditional method of contracting for public services through competitive tendering. They emphasize that PPPs are worthwhile because both the public and the private sector can benefit by combining their specific qualities (Rosenau, 2000). Therefore, some define PPPs as "an organised relationship between public and private organisations, which establishes common scope and objectives and uses defined roles and work methodology to achieve shared goals" (ENISA, 2011).

WHY IS A PPP VALUABLE FOR IMPROVING THE EFFECTIVENESS OF CIP?

As previously mentioned, CIs are responsible for providing basic services to society, which makes society dependent on the correct functioning of CIs. Without the provision of those basic services, vital societal functions could be in danger. Public entities are always the ones in charge of ensuring the well-being of society and in considering the problems that may cause a disruption in one or more CIs, critical infrastructure protection (CIP) becomes one of the most significant activities that governments must pay attention to.

However, CIs are usually managed by private entities (Boin & McConnell, 2007; Dunn-Cavelty & Suter, 2009; Rothery, 2005), and as a consequence both public and private companies are responsible for providing resources and sharing responsibilities in terms of CIP. This requires a significant effort to promote cooperation between parties through public-private partnerships (PPPs) to ensure critical infrastructure protection (Busch & Givens, 2012).

For instance, PPPs are very common in the energy sector, where private companies are usually in charge of these infrastructures (Yusta, Correa, & Lacal-Arántegui, 2011). In fact, it is responsibility of the public and private participants to coordinate their efforts to improve the robustness of CIP, meaning that cooperation between public entities and private companies in improving CIP is not only sensible but essential (Dunn-Cavelty & Suter, 2009).

In fact, maintaining CIs and emergency equipment is an important part of CIP. In those cases, in which both public and private entities are involved in CIP, establishing PPPs regarding maintenance activities could improve the protection level of the CI.

A meaningful collaboration between partners embraces not only sharing responsibilities and duties but also resources and information. In fact, sharing useful information provided by the stakeholders involved in the partnership not available anywhere else, help improving the decision making process of critical infrastructure owners, government representatives and crisis managers. Actually, the fact of having more information available improves the quality of the decisions taken. Sharing information among partners, helps learning from other's experiences before an unpredicted event affect ourselves. Having information regarding crises that have affected others in the past beforehand, is always useful to improve prevention, response and recovery activities in case a crisis occurs.

Table 4 illustrates how a PPP could have improved CIP and more precisely the efficiency of the maintenance activities in some of the real situations discussed above.

STEPS FOR SETTING UP A PPP

The European Network and Information Security Agency (ENISA) has designed a framework that helps in the design of PPPs related to CIP answering the following five key questions (ENISA, 2011):

1. Why is PPP needed?
2. Who should it involve?
3. How should it be governed?
4. What services and incentives should be offered?
5. When should the actions be taken to start and maintain sustainability?

Table 4. Reasons why a PPP may be valuable

Real Incident	Is a PPP valuable? Why?
Smoke incident	No. Washington D.C.'s subway system is public. Consequently, there is no need for collaboration.
Snow chaos at Heathrow airport	Yes. Providing a reliable air traffic system is responsibility of the government. However, when the crisis took place the airport was operated by Ferrovial, which is a private company. Therefore, the responsibility for maintaining snowploughs was shared and an effective PPP could have reduced the snow chaos of 2010.
Flood prevention system	No. Flood prevention systems are public. Consequently, there is no need for collaboration.
Energy supply for hospitals	It depends on the country. In some countries, private companies are either owners or managers of hospitals. Providing a reliable health service to their citizens is responsibility of government. Therefore, public companies need to ensure that generators are available at all times in private hospitals and force them to follow a set of maintenance activities.
Improper maintenance of traffic infrastructure	It depends on the country. If the road is public, the maintenance activities are the sole responsibility of public entities. However, in some countries, like in Spain, private companies manage some highways and in this particular example, the collaboration of both public and private companies will be valuable.
Eskom crisis in South Africa	No. Eskom is a public electricity utility. Therefore, the government has the sole responsibility of performing the required maintenance activities.

1. Why Is PPP Needed?

Identifying the main reasons why a PPP needs to be created, considering which is the scope of the partnership and thinking about which is the problem that is going to be addressed are the principal and most important steps need to be taken to design an effective and sustainable PPP. Giving an answer to all these questions will help defining the approach of a PPP related to CIP.

2. Who Should It Involve?

the next step in the path of establishing an effective ppp is deciding which will be the stakeholders involved and which will be the links with external organisations that are not directly members of the ppp but could have interest on the activities that are being performed.

3. How Should It Be Governed?

Establishing the rules to determine how the PPP will be governed is also essential. Giving an answer to questions such as, how will the PPP be organised or how will partners work together help defining the governance of an effective PPP. Moreover, establishing basic rules for the correct functioning of a PPP, such as how partners will communicate is also key to develop a successful PPP.

4. What Services and Incentives Should Be Offered?

Deciding which will be the services that will be provided by the PPP to all the partners is the next important step in the process of establishing an effective PPP. Good practice guides, technical evaluation or training exercises are just some of the services that PPPs could offer to all the participants.

Specifying the incentives that will be offered by the PPP to the partners is also important to develop a successful PPP. Access to unique information, cost savings by sharing efforts to solve critical problems or opportunities to contribute to strategic direction of national policies are examples of incentives that could be offered by a PPP.

5. When Should Action Be Taken to Start It and Maintain Sustainability?

The final step is to think about the evolution over time of the PPP should have how the PPP started and how it will grow. In fact, different structures could be used to develop a new PPP. A PPP could be created following bottom-up structure, what means that a community recognises the need to work together to solve a problem and then other members join to it. It could also be created the opposite way following a top down structure. Moreover, sometimes PPPs split or merge at some point and need a restructure. Thinking about all the internal structure and ensuring it is understandable for everyone is key to set up an efficient PPP.

THREE PPP TYPES IN MAINTENANCE

Political incentives often prejudice infrastructure expenditure towards new assets over maintaining existing assets. While the opening of a new bridge or of a new water supply system attracts the media, the maintenance of already existing assets is not interesting at all. Consequently, infrastructure assets are often poorly maintained due to poor planning or deferred maintenance. Regular maintenance maintains service standard and minimizes lifecycle costs by not allowing the asset to degrade to the point that costly rehabilitation is needed.

Protecting critical infrastructures against hazards requires ensuring that daily maintenance activities are properly conducted. The correct maintenance of an infrastructure is an everyday activity on any infrastructure manager's list of responsibilities. In this regard, CI managers are committed to ensuring that the assets of the company are correctly maintained. However, private companies are not the only responsible party. In the specific case of CIs, public bodies also need to be involved in supervising maintenance activities. As previously noted, public authorities are responsible for ensuring that CIs are working properly so that the welfare of society is not negatively affected.

Considering what has been discussed so far, we have identified three different ways in which public authorities are currently involved in maintenance activities with the principal aim of ensuring critical infrastructure protection.

The first case is related to grants. Sometimes public entities decide to lend public infrastructures to private operators during a certain amount of time. This way, although public entities still own the infrastructures themselves, private companies are the ones in charge of managing the infrastructures. Once the grant period ends, the deterioration level of the CI will be greater or lesser depending on how well it has been maintained. As a consequence, public entities, as owners of the infrastructures, need to ensure that the maintenance activities are properly carried out so they will still have a valuable infrastructure once the grant period is concluded. The improper road maintenance in Spain is a good example of how this type of PPP for road maintenance could have helped prevent the deaths and other damage that occurred due to a lack of government investment during the worst period of the economic crisis Spain.

The second case involves giving a private company not only the right to provide a public service to society as a whole but also the right to own a CI. In this case, private companies, as CIs owners, are also in charge of correctly managing those assets. Given that the welfare of society depends on the basic needs provided by CIs, regardless of whether CI owners are public or private, they will always have the responsibility of providing the effort and resources necessary to protect CIs. However, in those cases where CI owners are private, they must still allow public entities to supervise the management of CIs since it is of public interest to ensure that maintenance activities are correctly carried out and the CIs are correctly protected. For instance, when the owner of a hospital is a private company, as was the case in the two hospitals in Rhode Island, both it and the relevant public entities must ensure that the proper maintenance activities are being undertaken to prevent blackouts.

The third case is related to the responsibility that CI owners and managers have to ensure that the equipment needed to respond to a crisis is always available and properly maintained. Apart from the daily maintenance routines of the CI itself, it is important to bear in mind that there is also some equipment that will be only used in the event a crisis occurs. Ensuring that this equipment is correctly maintained is one of the basic pillars for being able to efficiently respond to a crisis. Public entities and private entities therefore need to collaborate in order to ensure that maintenance activities are being performed properly.

The availability of snowploughs at Heathrow and generators at Rhode Island Hospital are examples that illustrate the benefit of this type of PPP for CI maintenance.

FUTURE RESEARCH LINES

This chapter has explained why PPPs are valuable for CIP. The fact that private companies usually own CIs and that society is increasingly dependent on the services provided by CIs demonstrate the need for close collaboration between public bodies and private companies to ensure CIP. However, improving the efficiency of partnerships between public and private parties is not automatic; it is in fact a continuous process.

Although public entities are conscious of the importance of PPPs, there is still no model that systematizes the optimum way to develop PPPs for CIP. Having identified the need for a model that will help PPPs to develop successfully, our future work aims to develop such a model and produce a trustworthy guide that will make it easier to develop efficient PPPs in order to protect CIs against hazards.

In order to implement the model, firstly the stakeholders involved in PPPs for CIP will be defined. Then, the policies that will need to be implemented will be identified. Moreover, a set of indicators will be selected in order to monitor how the PPP is evolving over time and to measure changes in its efficiently level.

CONCLUSION

While governments are tasked with providing for the well-being of their citizens, which includes protecting CIs and emergency response in the event of a crisis, at the same time current economic and political trends mean that private companies are nowadays the owners or operators of most CIs. Although the interests of public and private entities usually diverge, in the case of CIs, as previously explained, collaboration between public and private companies needs to be very fluent and efficient in order to ensure that CIs, which provide basic services to an increasingly dependent citizenry, are properly maintained and thus protected against crises.

To achieve a sufficient level of protection for any CI, it is important to increase the resilience level of the system. In fact, when a CI disruption occurs, the overall resilience level of the company is what will eventually help to reduce the impact and decrease recovery time. To improve the resilience level of a system, effective resilience-building policies must be implemented; maintenance is one of those policies that helps to enhance resilience and ultimately protect CIs.

As the real cases cited in the chapter illustrate, in cases where CIs are the responsibility of both public and private companies, responsibility for performing the proper maintenance activities needs to be shared. To that end, an efficient PPP is valuable for undertaking maintenance activities that improve CIP.

Therefore, apart from introducing the terms Critical Infrastructure, Crisis Impacts and Resilience and justifying the urgent need to develop PPPs in nowadays society with real examples, this book chapter also provides helpful practical guidelines to CI owners and managers about how to develop successful PPPs in order to improve CIP. Moreover, three different types of PPP have been identified as a result of the research done. Finally, the book chapter also suggest interesting future research lines are proposed.

REFERENCES

Abkowitz, M. (2008, October 1). 10 lessons learned the hard way. *Vanderbilt Magazine*. Retrieved from http://www.vanderbilt.edu/magazines/vanderbilt-magazine/2008/10/lessons-learned-the-hard-way/

Alexander, D. (2005). Towards the development of a standard in emergency planning. *Disaster Prevention and Management: An International Journal*, *14*(2), 158–175. doi:10.1108/09653560510595164

Aratani, L. (2015, May 14). Metro resumes old strategy to prevent more smoke incidents in tunnels. *Washington Post*. Retrieved from https://www.washingtonpost.com/local/trafficandcommuting/metro-resumes-old-strategy-to-prevent-more-smoke-incidents-in-tunnels/2015/05/14/ee73c3b4-fa7b-11e4-9030-b4732caefe81_story.html?hpid=z2

Blanco, P. (2013, April 27). Carreteras sin cuidar o reparar. *El Pais*. Retrieved from http://politica.elpais.com/politica/2013/04/27/actualidad/1367084371_564522.html

Boin, A., & McConnell, A. (2007). Preparing for critical infrastructure breakdowns: The limits of crisis management and the need for resilience. *Journal of Contingencies and Crisis Management*, *15*(1), 50–59. doi:10.1111/j.1468-5973.2007.00504.x

Bureau of Transport Economics (BTE). (2001). Economic Costs of Natural Disasters in Australia.

Busch, N. E., & Givens, A. D. (2012). Public-Private Partnerships in Homeland Security Opportunities and Challenges.

Calder, S. (2010, December 19). Snow at Heathrow and Gatwick ruins plans of thousands. *BBC News*. Retrieved from http://www.bbc.com/news/uk-12030233

Chang, S. E., McDaniels, T. L., Mikawoz, J., & Peterson, K. (2007). Infrastructure failure interdependencies in extreme events: Power outage consequences in the 1998 Ice Storm. *Natural Hazards*, *41*(2), 337–358. doi:10.1007/s11069-006-9039-4

Chen, J., Chen, T. H. Y., Vertinsky, I., Yumagulova, L., & Park, C. (2013). Public–private partnerships for the development of disaster resilient communities. *Journal of Contingencies and Crisis Management*, *21*(3), 130–143. doi:10.1111/1468-5973.12021

EU Commission. (2005). Green Paper on a European Programme for Critical Infrastructure Protection COM (2005) 576.

Deshmukh, A., Ho Oh, E., & Hastak, M. (2011). Impact of flood damaged critical infrastructure on communities and industries. *Built Environment Project and Asset Management*, *1*(2), 156–175. doi:10.1108/20441241111180415

Dunn-Cavelty, M., & Suter, M. (2009). Public–Private Partnerships are no silver bullet: An expanded governance model for Critical Infrastructure Protection. *International Journal of Critical Infrastructure Protection*, *2*(4), 179–187. doi:10.1016/j.ijcip.2009.08.006

ENISA. (2011). Cooperative models for effective Public-Private Partnerships: Good Practice Guide. ISBN: 978-92-9204-054-3

Eskom. (n.d.). Eskom load shedding. Retrieved September 30, 2015, from http://loadshedding.eskom.co.za/LoadShedding/Description

Eusgeld, I., Nan, C., & Dietz, S. (2011). "System-of-systems" approach for interdependent critical infrastructures. *Reliability Engineering & System Safety*, *96*(6), 679–686. doi:10.1016/j.ress.2010.12.010

Fund, M. I. (2004). *Public-Private Partnerships*. Fiscal Affairs Department.

Hallegatte, S., & Przyluski, V. (2010). The economics of natural disasters: concepts and methods. *World Bank Policy Research Working Paper Series 5507.*

Harrington, M. (2002, August 1). EMERGENCY: The critical condition of hospital power. Retrieved from http://ecmweb.com/cee-news-archive/emergency-critical-condition-hospital-power

History.com Staff. (2009). Hurricane Katrina. Retrieved from http://www.history.com/topics/hurricane-katrina

Hodge, G. A., & Greve, C. (2007). Public–private partnerships: An international performance review. *Public Administration Review*, *67*(3), 545–558. doi:10.1111/j.1540-6210.2007.00736.x

Holdeman, E. (2015, May 18). Critical Infrastructure Maintenance: It's Not Sexy But It's Necessary. *Emergency Management.com.* Retrieved from http://www.emergencymgmt.com/emergency-blogs/disaster-zone/Maintenanceofcriticalinfrastructure.html

Jaffe, E. (2015, February 12). America's Infrastructure Crisis Is Really a Maintenance Crisis. *CityLab.com.* Retrieved from http://www.citylab.com/cityfixer/2015/02/americas-infrastructure-crisis-is-really-a-maintenance-crisis/385452/

Jimenez-Gálvez, J. (2015, July 11). El riesgo mortal de volver a las carreteras. *El Pais.* Retrieved from http://politica.elpais.com/politica/2015/07/11/actualidad/1436571260_237661.html

Labaka, L. (2013). *Resilience Framework for Critical Infrastructures* [Doctoral dissertation]. University of Navarra.

Labaka, L., Hernantes, J., Comes, T., & Sarriegi, J. M. (2014). Defining Policies to Improve Critical Infrastructure Resilience.*Proceedings of the 11th International ISCRAM Conference.*

Laugé, A., Hernantes, J., Labaka, L., & Sarriegi, J. M. (2012). Análisis y clasificación de los impactos en situaciones de crisis. *Revista Internacional de Desastres Naturales, Accidentes e Infraestructura Civil*, 12(2).

Lequeux, Q., & Ciavola, P. (2012). *Methods for estimating the costs of coastal hazards. Technical report.* CONHAZ.

Lewis, J. (2010, December 25). Heathrow bosses knew an inch of snow would cripple airport. *Telegraph UK.* Retrieved from http://www.telegraph.co.uk/finance/newsbysector/transport/8225006/Heathrow-bosses-knew-an-inch-of-snow-would-cripple-airport.html

Lorenz, J., Battiston, S., & Schweitzer, F. (2009). Systemic risk in a unifying framework for cascading processes on networks. *The European Physical Journal B*, *71*(4), 441–460. doi:10.1140/epjb/e2009-00347-4

Maraña, P., Hernantes, J., Labaka, L., & Sarriegi, J. M. (2015) CRISIS SCORECARD: How to reduce impacts increasing resilience and awareness.*Proceedings of The International Emergency Management Society Workshop (TIEMS)*.

Middelmann, M. H. (2007). *Natural hazards in Australia: identifying risk analysis requirements*. Geoscience Australia.

Nates, J. L. (2004). Combined external and internal hospital disaster: Impact and response in a Houston trauma center intensive care unit. *Critical Care Medicine*, *32*(3), 686–690. doi:10.1097/01.CCM.0000114995.14120.6D PMID:15090948

Nicolaides, G. (2015, January 15). Eskom: Bad maintenance led to load shedding. Retrieved September 30, 2015.

Norris, F. H., Stevens, S. P., Pfefferbaum, B., Wyche, K. F., & Pfefferbaum, R. L. (2008). Community resilience as a metaphor, theory, set of capacities, and strategy for disaster readiness. *American Journal of Community Psychology*, *41*(1-2), 127–150. doi:10.1007/s10464-007-9156-6 PMID:18157631

Pfurtscheller, C., Lochner, B., & Thieken, A. H. (2011). Costs of Alpine hazards. *WP8 report of the project ConHaz–Costs of Natural Hazards (D8. 1)*. Retrieved from http://conhaz. org

Rinaldi, S. M. (2004, January). Modeling and simulating critical infrastructures and their interdependencies. *Proceedings of the 37th annual Hawaii international conference onSystem sciences* (pp. 8-pp). IEEE. doi:10.1109/HICSS.2004.1265180

Rosenau, P. V. (2000). The strengths and weaknesses of public-private policy partnerships. In *Public-private policy partnerships* (pp. 217-242).

Rothery, M. (2005). Critical infrastructure protection and the role of emergency services.

Sapountzaki, K. (2007). Social resilience to environmental risks: A mechanism of vulnerability transfer? *Management of Environmental Quality: An International Journal*, *18*(3), 274–297. doi:10.1108/14777830710731743

Sarriegi, J. M., Sveen, F. O., Torres, J., & Gonzalez, J. J. (2008). Towards a research framework for critical infrastructure interdependencies. *International Journal of Emergency Management*, *5*(3), 235–249. doi:10.1504/IJEM.2008.025094

Sheffi, Y., & Rice, J. B. Jr. (2005). A supply Chain View of the resilient Enterprise. *MIT. Sloan Management Review*, *47*(1).

Stewart, G. T., Kolluru, R., & Smith, M. (2009). Leveraging public-private partnerships to improve community resilience in times of disaster. *International Journal of Physical Distribution & Logistics Management*, *39*(5), 343–364. doi:10.1108/09600030910973724

Swinford, S. (2010, December 27). Theresa Villiers: Government knew Heathrow was not prepared for snow. *Telegraph UK*. Retrieved from http://www.telegraph.co.uk/news/weather/8225790/Theresa-Villiers-government-knew-Heathrow-was-not-prepared-for-snow.html

Teisman, G. R., & Klijn, E. H. (2002). 10 Public-private partnerships in the European Union. In *Public-Private Partnerships: theory and practice in international perspective* (p. 165).

The Council of the European Union. (2008). COUNCIL DIRECTIVE 2008/114/EC of 8 December 2008 on the identification and designation of European critical infrastructures and the assessment of the need to improve their protection. *European Union*.

Turner, B. A. (1976). The organizational and interorganizational development of disasters. *Administrative Science Quarterly*, *21*(3), 378–397. doi:10.2307/2391850

World Bank. Asian Development Bank, & Inter-American Development Bank. (2014). Public-Private Partnerships: Reference Guide, Version 2.0. Retrieved from https://openknowledge.worldbank.com/handle/10986/20118

Yusta, J. M., Correa, G. J., & Lacal-Arántegui, R. (2011). Methodologies and applications for critical infrastructure protection: State-of-the-art. *Energy Policy*, *39*(10), 6100–6119. doi:10.1016/j.enpol.2011.07.010

KEY TERMS AND DEFINITIONS

Best Practices: Procedures or recommendations that are accepted or prescribed as being correct or most effective.

Corrective Maintenance: Unplanned maintenance activities performed after an incident occurrence in order to repair damage or strengthen infrastructures.

Crisis Emergency Equipment: The equipment that needs to be available when a crisis occurs.

Crisis Impact: Consequences a crisis generates.

Critical Infrastructure: An asset system which is essential for maintaining the social well-being of society.

Preventive Maintenance: Periodic activities performed before a crisis occurs.

Public-Private Partnership: Collaboration among public and private companies in which profits, risks and efforts are shared.

Resilience: The capacity of a system to avoid a crisis occurrence, and in the case where it cannot be avoided, the ability to reduce the impact and the recovery time needed to bounce back to the initial situation.

Chapter 4
Graphical Techniques and Methods:
Validating how they Improve Critical Assets Management

Adolfo Crespo Márquez
University of Seville, Spain

Khairy A. H. Kobbacy
Taibah University, Saudi Arabia

Luis Barberá
University of Seville, Spain

Samir M. Shariff
Taibah University, Saudi Arabia

ABSTRACT

GAMM (Graphical Analysis for Maintenance Management) is a method that supports decision-making in maintenance management through the visualization and graphical analysis of data. GAMM also allows the identification of anomalous behavior in equipment, derived from its own operations, maintenance activities, improper use of equipment or even as a result of design errors. As a basis for analysis, the GAMM method uses a nonparametric estimator of the reliability function using historical data, sometimes in very limited amounts. However, for successful results, experience and advanced knowledge in maintenance management are strictly necessary. In order to ease the interpretations of the GAMM method results, with the intention that the method becomes really amicable for managers, a set of basic rules have been developed. This set of rules leads to a proper and objective interpretation of GAMM results, improving the decision making.

DOI: 10.4018/978-1-5225-0651-5.ch004

INTRODUCTION

Maintenance management has grown in the past decade as a result of the continuous advancement of industrial organizations towards more efficient operations and improved productivity (Kobbacy & Jeon, 2001). Maintenance engineering has been fueling this continuous improvement of the maintenance management process by incorporating knowledge, intelligence and analysis (Barberá et al. 2012). Graphical tools supporting decision-making in the area of operational reliability can help to achieve an accurate and efficient management of assets and resources in organizations (Birolini, 2007), even when devices have functional configuration of high complexity. To obtain real applications of analytical models, practical, functional, innovative and simple tools must be generated (Barberá et al. 2013). This will help to make tactical and operational decisions easier. Additionally, the development of simple technical tools to facilitate the exercise of analysis and results outcome (Nelson, 1969) provides a framework for the control and monitoring of action plans implemented in terms of maintenance activities (Surucu1 & Sazak, 2013; Zhou et al., 2014).

The GAMM method (Barberá et al. 2013) is a graphical tool based on collected data related to the sequence of technical equipment revisions. This method of graphic analysis adds variables with useful information that allows the management to visualize support patterns and make decisions in the overall maintenance management, such as: number of technical revisions per-formed, type (corrective or preventative), duration, operational failure during the routine service, reliability of the equipment before the service, or the variability of equipment failure over a period of time. Moreover, GAMM provides a visual representation from a complete or partial historical record of the maintenance work performed, showing different patterns of analysis providing useful information for decision making and problem solving (Barberá et al. 2013).

After numerous cases where the method was applied, we realized that it was possible to develop a set of simple rules for preventive maintenance program evaluation using GAMM. With these rules, maintenance managers can simply refine refine their decision-making derived from this method (when generating Graphic 1 & Graphic 2, as we will see).

BACKGROUND AND PROBLEM STATEMENT

The GAMM method (Barberá et al. 2013) is a support tool for operational maintenance management and preventive maintenance program evaluation. The tool provides useful information regarding the reliability and maintainability of systems or equipment for analysis when considering variables such as type of intervention, duration of intervention or the existence of a stop during the intervention of the equipment/system. This broadens the spectrum display for maintenance management, acquiring new evaluation parameters and graphically determining possible areas of improvement.

GAMM is a quantitative and qualitative analysis method that intends to support decision making in maintenance management using the logic of a dispersion diagram which integrates the following variables: type of maintenance task, duration, and state of equipment/system during the maintenance intervention. The combination of these new variables with a graphic display (Graphic 1 and 2, Figure 1 and 2) of the sequence of interventions generates synergies with regard to the information given by the diagram, thus establishing new sources of information that can be analyzed. In addition, GAMM is able to estimate the reliability function, which is calculated using the algorithms on the basis of the

Figure 1. Examples of the graphs generated by GAMM (Graphic 1)

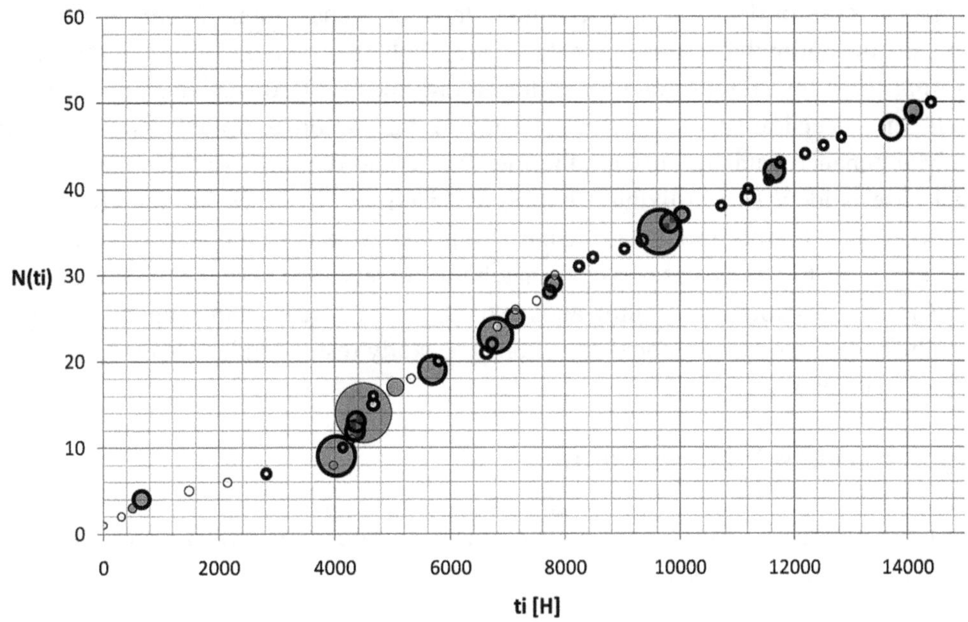

Figure 2. Examples of the graphs generated by GAMM (Graphic 2)

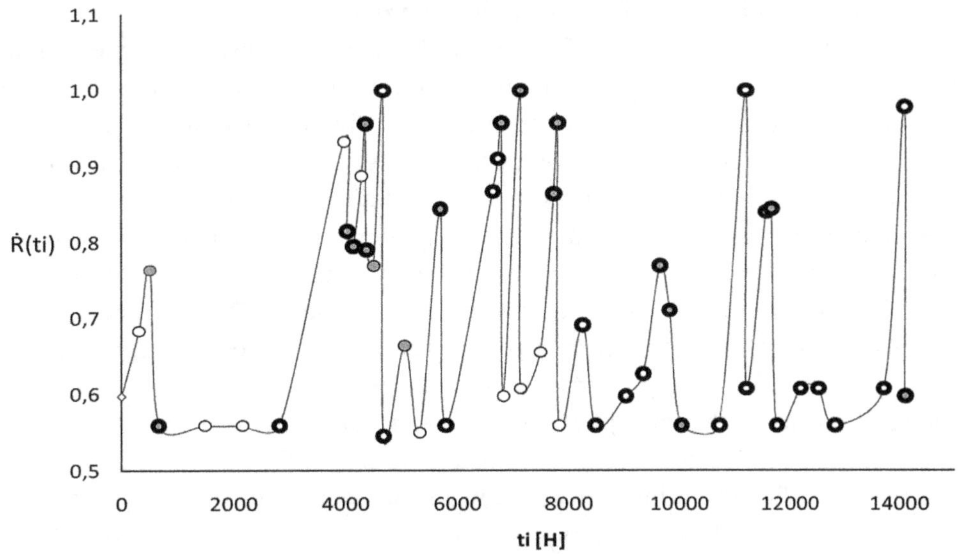

nonparametric method of Nelson–Aalen, and evaluated in a given time t_i, that is, in the moment before the maintenance intervention. The estimation method can consider both, the historic full data, and the censored historical data.

With the intention to test our set of rules for GAMM results interpretation, we will compare these rules results to those offered by the tool HIMOS (Hybrid Intelligent Maintenance Optimization System), for the same set of data (Kobbacy & Jeon, 2001). HIMOS (Hybrid Intelligent Maintenance Optimization System) is a system for decision support in maintenance management. More specifically, HIMOS helps to determine the optimal maintenance policy, essentially the frequency of Preventive Maintenance (PM) which minimizes the sum of down time due to PM and CO (Corrective Maintenance). HIMOS uses an 'intelligent' decision support system to carry out an automated analysis of the maintenance history data. Maintenance data are presented to the system and the most suitable mathematical model from a model-base is identified utilizing a hybrid knowledge/case based system (KBS/CBR) (KAH Kobbacy & J Jeon, 2001). Thus initially a rule base is applied to select a model. If no model is matched, the system reverts to its historical case-base to match the current case with a similar case that has been previously modelled. This double reasoning adds to the system's true learning capabilities (intelligence) and increases the rate of success of model selection. The study results include optimal PM intervals for a sample of industrial data sets.

Reviewing the GAMM Method Results

GAMM graphics allows appreciating different patterns in equipment behavior over time when exposed to maintenance in a given operational context, for example:

- Trend in the number of maintenance tasks.
- Preventive maintenance (PM) deviation: adding variable analysis as the type of intervention can display the frequency of completion of PM (programmed) and therefore can track and monitor the attainment of the established PM frequencies in the program.
- Quality of operations and/or quality of the PM tasks: a graphic display of interventions by type (preventive or corrective) allow the recognition of certain patterns offering information about the quality of operations and/or the performance of existing PM activities. By observing corrective interventions between preventive ones, sequences of corrective short term stoppages and/or accumulation of corrective interventions immediately after a PM intervention; the existence of problems in the maintenance execution, or in the equipment operation, can be evident.
- GAMM also shows the estimated reliability of the equipment/system prior to the maintenance task. This does really complement quality of operations and/or PM analysis. For instance, this may demonstrate the existence of corrective interventions in periods of high levels of expected reliability, indicating abnormal premature failure modes.
- Efficiency and quality in the implementation of maintenance can also be analyzed. The duration of each maintenance task can be represented using the size of the bubble in GAMM (Graphic 1). This variable identifies those interventions longer than expected times. This may question the maintenance schedule and workload, or could reveal the existence of other possible factors impacting maintenance, such as lack of tools, lack of parts, lack of staff training, and so forth.
- Impact on production can also be graphically appreciated with GAMM. The stop variable (Det) determines the status of equipment/system at the time the equipment is maintained. The bubbles with hard shaded border represent interventions in which the equipment stops, affecting their individual availability and possibly affecting the overall system availability.

In summary, GAMM is a quantitative and qualitative analysis that intends to support the decision making to reduce the high costs of systems unavailability as well as costs associated to unforeseen maintenance interventions. The method provides valuable information to directly implement action plans and to improve overall maintenance management.

Rules for the GAMM Method Results Interpretation

As mentioned above, in order to help maintenance decision making based on GAMM analysis and interpretation, a set of rules for preventive maintenance program evaluation will now be presented.

The maintenance managers will be able to make the best decision making according to these set of rules, derived from the vision of the GAMM graphics (Graphic 1, Graphic 2).

These rules provides easy access to certain variables patterns showing useful information for maintenance management and decision making in the short, medium and long term (see Table 1).

Validating the GAMM Rules

GAMM rules validation aims at substantiating that GAMM performs with an acceptable level of accuracy. In the literature, several validation methods are found. In (O'Keefe, 1987), there is a review of qualitative and quantitative validation methods for expert systems. In (O'Leary, 1987), a different framework

Table 1. Set of rules for GAMM method results interpretation

Rule 1	**If** the Graphic 1 shows a line of $N(t_i)$ (number of interventions) vs t (time) which has approximately constant slope and $R(t_i)$ ranging between $R(t_i)$ max = 1.00 and $R(t_i)$ min = 0.70, **then** you have an acceptable[R1.1] maintenance program. **If** the line of $N(t_i)$ vs t has not a constant slope or $R(t_i)$ has sudden or large fluctuations, **then** the maintenance program is not optimal.
	R1.1: regular number of maintenance events over an increasing operating time which means that a certain average reliability over time is guaranteed thanks to your maintenance activities.
Rule 2	**If** number of PM[1] interventions > 80% of the total interventions ($N(t_i)$), **then** it could be considered as an acceptable[R2.1] level of PM. **If** the number of PM interventions > 95%, **then** increase PM interval could be considered[R2.2]. **If** the number of PM interventions < 80%, **then** it could be necessary to evaluate the quality and frequency of the PM interventions. Decrease PM interval could be considered[R2.3].
	R2.1: according to many industry best practices. However, we should check the effects of existing failures and how they are distributed just in case they may become critical (size of the circle). **R2.2:** Number of PM interventions > 95% could also be a consequence of excessive maintenance to some non-critical failure modes. **R2.3:** Number of preventive interventions below 80% are scenarios showing room for potential improvement of PM policies.
Rule 3	**If** there are CO[2] interventions carried out in periods of high reliability ($R(t_i)$>0.85) or shortly after a PM intervention (less time than the average duration between PM interventions), **then** the PM task before that CO intervention could be poor (problems in the execution of PM intervention) or the failure could be the result of improper equipment operation, among others, and it is necessary to evaluate it.
Rule 4	**If** there is a logic/constant pattern of PM durations[R4.1] (downtimes, according to the tasks carried out on each intervention), **then** the PM program duration shows consistent maintainability. **If** there is a high variability of PM durations, **then** deviations in execution time due to external factors (lack of tools, waiting for parts, lack of staff training, accessibility problems, etc.) should be checked.
	R4.1: However, in some cases there are two or more different patterns of PM duration due to, e.g., overhauls.
Rule 5	**If** the temporal pattern of time between PM interventions is not constant, **then** the maintenance program is probably not well scheduled or its fulfillment should be reviewed.

[1] Preventive maintenance.

[2] Corrective maintenance.

for system validation is suggested, which includes content validity, criterion validity, objectivity and reliability. Typically, engineers have validated intelligent software system performance by running test cases through a system and comparing results against existing results or experts´ opinions. We will here adopt this method, and we will compare results for selected cases where GAMM rules offer recommendations to those recommendations offered by HIMOS for the same cases. This is done for specific cases that we have selected, which were previously validated by an expert panel, in the time that HIMOS was developed. More precisely, GAMM method will be applied to analyze a total of 5 sets of historical data (Figures 3 to 12), coming from the petrochemical industry (Kobbacy & Jeon, 2001).

Figure 3. Graphic 1 of GAMM for using Data set 1

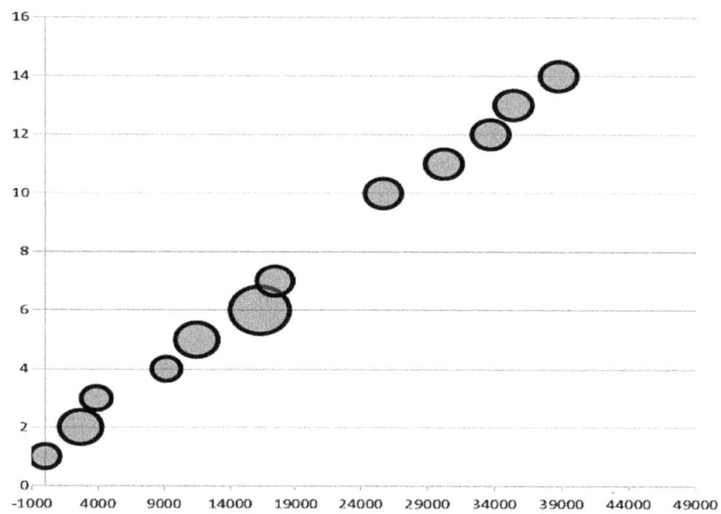

Figure 4. Graphic 2 of GAMM for using Data set 1

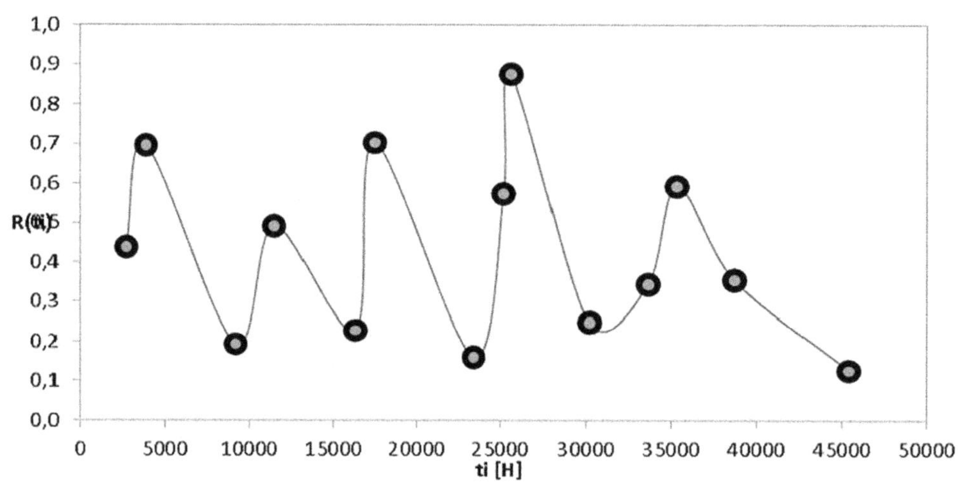

Figure 5. Graphic 1 of GAMM for using Data set 2

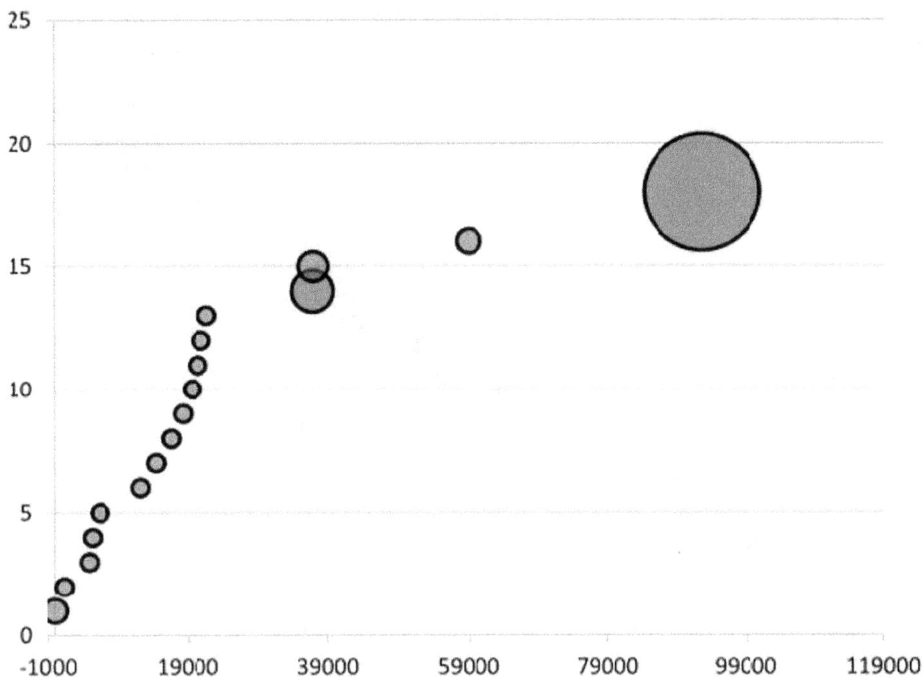

Figure 6. Graphic 2 of GAMM for using Data set 2

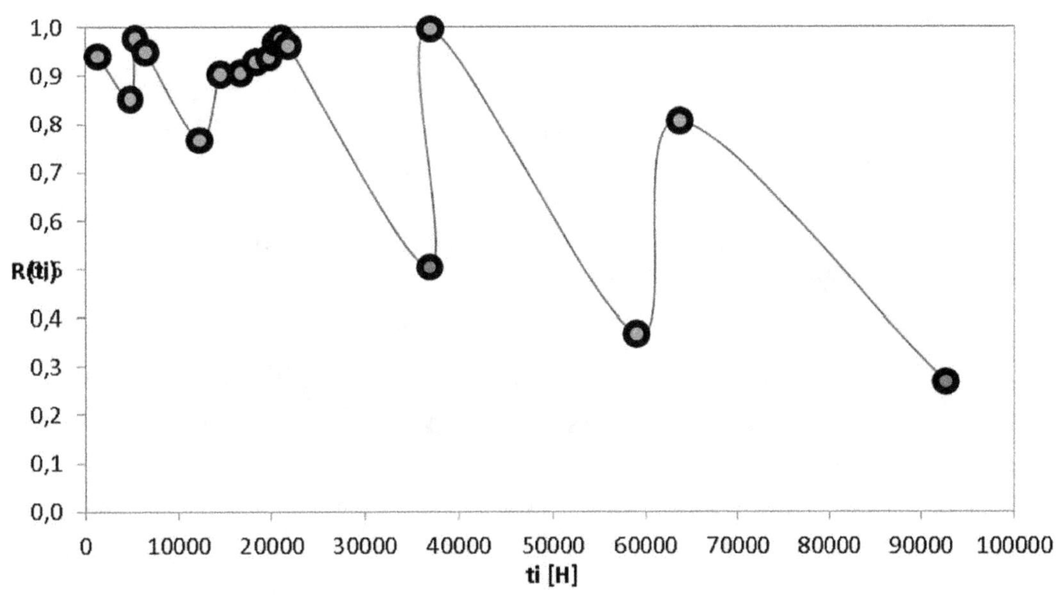

Figure 7. Graphic 1 of GAMM for using Data set 3

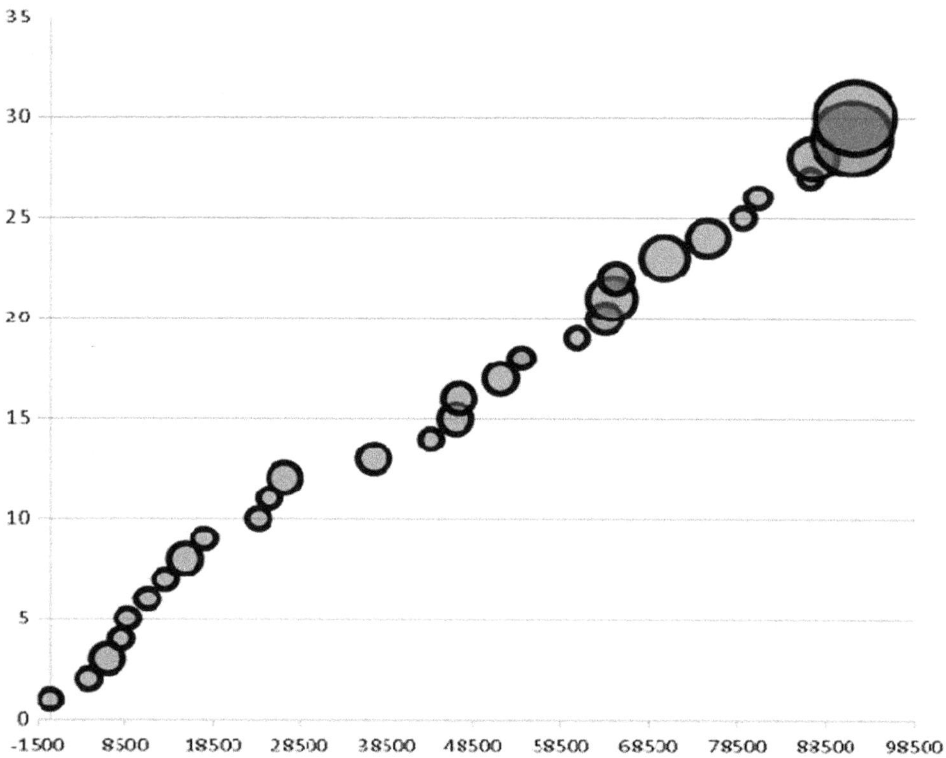

Figure 8. Graphic 2 of GAMM for using Data set 3

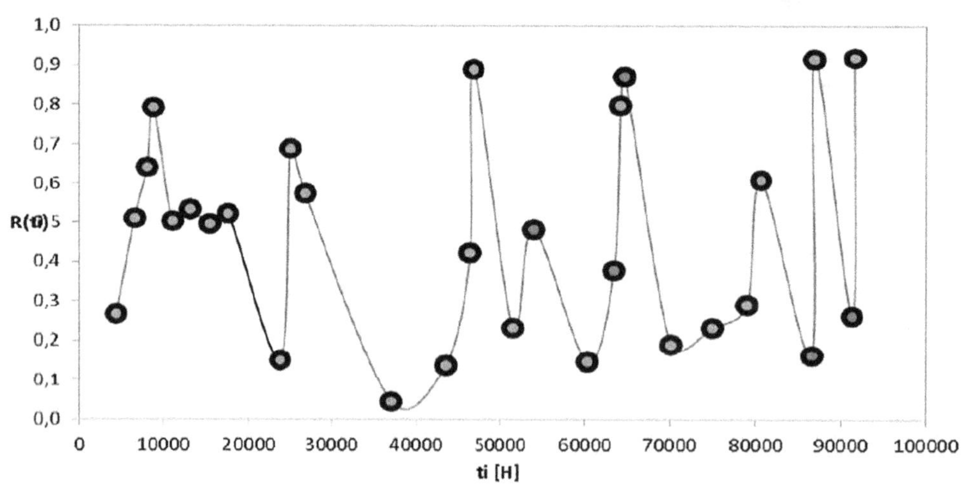

Figure 9. Graphic 1 of GAMM for using Data set 4

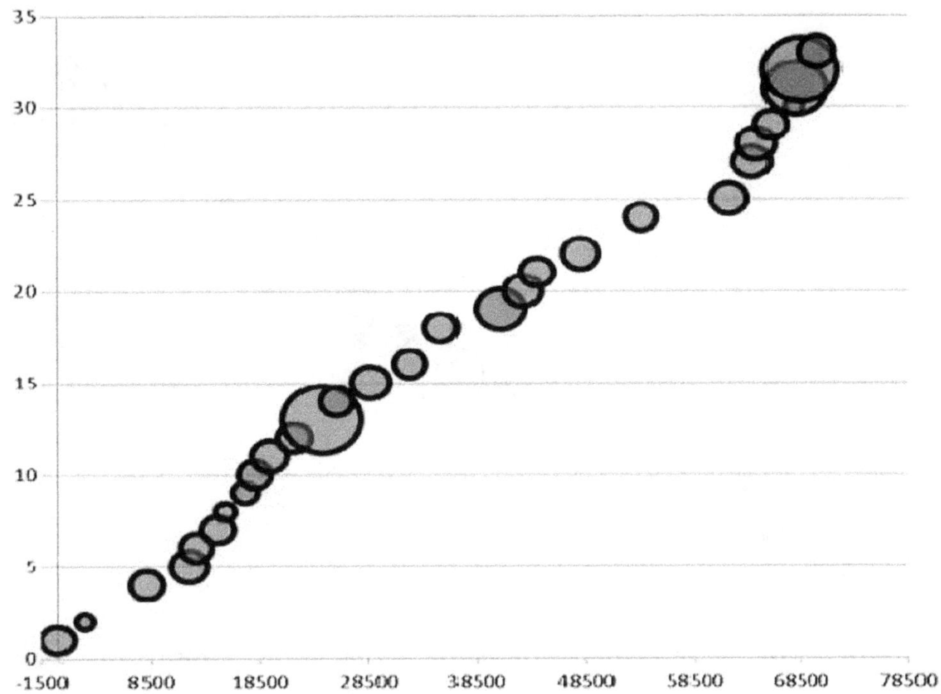

Figure 10. Graphic 2 of GAMM for using Data set 4

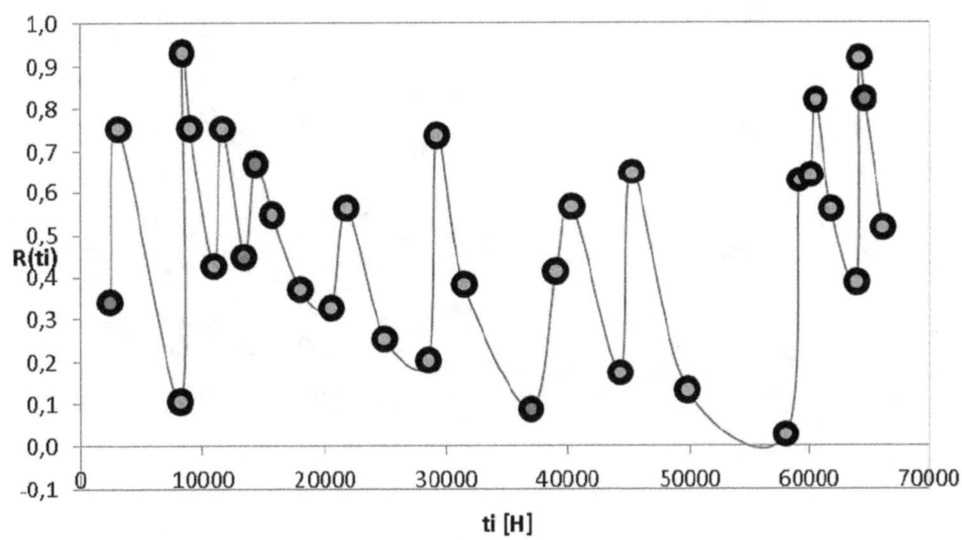

Figure 11. Graphic 1 of GAMM for using Data set 5

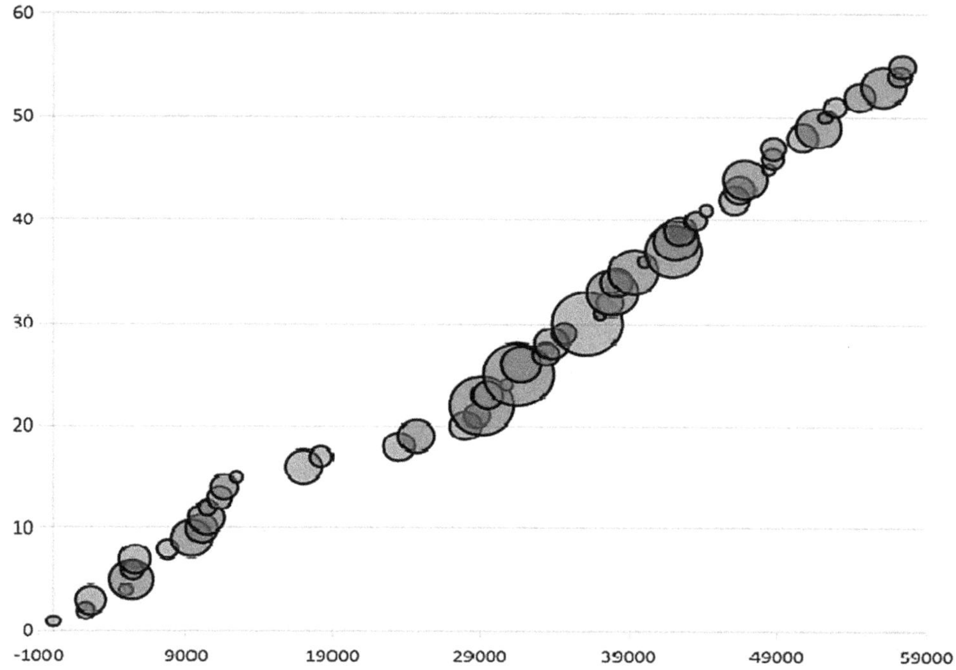

Figure 12. Graphic 2 of GAMM for using Data set 5

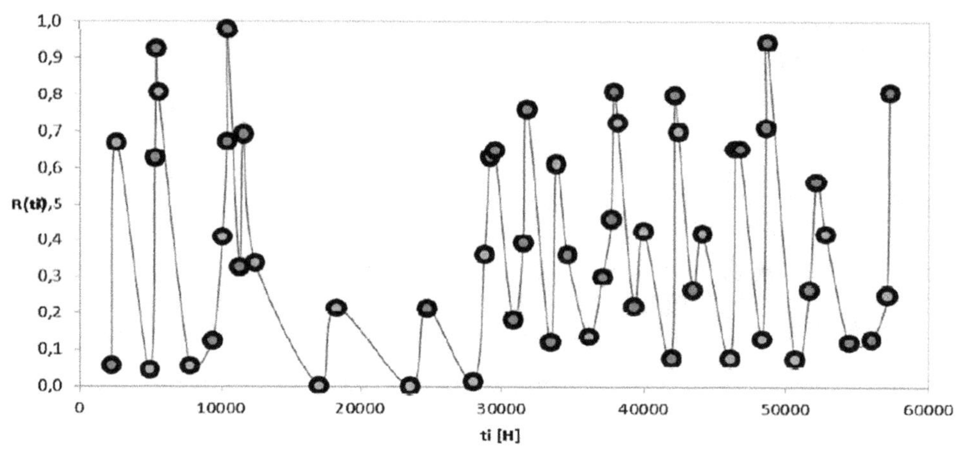

As previously mentioned, these data sets were analyzed by an expert panel (five experts in maintenance modeling) and HIMOS with the aim of comparing their recommendations of appropriate maintenance policies for each data set (KAH Kobbacy & J Jeon, 2001).

We have now summarized the recommendations offered by HIMOS and GAMM in Table 2.

Table 2. Comparison between HIMOS and GAMM' recommendations for 5 data sets

	Data Set 1
HIMOS	Increase PM interval from 135 to 176 days (CBR-Geometric I model)
GAMM	R(ti) has sudden or large fluctuations: maintenance program is not optimal. Number of PM interventions > 95%: increase PM interval could be considered. There is a logic/constant pattern of PM durations (downtimes, according to the tasks carried out on each intervention): PM program duration shows consistent maintainability. The temporal pattern of time between PM interventions is not constant: maintenance program should be reviewed.
	Data Set 2
HIMOS	Increase PM interval from 177 to 403 days (CBR-RPOW model)
GAMM	The line of N(ti) vs t has not a constant slope and R(ti) has sudden or large fluctuations: maintenance program is not optimal. Number of PM interventions > 95%: increase PM interval could be considered. There is a logic/constant pattern of PM durations (downtimes, according to the tasks carried out on each intervention): PM program duration shows consistent maintainability. The temporal pattern of time between PM interventions is not constant: maintenance program should be reviewed.
	Data Set 3
HIMOS	Increase PM interval from 159 to 223 days for the first sequence but gradually decrease to 86 days (tenth sequence) (NHPPScoSpm model)
GAMM	R(ti) has sudden or large fluctuations: maintenance program is not optimal. Number of PM interventions > 95%: increase PM interval could be considered. There are CO interventions carried out in periods of high reliability (R(ti)>0.85) and also shortly after a PM intervention (less time than the average duration between PM interventions): PM interventions before that CO interventions could be poor (problems in the execution of PM intervention) or the failure could be result of improper equipment operation and it is necessary to evaluate them. There is a logic/constant pattern of PM durations* (downtimes, according to the tasks carried out on each intervention): PM program duration shows consistent maintainability. The temporal pattern of time between PM interventions is not constant: maintenance program should be reviewed.
	Data Set 4
HIMOS	Increase PM interval from 108 to 140 (first sequence)–117 (tenth sequence) days (RPDWSco)
GAMM	R(ti) has sudden or large fluctuations: maintenance program is not optimal. Number of PM interventions > 95%: increase PM interval could be considered. There are CO interventions carried out shortly after a PM intervention (less time than the average duration between PM interventions): PM task before that CO intervention could be poor (problems in the execution of PM intervention) or the failure could be the result of, improper operation of equipment and it is necessary to evaluate them. There is a logic/constant pattern of PM durations* (downtimes): PM program duration shows consistent maintainability. The temporal pattern of time between PM interventions is not constant: maintenance program should be reviewed.
	Data Set 5
HIMOS	Decrease PM interval from 103 to 21 days (Delay-time Model, Weibull distribution, manual selection mode)
GAMM	Graphic 1 shows a line of N(ti) (number of interventions) vs t (time) which has partially constant slope and R(ti) has sudden or large fluctuations: maintenance program is not optimal. Number of PM interventions < 80%: it could be necessary to evaluate the quality and frequency of the PM interventions. Decrease PM interval could be considered. There are CO interventions carried out in periods of high reliability (R(ti)>0.85) and shortly after a PM intervention (less time than the average duration between PM interventions): PM task before that CO intervention could be poor (problems in the execution of PM intervention) or the failure could be the result of improper operation of equipment and it is necessary to evaluate them. There is a high variability of PM durations: deviations in execution time due to external factors (lack of tools, waiting for parts, lack of staff training, accessibility problems, etc.) should be checked. The temporal pattern of time between PM interventions is not constant: maintenance program should be reviewed.

It is interesting how, being the general HIMOS advice in accordance with GAMM recommendations, GAMM is able to show additional information considering different variables and using the same data sets. As can be seen in Table 2, for each data set there was general agreement between the results produced by HIMOS and GAMM recommendations. Thus, it was agreed that the proposed rules are suitable for the correct and objective interpretation of the GAMM method. In a knowledge-based system (KBS) or `intelligent' decision support system DSS (IDSS), this expertise is usually provided by the so called knowledge subsystem. In GAMM, this expertise concerns the right interpretation of the graphs (performed from a data set) based on the rule base proposed. The rules were developed by looking for patterns in different data sets for different equipment and considering all aspect of the GAMM method. Therefore, knowledge base system implemented in GAMM consists in the set of 5 rules as shown in Table1.

CONCLUSION

The aim of this paper was to present and to validate a proposed set of rules to help and support managers when implementing the GAMM method, which has been developed and published by the authors of this publication. Additionally, for validation of the proposed interpretation rules, GAMM method was applied to analyze a total of 5 sets of historical data (which are included in the appendix), from the petrochemical industry. This data was previously analyzed by HIMOS and HIMOS results were validated by a panel of experts. Comparing their recommendations, HIMOS and GAMM provided consistent results.

FUTURE RESEARCH DIRECTIONS

As a future research line, it is interesting that the joint application of HIMOS and GAMM could be considered as a quantitative and qualitative analysis that intends to support the decision-making in maintenance management and helps to reduce high costs of unavailability, as well as the costs associated with un-foreseen maintenance activities. The joint application of HIMOS and GAMM could be an excellent tool supporting maintenance management in short and medium terms. They provide useful information, clear and easy to interpret, which can be checked with one another in order to determine the optimal maintenance policy (frequency of preventive maintenance). This broadens the spectrum display for maintenance management, acquiring new evaluation parameters and determining possible areas of improvement.

REFERENCES

Barberá, L., Crespo, A., Viveros, P. & Arata, A. (2013). The Graphical Analysis for Maintenance Management Method: A Quantitative Graphical Analysis to Support Maintenance Management Decision Making. *Journal of Quality and Reliability Engineering International*, 29(1), 77–87. DOI:10.1002/qre.1296

Barberá, L., Crespo, A., Viveros, P., & Stegmaier, R. (2012, January-March). Advanced model for maintenance management in a continuous improvement cycle: Integration into the business strategy. *International Journal of System Assurance Engineering and Management*, *3*(1), 47–63. doi:10.1007/s13198-012-0092-y

Birolini, A. (2007). *Reliability Engineering* (5th ed.). Springer Berlin Heidelberg New York.

Crespo Márquez A., Gupta J. (2006). Contemporary Maintenance Management. Process, Framework and Supporting Pillars. *Omega: The international journal of management science*, 34(3), 325-338. DOI:.10.1016/j.omega.2004.11.003

EN 13306:2001. (2001). Maintenance terminology. *European Standard. CEN. European Committee for Standardization*, Brussels.

Kobbacy, K., & Jeon, J. (2001). The development of a hybrid intelligent maintenance optimisation system (HIMOS). *Journal of the Operational Research Society*, 2001, 762–778.

Mishra, R. P., Anand, D., & Kodali, R. (2006). Development of a framework for world-class maintenance systems. *Journal of Advanced Manufacturing Systems*, *5*(2), 141–165. doi:10.1142/S0219686706000807

Nelson, W. (1969). Hazard Plotting For Incomplete Failure Data. *Journal of Quality Technology*, *1*, 27–52.

O'Keefe, R. M., Balci, O., & Smith, E. P. (1987). Validating expert system performance. *IEEE Expert*, *2*(Winter), 81–87. doi:10.1109/MEX.1987.5006538

O'Leary, D. E. (1987). Validation of expert systems—with application to auditing and accounting expert systems. *Decision Sciences*, *18*(3), 468–486. doi:10.1111/j.1540-5915.1987.tb01536.x

Pintelon, L. M., & Gelders, L. F. (1992). Maintenance management decision making. *European Journal of Operational Research*, *58*(3), 301–317. doi:10.1016/0377-2217(92)90062-E

Surucul B., & Sazak, H.S. (2013). Graphical Methods for Reliability Data. In *Wiley Encyclopedia of Operations Research and Management Science*. John Wiley & Sons, Inc.DOI:10.1002/9780470400531.eorms1062

Van Horenbeek, A., Pintelon, L., & Muchiri, P. (2010). Maintenance optimization models and criteria. *Int J Syst Assur Eng Manag*, *1*(3), 189–200. doi:10.1007/s13198-011-0045-x

Zhou, Y., Kou, G. and Ergu, D. (2014). Analysing Operating Data to Measure the Maintenance Performance. *Journal of Quality and Reliability Engineering International.* DOI:10.1002/qre.1584

Chapter 5
Impact of the Knowledge Management in Maintenance Engineering:
Effects on Industrial Production

Javier Cárcel-Carrasco
Universitat Politècnica de València, Spain

Manuel Rodríguez-Méndez
ESeyPro S.L., Spain

María Carmen Carnero
University of Castilla – La Mancha, Spain & University of Lisbon, Portugal

ABSTRACT

Knowledge management has been analyzed in numerous areas of the industrial enterprise, especially in the areas of strategic management, innovation, trade, or administration. However in operational areas with technicians working mainly on the basis of its experience gained over the years, such as the departments of industrial maintenance, there are no deep analysis of the incidence of the knowledge management in these areas. The peculiarities in this type of activity on the inside of the company, knowledge of these people is strongly based on your experience (strong tacit component), difficult to measure and articulate, and however, on numerous occasions, this knowledge not transmitted, can be a high cost for the company (many times assumed as inevitable) due to the increase of production and services downtime, loss of efficiency, or time of coupling of new personnel to these areas.

INTRODUCTION

There is an abundant literature on knowledge management in various industrial activities and services, and the effects of their application (Bahoque, Gómez & Pietrosemoli, 2007; Colino, Martinez & Martinez, 2010; Colino & Riquelme, 2000; Chua & Gu, 2008; Ferrada & Serpal, 2009; Rivas & Flores, 2007; Ven-

DOI: 10.4018/978-1-5225-0651-5.ch005

tura & Ordoñez, 2007; Yang, 2006;), but normally these studies focus on global management (especially in the more explicit part), with an emphasis on trade, accounting or internal administrative management, or development activities, but dealing sparingly with the tactical actions of industrial trades, normally considered as an "expense for the company", but having the effect, however, of substantially reducing costs involved (often taken by management itself). Therefore, to manage knowledge in these areas of work means, in itself, not only an improvement in the efficiency of the processes within industrial trades, but also a reduction in the spending caused to the company (by stoppage of production, loss of energy, loss of efficiency or reliability of systems and facilities and longer adaptation time of new technicians).

It is necessary to analyze the personal knowledge to develop organizational knowledge (Martin, 2008; Pauleen, 2009; Volkel & Haller, 2009), that allows to make an analysis of cost-benefit on its application (Volkel & Abeckar, 2008). The internal activities of the industrial enterprise, maintenance needs deep technical knowledge, highly experienced staff and has traditionally been the structure within the company where there is major component of tacit knowledge. Since their duties directly affect the reliability of systems and installations (Sols, 2000), eliminating unwanted stops and performance to critical processes, the need for the appropriate management of such information/knowledge is given that you can have a great strategic value for the company.

Personnel policies, so prevalent today, that they include subcontracting of industrial services, the objectives of the staff, serving only for reasons of age, without submitting to the recently incorporated into a loyalty preprocessing to reduce the external rotation, it is possible to achieve, in some cases, successes in the short term but certainly the lack of sufficient time to create and transfer knowledge in a controlled way, and the lack of sedimentation cultural to assimilate it and implement it, will be the implementation of any model of knowledge management in the medium term (Muñoz, 1999), and to a large extent, the loss of control of the reliability and efficiency of production processes or industrial internal services.

It should be borne in mind that the operational impact of maintenance actions has an impact on the company (Table 1.), affecting most of the fundamental tactical actions, and given that this action can be strategic, it is necessary to check conditions for the capture of that knowledge (Cárcel, 2014a).

Table 1. Tactical aspects of the companies and their relationship with maintenance

Tactical In Business Aspects	Operational Impact of the Maintenance
Production	High incidence, directly affecting the levels of reliability and stops.
Fixed assets depreciation	Increases the operational life of fixed assets.
Repairs and maintenance	Direct responsibility
Fixed assets investment	Transfer to maintenance, once. Your point of view and experience should be in the election.
Staff	In regards to maintenance, is required high qualification and experience.
Training and education	In reference to the maintenance, the training must be integrated with its fundamental tactical functions.
External services and outsourcing.	All subcontractors for repairs or maintenance must be checked by maintenance.
Energy consumption	You must be one of the main functions of the Organization of the maintenance, control and monitoring of energy consumption.
R & D	In the actions of r & d for equipment, facilities and processes, should be the vision of the Department of maintenance

Source: Own elaboration

This chapter is intended to make an approximation of the factors of management knowledge in maintenance engineering, and the effects on industrial production, as well as mark knowledge management processes and identify barriers and facilitators that could have impact on such activities.

THE PRINCIPLES OF KNOWLEDGE MANAGEMENT

The knowledge is the ability to act, process and interpret information to generate more knowledge or to solve a particular problem. A momentous change, in which the growth of economies and enterprises is driven by knowledge and ideas, more, has occurred in recent years that traditional resources (Del Moral, 2007). We are moving towards a society driven by the knowledge, where traditional tangible assets are losing value in favour of intangibles (Peña, 2001; Sánchez, Chaminade & Escobar, 1999). So that it can be considered knowledge as the main intangible ingredient both in business and in the economy as a whole (OCDE, 1996).

Comes from the knowledge a person considered, interprets and uses information combined with its own experience and ability (Zapata, 2001).

The importance of knowledge in organizations depends on what can be done with it within a scope of business (Nonaka, 1999). That is, the knowledge itself is not relevant, so it cannot be used to give rise to actions of creation of value (Xiomara, 2009).

Knowledge management involves, therefore, difficult to observe and manipulate practices, and are sometimes even unknown to those who possess them. This presents a problem for businesses, more familiar with the management and accounting of fixed capital. Among the various categories of investments related to the knowledge (education, training, software, R&D, etc.), knowledge management is one of the least known both qualitative and quantitative, as well as in terms of costs and economic returns (OCDE, 2004).

Definitions of knowledge management that appear in the academic literature adopt various perspectives (Palacios & Garrido, 2006), reflecting the nuclear approach postulates theoretician who is part, are taken as a collection of processes, management of stocks, applying the knowledge to create competencies, or the management of the talent of the staff that creates knowledge.

All the perspectives of knowledge management, based on principles and practices, very rarely are assigned to the part that stores highest level of tacit knowledge within industrial enterprises and services, such as tactical maintenance, Assembly and projects activities aspects in general difficult documentation, which normally are based largely on our employees experience over the years difficult to capture and even more difficult to transfer.

The problem of transmission of knowledge in industrial maintenance operators occurs when the operator possessing that knowledge, leaves job, the company loses it, suffering economic and operational problems arising from this.

The top of Figure 1 shows the assimilation and experience curve according to the length of employment that is normally observed in maintenance organizations. It notes that before the change of operator the replacement needs a bedding-in period, with reduced operational efficiency until this period is over. A model of knowledge in maintenance management should be to work together to capture that experience or tacit knowledge, and so reducing these contact times (Figure 1B).

But on numerous occasions the knowledge management is underused and inefficiently deployed (Ordonez, 1999,2001), knowledge management can be defined as "structures, systems and interactions

Figure 1. Transformation of knowledge based on experience in maintenance process of retention curves
Source: Own elaboration

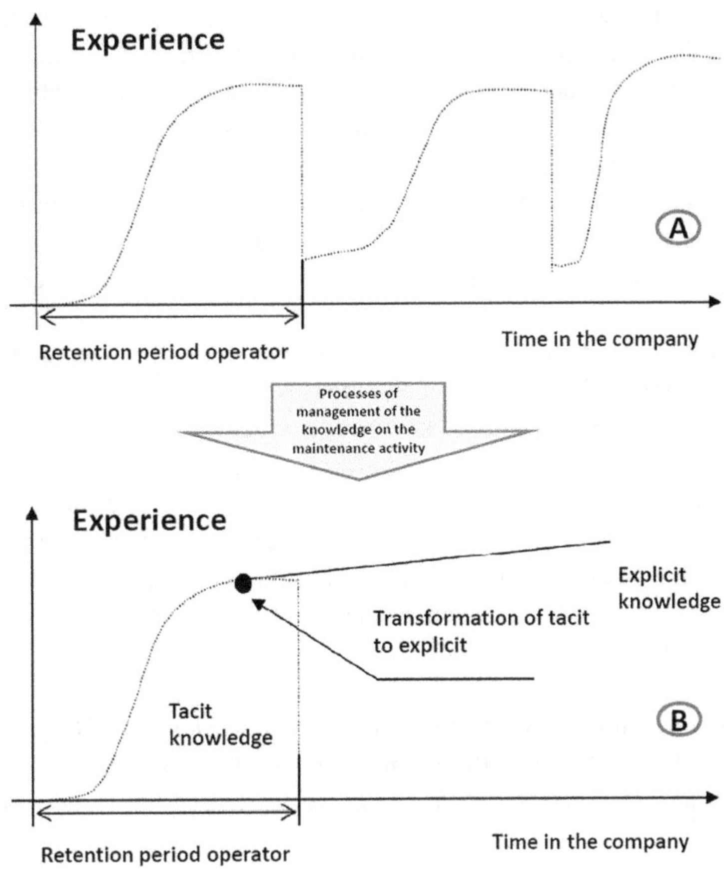

consciously integrated and designed to allow the management of the set of knowledge and skills of the company" (Tiemessen, Lane, Crossman & Yinkpen, 1997), turning it into a resource of fundamental strategic importance (Bueno, 1999, 2000; Hedlund, 1994; Nonaka & Takeuchi, 1995, 1994; Ventura, 1996; Wernerfelt, 1984).

Knowledge generation occurs in a transformation of tacit knowledge of individuals in explicit group and organizational level (Nonaka & Takeuchi, 1995), and they internalize each one of the members of such groups, turning it again into tacit. This process generates four phases, which are: socialization, externalization, combination and internalization.

Model of knowledge creation is shown in a perspective multilevel seen in the spiral of knowledge, which is not a linear, sequential, process but exponential and dynamic, that part of the human element and its need to compare and validate their ideas and premises. In this way, the individual through the experience creates tacit knowledge, which conceptualizes, transforming it into explicit single. It becomes social explicit knowledge to share with any of the persons involved in the organization. The next step is to internalize the common experiences, transforming the social explicit knowledge into tacit individual (Martínez et al., 2002). Having mechanisms for structuring and use past within the industrial company experiences possible, members not improvise continuously on the same experience (Cegarra & Moya,

2003). The focus of the knowledge management is to leverage and reuse resources that already exist in the Organization, in such a way that people can select and apply best practices (Wah, 1999).

Knowledge cannot be created regardless of the action, changing the notion of knowledge as a commodity which individuals or organizations can be purchased, to the study of knowledge as something that players develop through the action. Polanyi work has been very influential in the definition of knowledge as something dynamic, and whose tacit dimension hinders its transmission, which is largely in the fundamental activities of industrial maintenance.

According to Rodriguez in his study on the creation and management of knowledge in the business sector (Rodríguez, 2006), some of the facts that justify the importance of the management of the knowledge, run between capture and share best practices to strengthen the chain of command and improve communication.

Knowledge structures are built on past experience and are used to sort data to the following interpretation and action. Hence, the individual knowledge focuses on individual knowledge structures, while the group knowledge relates to organizational knowledge structures (Pérez, 2007).

Generating knowledge organization must design forms of labour and establish policies that lead the company into a situation that can be characterized by three conditions or basic facilitators (Peris, Loria & Mendez, 2002). The postulates of Peris, largely approaching the appropriate measures that should be in the internal activities of the industry (operation and maintenance), due to the characteristics of operation of these areas (team work, strong component of tacit knowledge, actions of emergency and resolution of faults that can involve all production of the company):

1. Equity and recognition of the effort and the merits of each Member of the organization.
2. Promotion of trust in and between the members of the Organization, such as a requirement to delegate to them the necessary autonomy in skilled work, and obtain cooperation in knowledge creation.
3. Congruence between the company's objectives and the objectives and interests of the workers engaged in knowledge creation.

If these large basic conditions are reached, other aspects of the Organization are largely technical issues of design. Some of the main features that should have the design of the creative organization of knowledge, are:

1. Adequate levels of formalization and centralization of decision-making.
2. Policies and practices of human resources. Recruit staff with the necessary qualifications, appropriate training (internal and external), evaluation based on group results, and ensure retention and tenure at the company.
3. Importance of the work teams. It is Basic for the context created for share tacit and explicit knowledge. It is advisable that they are multifunctional, is auto-managed and formed parallel organizations.

For this reason, an organizational space has been created in which members of the organization, founding their relations on the basic conditions set forth, can share information, goals and interests. In this way, we ensure the voluntary cooperation of the members of the organization and the contribution of their information, which is necessary for the creative organization of knowledge.

The literature identifies many roles and values in knowledge management at the level of how organizational staff (Cheong & Tsui, 2010a, 2010b) handle the knowledge management process, and it should be established as a team mission (by humans), encouraging awareness and proactivity to capture the knowledge by the group. As in all human groups, there is relationship between competitiveness and cooperation among its members. The conditions of these relationships must be established (Hooff & Huysman, 2009; Kulkarni, Ravidran & Freeze, 2007; Lin, 2007; Tirpak, 2005), to succeed in the internal exploitation and maintenance units of the company (due to the intrinsic conditions of these workers, there is usually a high component of stress, reduced templates, and a feeling that "what I know, is what gives me value within the company, and I do not care to share it". It is this human factor that makes the integration of group knowledge so difficult.

PROCESSES, DIMENSIONS, AND TOOLS IN KNOWLEDGE MANAGEMENT

Barragan analyzes the taxonomies of models of knowledge management (Barragan, 2009) where you can find points in common that allow summarize them and reassemble them to homogenize criteria in areas where the study and development of knowledge management have had an important development; These criteria include theoretical, conceptual, philosophical, technical, scientific, cognitive, intellectual capital, knowledge management and social aspects.

- The dimensions in the management of the knowledge according to empirical studies and theoretical (Tari & Garcia, 2009), that are directly related to the quality control, can be enumerated in the following:
- Creation of knowledge (organizational learning): acquisition of information, dissemination of information and shared interpretation.
- Transfer and storage of knowledge (organizational knowledge): store knowledge and knowledge transfer.
- Application and use of knowledge (learning organization): teamwork, promote dialogue, establish systems to capture and share learning, relationship between different departments or functional areas and commitment to learning.

Within the context of a company, if we define knowledge as a process management, an approach to this process could be composed basically of the generation, coding, transfer and use of knowledge (Wiig, 1997).

The dimensions in the management of the knowledge according to empirical studies and theoretical (Tari & Garcia, 2009), that are directly related to the quality control, can be enumerated in the following:

- Creation of knowledge (organizational learning): acquisition of information, dissemination of information and shared interpretation.
- Transfer and storage of knowledge (organizational knowledge): store knowledge and knowledge transfer.
- Application and use of knowledge (learning organization): teamwork, promote dialogue, establish systems to capture and share learning, relationship between different departments or functional areas and commitment to learning.

Within the context of a company, if we define knowledge as a process management, an approach to this process could be composed basically of the generation, coding, transfer and use of knowledge (Wiig, 1997).

The generation and transfer of knowledge are processes with a greater amount of tacit knowledge. Both the stage of coding in the use stage, tacit knowledge is converted into explicit knowledge to understand and available to all members of the company itself. Bueno (2002) points out that the fundamental aspects of the knowledge management are the creation and distribution of knowledge, although there are qualifications expressed by other authors surplus (Table 2.) processes involved in the management of knowledge (Quintana, 2006).

It should be defined, by the relevance that can be taken, the concept of loss or leakage of knowledge, where we find the extensive problem of the loss of important personnel in the Organization (factor that occurs with high incidence among the staff of office, due to high personnel rotation or by outsourcing of business policies). Clearly demonstrates that the abandonment of key individuals it is a net loss of knowledge, limiting the extent access to knowledge and learning for employees who replace them not to hire a new worker equally profitable. A high abandonment rate breaks the continuity in the Organization and leads to a social environment in which workers are suspicious of fellow (Perez de Miguel, 2006).

For the proper management of knowledge, a set of tools and technologies needed (Gray & Meister, 2006; Kim, Breslin & Decker, 2009; Davenport, 1997; Sher & Lee, 2004; Wong & Aspinwal, 2004), which produce cheaper and it confers an apparent reliability and efficiency in the difficult task of cap-

Table 2. Knowledge management processes by different authors

Author	Process Knowledge Management
Wiig (1997)	Generation. Acquisition. Creation. Encryption, storage or integration. Transfer. Use
Hernan (1998)	Creation. Development. Broadcasting.
Revilla & Pérez (1998)	Creation. Development. Broadcasting. Exploitation
Ditzel (2005)	Planning. Definition of the desired state or of what should be. Analysis of the real situation. Comparison of the actual situation and the desired situation. Planning of actions. Development. Generate. Purchase. Order. Archiving knowledge. Transfer. Use. Evaluation and review
Alavi & Leiden (1999)	Creation. Share. Distribution
Bhatt (2001)	Creating validation. Presentation. Distribution. Application
APQC cited by Luan/Serban (2002)	Identification. Capture. Transfer
Alavi &Leiden (1999)	Creation. Share. Distribution
Lai & Chu (2002)	Generation. Repository model. Sharing and transfer. Use
Lee & Hong (2002)	Capture. To develop. Share. use
Petrides & Nodine (2003)	Generation. Storage. Distribution. utilization
ARIS Manual (2003)	Acquisition. Presentation. Transfer. Use. elimination
Corrêa da Silva & Agustí-Cullel (2003)	Generate and acquire. Store and save. To access. use
Blázquez (2004)	Generation. Capture. Processing. Storage
McCann & Buckner (2004)	Purchase and build. Share and retain. Apply
Peluffo & Beatriz (2005)	Generation. Share or distribute. Use

Source: own elaboration based Quintana, 2006

turing the strategic knowledge and that generates value for the organization. The Japanese industrial companies were pioneers in the study and implementation of its management, especially the automobile sector, as defined by Rivas (Rivas & flowers, 2007), or the definition of Binney (Binney, 2001), of the spectrum of management of knowledge, for different areas of industrial enterprise.

Are numerous authors, indicating relevant values, know what you want to manage (knowledge), through audits of maintenance, and clarify as spreads such information-knowledge in the heart of the Organization, through simple, Visual, and intuitive tools, such as the maps of information or knowledge maps.

Must be the identification of the knowledge residing in the bosom of the same as well as characteristics or identifying elements, using knowledge maps (Hansen & Kautz, 2004; Quintana, 2006; Seeman & Coen, 1997).

A map of knowledge is a tool for the localization of knowledge within an organization. It is similar to a map of information oriented to knowledge rather than information. You can have a pictorial representation in the form of a network of knowledge (Gil, Perez & Lopez, 2008), and the reasons for making it can be varied (Pérez, 2005), from finding key sources and restrictions in the creation of knowledge and its flows and improve customer response times, takes it decisions and solving problems, providing access to the required information.

Through the audit of knowledge (Table 3), it is got the first step that must be performed in knowledge management projects.

The audit identifies opportunities and help to locate and evaluate the sources where is stored the knowledge, activities that transform the knowledge, and the factors involved, and finally allows to establish patterns of solution (Perez, 2007). Also, it helps to get the requirement for the design of an efficient system of knowledge management, exploration of the first activities to be carried out in the knowledge chain (Holsapple et al., 2004).

Table 3. Fundamental features of knowledge audit

Fundamental Features of Knowledge Audit			
The Objectives That Pursues the Knowledge Audit (Debenham, 1994)	**Results that Must Include a Report of Audit**	**You Must Respond to the Following Questions (Pérez, 2007)**	**Methods of Analysis**
• A comprehensive and structured view of knowledge contained in a particular section of an organization • Identify sources of knowledge • Qualitatively and quantitatively characterize the knowledge in the sources	• Description of the audited process • Sensitivity analysis of found and their respective conclusions • Diagram of the auditee knowledge blocks, relations between the blocks and the sources where knowledge resides • Description of the means used to record information related to the qualitative and quantitative characteristics of knowledge	• What knowledge does need the Organization to support your business? • Where is the knowledge in the Organization? • How flows the knowledge within the Organization? • How it is captured, stored and exchanged knowledge? • How are did visible that knowledge? • How people keep updated such knowledge? • How can is defined in the Organization knowledge? • How to create knowledge in the Organization?	• Questionnaires based on surveys of knowledge: these help to obtain extensive descriptions about the status of operation of knowledge • environmental analysis of the tasks: helps to understand that knowledge is present in roles • verbal Protocol Analysis: identifies knowledge • knowledge mapping elements: used to develop conceptual maps or hierarchical • analysis of the critical role of knowledge • Preparation of an inventory of knowledge and requirements analysis

Source: own elaboration from various authors

In the literature reviewed has been found a use of ontologies as a formalism to represent knowledge as support audit of knowledge processes in organizations. The necessity of implementing ontologies in process modelling is a topic of interest (Rojas & Montilva, 2009). Process modelling describes the domain of the application in terms of a system consisting of a set of related elements: objective, processes, actors, activities (workflow), objects, organizational structure, business rules and events. Sometimes, modelling each one of these elements is not supported a method clear allow doing so. In this sense, an ontology might contribute to achieving such conceptual clarity.

Ontologies also help process modelling groups to establish, differentiate and relate objectives, processes, activities, resources, rules, players, technologies and events that characterize business systems. It also facilitates communication between the actors involved in the development of software, to provide a unique and consensual definition of the concepts of the application domain.

The results of an audit of knowledge, are the requirements to develop properly (Paniagua, 2007) knowledge management project.

The audit identifies opportunities and help to locate and evaluate the sources where the knowledge, activities that transform the knowledge is stored and the factors involved in these activities, and finally allows establishing patterns of solution (Perez, 2007). Also, it helps to get the requirement for the design of an efficient system of knowledge management, exploration of the first activities in the chain of knowledge (Holsapple & Jones, 2004).

INDUSTRIAL MAINTENANCE AND THE KNOWLEDGE MANAGEMENT

Given that the idea is to measure an intangible asset (knowledge management processes), it makes the process very difficult. This is why some authors propose measures to assess qualitative rather than quantitative knowledge (Baez, 2007; Edvinsson & Malone, 1997; Norton & Kaplan, 1996).

Consequently, the effective management of the maintenance is one of activities crucial for most of the companies with physical assets. Are therefore logical efforts to optimize its performance, engaging for this purpose human resources technicians and with the model and strategies most appropriate to each company (Zhou et al., 2009; Sun et al., 2007; Pongpech et al., 2006; Oke, et al., 2005).

Such an operational definition of Industrial maintenance could be the set of techniques that are intended to achieve an optimal use of productive assets, keeping them in the State that requires an efficient production.

The following items may be removed from this definition:

- State of the maintenance.
- Requirements of availability or preservation of that State.
- Set of techniques and procedures that conservation-oriented.
- Activity of replacement, repair or modification of units, components, assemblies, equipment or industrial plant systems.

It is shown by the very nature of the maintenance elements linked to knowledge, since the technique can be defined as the form or means to carry out an activity, involving, as a result, the presence of intellectual capital incorporated or not into manufacturing assets or personnel. The special action or maintenance activity required techniques or knowledge very specific and contingent, of high strategic value, involving complexity and great efforts in your registration, transmission and application.

In terms of the expression of its goal: the achievement of availability in equipment and facilities requirements, involves the location of maintenance activities in high contingency scenarios and uncertainty, which contained very dynamic, perishable and specific information, and its application procedures, reveal themselves as essential for an efficient functioning of the plant. In another case, the maintenance of the plant should respond of high costs of intervention, based on a repetitive and inconsistent search for information in phases of detection, diagnosis, prevention, and repair of the fault.

The maintenance function therefore meets two major objectives: firstly, retain the State of assets, in second, improve levels of availability at a lower cost.

It derived the following conclusions with regard to the basic knowledge necessary for the actions carried out by the maintenance function:

- Knowledge necessary for the actions of selection, acquisition, installation and commissioning of the equipment and General facilities:
    - Knowledge of the requirements of the plant in relation to the needs of capacity and normal operation of the equipment.
    - Knowledge of real team response to these requirements.
- Knowledge necessary to carry out actions of prediction, prevention and operative correction of equipment:
    - Knowledge of the requirements of the plant in relation to the needs of equipment operation.
    - Knowledge of real team response to these requirements.

Some of the most frequent issues critical, tacit knowledge and knowledge management, with the specialists and maintenance technicians are:

- Changes in the staff.
- Inexperienced operators.
- Lack of information on measures to be taken and steps to be followed before certain breakdowns or incidents.
- Dependence on knowledge and tacit experience of operators.
- Disruption of information about facilities.
- Lack of learning and the personal recycling systems.

The problems of personnel changes in the maintenance template translates into economic losses due to the lack of knowledge by the operator of: existing facilities, typical failures and measures to adopt to them, time of filming and adaptation to the shape and systems labour, etc. The limited experience of the operator requires others to abandon their tasks to be able to teach you the locations, types of installations, mode of work, etc., with the consequent loss of productivity and performance.

In larger companies, the problem is acute and the cost of these changes is increased considerably, since the facilities to learn, the work to be carried out, etc., are much higher. It must also take into account to analyse these costs, the failures (the increase in the average of troubleshooting time).

There are costs of outage or inefficiency, posed to the company by adding new operators to equipment maintenance, should be stressed also other induced costs.

These induced costs arise from the failure of the operator to solve a critical fault in a given time. These critical faults, as opposed to non-critical faults, differ in that these pose a cost high to the company as, for example, the cessation of production until this fault is not rectified.

Other relevant problems when it comes to good maintenance of facilities is the lack of information on measures to adopt, and order of sequential execution to faults which have not been produced before, or you have not occurred in the presence of the operator.

In the majority of cases, are older workers who know better facilities and equipment, as well as their specific behaviour, measures to take to any incident, review what and how to do it, in particular, for each machine, etc.

This experience gained over the years, it is not committed or competition of the educational system, however, is of vital importance for the proper functioning of the company.

The problem is that if the operator possessing that knowledge, leaves job, the company loses it, suffering economic and operational problems arising from them.

FACTORS IN THE MANAGEMENT OF KNOWLEDGE IN INDUSTRIAL MAINTENANCE

Reviewing the bibliography on analysis of the situation in industrial sectors (Cárcel 2014a, 2014b), the main factors involved in knowledge management in maintenance have been identified, and they have an exponential influence on the overall efficiency of the company (Table 4), given that they affect operational reliability and have economic repercussions. Although the data shown suggest great similarity between what is seen by technicians and maintenance managers, there are some differences due to the specific nature of the jobs, since maintenance managers are normally educated to University level, are used to using computer-based tools, and have a higher level of organization in their role, as compared to technicians.

The following categories or dimensions to describe the phenomena of the knowledge management in maintenance personnel can be assigned:

- **Acquisition and Generation of Internal and External Knowledge:** Activities from the outside or inside that support organizational processes and provide new knowledge on the maintenance activity.
- **Important Elements of the Knowledge Generation:** Individual and organisational actions that support or limit the generation of knowledge.
- **Transfer of Knowledge through Formal and Informal Mechanisms:** Formal mechanisms encoded or stored knowledge in databases, documents or software tools, which you can access and easily used by any member of the Organization (Hansen, Nohria & Tierney, 1999). Informal mechanisms that help transfer knowledge through direct contact from person to person.
- **Important Elements of Knowledge Transfer:** Individual and organisational actions that support or limit the transfer of knowledge.
- **Use of Knowledge:** The application of newly acquired knowledge in routine activities in the company.
- **Tools for Knowledge Management:** Use of audits and knowledge maps that improve the processes in which the maintenance activity is involved.

Table 4. Factors in knowledge management observed according to its dimensions

Phenomenon Under Study	Technical Maintenance Operations	Controls or Heads of Maintenance
Acquisition and generation of knowledge	**External:** Suppliers of materials and equipment. Catalogues and guides of manufacturers. External installation and assembling companies. The internal client (the rest of the industry). **Internal** Self-learning. Training courses Formal meetings in the area. Informal meetings with colleagues.	**External:** Contact with companies in the sector (production and maintenance Areas) Suppliers of materials and equipment. Catalogues and guides of manufacturers External installation and assembling companies. The internal client (the rest of the industry). Consultations on the internet. Attendance to congresses and fairs. **Internal:** Self-learning. Training courses Meetings in the area
Elements in the acquisition and generation of knowledge	Proactive attitude of the address. Staff motivation. Opportunity to learn. Take part in the decision-making process. Specific training in the environment. Agile access to external sources.	Proactive attitude of the address. The size of the company. Staff motivation. Opportunity to learn. Take part in making investment decisions. Specific training in the environment. Agile access to external sources.
Knowledge transfer	**Formal mechanisms:** Documents Intranet Meetings of the maintenance area. **Informal mechanisms:** Face-to-face communication Informal conversations	**Formal mechanisms:** Documents Intranet Meetings of the maintenance area. Analysis of quantitative data. **Informal mechanisms:** Face-to-face communication Hall talks. Meetings with colleagues from other companies. Electronic mail. Intranet.
Elements in the transfer of knowledge	Working environment. Staff motivation. Take part in the decision-making process. Simple tools of knowledge acquisition. Availability of time.	Management style. Staff motivation. Take part in the decision-making process. Simple tools of knowledge acquisition. Availability of time.
Use of knowledge	Resolution failures. Knowledge of the environment. See opportunities for action.	Maintenance planning. Mark priorities. Optimize technical resources. Economic optimization. Improvement of reliability and response times.
Elements for the use of the knowledge	Involve employees in the improvement of the work through the suggestions system. Agile and simple system as container knowledge. Group work. Support of top commanders.	The orientation of the system of continuous improvement to work in teams. Encourage the Working Group to eliminate islands of knowledge. Continuity in time.

Source: Own elaboration

With greater detail are might describe categories influencing knowledge management:

Acquisition and Generation of Knowledge

Relations with specialized outsourced companies, installation companies, manufacturers of materials and equipment and relations with departments of maintenance of other companies, have been expressed as the fundamental sources of acquisition and generation of external knowledge. As for the generation of knowledge for internal media, meetings organization and informal conversations with colleagues, as well as self-learning are the main sources of acquisition and generation. These activities acquired from the outside or inside are that support greater organisational processes and provide new knowledge to the maintenance organization and therefore the company.

Similarly, a proactive attitude of the leadership, the existence of organizational culture in the maintenance department and motivation of the staff involved that creates opportunities to learn, are important elements in the generation of knowledge. As soon as the barriers, the availability of time, for use in actions which are not characteristic of maintenance activity, dispersion or not update with the information you need.

In relation to the style of leadership, are aligned to the model proposed by Pedraja (Pedraja, Rodríguez & Rodriguez, 2006), where the style of leadership is considered determinant of organizational effectiveness, as defined Robbins, influences in individuals and in groups, to achieve success (Robbins & Judge, 2009), noting that the power of the expert is the most efficient form of leadership. This last is special consideration between the maintenance personnel, where the top experts in each area are those people with many years of experience and used to solve vital technical problems for the company, and that they normally act as leaders.

Therefore, that leadership style has influence on the results of the organization, confirming studies from different authors (Pedraja, Rodriguez, & Rodriguez, 2006; Pedraja & Rodriguez, 2004; Robbins & Judge, 2009; Yukl, 2008), and the behavior of the leader influences people, generating effects in the conduct of the individuals in the organization (Yukl, 2008). In this respect, Robbins and Judge (2009) indicate that effective leadership promotes and facilitates mechanisms to improve quality, productivity and use of resources.

Relations with material and equipment manufacturers, subcontractors and relations with departments of maintenance of other companies in order to learn technological know-how specific, necessary for the proper performance of their activity, have revealed as key factors in the acquisition of external knowledge, which allows the creation of individual knowledge that will subsequently generate organizational knowledge. Commanders also give incidence queries by internet and attending conferences and fairs, as well as informal contacts with companies in the sector.

Which employees are part in the decision-making, the size of the company and a participatory management, they have a positive influence on the acquisition and generation of external knowledge, technical decisions in the field of their daily work, as valued by managers being part of strategic decision-making.

In terms of internal knowledge acquisition, information from internal customers (production and other areas in the case of industrial enterprises, and external clients in the case of services companies such as hotels, etc.), allows to offer these services they really need, and gives the Organization the opportunity to know what is what are requiring customers who make use of the maintenance activity.

This acquisition of knowledge promotes human talent in the maintenance organization encouraging the circulation of knowledge and the flow of ideas (Garcia & Leal, 2008). Therefore, most of the company's value lies in its intellectual capital, and is essential to the permanent updating of skills and knowledge, training and continuing education, as well as the constant evaluation of persons (Marvel, Rodriguez & Núñez, 2011).

The creation of internal knowledge is based on self-learning of the employee and in the meetings held between the members of the teams of maintenance. It comes on the basis of personal motivation and is key in such activity, in a highly technological environment, which helps to provide a better service. Self-learning depends on individual tacit knowledge developed through their experience, learning through action, social interaction and communication within the company, that favours it a participative management style (Zapata, 2001).

To Olaz (2003), training is a key element to improve the productivity of people, based on optimizing the suitability of the person to the job, something which is vital within the maintenance organization, where periods of personal new mesh are high. In this regard, notes Olaz (2003), which is indicative of higher productivity, according to the theory of human capital, a high educational level of staff, so any increase in the level of education would be transformed into an increase in productivity (Garrido, 2007; Mungaray & Ramírez, 2007).

As a tool of socialization that supports the creation of organizational knowledge, the meetings are used in a fundamental way. This promotes learning among all members and supports that it can trigger new ideas. This form of internal creation of knowledge is supported by the organizational culture and the personal motivation of each one of the members of the Organization of maintenance, detected among practitioners in greater degree, the use of meetings or informal discussions.

Transfer of Knowledge through Formal and Informal Mechanisms

The formal mechanisms that encode or stored knowledge, documents (reports, manuals, drawings, databases, etc.), or tools computer, which you can access and easily used by any member of the Organization (Hansen, Nohria & Tierney, 1999). Informal mechanisms will help transfer knowledge through direct contact from person to person.

The documents are those who have a greater impact in the Organization at the time of transferring the knowledge, but unfortunately there is plenty of time to document activities or important actions for the development of the services provided. The meetings are relevant to the Organization as a means to transfer knowledge. These are frequently, most often between commanders and heads of maintenance, allowing you to know the global strategies.

Intranet as well as the use of electronic mail are used in the transfer of knowledge by commanders of maintenance. Maintenance technicians, given that they typically run hand crafts works, usually do not have an individual computer, using a collective service where the data of executed actions, or parts of work and with access to the intranet as a whole are inserted. In many cases there is a clear definition of what home Intranet and is easier to access their own sources of knowledge for some employees.

It has been seen that greater use is made of informal mechanisms. Informal personal mechanisms, such as face-to-face communication and chatting around the water cooler are commonly used by technicians.

Technology can increase access and simplify the problem of knowledge appropriate to the person at the right time, however, since the technical operations, indicate the importance that should be an agile, simple and schematic way in use.

Use of Knowledge and Tools for Their Management

The appropriate management of knowledge and the application of the knowledge acquired in the routine activities of maintenance in the company, and its improvement, is assumed by all components, as an important factor that can positively influence different actions that strategically affect enterprise-wide, such as:

- Resolution failures.
- Knowledge of the environment.
- See opportunities for new shares.
- Maintenance planning.
- Mark priorities for investment, reliability and energy efficiency.
- Optimize technical resources.
- Economic optimization.
- Improvement of reliability and response times.

An improvement in the management of information and knowledge, is positively in all these actions, and especially in the resolution of large malfunctions, or failures not cyclic spaced in time and normally not recorded her performance.

With regard to the tools to be used to gather strategic information which helps to improve knowledge management, they are habitually little used in the area of maintenance. The limited use of internal action audits, and information and knowledge maps, is recognized, as criticity diagrams are only made in specific facilities and equipment which are essential to the activity of the company.

Facilitators and Barriers to the Generation and Transfer of Knowledge in Maintenance Organizations

The elements that facilitate the generation and transfer of knowledge are observed to an organisational and individual level (Chen & Chen, 2006; Goh, 2002; Ajmal & Koskinen, 2008; Geisler, 2007; Wright, 2005).

A flexible proactive organizational culture together with a participatory style of leadership, are elements that allow to develop both the generation and transfer of knowledge within the organization. Individual level personal motivation and opportunity to learn it facilitates generating the knowledge that leads to organizational knowledge to be shared with other members of the company. The empirical evidence from the present study shows that open organizational culture encourages engineers to generate and share their knowledge in a more successful way and at the same time, supports communication between the members of the company. In the control of maintenance, a clustered physical distribution of their jobs facilitates the transfer of knowledge.

The need for the figure of a "Manager of knowledge in the activity of maintenance", can be an important facilitator in the catchment of the transfer and use of knowledge, and to help overcome resistance to change (Garcia & Marin, 2009). This figure should be a person with training technical, organisational and notions of knowledge management, with extensive experience in the operational area (who knows in depth of first-hand the factors influencing their work), and which brings together all the efforts of the maintenance organization to manage a strategic knowledge that can be used by the entire organization.

Their dedication might be partial or total (according to the characteristics of the company), sharing it with the dedication in other facets of the holding area, and you could meet at the same time a link's link with the rest of the Organisation (production, management, etc.), which would help the greater quality of the service of maintenance. This suggests that the knowledge that you want to transfer needs to be a priority within the organization where the transfer required to be planned as the rest of the company's strategic activities.

The availability of time to properly document important actions, cultural barriers with a culture based on "own knowledge", not shared, all in technical operations, as well as getting the total involvement of the staff, are barriers in the present study.

Similarly, the massive use of informal knowledge transfer mechanisms, which make that information is in "Islands" within the Organization itself has been identified.

Participants in the study considered that the possibility of applying their expertise in the activities of the Organization motivates them in self-learning, learning new tools and create new ways of doing things. When this personal motivation is enhanced in the knowledge that their opinions and suggestions to acquire external knowledge cannot be taken into account, the transfer and use of knowledge processes are strengthened.

Table 5 shows a summary of the main features that have been identified in tools, barriers and facilitators and implications in knowledge management processes.

CONCLUSION

This chapter focuses on the ways that maintenance organizations of companies generate, transfer and use their knowledge, and the impact it can produce on the organization as a whole. These processes are characterized by a Kantian process (people, means and environment).

Both the acquisition of external knowledge and internal knowledge creation are important activities to generate that knowledge to critical actions (breakdowns, emergencies, etc.) not cyclical, can assume an important strategic value. All of them are aware that having a proper knowledge management, positively affects the following actions performed by maintenance:

- Capture of strategic tacit knowledge of technical maintenance operations.
- Resolution of critical faults in shortest time (especially the non-cyclic).
- Reduction of the times of operational maneuvers.
- Facilitate area change or replacement of staff.
- Decrease of coupling again personal times.
- Capture of information and transfer of subcontractors.
- Share knowledge of employees who can be used by others that can detect new opportunities for improvement.
- Improvement of the knowledge of the reliability of the equipment and facilities.
- Improvement of the knowledge for the detection and improvement of energy efficiency actions.
- Optimization of the time, which is again in the knowledge management and the reduction of the maintenance costs.

Table 5. Tools, barriers and facilitators in the Knowledge management in the activity of maintenance

Category of the Phenomenon Under Study	Technical Maintenance Operations	Controls or Heads of Maintenance
Tools for knowledge management	Maps of information and knowledge. Agile and simple systems to capture experiences	Audits of maintenance. Energy audits. Audits of knowledge. Maps of information and knowledge. Diagrams of criticality.
Barriers in knowledge management	Lack of time to document properly important actions. Cultural barriers. “Own knowledge”-based culture, not shared. Involvement of staff. Greater use of informal mechanisms for knowledge transfer.	Lack of time to document properly important actions. Cultural barriers. Involvement of staff. Greater use of informal mechanisms for knowledge transfer.
Facilitators in the knowledge management	Open and flexible proactive organizational culture. Participatory style of leadership. Personal motivation of the employee. Opportunity to learn. Organizational culture of the holding area. Management style Means of communication. Use of a manager’s knowledge of the activity of maintenance.	Open and flexible proactive organizational culture. Participatory style of leadership. Personal motivation of the employee. Opportunity to learn. Organizational culture of the holding area. Physical space Management style Means of communication. Use of a manager’s knowledge of the activity of maintenance.
Observations	Lots of strategic information, collection of handwritten way kilter notes and own notebooks, drawing endorsements, not shared with the rest of the Organization, hindering the transmission and use of knowledge to the rest of the organization.	All considered that an awareness and knowledge of the directorate-general is essential to get the media and promote better management of knowledge and maintenance optimization, with a medium and long-term vision.
Involvement of management knowledge in the activity of maintenance.	-Capture of strategic tacit knowledge of technical maintenance operations. -Resolution of critical faults in shortest time (especially the non-cyclic). -Reduction of the times of operational maneuvers. -Facilitate area change or replacement of staff. -Decrease of coupling again personal times. -Capture of information and transfer of subcontractors. -Share knowledge of employees who can be used by others that can detect new opportunities for improvement. -Improvement of the knowledge of the reliability of the equipment and facilities. -Improvement of the knowledge for the detection and improvement of energy efficiency actions. -Optimization of the time, which is again in the knowledge management and the reduction of the maintenance costs.	

Source: Own elaboration

Self-learning is key in this kind of activity in a technological environment and with demand for rapid and efficient action. The knowledge management is enhanced by a proactive and participatory management style that promotes the emergence of new ideas and work processes. Likewise, this organizational culture must be opened, allowing the address encourage employees to share their knowledge and to facilitate communication between the members of the company. These findings are supported by studies (O’Dell

& Grayson, 1998; Ruggles, 1998) where it was noted that companies with an open culture that motivates to generating and sharing knowledge will have more success in the implementation of these processes.

A greater use of informal meetings as a means of generation and transfer of knowledge, especially among the groups of technical operations, with a minor organizational culture that managers or heads of maintenance is detected.

Audits (maintenance, knowledge, energy, etc.), tend not to be used, what shows that the application of these techniques would enhance a first process in the elaboration of a global strategy of knowledge management. To the maintenance organization perform successfully the replica of his know-how, by technical operations, require agile and simple mechanisms that allow them to share their experiences, which generate knowledge quickly and efficiently.

Similarly, has been detected that on numerous occasions, the documentation for use in their activities, tends to be disaggregated, and often not up-to-date.

The key barriers identified for the proper management of knowledge have been the availability of time to properly document important actions, cultural barriers with a culture based on "own knowledge", not shared, especially in technical operations (with a high component of knowledge, tacit and therefore not registered), as well as getting the total involvement of the staff they are barriers in the present study.

The need for the figure of a "Knowledge Manager", as an important facilitator it confirms in the catchment of the transfer and use of knowledge within maintenance departments. This figure should be a person with training technical, organisational and notions of knowledge management, with extensive experience in the operational area of maintenance (who knows first-hand deep in the factors influencing their work), and which brings together all the efforts of the maintenance organization to manage a strategic knowledge that can be used throughout the Organization not being an accidental Manager (Dorfman, 2001), but with a partial or total dedication that mark the continuity of the project. Since innovation is a knowledge intensive process, this new concept will be necessary to rethink many of the management decisions associated with him (Gonzalez & Garcia, 2010).

The foregoing suggests that the knowledge that you want to transfer needs to be a priority in the activity of the industrial maintenance, i.e. must be planned and included in the strategic planning of the company. The power of tacit knowledge is combined with explicit knowledge and objective continuously to go to generating new knowledge (Arbonies & Calzada, 2004).

REFERENCES

Ajmal, M. M., & Koskinen, K. U. (2008). Knowledge transfer in project-based organizations: An organizational culture perspective. *Project Management Journal*, *39*(1), 7–15. doi:10.1002/pmj.20031

Arbonies, A., & Calzada, I. (2004). El poder del conocimiento tácito: Por encima del aprendizaje organizacional. *Intangible Capital*, *6*(0), 1–17.

Baez, J. (2007). *Investigación cualitativa*. Madrid: Esic Editorial.

Bahoque, E., Gómez, O., & Pietrosemoli, L. (2007). *Gestión del Conocimiento en la Industria de la Construcción: Estudio de un caso. Revista Venezolana de Gerencia, 12(39), 393 – 409.*

Barragán, A. (2009). Aproximación a una taxonomía de modelos de gestión del conocimiento. *Intangible Capital*, 5(1), 65-101. doi:10.3926/ic.2009.v5n1.p65-101

Binney, D. (2001). El espectro de la gestión del conocimiento: El entendimiento del panorama de la GC. *Diario de la Gestión del conocimiento*, 5(1), 33-42.

Bueno, E. (1999). *Gestión del conocimiento y capital intelectual' Experiencias en España*. Comunidad de Madrid- Euroforum Escorial Madrid.

Bueno, E. (2000). *La dirección del conocimiento en el proceso estratégico de la empresa: información, complejidad e imaginación en la espiral del conocimiento* (pp. 55–66). Euroforum Escorial.

Bueno, E. (2002a). Enfoques principales en Dirección del Conocimiento (Knowledge Management) y tendencias. En R. Hernández (Ed.), Gestión del Conocimiento: Desarrollos teóricos y aplicaciones. Cáceres: Ediciones La Coria, Fundación Xavier de Salas.

Bueno, E. (2002b). *La sociedad del conocimiento: un nuevo espacio de aprendizaje de las personas y organizaciones, en La Sociedad del Conocimiento*. Presidencia de la Generalitat Valenciana, Valencia: Monografía de la Revista Valenciana de Estudios Autonómicos.

Cárcel Carrasco, F. J. (2014a). *La gestión del conocimiento en la ingeniería del mantenimiento industrial: Investigación sobre la incidencia en sus actividades estratégicas. Omnia Science*. doi:10.3926/oms.197

Cárcel Carrasco, F. J. (2014b). *Planteamiento de un modelo de mantenimiento industrial basado en técnicas de gestión del conocimiento. Omnia Science*. doi:10.3926/oms.198

Cegarra, J. & Moya, B. (2003). Orientadores del capital relacional. *Cuadernos Administración*, 16(26), 79-97.

Charmaz, K. (2006). *Constructing grounded theory. A practical guide through qualitative analysis*. London: SAGE.

Chen, M. Y., & Chen, A. P. (2006). Knowledge management performance evaluation: A decade review from 1995 to 2004. *Journal of Information Science*, *32*(1), 17–38. doi:10.1177/0165551506059220

Cheong, KFR. & Tsui, E. (2010a). The roles and values of personal knowledge management: an exploratory study. *VINE: The journal of information and knowledge management systems*, 40(2), 204–227.

Cheong, K. F. R., & Tsui, E. (2010b). Exploring the synergy between business process management and personal knowledge management. *Cutter IT Journal*, *23*(5), 28–33.

Chua, A. Y. K., & Goh, D. H. (2008). Untying the knot of knowledge management measurement: A study of six public service agencies in Singapore. *Journal of Information Science*, *34*(3), 259–274. doi:10.1177/0165551507084139

Colino, J. & Martinez, J; Martínez-Carrasco, F. (2010). La gestión de la innovación en la industria. El caso de la región de Murcia. *Economía Industrial*, 377, 150-158.

Colino, J. & Riquelme, P. (2000). Estructura industrial y desarrollo tecnológico en la región de Murcia. *Economía Industrial*, 335/336, 171-183.

Dataware Technologies. (1998). *Seven Steps to Implementing Knowledge Management in Your Organization*. Manual de trabajo.

Davenport, H. (1997). Ten principles of knowledge management and four case studies. *Knowledge and Process Management*, *4*(3), 187–208. doi:10.1002/(SICI)1099-1441(199709)4:3<187::AID-KPM99>3.0.CO;2-A

Del Moral, A. (2007). *Gestión del Conocimiento*. Thompson Editores. España.

Ditzel, B. (2005). *Desarrollo de un modelo de gestión del conocimiento para un departamento universitario. Tesis doctoral.* Campus Tecnológico de la Universidad de Navarra. Escuela Superior de Ingenieros de San Sebastián.

Dorfman, P. (2001). The accidental knowledge manager. *Knowledge Management*, *4*(2), 36–41.

Douglas, D. (2004). Grounded theory and the 'And' in entrepreneurship research. *Electronic Journal of Business Research Methods*, 2(2).

Edvinsson, L., & Malone, M. S. (1997). *Intellectual Capital: Realizing Your Company's True Value by Finding its Hidden Brainpower*. New York: Harper Business.

Eich, D. (2008). A Grounded Theory of High-Quality Leadership Programs: Perspectives From Student Leadership Development Programs in Higher Education. *Journal of Leadership & Organizational Studies*, *15*(2), 176–187. doi:10.1177/1548051808324099

Eisenhardt, K. (1989). Building theories from case studies research. *Academy of Management Review*, *14*(4), 532–550.

Ferrada, X., & Serpell, A. (2009). La Gestión del Conocimiento y la Industria de la Construcción. *Revista de la Construcción*, *8*(1), 46–58.

García, M., & Leal, M. (2008). Evolución histórica del factor humano en las organizaciones: De recurso humano a capital intelectual. *Omnia*, *14*(3), 144–159.

Garrido, C. (2007). La educación desde la teoría del capital humano y el otro. *Educere*, *11*(36), 73–80.

Geisler, E. (2007). A typology of knowledge management: Strategic groups and role behavior in organizations. *Journal of Knowledge Management*, *11*(1), 84–96. doi:10.1108/13673270710728259

Gil, M.; Pérez, A. & López, G. (2008). La auditoría del conocimiento como etapa previa a la gestión del conocimiento en una institución educativa Mexicana. *Revista Ciencia Administrativa*, 2008(2), 17-27.

Glaser, B. G., & Strauss, A. L. (1967). *The discovery of grounded theory*. New York: Aldine deGruyter.

Goh, S. C. (2002). Managing effective knowledge transfer: An integrative framework and some practice implications. *Journal of Knowledge Management*, *6*(1), 23–30. doi:10.1108/13673270210417664

González, R., & García, A. (2010). Innovación abierta: Un modelo preliminar desde la gestión del conocimiento. *Intangible Capital*, *7*(1), 82–115. doi:10.3926/ic.2011.v7n1.p82-115

Gray, P. H., & Meister, D. B. (2006). Knowledge sourcing methods. *Information & Management*, *43*(2), 142–156. doi:10.1016/j.im.2005.03.002

Hansen, B., & Kautz, K. (2004). Knowledge zapping: A technique for identifying knowledge flows in software organisations. *EuroSPI.*, *2004*, 126–137.

Hansen, M. T., Nohria, N., & Tierney, T. (1999). What's your strategy for managing knowledge? *Harvard Business Review*, (March-April): 106–116. PMID:10387767

Hardy, M., & Bryman, A. (2004). *Handbook of data analysis*. London: Sage Publications. doi:10.4135/9781848608184

Hedlud, O. (1994). A model of knowledge management and the N-form corporation. *Strategic Management Journal, 15*(Suppl. 2), 73–90. doi:10.1002/smj.4250151006

Holsapple, C., & Jones, K. (2004). Exploring primary activities in the knowledge chain. *Knowledge and Process Management, 11*(3), 155–174. doi:10.1002/kpm.200

Hooff, B., & Huysman, M. (2009). Managing knowledge sharing: Emergent and engineering approaches. *Information & Management, 46*(1), 1–8. doi:10.1016/j.im.2008.09.002

Kim, H., Breslin, J. G., Decker, S., Choi, J., & Kim, H. (2009). Personal knowledge management for knowledge workers using social semantic technologies. *International Journal Intelligent Information and Database Systems, 3*(1), 28–43. doi:10.1504/IJIIDS.2009.023036

Kogut, B., & Zander, U. (1992). Knowledge of the firm, combinative capacities, and the replication of technology. *Organization Science, 7*(3), 502–517.

Kulkarni, U., Ravidran, S., & Freez, E. R. (2007). A knowledge management success model: Theoretical development and empirical validation. *Journal of Management Information Systems, 23*(3), 309–347. doi:10.2753/MIS0742-1222230311

Lin, H. F. (2007). A stage model of knowledge management: An empirical investigation of process and effectiveness. *Journal of Information Science, 33*(6), 643–659. doi:10.1177/0165551506076395

Martin, J. (2008). Personal knowledge management: the basis of corporate and institutional knowledge management. In J. Martin & K. Wright (Eds.), *Managing Knowledge: Case Studies in Innovation*. Alberta: Spotted Cow.

Martinez, I., & Ruiz, J. (2002). Los procesos de creación del conocimiento: El aprendizaje y la espiral de creación del conocimiento. *Proceedings of theXVI Congreso Nacional de AEDEM*, Alicante.

Márvel, M., Rodríguez, D., & Núñez, M. A. (2011). La productividad desde una perspectiva humana: Dimensiones y factores. *Intangible Capital, 7*(2), 549–584. doi:10.3926/ic.2011.v7n2.p549-584

Miles, M., & Huberman, A. (1994). *Qualitative Data Analysis: A sourcebook of new methods*. California: Sage Publications.

Mungaray, A., & Ramírez-Urquidy, M. (2007). *Human capital and productivity in microenterprises*. Universidad Autónoma de Baja California.

Muñoz, J. (1999). Sobre gestión del conocimiento, un intangible clave en la globalización. *Economía Industrial*, 330.

Nonaka, I. (1991). The knowledge-creating company. *Harvard Business Review*, November-December, 96–104.

Nonaka, I. (1994). A dynamic theory of organizational knowledge creation. *Organizatioll Saence*, *5*(1), 14–37. doi:10.1287/orsc.5.1.14

Nonaka, I., & Konno, N. (1998). The concept of "Ba": Building a foundation for knowledge creation. *California Management Review*, *40*(3), 40–54. doi:10.2307/41165942

Nonaka, I., & Takeuchi, H. (1995). *The knowledge-Creating Company: How Japanese Companies Create the Dynamics of Innovation*. New York: Oxford University Press.

Nonaka, I., & Takeuchi, H. (1999). *La Organización Creadora de Conocimiento*. Oxford, México.

Norton, D., & Kaplan, R. (1996). *The Balanced Scorecard: Translating Strategy into Action*. Boston: Harvard Business School Press.

O'Dell, C., & Grayson, C. J. (1998). If only we knew what we know: Identification and transfer of internal best practices. *California Management Review*, *40*(3), 154–170. doi:10.2307/41165948

OCDE. (1996). The Knowledge-Based Economy.

OCDE. (2004). *Medición de la gestión de conocimientos en las empresas: primeros resultados*.

Olaz, A. (2003). *Análisis de los procesos de transformación en las cualificaciones profesionales en el sector financiero* [Thesis Doctoral]. Universidad de Murcia. Facultad de Economía y Empresa.

Ordoñez de Pablos, P. (1999). *Gestión del conocimiento y medición del capital intelectual en la empresa internacional*. Proyecto de Doctorado, Universidad de Oviedo.

Ordoñez de Pablos, P. (2001). *Capital intelectual, gestión del conocimiento y sistemas de gestión de recursos humanos: Influencia sobre los resultados organizativos* [Thesis Doctoral]. Universidad de Oviedo.

Ordoñez de Pablos, P., & Parreños, J. (2003). Aprendizaje organizativo en el contexto internacional: Implicaciones para la gestión del conocimiento. *Investigaciones Europeas de Dirección y Economía de la Empresa*, *9*(2), 205–216.

Ordoñez de Pablos, P., & Parreños, J. (2005). Aprendizaje organizativo y gestión del conocimiento: Un análisis dinámico del conocimiento en la empresa. *Investigaciones Europeas de Dirección y Economía de la Empresa*, *11*(1), 165–177.

Pace, S. (2004). A grounded theory of the flow experiences of Web users. *International Journal of Human-Computer Studies*, *60*(3), 327–363. doi:10.1016/j.ijhcs.2003.08.005

Palacios, D., & Garrigos, F. (2006). Propuesta de una escala de medida en la gestión del conocimiento en las industrias de biotecnología y telecomunicaciones. *Investigaciones Europeas de Dirección y Economía de la Empresa*, *12*(1), 207–224.

Paniagua, E. (2007). *La Gestión tecnológica del conocimiento*. Murcia: Universidad de Murcia, Servicio de Publicaciones.

Partington, D. (2000). Building grounded theories of management action. *British Journal of Management*, *11*(2), 91–102. doi:10.1111/1467-8551.00153

Pauleen, D. (2009). Personal knowledge management: Putting the "person" back into the knowledge equation. *Online Information Review*, *33*(2), 221–224. doi:10.1108/14684520910951177

Pavez, A. S. (2000). *Modelo de implantación de Gestión del Conocimiento y Tecnologías de Información para la Generación de Ventajas Competitivas*. Valparaíso, Chile: Universidad Técnica Federico Santa María.

Pedraja, L. & Rodríguez, E. (2004). Efectos del estilo de liderazgo sobre la eficacia de las organizaciones públicas. *Revista facultad de ingeniería*, 12(2), 63-73.

Pedraja, L., Rodríguez, E., & Rodríguez, J. (2006). Liderazgo y decisiones estratégicas: Una perspectiva integradora. *Asociación Interciencia*, *31*(8), 577–582.

Peña, P. (2001). *To know or no to be. Conocimiento, el oro gris de las organizaciones*. Madrid: DINTEL.

Pérez, A. (2007). *Modelo para la Auditoría del Conocimiento Considerando los Procesos Clave de la Organización y Utilizando Tecnologías Basadas en Conocimientos. Tesis doctoral*. Universidad de Murcia.

Pérez de Miguel, A. (2006). Hacia el Índice de Conocimiento Roto: Introducción a los coeficientes de rotura de conocimiento. Documentos de trabajo "Nuevas tendencias en dirección de empresas". DT 06/06.

Peris, F.; Loria, B. & Méndez, M. (2002). Creación de conocimiento y diseño de organizaciones: Equidad, confianza y objetivos compartidos como reto de la gestión del conocimiento. *Cuadernos de CC.EE. y EE.*, 43, 41-56.

Petrides, L. A., & Nodine, T. R. (2003). *Knowledge Management in Education: Defining the Landscape*. Half Moon Bay, California: March Institute for the Study of Knowledge Management in Education.

Polanyi, M. (1958). *Personal Knowledge: Towards a Post-Critical Philosophy*. University of Chicago Press.

Polanyi, M. (1966). *The tacit dimension*. London: Routledge & Kegan Paul.

Quintana Fundora, Y. (2006). *Gestión por el conocimiento en la carrera de Ingeniería Industrial. Administración de operaciones. Tesis en opción al título de Master en Ciencias*. Cuba: Facultad de Industrial Economía. Universidad de Matanzas.

Rivas, L., & Flores, B. (2007). La gestión de conocimiento en la industria automovilística. *Estudios Gerenciales*, *23*(102), 83–100. doi:10.1016/S0123-5923(07)70003-7

Robbins, S., & Judge, T. (2009). *Comportamiento Organizacional. Decimotercera edición*. México: Pearson Educación.

Rodríguez, D. (2006). Modelos para la creación y gestión del conocimiento: Una aproximación teórica. *Revista Educar*, *37*, 25–39.

Rojas, C. M., Montilva, C. J., & Barrios, A. J. (2009). *Revista Colombiana de Tecnologías Avanzadas*, *1*(13), 72-80.

Ruggles, R. (1998). The state of the notion: Knowledge management in practice. *California Management Review*, *40*(3), 80–89. doi:10.2307/41165944

Sánchez, M., Chaminade, C., & Escobar, C. (1999). En busca de una teoría sobre medición y gestión de los intangibles en la empresa: Una aproximación metodológica. *Ekonomiaz*, 45, 188-213.

Seemann, P., & Cohen, D. (1997). The Geography of Knowledge: From Knowledge Maps to the Knowledge Atlas. *Knowledge and Process Management*, *4*(4), 247–260. doi:10.1002/(SICI)1099-1441(199712)4:4<247::AID-KPM107>3.0.CO;2-1

Sher, J. P., & Lee, C. V. (2004). Information technology as a facilitator for enhancing dynamic capabilities through knowledge management. *Information & Management*, *41*(8), 933–945. doi:10.1016/j.im.2003.06.004

Sols, A. (2000). Fiabilidad, Mantenibilidad, Efectividad, un enfoque sistémico, Comillas, Madrid.

Tari, J., & Garcia, M. (2009). Dimensiones de la gestión del conocimiento y la gestión de la calidad: Una revisión de la literatura. *Investigaciones Europeas de Dirección y Economía de la Empresa*, *15*(3), 135–148. doi:10.1016/S1135-2523(12)60105-1

Tiemessen, L., Lane, H. W., Crossan, M. M., & Yinkpen, A. C. (1997). Knowledge management in international joint ventures. In P.W. Beamish, & J.P. Killing (Eds.), Cooperative strategies. North American perspectives. The New Lexington Press.

Tirpak, T. M. (2005). Five steps to effective knowledge management. *Research Technology Management*, *48*(3), 15–16.

Ventura, J. (1996). *Análisis dinámico de la estrategia empresarial: Un ensayo interdisciplinar. Servicio de publicaciones*. Universidad de Oviedo.

Ventura, J. & Ordóñez de Pablos, P. (2007). Análisis de estrategias de conocimiento en la industria manufacturera española: Evidencia empíricas. *Tribuna de Economía*, 836, 141-161.

Volkel, M., & Abecker, A. (2008). Cost-benefit analysis for the design of personal knowledge management systems. Proceedings of the International Conference on Enterprise Information Systems ICEIS '08 (pp. 95–105).

Volkel, M., & Haller, H. (2009). Conceptual data structures for personal knowledge management. *Online Information Review*, *33*(2), 298–315. doi:10.1108/14684520910951221

Wah, L. (1999). Behind the Buzz. *Management Review*, *88*(4), 16–19.

Wernerfelt, B. (1984). A resource based view of the firm. *Strategic Management Journal*, *5*(2), 171–180. doi:10.1002/smj.4250050207

Wiig, K. M. (1997, June). Integrating Intellectual Capital and Knowledge Management. *Long Range Planning*, *30*(3), 399–405. doi:10.1016/S0024-6301(97)90256-9

Wong, K. Y., & Aspinwall, E. (2004). Knowledge management implementation frameworks: A review. *Knowledge and Process Management*, *11*(2), 93–104. doi:10.1002/kpm.193

Wright, K. (2005). Personal knowledge management: Supporting individual knowledge worker performance. *Knowledge Management Research and Practice*, *3*(3), 156–165. doi:10.1057/palgrave.kmrp.8500061

Xiomara, P. (2009). La gestión del conocimiento y las Tics en el siglo XXI. *CONHISREMI, Revista Universitaria de Investigación y Diálogo Académico*, 5(1).

Yang, J. (2006). La estrategia de gestión del conocimiento y su efecto en el crecimiento corporativo. *Economía Industrial*, 362, 123-133.

Yin, R. K. (1995). *Case study research: design and methods*. California: Sage Publications.

Yukl, G. (2008). Liderazgo en las organizaciones (6th ed.). Madrid: Pearson Educación S.A.

Zapata, L. (2001). *La Gestión del Conocimiento en Pequeñas Empresas de Tecnología de la Información: Una Investigación Exploratoria. Document de treball núm. 2001/8.* Universitat Autònoma de Barcelona. Facultat de Ciències Econòmiques i Empresarials.

KEY TERMS AND DEFINITIONS

Audits of Maintenance: Maintenance assessment applied to an organization to identify which aspects are in need of improvement. There is a distinction between technical audits, which assess the condition of the facilities, and management audits, which identify the level of excellence of the maintenance department and its management.

Experience Curve: Assessment of improvement in efficiency or productivity through learning as an activity is repeated.

Explicit Knowledge: Knowledge that can be processed or transmitted easily through formal language.

Industrial Maintenance: Production support system whose aim is to guarantee optimum working of facilities and productive systems at minimum cost.

Know-How: Pre-existing knowledge in an organization, in the form of techniques, procedures, information, etc. which has in many cases been obtained by experience or repetition in carrying out activities, processes or decision making.

Knowledge: Stored information which has been acquired by experience, training, practice or deduction (introspection) and which can produce more knowledge or solve a problem.

Tacit Knowledge: Knowledge acquired through habit, experience or culture, including subjective elements such as personal beliefs, intuition, etc. this kind of knowledge is difficult to process or transmit systematically.

Chapter 6
Asset Management for Buildings within the Framework of Building Information Modeling Development

Antonio Jesús Guillén López
University of Seville, Spain

Adolfo Crespo Márquez
University of Seville, Spain

Jose A. Sanz
University of Seville, Spain

Khairy A. H. Kobbacy
Taibah University, Saudi Arabia

Samir M. Shariff
Taibah University, Saudi Arabia

Etienne Le Page
École Centrale de Marseille, France

Vicente González-Prida
University of Seville, Spain

ABSTRACT

Buildings lifecycle management is an area of great interest. Due this, the R&D is being promoting worldwide looking for new effective maintenance tools and methodologies. In this scenario there are two development lines whose convergence can bring great advances in this area: Asset Management (AM) and Building Information Modeling (BIM). BIM models are transforming the way buildings are conceived, designed, constructed and managed. But current use of BIM concentrates on preplanning, design, construction and integrated project delivery of buildings and facilities, rather than maintenance and building operation management. Asset Management tools, including Facilities Management (FM), and application frameworks provide the approach and required elements to get more efficiency and efficacy in the building lifecycle management. This chapter introduces the application of AM for building and how the development of BIM models is the key element to allow its effective implementation.

DOI: 10.4018/978-1-5225-0651-5.ch006

INTRODUCTION

Nowadays, buildings are generally required to operate efficiently. A building that operates efficiently is that one that most of its characteristics can be measured, monitored and recorded in reference to a relevant actuation. However, operation and maintenance productivity is decreasing due to the current complexity of buildings. In addition, it is estimated that operation and maintenance phase constitutes approximately 60% of the total lifecycle cost of a facility or building. Within the components of this cost is interesting to highlight those related to inadequate interoperability between different roles and stakeholders involved in the building lifecycle. A study by the US National Institute of Standards and Technology (NIST) showed that the annual costs associated to inadequate interoperability among software systems was $15.8 billion (Gallaher et al. 2004). Two thirds of this cost was incurred as a result of on going facility operation and maintenance activities (Shen et al. 2010). The sector has realized that information management is one of the main reasons of this over cost. BIM (Building Information Modelling) is a holistic approach to the design, construction and management of those facilities used in the built environment. The introduction of BIM can addressed this challenge. If appropriate operational information can be incorporated into this model, all the stakeholders will have all the information needed to take decisions.

Facilities Management may cover a wide range of activities, from ordinary maintenance task and operation control to performance monitoring (both views: operation and sustainability), or outsourcing service management and financial asset management. Facilities Management (FM) tasks require gathering and sharing large amounts of information related to facilities components. This information covers historical inspection data and operation information. Despite the availability of Computerized Maintenance Management Systems (CMMSs) or Computerized Aided Facilities Management (CAFM), these systems focus on the data management aspects (i.e. work orders, resource management and asset inventory), the lack of functions required to facilitate data collection and data entry, as well as the data retrieval and visualization when and where needed. Building Information Modelling (BIM) provides opportunities to improve the efficiency of FM software by sharing building information between different applications/users throughout the facility lifecycle. Moreover, maintainability issues are poorly treated at the design and construction stages (Liu, 2013), such as accessibility of elements for maintenance purpose, lack of maintainability information and lack of communication between stakeholders and particularly with the FM team. Many studies intent to define the data needed for FM team, as well as the way they are provided and delivered. Nevertheless, there are lacks of guidelines and tools for capturing BIM models of existing facilities, coping with non-consistent terminologies and taxonomies, requirements specification in BIM applications, and identifying which information and what levels of detail are desired by the FM teams (Parsanezhad et al., 2014).

Therefore, buildings maintenance management has become a research area of great interest. One of the main aims is to eliminate delays, confusion, and inaccuracy generated by traditional O&M documents namely as-built drawings, O&M manuals, maintenance schedules, room datasheets, asset performance data documents, and cost datasheets in either paper format or static digital formats such as Portable Document Format (PDF) and scattered spreadsheets (Schevers et al., 2007). This drives to the information system application development, where BIM models and Asset Management information system (AIM, Asset Information Model) are the main references.

In the past decades, there had been a growing interest in the use of Building Information Models (BIM) by the construction sector due to many benefits and resource savings during design, planning, and construction of new buildings (Eastman et al 2013). Ashworth et al. (2015) highlight the values added by the implementation of BIM for FM sector, such as Cost and time savings, Real time data (Learning cycle, Informed decision making, Buildings improvement), Cultural impact (Improve users' experience, Assist procurement process), Maximising operational aspects and Understanding business journey. The use of BIM concentrates on preplanning, design, construction and integrated project delivery of buildings and infrastructure but, recently, research focus shifts from earlier life cycle (LC) stages to maintenance, refurbishment, deconstruction and end-of-life considerations especially of complex structures (Volk et al., 2013). BIM has proved over time to be a valuable asset in contributing to the management of relevant building data (Cahill et al 2012; Volk et al., 2013).

An extensive use of BIM for all FM/AM sector does not exist yet. The literature shows that, despite many attempts to BIM-FM implementation exist and apparently provide good results, they are still piecemeal and FM cases showing extensive use of BIM are lack (Kassem et al., 2015; Volk et al., 2013). Moreover, case studies generally are focused on specifics FM area and still present many gaps, issues and challenges to be implemented correctly (Parsanezhad et al., 2014). Perception of BIM (lack of benefit evidence, lack of FM engagement), standards and policy (interoperability, tools integration), risk and uncertainty (potential lack of interest from investors, creation new market and new roles, validate BIM data), building physical aspect (create models of existing facilities, different management strategies), information management and technology (data repetition and loss, overload of data, use of different software, information update), education and skills… are few examples resumed by Ashworth (2015). However, one of the most relevant changes appears in the attitude of the FM team that becomes more process oriented and seeks to identify process inefficiencies that could be mitigated by the adoption of BIM (Fukuda et al., 2014).

Are AM and BIM a natural partnership? The data-rich environment of BIM has the potential to assist in the fulfilment of AM requirements. For example, employer's requirements or specifications could require relevant information to be included into each designer's BIM model in order to facilitate requirements. Technology alone, in this case BIM, cannot deliver business outcomes. The integration of BIM tools within AM approach ensure that the organization obtains the results that expects (Love et al. 2014). AM specifically connects BIM technology application with the fact of how an asset owner obtains and manages value from their assets. With this view, the implementation of BIM should not be considered an isolate IT effort for an organization but integrated within the asset information system in the asset management system. AM approach allows taking the decision to implement BIM based on overall strategy and business value rather than the use of ROI or other financial methods as unique reference.

The interest of including this chapter in this book is to present the AM application to the construction and facilities sector. The attractions for asset management in this area can be summarized in two aspects:

- On one hand, the potential that AM has in this sector. This is a sector where: there is much room for improvement and cost reduction over the lifecycle of assets; in most cases, they are complex assets that require the use of management tools; and finally, the volume of business in the sector is huge. In addition to new building business, the major factor that has contributed to overruns in construction projects has been the creation and transfer poor quality information, which has manifested as errors, omissions and information redundancy.
- BIM technology development is a tool that aids the effective application of AM. Its current process of generalization (among organizations and technicians), not only as a tool but a requirement (even legal requirement, see UK BIM level 2 requirement from 2016) makes up an scenario where asset management can provide the necessary support for an effective implementation.

The growing rate of BIM users can pull the AM application development. Following similar approach standards as PAS 1192-2 and 1192-3, they have been recently developed where the AM principles are connected and used together with BIM.

BIM BACKGROUND

The term BIM (Building Information Modelling) can indicate a process, a discipline or a technology. It is focused on delivering and using a digital model of a building or facility, i.e. an information model of such assets. This model is much more than a 3D representation of the building. 3D modelling is just one aspect of BIM. Its real strength and power lies in the knowledge database, which can be used in conjunction with other software to deliver quick and reliable information in areas of sustainability, estimating, structural analysis, demolition and reconstruction (Zhang 2009).

The given definition of BIM by the Building Smart Alliance introduces the following aspects that serve to understand the holistic view of BIM conception and its potential:

- A digital representation of physical and functional characteristics of a facility. Not only graphical information of the building elements but also the rest of information types that can be used to manage all the lifecycle phases: manufacture and vendor data, service and use requirements, operation and maintenance data, performance parameters, energy consumption, etc.
- A shared knowledge resource for information about a facility, forming a reliable base for decision during its lifecycle.
- A platform for collaboration by different stakeholders at different phases of the facility lifecycle in order to insert, extract, update or modify information in the BIM support reflecting different roles according to each stakeholder's interest.
- The BIM is shared digital representation founded on open standards for interoperability. In addition to the standardization needs, this point highlights the open character of BIM conception, in order to allow the combined use of different software an application (3D design, FM software and others) and to support the successive software updates.

The interest of the construction sector in BIM applications has grown in the last decades. The benefits and the resource saving during design, planning, and construction of new buildings are the main reasons of this evolution. Despite the development of 3D modelling started in the 1970s, BIM modelling was not introduced in pilot projects until early 2000s to support building design of architects and engineers (Volk et al. 2014). Comparing BIM with 2D CAD, this uses views and representations that are difficult to check and update considering that, if one view is modified, the rest should be revised too. Traditional 2D CAD only includes graphical entities in contrast to the intelligent contextual semantic of BIM models that include geometric and non-geometric data, characterizing, among others: the geometry, spatial relationships, geographic information, quantities and properties of building elements, cost estimates, material inventories, project schedule and operation and maintenance. BIM is an "object oriented" approach to CAD where designers and stakeholders work primarily with architectural elements instead of vector based graphics. Objects have "intelligence" that provide a certain degree of automation

In order to illustrate the progressive introduction of capability or functionalities within the models, the following terms have been employed:

- **2D Models:** Classical 2D CAD model application.
- **3D Models:** 3D design and tools. Parametric and object oriented approach. 3D digitalization.
- **4D Models**: Scheduling and project sequencing. The capabilities of capture data and information Representation is used for planning and control project and construction execution.
- **5D Models:** Cost estimating. Budget estimation and control of the construction phase.
- **6D Models:** Sustainability. Control of the impact of construction and operation.
- **7D Models:** Facilities Management. Including operation and maintenance planning and execution and the rest of life-cycle considerations.

This view can be interesting in order to show the BIM potential. It also provides a sequential way to implement BIM solutions by the organizations. Nevertheless, given a more holistic and accurate view of BIM, this model can be seen as a unique model of the building, a common and general digital data base of building and facilities. This model can present different applications. In the same way, the Computer Integrated Construction Research Program (2013) defined a BIM use as a method or strategy applied during a facility's lifecycle in order to achieve one or more specific objectives. For example, Azhar et al. (2011) list the following BIM applications:

- **Visualization:** 3D renderings can be easily generated in-house with little additional effort.
- **Fabrication/Shop Drawings:** It is easy to generate shop drawings for various building systems, for example, the sheet metal ductwork shop drawing can be quickly produced once the model is completed.
- **Conflict, Interference and Collision Detection:** Because BIM models are created, to scale, in 3D space, all major systems can be visually checked for interferences. This process can verify that piping does not intersect with steel beams, ducts or walls.
- **Code Reviews:** Fire departments and other officials may use these models for their review of building projects.
- **Forensic Analysis:** A building information model can easily be adapted to graphically illustrate potential failures, leaks, evacuation plans, etc.
- **Facilities Management:** Facilities management departments can use BIM for renovations, space planning, and maintenance operations.
- **Cost Estimating:** BIM software(s) have built-in cost estimating features. Material quantities are automatically extracted and changed when any changes are made in the model.
- **Construction Sequencing:** A building information model can be effectively used to create material ordering, fabrication, and delivery schedules for all building components.

Despite the operational phase, (Figure 1) it is estimated that operation and maintenance constitutes approximately 60% of the total lifecycle cost of a facility or building, in the list of applications above. It is observed again how O&M and FM play a discrete role. The "nD" representation also shows the lack of an appropriated treatment of the phases after construction. This is one of the principal lacks in the current BIM models as it will be presented in next sections. Zhang et al. 2009 discuss how sustainable design and lifecycle management of buildings need new type of information compared to traditional building process.

The availability of a single BIM, with a unique well-structured solution, is possible by capturing information throughout design, procurement and construction of a building. It serves as a living record

Figure 1. Building and Facilities life-cycle phases: extension and BIM application maturity by phases

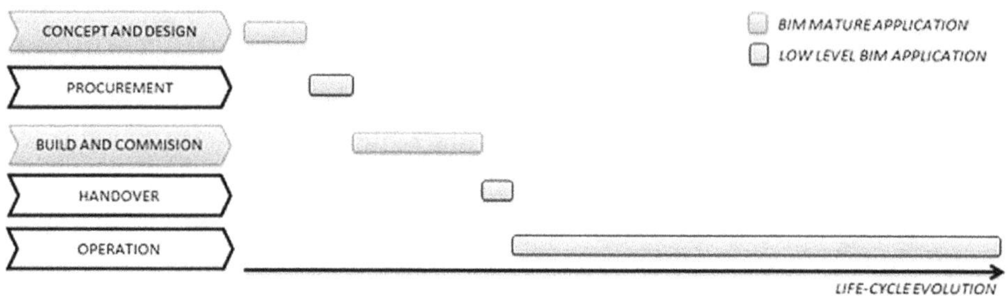

of the building for operations and maintenance throughout its lifecycle where the information data base will be integrated with its own O&M's elements and asset management. All of the non-graphic asset information contained in BIM models (manufacturer, serial number, model number, and so on) can be migrated/integrated into asset management system. In this sense, the conception of BIM is closed to Asset Management conception in the terms presented by ISO 55000.

BIM Benefits and Challenges in Operational and Maintenance Phase

Governments around the world have recognized in general terms the inefficiencies affecting the construction industry. Besides the components of these costs, it is interesting to highlight those related to inadequate interoperability between different roles and stakeholders involved in the building lifecycle (Gallaher et al 2004). The table bellow clearly shows this problem, and how these cost are specially relevant in the O&M phase.

As a result, different regulatory initiatives have appear that either recommend or mandate the use of building information modelling as a strategy to address a declining productivity (Kassem et al. 2015). In this case, the most significant reference is probably UK. UK has mandated BIM level 2 on all centrally procured projects from 2016, including the handover of digital data required for the operational phase. Although this mandate prescribes an operational handover specification, there is still a limited amount of research on the FM industry with regards to BIM. In this moment new researches are being focused on use cases for application of BIM for FM. The jointly application of FM and BIM is less explored compared to design, planning and construction phases. As it will be argued in the section bellow, this can be considered within the asset management development field.

Table 1. Costs of inadequate interoperability by stakeholder group, by lifecycle phase ($Millions)

Stakeholders Group	Planning, Design and Engineering Phase	Construction Phase	O&M Phase	Total
Architects and Engineers	1007.2	147	15.7	1169.8
General Contractors	485.9	1265.3	50.4	1801.6
Specialist Fabricators and Suppliers	442.2	1762.2	-	2204.6
Owners and Operators	722.8	808.0	9027.2	10648.0

(Gallaher et al 2004)

The current evolution of the sector is promoting a great increment in the number of BIM users: companies and technician (architects, engineers, etc.). The expertise is growing at the same rate. Nevertheless, this application cases are mainly focus on design and construction. The use of BIM models for operation and maintenance phase is not so extended. For this reason, although there are many references about the potential benefits of BIM use, there are few references about the quantification and realization of these benefits in practical cases for operational phases. In conclusion, BIM benefits for FM or AM are not well characterized.

The benefits of implementing BIM processes and technologies have been widely exposed. Following Love et al. (2014), these benefits include: reduction in construction costs; quality improvement regarding design information; integration of project systems, data and teams; reduced trend to change orders; interoperability improvement; asset management implementation to the whole asset lifecycle. Similar aspects are highlighted by IFMA (2013). Analysing the use of BIM within Facilities Management, the benefits are summarized as follows:

1. It reduces cost: accurate data management with lower data capture and O&M costs;
2. It improves performance: bets data management allow preventing problem and faster analysis and correction;
3. It integrates system: BIM can be integrated with management software as CMMS, CAFM, BAS, ERP.

Indeed, despite BIM is nowadays mainly used in design and construction phases, the operational phase of a building can be the most relevant area for BIM application as it is the main contributor to the building lifecycle cost. Estimations show that lifecycle cost is five to seven times higher than the initial investment costs and three times the construction costs (Kassem et al 2015). Taking this into account, these authors describe the following benefit of BIM application in FM:

- Improvement to the current manual processes of information handover; improvement to the accuracy of FM data.
- Efficiency increment of work orders execution, in terms of speed to accessing data and locating interventions. Such value is derived from the capability of BIM to provide a data-rich visual and integrated data environment.
- Improvement in the accessibility of FM data that can be found within the model.
- Efficiency increment for creating bespoke plans, elevations and visual renders all from the same model.
- Ability to attach legislative/statutory compliance data, which can be reported and scheduled out of the one model.
- Potential for finding an accurate fault reporting through the model interrogation.
- Ability to plan refurbishment projects scenarios in a 3D environment.

A very interesting approach is presented by Carbonari et al. (2015) that allow understanding the capabilities of BIM, especially for FM and AM application. In this chapter, the conclusions of a workshop with different FM professionals are shown, distinguishing and describing barriers, opportunities and added value of BIM and FM integration.

The other side of the potential benefits are the challenges to face in order to access them. MacArthur (2015) identifies four key challenges that must be overcome when developing BIM models suitable for sustainable operations management:

1. Identification of critical information required to inform operational decisions,
2. The high level of effort to create new or modify existing BIM models for the building(s),

Table 2. Barriers, opportunities and added value of integration of BIM and FM

	Definition	Description
Barriers	Perception of BIM	BIM still a new topic and its potential is not fully understood. Still there is no evidence to convince facility managers to adopt this technology.
	Physical aspects	Specially related to BIM for existing buildings. There is need to understand the extension of the model, what are the necessary data what is the effort required to get the model
	Standards and policy	There is a need for a unique BIM international standard.
	Education and skills	The construction industry is in the middle of the BIM learning curve. Training not only provides new skills for new technicians but also helps to understand what can be achieved.
	Risk and uncertainty	Resistance to change of stakeholders. Investor will not probably be interested in FM-BIM at design stage or BIM of buildings that will be sold once the building is completed.
	Information Management and technology	It is necessary to decide what information of BIM is useful for FM, how integrated the different software and application and who will be in charge of information changes during the life-cycle.
Opportunities	Knowledge capture	Share information and knowledge. Increased asset value capturing FM know-how
	WLC at early stages	More accurate tenders. Long terms savings
	Assist business journey	BIM can be used as a tool to increase flexibility and adaptability of a building, improving the quality of the work environment and therefore having positive effects on productivity.
	FM early engagement	Opportunity for facility management to be involved and influence the design process. Feedback loop for continuous improvement and less rework. Calculation of total expenditure during whole life.
	Data strategy	Identifying data needed by FM during predesign stage providing and single source data. Allow FM and owners test the outcomes in the building use phase. Facilitate the handover.
	Support soft/hard services and strategic FM	Informed decisions. Improve maintenance strategy. Cultural behaviours. Facilitate data location.
Added value	Cost and time savings	BIM has a positive impact in cost and time savings, reducing errors, omissions and rework.
	Real time data	Update the model with changes. This allow decision based on real time data about the building condition. Learning cycles and improvements.
	Cultural impact	Improve user experience. Assist procurement process
	Maximise operational aspects	FM can use BIM not only for location and visualisation purpose but also as an O&M tool. Model data can be used for performance analysis and planning future strategies
	Understanding business journey	Use BIM to accommodate building business needs.

(Carbonari et al 2015)

3. The management of information transfer between real-time operations and monitoring systems and the BIM model, and
4. The handling of uncertainty based on incomplete building documentation.

Kassem et al. (2015) summarize the following challenges that are hindering the exploitation of BIM in FM:

- The lack of methodologies that demonstrate tangible benefits of BIM in FM, which is reflected by limited demand of BIMs for FM by clients and operators; the need for rigorous BIM specifications for modelling requirements;
- The interoperability between BIM and FM technologies and the difference in their lifespan;
- Limited knowledge of requirements for the implementation of BIM in FM (e.g. what information is to be provided, when and by whom);
- The lack of open systems and standardised data libraries that can be utilised as a bridge between BIM and CAFM technologies;
- The current number of operational systems, managing the same building;
- The lack of clear roles, responsibilities, contract and liability framework; the shortage of BIM skills in the FM industry; and
- The rigid industry cultural approach to adopt new processes and technologies.

Comparing challenges describing with potential benefits, one realizes the great number of coincidences between them. This mean that, in fact, we are at the very beginning of BIM application for FM/AM. The AM application prospect of using BIM in building operation and maintenance increases as the adoption of BIM solutions within design and construction firms grows (Zhang et al 2009). As architects move more rapidly to BIM, the opportunities to leverage digital design data for downstream facilities/asset management become more prevalent. Owners and operators can improve the obtained value from their assets by using the high-quality building information from BIM, integrating this within de asset management information system and the decision-making process within an asset management strategy.

Information and Data Management and Technologies Integration

From all the above mentioned aspects, one of the key points of BIM application is the information and data management. BIM development provides the opportunity to get and manage more information with more quality, incrementing its use and applicability. At the same time, it is a great challenge to obtain such a powerful information system, including the integration of other software and technologies that can be information sources.

A first step in this process is the determination of information needs. Two dimensions have to be managed at the same time: the election of the element breakdown structure for the organization (until the so called intervention level) and the level of development of elements (LOD). Both dimensions are connected but express different aspect of the system and its elements. O&M task management is very sensible to final description of the system and, by extension, FM and AM.

Jointly with this last consideration, there are two problems, totally opposite one to the other, that must be taken into account:

- **Information Filtering:** Much of the information typically included in models is unnecessary for day-to-day operations.
- **New Information Needs**: Relevant information for maintenance and reliability management is not present or is not reliable for advanced approaches. In some cases it is possible to find CMMS or CAFM, but this can result an additional complication.

Determining what Level of Development (LOD) is necessary to achieve the benefit for that specific model element. The Level of Development describes the level of completeness to which a Model Element is developed. There are several ways that Level of Development can be documented. The most widely accepted by the industry is LOD defined in the model progression specification and adopted in AIA E202

Regarding what information should include a specific BIM model; one important reference is about the BIM maturity levels. PAS 1192-2 and PAS 1192-3 are both known as BIM Level 2, in response to the UK Government's Construction Strategy published in 2011. Following this approach, four maturity levels are distinguished, from level 0 to level 3. This level 2 includes:

Managed 3D environment held in separate discipline "BIM" tools with attached data. Commercial data managed by an enterprise resource platform. Integration on the basis of proprietary interfaces or bespoke middleware could be regarded as "pBIM" (proprietary). The approach may utilize 4D programme data and 5D cost elements as well as operational systems.

In addition to this, collaboration at BIM Level 2 is file-based in opposition to paper-based (BIM Level 0) or through integrated web services (envisaged as BIM Level 3 at the time of writing).

Interoperability is one of the keystones of BIM models and its uses. Different organizations and roles take part along the building lifecycle. They may apply different software or may demand different types of data. The Industry Foundation Classes (IFC) is a standardized non-proprietary data format for sharing and accessing construction and facility management data, enabling interoperability between heterogeneous software applications. This format is described by the ISO 16739:2013 "Industry Foundation Classes (IFC) for data sharing in the construction and facility management industries". It is an open

Table 3. An example of the Level of Development descriptions

Level of Development (LOD)	Description
LOD 100: Schematic Design Model	Overall building massing indicative of area, height, volume, location and orientation may be modelled in three dimensions or represented by other data
LOD 200: Design Development Model	Model Elements are modelled as generalized system or assemblies with approximate quantities, size, shape, location and orientation. Non-geometric information may also be attached to model elements.
LOD 300: Construction Documentation Model	Model Elements are modelled as specific assemblies accurate in terms of quantity, sixe, shape, location and orientation. Non-geometric information may also be attached to model elements
LOD 400: Construction Model	Model Elements are modelled as specific assemblies accurate in terms of quantity, sixe, shape, location and orientation with complete information assembly, and detailing information. Non-geometric information may also be attached to model elements
LOD 500: Record Model	Model Elements are modelled as constructed assemblies actual and accurate in terms of size, shape, location, quantity and orientation. Non-geometric information may also be attached to model elements

(Computer Integrated Construction Research Program, 2013)

international standard for BIM data that is exchanged and shared among software applications used by the various participants in a building construction or facility management project.

The Construction Operations Building Information Exchange (COBie) is a data exchange model for the publication of building information subsets focused on delivering non-geometric information and asset data, rather that geometric information. It is a subset of the IFC, but may also be conveyed using spreadsheet. Initially COBie was developed by the US Army Corps of Engineers as a method of delivering O&Ms specific data in a standardized format, delivered in a spreadsheet format.

COBie helps capturing and using information essential to support operations, maintenance and asset management. The COBie file may contain data from consultants, the contractor, sub-contractors, suppliers, and even the client. Ultimately, the data will provide information for the efficient operation and management of the facility. COBie consists of multiple sheets documenting attributes of the facility, its systems and assets and details of their product types, warranties, maintenance requirements etc. While the project is developing, additional attributes, issues and documentation can be associated to specific items. UK government has introduced de COBie format as principal part of Level 2 BIM requirement that will be mandatory for all centrally procured Government contracts from 2016. Level 2 BIM includes 3D BIM models and project and asset information available in COBie format (Carbonari et al 2015)

Particular technologies, such as the use of cloud computing, passive RFID tags and 3D scanning, have facilitated added value to BIM in a Facility Management/Operations context (McArthur et.al 2015). Vice versa, the BIM potential aids in the introduction and effective use of these technologies and others as condition monitoring and predictive maintenance, energy consumption control, etc. Actually, these technologies can be seen as information sources that are integrated and used in the building management through the BIM implementation and its interoperability with rest of software packages, which are linked to the operation and maintenance of the building (Figure 2).

Figure 2. Software interoperability

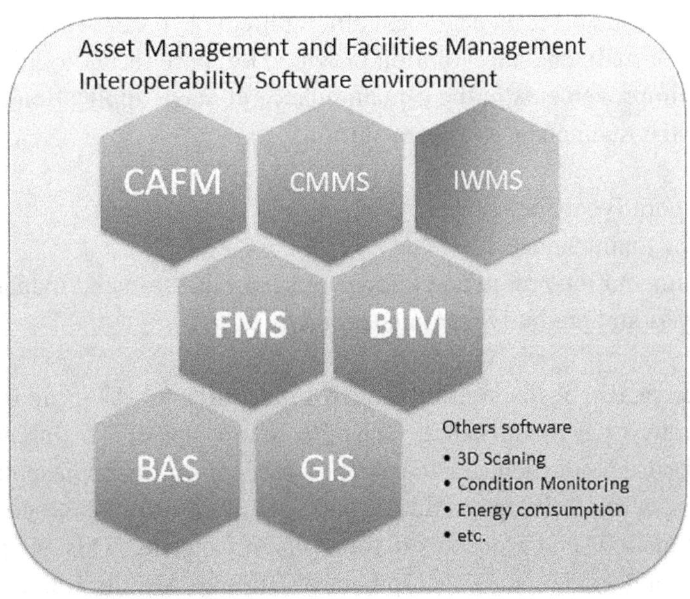

Facility Management Systems are software packages that support the maintenance and management of a facility. It helps to manage work orders, assets, inventory, and safety. Some names that can be classified under Facility Management Systems include Computerized Maintenance Management Systems (CMMS), Computer-Aided Facility Management (CAFM), and Computerized Maintenance Management Information System (CMMIS). If an organization has a FMS, it should be evaluated for its ability to support the BIM Data and Uses as defined in the previous steps. If an organization does not have a FMS, it is important that a proper one is selected. According to the interoperability requirements of the building asset management approach, the FMS should support the importing and/or exporting of data to other systems. It is especially important to consider the integration and acceptance of BIM Data, i.e. FMS should have the ability to import and/or export BIM data directly from the BIM Model. For this purpose, FMS have to support open standards for data transfer such as COBie. Finally, FMS should be able to handle other graphical data such as photographs and plans and the interface with others technologies (BAS, Condition Monitoring, etc.). Comparing with the simple application of CAFM or CMMS software, the use of BIM models aids in two aspects:

- Allow introduce AM and FM from the very beginning of the building lifecycle (designing and constructions).
- BIM software as interface: intelligent and detailed representation of physical and functional characteristics.

Building Automation Systems (BAS) have a main role in new intelligent buildings. They conform the control system of the building and include facilities control and the interaction with users. These systems are going to produce relevant information and will be one of the main sources for FM. Once again, the problem to use these data is the interoperability of BAS hardware/software with FMS and BIM.

This interoperability software environment for AM, with BIM models as a central element, is totally aligned with the E-maintenance principles, term that serves as a conceptual support to general use and applications of ICTs in maintenance. This is related to the effective introduction of new capabilities at the service of maintenance and asset management among lifecycle product stages. This approach helps to understand the scope of maintenance evolution provided by these technologies. Muller et al. (2008), analyzing the potential improvements in the e-maintenance concept application context, introduce the following references to the maintenance tasks evolution:

- Cooperative/collaborative maintenance.
- Remote and on-line maintenance.
- Maintenance documentation/record and knowledge capitalization and management.
- Fault/failure analysis and predictive maintenance.

The current perspective, established within the BIM discipline for AM, can be considered complementary to the vision drawn by E-maintenance. One of the main gaps in E-maintenance application is its contribution, which is focused on structuring the technological discipline represented by E-maintenance, without discussing the specific techniques and methods required to fulfill the goals envisioned for the maintenance business functions and processes in future manufacturing. This gap precisely can be completed by the evolution supported by the development of BIM for AM discipline.

ASSET MANAGEMENT FOR BUILDING AND FACILITIES

The optimization of building operational phase is a complex issue. Together with its intrinsic complexity of such facilities, it has to be considered the legislation and sustainability demands, fitting with usage requirements. (Zhang et al., 2009). This complex scenario is completed with the trend to outsourcing services, and the introduction of procurement routes that include operation and maintenance in integrated supply contracts. FM as discipline provides a holistic view of the building operation and maintenance, an overall management of the resource available towards the strategic objectives of facilities' users and owners. Here is located the potential of FM rather than the accumulation of maintenance task or new software applications. FM is a broad concept, covering everything from real estate and financial management to maintenance and cleaning (Atkin and Brooks, 2009). FM is defined by EN 15221-1 as "the integration of processes within an organization to maintain and develop the agreed services which support and improve the effectiveness of its primary activities."

AM and FM are overlapped disciplines. The terms AM and FM are often used interchangeably, although there are differences in approach between the asset management and facilities management disciplines. FM is focused on the infrastructure and building sectors while AM has a broader application field. Especially in the building sector FM considered a consolidated profession (IFMA 2013). Both have generated their own standards or specifications and both have evolved their own language of preferred and defined terms. The ISO 55000 lists FM among the asset management activities. With this view FM can be considered as a part or tool of AM, since AM goes beyond FM providing a more comprehensive view and more potential benefits.

The benefits of AM are proven in many industries and business environments, improving the performance along the lifecycle and the contribution to safety, health and protection of the environment. Together with organizational commitment to quality, performance or safety, it helps to mitigate the legal, social and environmental risks associated to accidents in industrial facilities. Asset Management, as a discipline, allows organizations to optimize the whole life value of managing assets portfolios. For a

Figure 3. Integration of FM and AM in buildings and facilities management

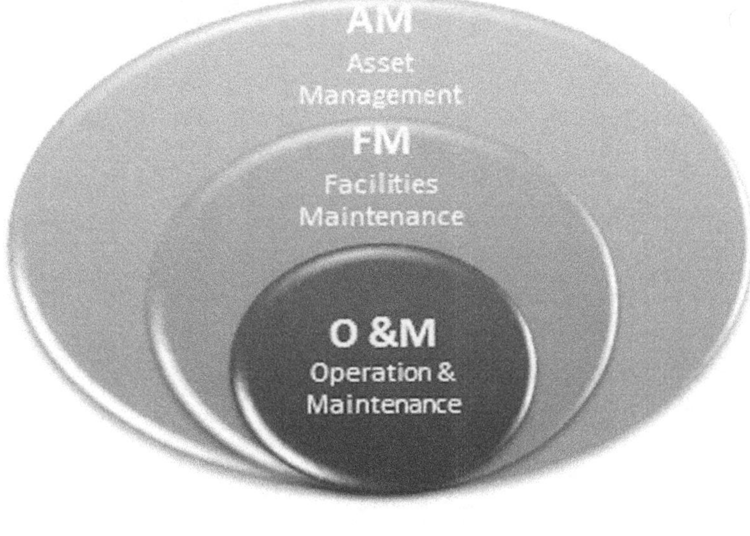

single organization, the list of assets or portfolio may contain diverse assets in nature, distributed over extensive geographical areas, and may be subjected to differing demand/utilization requirements. In agreement to ISO 55000, AM can be applied to all kind of assets, including physical assets (elements, inventory and properties) and intangibles assets (leases, brands, digital assets, use rights, licenses, intellectual property, etc.)

The general principles of AM have been defined by the family of standards ISO 55000, ISO 55001, ISO 55002. Among the aspects contained in this approach, those who can be described as more innovative, and representing a significant advance in the optimization of asset management, they may include:

- Managing the value of the asset. AM supports the realization of value while balancing financial, environmental and social costs, risk, quality of service and performance related to assets.
- Risk based decision-making. Effective control and governance of assets by organizations is essential to realize value through managing risk and opportunity, in order to achieve the desired balance of cost, risk and performance.
- Integrating the longer term activity of asset management with the shorter term activity of asset
- Stakeholders' treatment. The stakeholders' requirement but also the information exchange and the suitable access to the information (right information, to right person at the right moment)
- The information requirement and its treatment. AM is intensive in data/information management. Asset information system can be extremely large and complex and creating, controlling and documenting this information is a critical function of the asset management system.

This last point reinforces the importance of information management in AM. The Information System is one of the key elements in asset management systems (AMS) defined by ISO 55000-1 (Figure 4). An asset management system is a set of interrelated and interacting elements of an organization,

Figure 4. AMS elements

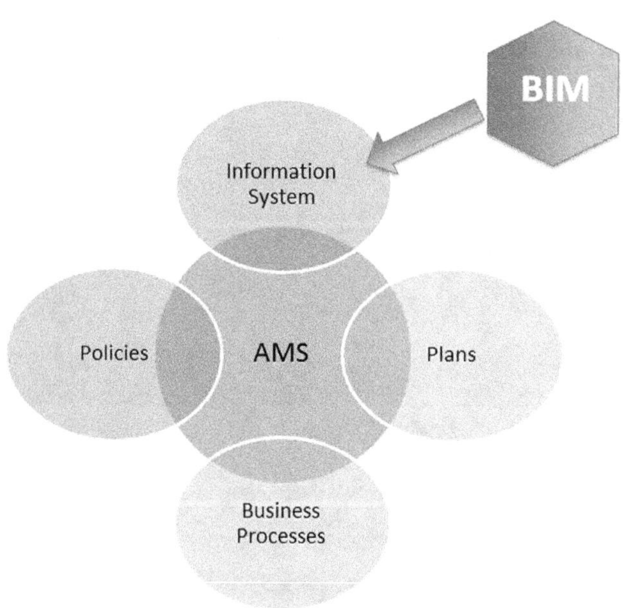

whose function is to establish the asset management policy and asset management objectives, as well as the processes needed to achieve those objectives. In this context, the elements of an asset management system should be viewed as a set of tools, including the information systems, which are integrated to give assurance that the asset management activities will be delivered. For building asset management this required information system could be implement from the implementation of the BIM model.

BIM and AMS: PAS 1192-2 and PAS 1192-3

PAS 1192-2:2013 "Specification for information management for the capital/delivery phase of construction projects using building information modeling" and PAS 1192-3:2014 "Specification for information management for the operational phase of assets using building information modeling", are complementary documents that specified an information management process to support building information modelling (BIM) Level 2, referred previously in this chapter in:

- The capital/delivery phase of projects, PAS 1192-2, (PIM, project information model)
- The O&M phase PAS 1192-3. (AIM, Asset Information Model)

PAS 1192-2, PAS 1192-3 applies to both, building and infrastructure assets, and the intended audience for these documents includes organizations and individuals responsible for the procurement, design, construction, delivery, operation and maintenance of buildings and infrastructure assets. These standards cross-reference with other existing standards concerned with the management of assets. In particular, are closely related to the ISO 55000 series of standards that provide one overarching framework for the adoption and implementation of PAS 1192-2 and PAS 1192-3.

PAS 1192-2 focuses specifically on project delivery, where the majority of graphical data, non-graphical data and documents, known collectively as the project information model (PIM), are accumulated from design and construction activities. PAS 1192-3 focuses on the operational phase of assets irrespective of whether these were commissioned through major works, acquired through transfer of ownership or already existed in an asset portfolio. The operational phase of an asset is deemed to commence at handover, but the requirements within PAS 1192-3 may also be helpful during major works. Progressively working through the various stages of the information delivery cycle, the requirements within this PAS culminate with the delivery of the as-constructed asset information model (AIM). AIM is handed over to the employer by the supplier once the PIM has been verified against what has been constructed and it is used to support the portfolio management activity for the life of the asset, which include the maintenance management.

CONCLUSION

At this moment, it is possible to affirm that the technological advances may provide a qualitative jump in building asset management. The current evolution of the sector is promoting a great increment of BIM applications. But these applications are mainly focus on design and construction. The use of BIM models for operation and maintenance phase is not so extended. For this reason, although there are a lot of references about the potential benefits of BIM use, there are few references about the quantification and realization of these benefits in practical cases for operational phases. In conclusion, BIM benefits

Figure 5. AIM and PIM: The information delivery cycle
(PAS 1192-2:2013)

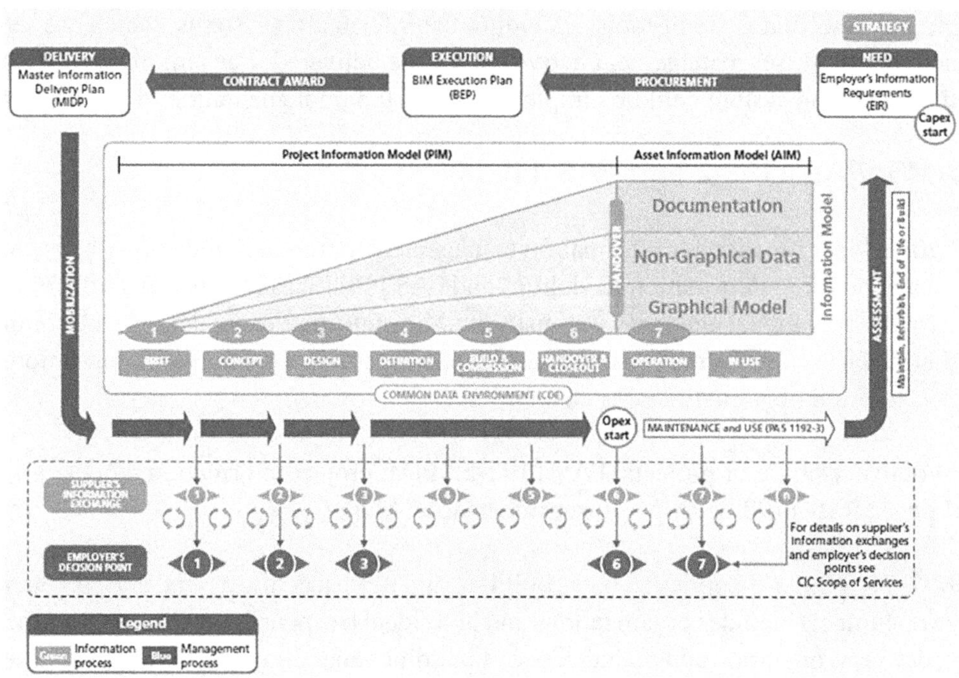

for AM are not well characterized. Methodologies and application frameworks have to be proposed and tested to develop reference use cases that will allow extending the knowledge about AM supported by BIM model as the base for an asset management information system. In this sense, the PAS 1192 standards series jointly with the most general ISO 55000, are the main references for this development process and for any particular implementation attempt.

REFERENCES

Azhar, S. (2011). Building Information Modelling (BIM): Trends, Benefits, Risks, and Challenges for the AEC Industry. *Leadership and Management in Engineering*, *11*(3), 241–252. doi:10.1061/(ASCE)LM.1943-5630.0000127

Cahill, B., Menzel, K., & Flynn, D. (2012). *BIM as a centre piece for optimised building operation. eWork and eBusiness in Architecture, Engineering and Construction*. Taylor & Francis Group.

EN 15221-1. (2006). European standard in facility management–Part 1: Terms and definitions.

Fukuda, S. (2014). BIM for FM: A Case Support for Business Life Cycle. *PLM 2014. IFIP AICT*, *442*, 63–74.

Gallaher, M. P., O'Connor, A. C., Dettbarn, J. L., & Gilday, L. T. (2004). Cost Analysis of Inadequate Interoperability in the U.S. Capital Facilities Industry. *National Institute of Standards and Technology*. Retrieved from http://fire.nist.gov/bfrlpubs/build04/PDF/b04022.pdf

IAM. The Institute of Asset Management. (2013). What is asset management? Retrieved from http://theiam.org/what-asset-management

IFMA, International Facilities Management Association. (2013). What is FM? Retrieved from http://www.ifma.org/know-base/browse/what-is-fm

ISO. (2013). ISO 55000 Asset Management - Overview, principles and terminology.

ISO. (2014). ISO 55001 Asset management — Management systems — Requirements.

ISO. (2014). ISO 55002 Asset management — Management systems — Guidelines on the application of ISO 55001.

Kassem, M., Kelly, G., Dawood, N., Serginson, M., & Lockley, S. (2015). BIM in facilities management applications: A case study of a large university complex. *Built Environment Project and Asset Management*, *5*(3), 261–277. doi:10.1108/BEPAM-02-2014-0011

Liu, R., & Issa, R. R. A. (2013). Issues in BIM for facility management from industry practitioners' perspectives. *Computing in Civil Engineering*, 2013, 411-418.

Love, P. E. D., Matthews, J., Simpson, I., Hill, A., & Olatunji, O. A. (2014). A benefits realization management building information modelling framework for asset owners. *Automation in Construction*, *37*, 1–10. doi:10.1016/j.autcon.2013.09.007

Love, P. E. D., Simpson, I., Hill, A., & Standing, C. (2013). From justification to evaluation: Building information modelling for asset owners. *Automation in Construction*, *35*, 208–216. doi:10.1016/j.autcon.2013.05.008

McArthur, J. J. (2015). A building information management (BIM) framework and supporting case study for existing building operations, maintenance and sustainability.*International Conference on Sustainable Design, Engineering and Construction. Procedia Engineering 00* doi:10.1016/j.proeng.2015.08.450

Muller, A., Marquez, A. C., & Iung, B. (2008). On the concept of e-maintenance: Review and current research. *Reliability Engineering & System Safety*, *93*(8), 1165–1187. doi:10.1016/j.ress.2007.08.006

Parsanezhad, P., & Tarandi, V. (2013). Is the age of facility manager's papers boxes over? *Proceedings of theCIB World Building Congress.*

PAS 1192-2:2013, Specification for information management for the capital/delivery phase of construction projects using building information modeling. (2013).

PAS 1192-3:2014, Specification for information management for the operational phase of assets using building information modeling. (2014).

Schevers, H., Mitchell, J., Akhurst, P., Marchant, D., Bull, S., McDonald, K., & Linning, C. (2007). Towards digital facility modelling for Sydney opera house using IFC and semantic web technology. *ITcon*, *12*, 347–362.

Shen, W., Hao, Q., Mak, H., Neelamkavil, J., Xie, H., Dickinson, J., & Xue, H. et al. (2010). Systems integration and collaboration in architecture, engineering, construction, and facilities management: A review. *Advanced Engineering Informatics*, *24*(2), 196–207. doi:10.1016/j.aei.2009.09.001

Volk, R., Stengel, J., & Schultmann, F. (2014). Building Information Models (BIM) for existing buildings –literature review and future needs. *Automation in Construction*, *38*, 109–127. doi:10.1016/j.autcon.2013.10.023

Zhang, X., Arayici, Y., Wu, S., Abbott, C., & Aouad, G. (2009). Integrating BIM and GIS for large scale (building) asset management: a critical review. *Proceedings of theTwelfth International Conference on Civil, Structural and Environmental Engineering Computing*. doi:10.4203/ccp.91.93

Chapter 7
Service 4.0:
The Reasons and Purposes of Industry 4.0 within the Ambit of After-Sales Maintenance

Eduardo Castellano
IK4-IKERLAN, Spain

Gerardo Pagalday
IK4-IKERLAN, Spain

Patxi X. Zubizarreta
IK4-IKERLAN, Spain

Jone Uribetxebarria
IK4-IKERLAN, Spain

Adolfo Crespo Márquez
University of Sevilla, Spain

ABSTRACT

In recent years, the concept of Industry 4.0 has been significantly advanced in industrial circles as an aspect that provides a competitive differential. Through the technologies involved, machines can now monitor and relay information on their operating conditions for analysis and decision-making, as well as for prompting action. These new functions generally involve the development of technological projects and significant investments. This renders it expedient to explain why certain systems should be monitored, but not others, as well as the use to be given to the data gathered as a way of generating income for a firm. This approach is especially important in certain corporate operations, such as after-sales maintenance. This article introduces a reference framework that permits the effective and efficient management of after-sales maintenance services. This framework relates after-sales service technologies with product technologies (Industry 4.0), and therefore covers the reasons and purposes of Industry 4.0 within the ambit of after-sales service.

DOI: 10.4018/978-1-5225-0651-5.ch007

INTRODUCTION

Until recently, the bulk of the turnover for many capital goods manufacturers was generated by the sale of new machines, with their after-sales service being an obligatory extra to be provided with the new sale. For these kinds of firms, therefore, an after-sales service has been seen as a necessary evil and, in short, as an added cost.

As a result of the current crisis and the greater difficulties facing the sale of new machinery, together with the tighter margins in new sales, this viewpoint has gradually changed, and many capital goods manufacturers have begun to consider their after-sales service as both a business opportunity for generating recurring income from the pool of machines sold and a driver of customer loyalty.

The services that may be provided at this stage of the life cycle for machinery and equipment are linked to the actual servicing of the items sold (e.g., repairs, spare parts, preventive maintenance, and retrofitting), contracts for guaranteeing machine availability, contracts on the use of machinery (e.g., renting and leasing), services related to energy efficiency, process enhancement, and training, amongst others (Cohen, Agrawal & Agrawal, 2006).

The following figure provides a schematic view of the reference framework for industrial maintenance management proposed by Crespo (2007), which has been adapted to apply to the after-sales ambit.

The reference framework for after-sales maintenance described in this chapter consists of a number of initial phases that use the cost incurred by the unreliability of a machine's different parts to identify which ones should be remotely monitored and supervised by means of *Condition Based Maintenance (CBM)*. In other words, identification is made of those systems in the machine that require investment in technological development, and those which do not; hence the reason for Industry 4.0 in the after-sales maintenance service. Later phases, on the other hand, permit acting upon the alerts generated by these Industry 4.0 technologies for fine-tuning the necessary resources (i.e., management of stocks of spare parts, and dynamic scheduling of maintenance tasks) for the service's deployment; hence the purpose of Industry 4.0 in the after-sales maintenance service. The after-sales maintenance framework thereby provides information on the return on investment of Industry 4.0 for machinery, through the after-sales maintenance service, which is of particular interest for a firm's business outlook (e.g., funds spent vs. funds generated).

The next section describes the types of firms that have taken part in the case studies. A comparison between the reference framework for industrial maintenance (Crespo, 2007) and those firms' after-sales operations and specific techniques identified in the literaturehas been used to constitute the reference framework for after-sales maintenance services that is described in the third section. The fourth section identifies three typical levels of maturity for the after-sales maintenance service identified based on an analysis of the case studies, and lays out a roadmap for advancing toward higher levels of development.

CASE STUDIES

Based on the industrial maintenance management framework propounded by Crespo (2007), as shown in Figure 1, and the *Maintenance Effectiveness Survey* (Marshall Institute, 1999), a semi-qualitative questionnaire (Patton, 2001) has been compiled for each one of the phases. The questionnaire has been designed for use in industrial firms in the capital goods sector to identify those more salient differences existing between the approach to industrial maintenance and the consideration of the after-sales mainte-

Figure 1. Crespo (2007) industrial maintenance management framework

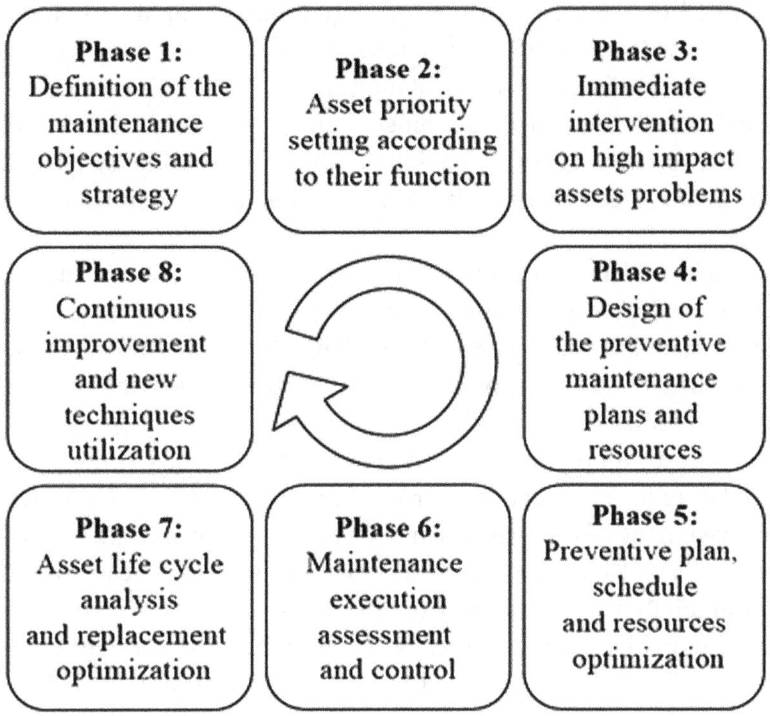

nance of the equipment sold, as well as the extent to which after-sales maintenance has been developed in those firms.

The next step involved interviewing CEOs, chief industrial officers, managers of after-sales services, and maintenance technicians. These interviews were supplemented by the evidence gathered from each one of the firms (e.g., documents and applications), whereby the answers provided could be compared with their daily practices (Patton, 2001).

The firms chosen for the study are located in the Basque Country (Spain). The machinery and capital goods manufacturing industry in this region has the third largest output in the European Union (SPRI, 2015). The sample firms represent different echelons in the capital goods sector's value chain:

- Three[1] manufacturers of subsystems and components for final equipment makers; electric motors, data gathering systems and CNCs, and remote controls.
- Three[2] manufactures and installers of small machinery; industrial washing machines, vending machines, and industrial boilers.
- Twelve[3] manufacturers and installers of medium- and large-sized machinery; compressors, machinery for core-making, blowing machines, presses, levellers, machining centres, automatic lines, etc.
- One[4] firm marketing industrial products; fork-lift trucks.
- One[5] firm providing multi-brand industrial after-sales services.

The importance of the after-sales maintenance service in each one of these groups of firms is as follows:

- Manufacturers of subsystems and components for final equipment makers; the after-sales service is generally provided by the maker of the final equipment or its distributor. Accordingly, these components are supplied with technical maintenance manuals, along with technologies that facilitate the service for the end installer or distributor.
- Manufactures and installers of small machinery; this type of manufacturer usually provides an after-sales maintenance service for its equipment during the warranty period. Once this period has expired, they are replaced by multi-brand after-sales service companies, which in turn act as consumer motivators.
- Manufacturers and installers of medium- and large-sized machinery; they begin providing an after-sales service on different operating levels (e.g., technical service, repairs, preventive and predictive maintenance, spares, and retrofitting). The after-sales service these firms provide is a major business opportunity in terms of both the potential income from the pool of machines sold and the reinforcement of customer loyalty.
- Firms marketing industrial products; many of these firms do not sell the product to the end customer, but instead offer the product under terms of renting or leasing. The efficiency and effectiveness of their management of the maintenance of their pool of machines is therefore crucial for controlling the costs involved. Specifically, it is vital to accurately estimate the costs associated with a product's unreliability over the course of its life cycle.
- Providers of multi-brand industrial after-sales services; these are services companies that provide an after-sales service to manufacturers and installers of small machines. The after-sales service is the core business of these firms. It is therefore essential for them to fine-tune the mix and the maintenance plan (corrective, preventive, and predictive), as well as the frequency of the service. Furthermore, it is important for these firms to know the unreliability costs over the life cycles of the equipment they manage, thereby enabling them to properly schedule the inclusion of upgrades and retrofitting, ensuring that maintenance remains profitable.

MANAGEMENT MODEL FOR THE AFTER-SALES MAINTENANCE SERVICE

This section describes the management model for the after-sales maintenance service designed according to the industrial maintenance reference framework (Crespo, 2007), after-sales experiences across the 20 firms taking part in the study, and specific analyses of the literature on each phase.

As can be seen on the left-hand side in Figure 2, the management framework for the after-sales maintenance service consists of eight phases related to service technologies (Service 4.0), linked through phases 4 and 5 to Industry 4.0 technologies (right-hand side of the figure).

There follows a description of each one of the model's eight phases.

Phase 1: Definition of the Strategy and Key Indicators of the After-Sales Business

Based on the case studies considered, it has been noted that those capital goods manufacturers that view their after-sales maintenance service as a source of costs, and not of income, face several common issues. This reactive approach means they organise the service as an operating unit for the urgent repair of failures. The operations are therefore corrective in nature, with no foresight or prior planning, and the

Figure 2. After-sale maintenance management framework

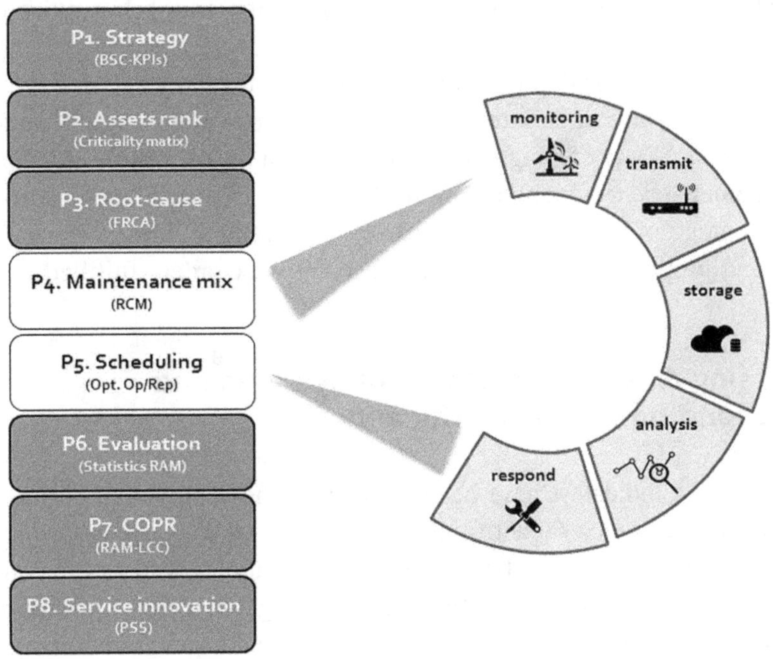

resources available in terms of technical customer care staff are few, which often means that production and commissioning personnel need to be called upon, thereby reducing productivity in new sales. In these kinds of firms, the after-sales maintenance service is seen as a problem, and not as a business opportunity.

By contrast, there are capital goods manufacturers that are turning after-sales maintenance into one of their firm's main sources of revenue (e.g., a 40% share of the firm's turnover and 60% of its operating profit). These firms have a unitary approach to service/business, where the service is proactive (e.g., preventive and remote predictive maintenance contracts, advanced sale of high-margin spares), and operations are planned (e.g., dimensioning of resources –people and stocks- to cater for the pool on a preventive basis with the required level of availability and service); in short, they understand the after-sales maintenance service to be an integral part of the sale of the product-machine within a *Product Service System* framework (Baines et al., 2007).

This first phase involves the definition of indicators, objectives, strategies and responsibilities in after-sales maintenance. The aim is to ensure that the strategy and operating objectives of after-sales maintenance are consistent with the business strategy and its overall objectives.

A highly valid tool for the consideration of after-sales maintenance as a business driver is the Balanced Scorecard (BSC) proposed by Kaplan and Norton (1992). The BSC is a model that translates a business's mission and strategy into a series of indicators with quantifiable targets centred around four perspectives: (1) Financial – investor perspective; (2) Customers – performance attributes valued by customers; (3) Processes – procedures and means provided by the business for achieving financial and customer objectives; (4) learning and growth – ability to improve and create value.

An analysis of the case studies has led to the identification of the following key indicators of after-sales maintenance:

- **Financial:** Ratio between the invoicing of the after-sales service and the invoicing from the sale of new products, and the percentage of the invoicing of the after-sale service over the operating profit.
- **Customers:** Percentage of availability of the machines sold, and level of service.
- **Processes:** Mean time between failures (MTBF), ratio between the mean time to repair and mean downtime (MTTR/MDT), percentage of corrective measures over preventive ones, level of occupation of the maintenance team, and turnover of the stock of critical spare parts.
- **Learning:** Percentage of corrective measures over preventive ones performed by maintenance staff in a due and proper manner, and percentage of work orders fulfilled in a due and proper manner.

Phase 2: Prioritising Assets, Systems, Subsystems and Components Depending on Their Criticality

Once the key objectives and indicators have been defined for the deployment of the after-sales maintenance service, the next phase involves identifying the criticality of the various systems, subsystems and components of the machines sold on the basis of their reliability, as well as the consequences arising from any shortfall in the same within an operational context (e.g. cost of the unavailability of machinery in the customer's processes, corrective costs for each failure, and safety and environmental costs). A prioritisation process based on the axes of reliability vs. impact leads to a *criticality matrix* type representation (Crespo et al., 2015).

This prioritisation allows identifying, on the one hand, those components whose more frequent and costly failures require greater attention and more resources and, on the other, those that require less attention and fewer resources for their maintenance. An initial objective at this stage, therefore, is to fine-tune the effective distribution of maintenance resources according to the importance and ramifications of potential failure events affecting equipment/sub-systems within the operational context in which they may occur. This will avoid overspending on the maintenance of non-critical (non-priority) equipment/sub-systems, and under-spending on the maintenance of critical (priority) equipment/subsystems, which therefore have a high potential for generating extra costs (e.g., penalties) if they are not properly addressed.

Furthermore, the *criticality matrix* is the starting point for identifying the type of CPS signals and parameters that need to be monitored for the prevention or early detection of highly critical failure modes. The capture of these parameters and signals will allow, on the one hand, identifying a machine's current operational contexts, and on the other, knowing which analogue or digital signals needed to be captured to verify the status of the different subsystems.

The case study has highlighted the fact there is not just one *criticality matrix* per product-machine model, but rather that each one's operating conditions in the customer's process may significantly alter the rate of occurrence of certain failure modes. It would therefore be more appropriate to refer to a family of *criticality matrices* for each model of machine.

There are different techniques that may be applied depending on whether the type of data available for the prioritisation is more qualitative (e.g., expert opinions) or quantitative (e.g., log records of coded failure modes and documented registers of the associated measures). Thus, the case studies have shown for qualitative data sources that the most appropriate technique for identifying an initial version of the prioritisation of criticality is the *Criticality Flowchart* (Crespo, 2007). This technique allows classifying a machine's systems, subsystems and techniques into three categories, A, B, C, with A being the most

Figure 3. Example of a criticality matrix

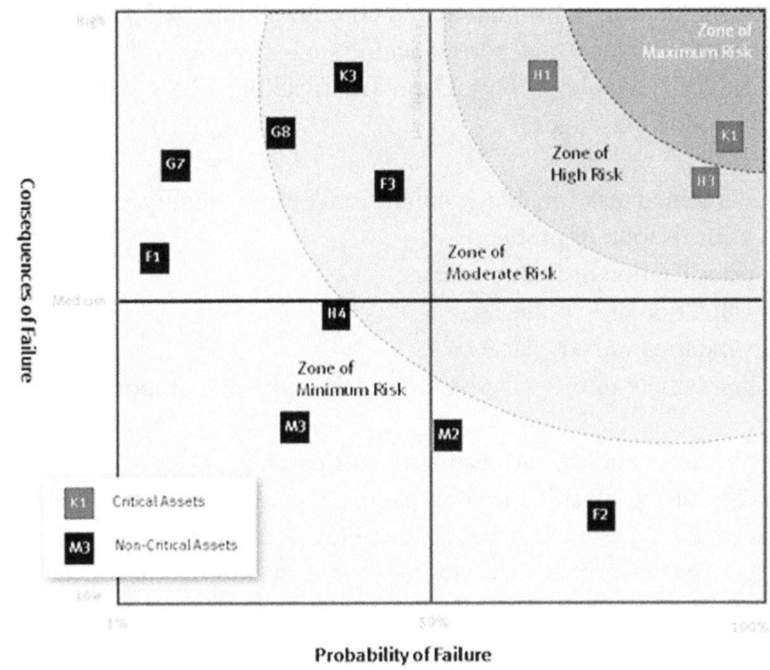

critical one. In the case of mixed sources involving both qualitative and quantitative data, the technique that provides the highest quality data is the *Overall Criticality by Risk* (Parra and Omaña, 2001). This method is based on the notion of risk, understood to be the result of multiplying the frequency of a failure by its severity (Jones, 1995). Both in this technique and in the previous one, risk is estimated through the frequencies and consequences of failures. The section on the consequences of failures takes into account impacts related to production losses linked to a lack of reliability, maintenance costs, maintainability, and potential damages involving the safety of the operators working the machines, and any environmental discharges. Finally, when the data source involved is of a quantitative nature, the technique that best exploits them in terms of asset prioritisation is the *Analytic Hierarchy Process* (Saaty, 1990). This technique helps analysts to arrange a problem's critical aspects into a hierarchical structure, reducing complex decisions to a series of comparisons that permit prioritising the different aspects (criteria) assessed.

Phase 3: Root-Cause Analysis of Failures

This phase involves analysing the reasons for the lack of reliability affecting the different systems, subsystems and components of high criticality identified in the previous phase, with solutions being proposed to eliminate or diminish their effects. The aim is therefore to identify the root cause of these critical failures in order to act upon them and reduce their possible impact on failures affecting operations, safety and the environment, thereby providing greater operational reliability.

As revealed in the case studies, the origin of this lack of reliability may be due to both technical and human shortcomings; errors in design, unsuitable manufacturing and installation technologies, incorrect procures for the use of manufacturing and installation technologies, or a lack of training in these technologies among manufacturing, installation and commissioning personnel.

There are numerous techniques that permit addressing a root-cause analysis of critical failures (Campbell and Jardine, 2001; Mobley, 1999). Generally speaking, all these techniques follow a logical process that covers the following stages:

1. Setting-up of a combined task force; personnel from design, quality, operations and maintenance.
2. Definition and prioritisation of problems.
3. Definition and prioritisation of failure modes.
4. Definition and validation of hypotheses.
5. Definition and validation of root causes.
6. Definition and assessment of the effectiveness of the solutions proposed.

As derived from the case studies, the results from this phase are not only beneficial in terms of improving the reliability of the machine product analysed (e.g., *Design for Reliability*) and therefore reducing critical elements in the *criticality matrix*, but they also help to encourage designers to consider criteria from a design perspective that facilitate the future maintainability of equipment (i.e., *Design for Maintainability*).

Phase 4: Design of the Mix (Corrective, Preventive, Predictive) and Maintenance Plan

This phase will involve drawing up the maintenance plan for the model of machine whose pool is to be attended to through the after-sales service.

The prioritisation of critical elements made in Phase 2 is a key input for orchestrating the maintenance mix:

- *Corrective maintenance* for components with failure modes that are not particularly critical. These kinds of failure modes are the ones that occur in secondary subsystems, and do not affect the asset's main functions.
- *Preventive maintenance* according to a specific frequency, based on the log record of failures, for failure modes of medium criticality.
- *Condition Based Maintenance* (CBM) for components with failure modes of high criticality.

Figure 4. Chart of maintenance strategies present in the mix

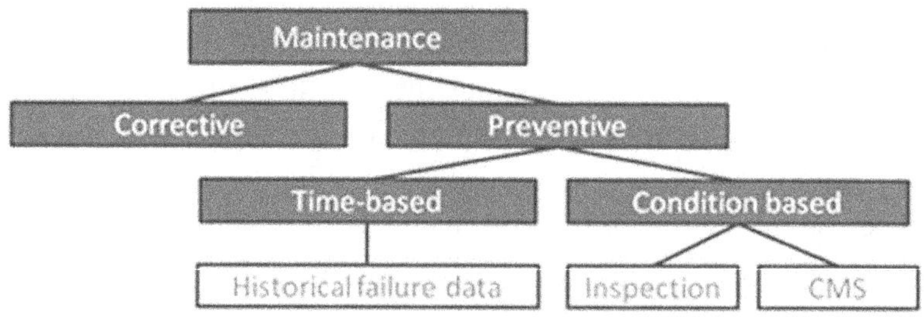

Preventive maintenance based on a log record of failures requires a huge volume of data to identify the optimum frequency for performing preventive maintenance tasks on the components of medium criticality on the machine in question. In practice, this requires a minimum of between ten and fifteen consecutive failure logs per component, or better still, per failure mode for each component, for modelling their distribution type (e.g., Weibull, Gamma, and Exponential). This distribution (i.e., probability of failure events over time), plotted against the preventive and corrective costs involved in operating in different time horizons, identifies the frequency of the preventive maintenance that optimises the intervention costs. As the following figure shows, this frequency is linked to a probability of failure that may be read the other way round in terms of availability. Therefore, improvements in availability will, in practice, mean reductions in the rate of preventive maintenance, an increase in preventive costs, and a reduction in the corrective costs associated with untreated failures.

Once the frequencies with which preventive maintenance is to be performed on a machine's various components have been established, a master plan is designed for preventive maintenance operations (e.g., personnel, specialists, tools, and critical spare parts), and the work orders to be followed in each case are drafted. These work orders are to contain the necessary fields for gathering the pertinent data for conducting the analysis required both in this phase (e.g., failure mode, consequence of the failure, date of occurrence, time and cost of the intervention) and in the preceding ones.

In addition, those components with highly critical failure modes will be the main candidates for a CBM, either through direct inspections (e.g., thermographs and the analysis of lubricants) or remotely (e.g., remote monitoring of critical parameters). These latter ones, and not others, will require the incorporation of sensorics, infrastructures for communication and data storage, and their subsequent analysis, either through mathematical or statistical models (e.g., data mining), which allow identifying operating ranges that comply with operating standards and generating alerts for any deterioration. This will give the reason for the inclusion of Industry 4.0 technologies into some of the machine's components, and not into others, in support of the after-sales maintenance service, and therefore where these investments in Industry 4.0 need to be focused the most.

Figure 5. Availability-cost ratio for different periods of preventive maintenance

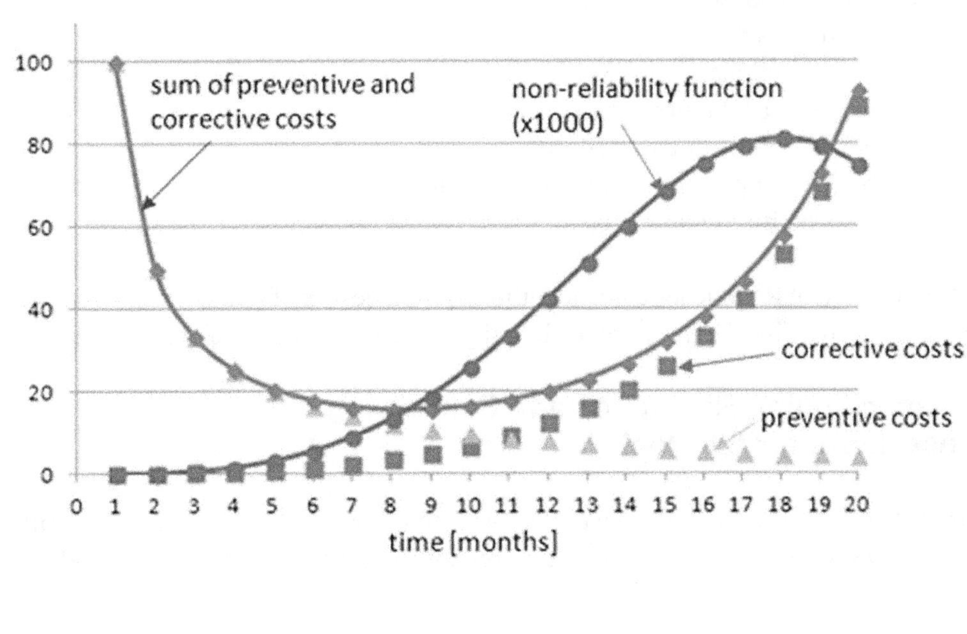

Based on the case studies, identification has been made of the different situations that favour the presence of remote monitoring systems for the condition as opposed to physical inspections by personnel when performing CBM:

- When the aim is to manage the after-sales maintenance of a machine pool, the actual location of the machines is a key issue. When the machines are installed in geographical regions that are very far apart, the remote monitoring of their critical components may be more economical than physical inspections.
- On some occasions, the machine's very design may render it extremely complicated to monitor the state of certain components through physical inspections because of their difficult access.
- The impaired performance of certain critical components may advance exponentially, whereby continuous remote monitoring is necessary in order to have enough time to react.

The case studies also reveal other issues that make it impossible to undertake preventive maintenance based on the log record of failures in favour of CBM, namely, the absence of that log record. This may be the case for new machines and those that have been subject to highly conservative preventive maintenance processes, and do not therefore have a log record of failures. In this latter case, CBM will allow tracking the advance of any deterioration in the components' performance, which may be used to estimate the potential failure.

Thirdly, the case studies have also revealed that within the field of capital goods and machine tools there is a very high degree of customisation across the board. This means that the master plan for the preventive maintenance of a certain model of machine needs to be taken as a starting point, and then adapted in each case to the idiosyncrasies of each individual machine and its operating conditions. This task makes it advisable to use configuration tools for maintenance plans.According to the literature, the main technique for the formal development of this phase is Reliability Centered Maintenance (Bloom, 2006; Moubray, 1991). Reliability Centered Maintenance (RCM) involves a procedure that allows identifying maintenance requirements and strategies based on the following seven questions:

1. What are the functions of the asset?
2. In what way can the asset fail to fulfil its functions?
3. What causes each functional failure?
4. What happens when each failure occurs?
5. What are the consequences of each failure?
6. What should be done to prevent or predict the failure?
7. What should be done if a suitable proactive task cannot be found?

A correct application of RCM leads to the identification of the mix of effective maintenance strategies linked to the product-machine for different availability objectives vs. cost of the service.

Phase 5: Dimensioning Resources and Scheduling Maintenance Tasks for the Pool

Once the preventive after-sales maintenance plan has been drawn up as required for fulfilling certain specific objectives as regards level of service (e.g., reliability-availability-cost) for a model of machine,

Figure 6. RCM logic

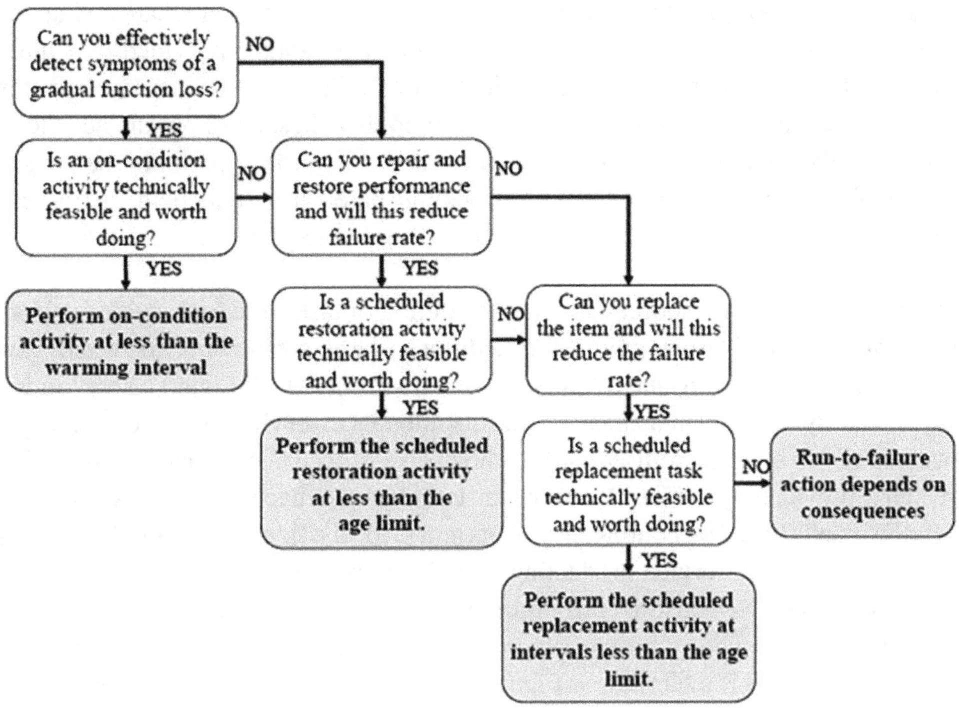

this next phase involves dimensioning the resources of both a proprietary and subcontracted nature (e.g., multi-brand services companies), as needed for attending to the global pool of machines sold, as well as the scheduling of tasks for responding with the required level of service.

The case studies have shown that, in contrast to industrial maintenance, after-sales maintenance poses a major challenge in this phase, namely, the geographical location of the machines. Accordingly, the dimensioning of the resources required for attending to the pool should consider not only the volume of maintenance personnel and spare parts needed for providing the necessary level of service, but also their more suitable location (i.e., siting of SAT centres and warehouses for spare parts). Likewise, the scheduling of tasks, and especially their grouping and sequencing, should take into account logistics aspects that are significantly different to those of industrial maintenance.

The literature reports different ways of addressing these challenges:

- Maintenance staff planning and scheduling analytical models as queue theory models (Shenoy and Bhadury, 1998), linear programming models (Duffuaa, 2000), Critical Path Method (CPM) and Programme Evaluation Review Technique (PERT), or simulation models (Persson and Saccani, 2009; Baker et al., 1997).
- The planning of spare part requirements using statistical inventory models (Tersine, 1994; Wilson, 1934), selective control procedures along with some heuristics (Huiskonen, 2001; Ramani and Krishnan Kutty, 1985), or using material requirements planning/manufacturing resources planning (MRP/MRPII) techniques (Shenoy and Bhadury, 1998).

- Different simulations techniques for quantitative economical analysis of maintenance policies (A. Berdinyazov et al., 2009; Enrico Zio et al., 2013), or predictive maintenance strategies for cost effectiveness analysis (E.Gilabert et al., 2015)The main advantage provided by simulation techniques (e.g., AnyLogic), compared to analytical mathematical approaches, is their flexibility for systemically modelling the multiple views of reality in this phase, without the need to incorporate unrealistic hypotheses, and for considering, in a straightforward manner, alternative scenarios for the location of resources and strategies for restocking spare parts, sequencing activities, assigning resources and their supervision, as well as their fine-tuning (Sagarna et al., 2015).

Beyond the dimensioning of resources and the scheduling of tasks, the case studies evidence that the practical management of this phase requires, for medium-to-large-sized pools, the use of some form of *Computerized Maintenance Management System* (e.g., Maximo, SAP-PM, etc.) These kinds of systems pave the way for the dynamic management of the maintenance service, in which the deployment of preventive tasks needs to be synchronised with ad hoc tasks of a corrective nature, as well as with potential failure alerts provided by the CBM systems installed on the machines. This will therefore require continuous readjustments based on contingency plans and prioritisation criteria in order to guarantee the robustness of the maintenance service for dealing with different scenarios of uncertainty.

Finally, the case studies have also identified the potential of CBM over the management of stocks of critical spare parts. Thus, when the time interval between the detection of a potential failure and its actual occurrence (i.e., the P-F interval, see Figure 8) is longer that the time needed to deliver its associated spare part, there will no need to keep that part in stock. These types of practices, although still infrequent in the management of after-sales maintenance, may lead to significant reductions in the costs involved in warehousing spare parts.

Both the preventive alerts flagging a potential failure in critical components, which may avoid high-impact interruptions of the customer's processes, as well as the potential saving in the warehousing of critical spare parts described in the previous paragraph, are arguments backing the installation of CBM systems, as well as the return on their investment; they are the purpose of Industry 4.0 within the field of after-sales maintenance.

Figure 7. Image of the simulator of resource dimensioning strategies and task scheduling in an AnyLogic simulation environment
Source: Sagarna et al. (2015)

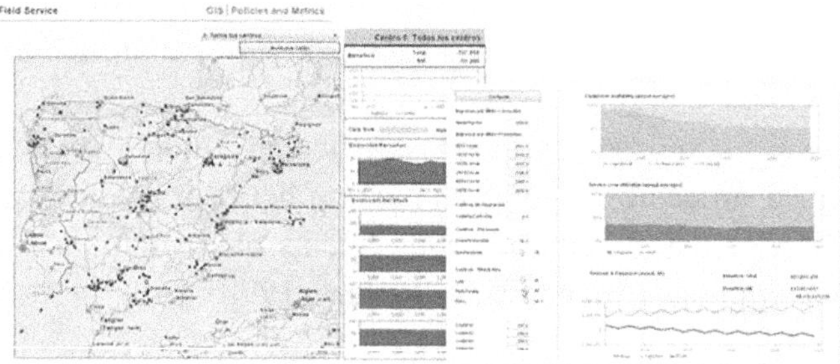

Figure 8. Behaviour curve of potential faults

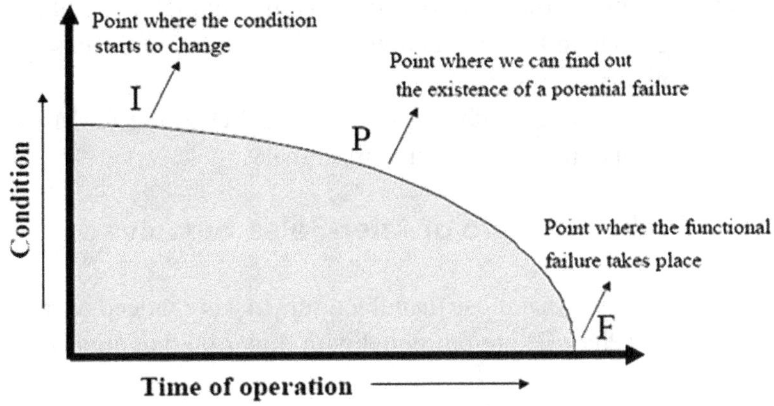

Phase 6: Assessment and Supervision of the After-Sales Maintenance Service

Based on the data recorded in the management systems of the after-sales maintenance service (e.g., CMMS in Phase 5), an assessment will be made of its performance in order to realign the resources and activities needed to attain, or improve, the goals set for the business indicators of after-sales maintenance (Phase 1).

In practice, this involves an intensive analytical procedure to convert the data from the CMMS's logs of work orders into meaningful information for assessment and supervisory purposes. Some of the indicators and basic statistical formulations commonly used in this phase are the following; Mean Time Between Failures (MTBF), Mean Up Time (MUT), Mean Time To Failure (MTTF), Mean Downtime (MDT), Mean Time To Repair (MTTR), Reliability function, Availability function, Maintainability function, Risk function, etc. (Rausand and Høyland, 2004).

The decision-making process within the management of after-sales maintenance should include technical indicators of reliability and maintenance, with cost indicators that include the consequences of failure events, as well as service level indicators. This consideration will allow maximising the service's cost-risk-profit relationship (Crespo, 2007).

Phase 7: Life Cycle Cost Analysis Based on Reliability

In the capital goods sector, the cost of purchasing a machine (i.e., CAPEX; design, development, acquisition, installation, staff, training, manuals, documentation, tools and facilities for maintenance, replacement parts for assurance, withdrawal) is often lower than the operating costs it will generate over the course of its life cycle (i.e., OPEX; manpower, operations, planned maintenance, storage, recruiting and corrective maintenance — penalizations for failure events/low reliability) (Crespo, 2007).

As regards OPEX, the analysis of costs due to unreliability, or the maintenance costs due to the asset's gradual deterioration, is referred to as RAM-LCC, which stands for Reliability, Availability, Maintainability Life Cycle Cost (Parra et al., 2012; Woodward, 1997; Woodhouse, 1993). Given the importance these types of costs have in a machine's life cycle, increasingly more buyers are asking for this information at the time of purchase.

In practice, this poses a major challenge for equipment manufacturers, as estimating the RAM-LCC requires them to know their machines' functions of reliability, maintainability and availability over time; with these calculations being made in Phase 6, and which, depending on the case study involved, very few manufacturers seem to master.

Therefore, if the equipment a firm manufactures is highly reliable, the RAM-LCC may be a major selling point compared to a competitor's less reliable machinery.

Phase 8: Innovation in the Portfolio of After-Sales Services

An analysis of the case studies shows that those manufacturers that are indeed capable of calculating the RAM-LCC of their equipment (Phase 7) are presented with major market opportunities:

- Firstly, knowing how the RAM-LCC develops over time enables a manufacturer to advise its customers about the best moment for carrying out the retrofitting of the machine's different systems, turning this is into a truly key moment in the sale.
- Secondly, for those manufacturers that provide renting or leasing services for their machines, the fact they know the RAM-LCC of their equipment ensures they are aware of the implicit costs these services involve, which means they can control their margin.

The same may be said for maintenance contracts, availability contracts, or the costs of warranties over different time horizons.

Therefore, establishing the equipment's RAM-LCC opens up a whole range of servitization possibilities (Baines et al., 2007; Vandermerwe and Rada, 1988) for those machines and their respective pools.

This phase involves the identification and design of different service mixes associated with the product and applicable to the pool, dimensioning the business linked to each mix for different levels of servitization. This is achieved by first defining different service packs depending on the firm's objectives, and the characteristics of the pool's asset types and existing segments. Based on a series of available services, they will be grouped according to those that can be offered for the different types of assets, estimating the value added in each case. This will provide the potential for offering those packs to the whole pool.

Figure 9. Example of RAM-LCC for two compressors

ECONOMIC COSTS (WITHOUT COSTS OF POOR RELIABILITY)

	Option A	Option B	
Acquisition costs	450.000	300.000	USD
Operational costs	20.000	10.000	USD/year
Preventive costs	3.120	3.400	USD/year
Overhaul cost	10.000	5.000	USD each 3 years
i - money rate	0,10	0,10	‰
T - asset life	15	15	Years

	Option A	Option B	
Acquisition costs	450.000	300.000	USD
Operational costs	152.122	76.061	USD NPV for T years
Preventive costs	23.731	25.861	USD NPV for T years
Overhaul cost (year 3)	7.513	3.757	USD NPV T = 3 years
Overhaul cost (year 6)	5.645	2.822	USD NPV T = 6 years
Overhaul cost (year 9)	4.241	2.120	USD NPV T = 9 years
Overhaul cost (year 12)	3.186	1.593	USD NPV T = 12 years
CTCV(P)	646.438	412.214	

COSTS OF POOR RELIABILITY

	Option A	Option B	
Failure freq.	1,5	6	Failure/year
MTTR	15	10	Hours/failure
Penalty costs	673	1.010	USD/hour
Cf	10.100	10.100	USD/failure
TCPf	15.150	60.600	USD/year
PTCPf(P)	115.232	460.928	USD NPV

TOTAL COSTS

	Option A	Option B	
CTCV(P)	646.438	412.214	USD NPV
PTCPf(P)	115.232	460.928	USD NPV
TOTAL(P)	761.670	873.142	USD NPV

This will be completed with each segment's characteristics, which will delimit the packs that may be offered in each case (Van Ostaeyen, 2014):

- **Product-Oriented:** Ownership of the asset is conveyed to the customer, which is offered additional services;
- **Use-Oriented:** The manufacturer retains ownership of the asset, and the customer is offered services related to the asset according to a renting or leasing type of arrangement;
- **Result-Oriented:** The customer is provided with the asset's functional results, which meet its requirements.

Secondly, business scenarios will be defined and assessed based on the service packs. This involves the definition of different business scenarios with which to cater for the actual pool. These scenarios will be addressed through different business models (Osterwalder and Pigneur, 2010). The KPIs will then be calculated for each one of the scenarios and models.

MATURITY LEVELS AND ROADMAP

Based on an analysis of the questionnaires used in the case studies, three maturity levels have been identified in relation to the after-sales maintenance model proposed, as shown in Figure 10.

Level 1: Intervention Unit

The first level corresponds to a basic level that is required in capital goods firms in order to sell new machinery. Its main characteristics are as follows:

- The firms provide an after-sales service solely when asked to do so, e.g., failures.
- The after-sales service is reactive; there are no preventive services.

Figure 10. After-sales maintenance maturity levels

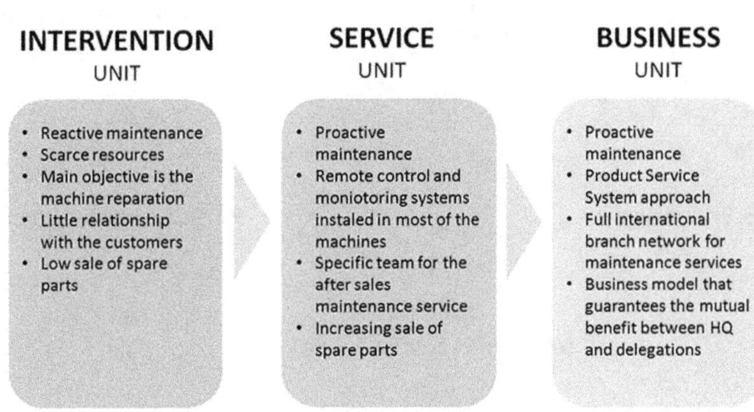

- Few resources are available for offering the service, which often means having to resort to resources both linked to production or commissioning of new machinery, with all the problems that this entails in terms of management and productivity..
- There is no liaising with the customer regarding the operating conditions of the machines installed in the latter's production processes.

The capital goods firms in this position (55% of those taking part in the study) look upon their after-sales service as a "necessary evil" when selling new products, considering that this activity belongs to a cost centre. The following spider chart provides a typical depiction of this level.

The diagram in Figure 11 contains a qualitative evaluation, on a scale of 1 to 5, for each one of the eight phases described in the previous section, rating their degree of development in the firm (i.e., what is done and how) as regards the objectives and techniques presented.

As Figure 11 shows, Phase 3 (i.e., Root-cause Analysis) is the one pursued the most in this group of firms. The reason for this is that, to a greater or lesser extent, all capital goods manufacturers have some kind of organisational mechanism (e.g., quality committees) that uses techniques for improving, through design, the operating reliability of their products, and therefore, reports of failures are reduced.

Level 2: Service Unit

The firms classified in this second level (30% of those taking part in the study) realise that their after-sales service may be a major business. Accordingly, their strategy includes taking major steps toward offering additional after-sales maintenance services to the pool of machines already sold. These services

Figure 11. Typical profile of an after-sales service in Level 1

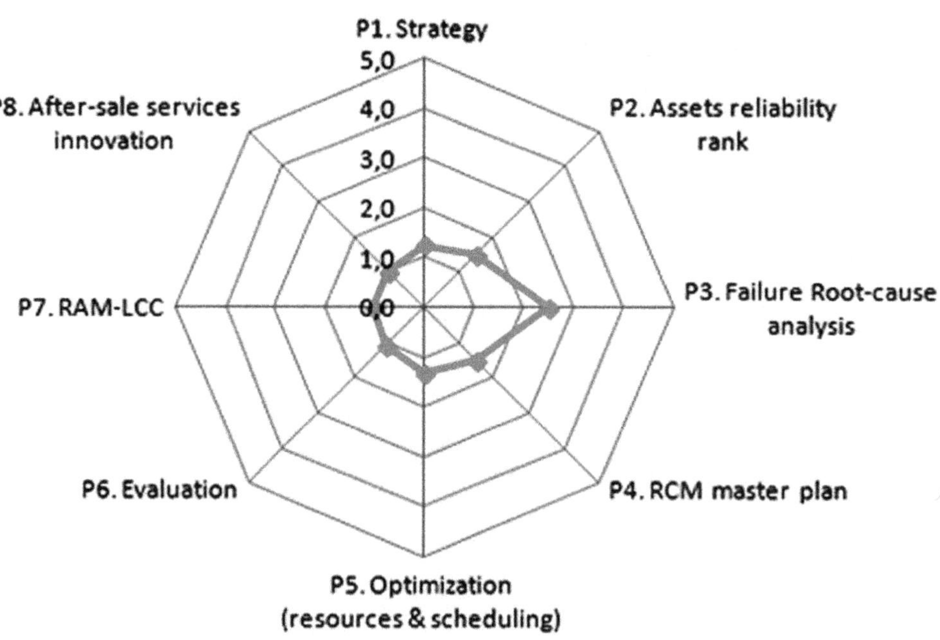

are both a recurrent income and a reinforcement of customer loyalty driver, which will impact on an increment of the future sales of new machines. The main characteristics of these firms are as follows:

- The after-sales service has a proactive approach.
- Remote control and monitoring systems are installed in machines bound for main and distant markets.
- There is a team responsible for developing the relationship with the customer.
- Spare parts are systematically offered before failure events occur.

As this figure shows, these kinds of manufacturers have chosen, from a strategic perspective (Phase 1), to generate recurring incomes through the after-sales service provided for the pool of machines they have sold. Nevertheless, given that this process is still in its infancy, the data used to prioritise the critical systems in their equipment (Phase 2) are still scarce, and so their approach is generic, qualitative, and based on the subjective experience of the maintenance staff. This situation extends to the elaboration of master plans for their machines (Phase 4). Nevertheless, significant progress is indeed being made in the systemisation of operating rules-procedures, as well as in the generation of work orders and in the incorporation of technology for the remote monitoring of systems, sub-systems and critical components. Concerning the dimensioning of resources and the scheduling of tasks (Phase 5), they tend to be found in the installation levels of a CMMS and in the expansion of a worldwide collaborating network with service companies. The subsequent phases (i.e., Phases 6, 7 and 8), which depend on the preceding ones, tend to be in their preliminary stages.

Figure 12. Typical profile of an after-sales service in an initial Level 2

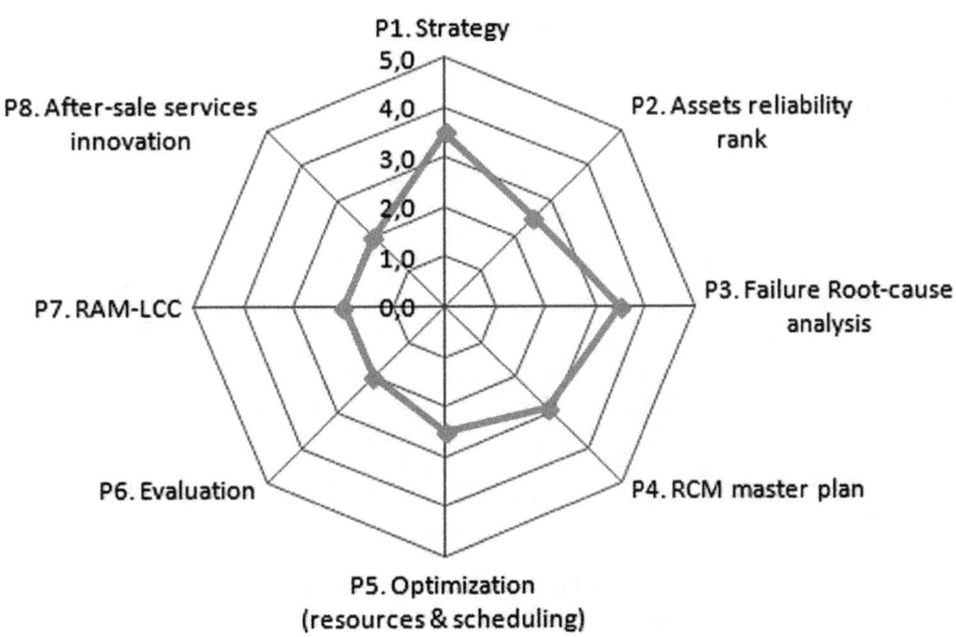

Level 3: Business Unit

The manufacturers in this level (15% of those taking part in the study) use their after-sales service to provide, in addition to those services on demand, other ones on a proactive basis (e.g., spare parts, preventive maintenance, retrofitting, contracts on the availability of machines, renting and leasing contracts, and enhanced process efficiency). The main characteristics of organisations belonging to this level are the following:

- The after-sales service is proactive.
- The sale of services is an integral part of the sale of machinery.
- There is an international technical, and commercial, service network that caters all markets.
- There is a proprietary business model that guarantees the joint benefit of the parent company and its delegations.

The following spider chart depicts the initial moments in this level.

These manufacturers have a clear strategy for their after-sales services defined as a business (Phase 1), with the corresponding income and operating profit being on a par with, or even exceeding, the figures obtained from the sale of new machinery. The automated gathering of data through work orders in electronic format enables many of these manufacturers to have updated families of criticality matrices segmented by operating conditions (Phase 2). The previous point permits drafting maintenance master plans (Phase 4) with a greater degree of customisation, with some of them having configuration tools for improving the maintenance tasks personalization. Regarding Phase 5, each CMMS is continuously fed with service data, which means that the scheduling of tasks and the level of stock of critical spare

Figure 13. Typical profile of an after-sales service in an initial Level 3

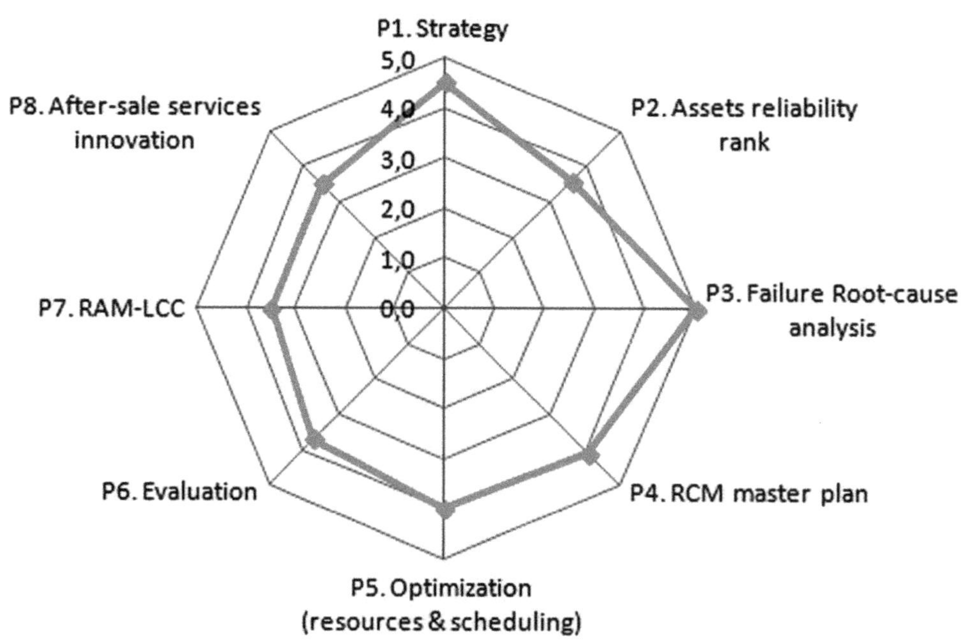

parts may be optimizing by the sequencing and grouping of coordinated activities with subcontracted services companies, as well as stock policies that besides the economic batch take into account preventive alerts for on-demand procurement. The availability of the latest data on the service allows conducting a performance analysis (Phase 6), as well as calculating the cost of unreliability for different time horizons (Phase 7). As described in Phase 8, in the previous section, this enables manufactures to open their catalogue of after-sales services to a broad range of contracts involving preventive maintenance, spare parts, availability contracts, retrofitting, renting and leasing services, etc. for the entire pool of machines sold. In fact, those manufacturers that have advanced the most in this level are even receiving requests for after-sales services (e.g., preventive maintenance) for their competitors' machines operating in their customers' production processes.

Roadmap

The three levels described above have been used to draw up a road map that identifies the progressive developments to be undertaken to enable capital goods manufacturers in lower levels to evolve, and thus be able to exploit the major business opportunities involved in after-sales services.

As noted previously, those manufacturers in *Level 1*, or just entering *Level 2*, lack any quantitative and objective prioritisation of their machine's failure modes. The reason for this is that these manufacturers do not gather structured information of the services they provide, but this it is scattered around in different paper folders, notes, unlinked spreadsheets, etc. This means that these kinds of manufacturers are not in a position either to draw up after-sales maintenance plans that are specifically customised for their own machines (i.e., mix of corrective, preventive and predictive maintenance), and likewise lack the necessary inputs for properly dimensioning the required resources (e.g., maintenance staff, specialists, spare parts, collaborations with services companies worldwide) for deploying the service effectively and efficiently.

The first logical step proposed is therefore to begin defining the criticality matrix of those machines sold to preferred customers that are in a position to pay for after-sales maintenance services. This will require starting to code and classify the failure modes of a machine's sundry subsystems according to their severity-impact, based on the log record of data available (e.g., physical and digital files), as well as on the tacit expertise of after-sales technicians. This first version of the criticality matrix will be fine-tuned after implementing more systematic methods of failure and intervention data recording.

This initial version of the criticality matrix may in turn be used to design a first version of the after-sales maintenance plan (i.e., mix of corrective-preventive-predictive tasks according to the criticality of the failure modes) for those sold machines in the pool that are to be serviced. In addition, the parameters to be monitored remotely are to be defined, along with the technological infrastructures and requirements for doing so. The initial draft of the after-sales maintenance plan will also be fine-tuned over time in response to any restatements of the criticality matrix, with the outcome being a catalogue of plans tailored to the machine's operating conditions.

These after-sales maintenance master plans may be used for an initial dimensioning of the resources (i.e., maintenance technicians and spare parts) required for attending to the target pool of machines for a pre-set level of service. The next step will involve drawing up the schedule of tasks to be performed on the pool, with specification of the policies on restocking critical spare parts, to ensure the utmost efficiency of the resources available.

The next step will involve the structured collection (e.g., using CMMS applications) and processing of the data generated once the service has been deployed: both the data gleaned from the work orders generated by these operations and the alerts triggered automatically by remote monitoring systems of the impaired performance of critical components.

The processing of these data will be the key for generating information that will inform the decisions to be made for improving the service indicators, as well as for analytically detailing the unreliability costs generated by the machines in the target pool. As noted in the description of the third maturity level (i.e. *Business Unit*), this will enable capital goods manufacturers to offer high value added after-sales services, generate recurring incomes, as well as reinforce customer loyalty, which facilitates future machines sales when they come to be obsolescent.

The next figure provides a schematic representation of the timeframe for the developments described here. The time horizons specified in the figure are based on the analysis of the case studies, and should be considered no more than a guideline.

CONCLUSION

The case studies conducted have led to the identification of an as-yet unexploited high business potential within the field of after-sales services for manufacturers of machinery and capital goods in the Basque Country (Spain). In terms of the sector as a whole, this region is the EU's third largest producer (SPRI, 2015).

These case studies have allowed adapting a reference framework for industrial maintenance (Crespo, 2007) to the field of after-sales maintenance. Furthermore, this new framework has shown how to profit from investments in Industry 4.0 technologies through the ambit of industrial after-sales services; Service 4.0.

Nevertheless, there have also been identified certain challenges. For instance, from a *cultural viewpoint*, some capital goods manufacturers are completely focus on the continuous technical improvement of their machines and just view their after-sales maintenance service as a problem and source of costs, and not as a business opportunity and potential source of incomes. Secondly, from a *processes and organisational*

Figure 14. Project roadmap based on the phases of the after-sales model

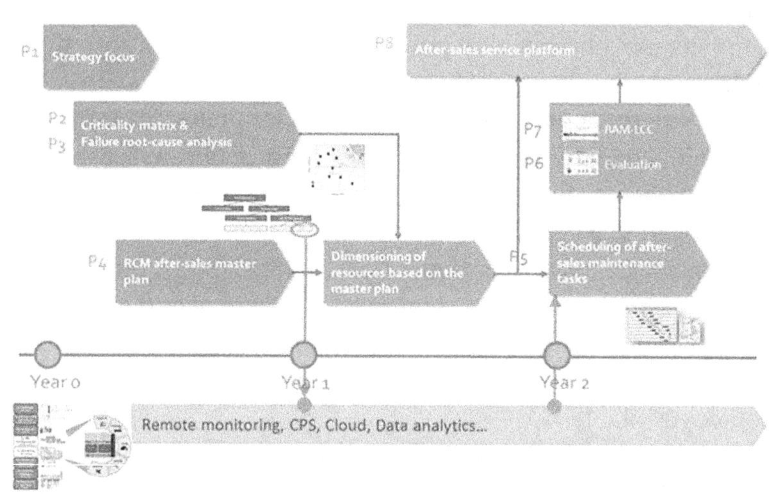

viewpoint, capital goods' machinery usually shows very high levels of customisation, hence, preventive maintenance master plans of a certain models need to be taken as a starting point, and adapted in each case to the idiosyncrasies of each individual instance and operating conditions. Also, the dimensioning of the resources required for attending to the pool must consider not only the volume of maintenance personnel and spare parts needed for providing the necessary level of service, but also the more suitable location of service sittings and spare parts warehouse. Likewise, the scheduling of tasks, and especially their grouping and sequencing, should take into account logistics aspects that are significantly different to those of industrial maintenance. In third place, and from a *technical viewpoint*, the case studies have also revealed that for some critical parts of the machine, when there isn´t historical failure data and the physical degradation model isn´t well known, a reliable failure prediction is quite difficult to perform.

The research performed has identified three maturity levels and a road map that will help manufacturers of machinery and capital goods on the path to making value proposals that will go beyond merely selling the physical product itself. As propounded, this expansion would begin by incorporating processes associated with the design, deployment and enhancement of after-sales maintenance services (e.g., sale of spare parts and preventive maintenance). The information generated would then open the door to other types of industrial after-sales services for the entire pool of machines sold, which may constitute major sources of recurring income and reinforce customer loyalty for the future sale of machinery, e.g., availability contracts, retrofitting of machines, and renting and leasing contracts.

REFERENCES

Baines, T., Lightfoot, H. W., Evans, S., Neely, A., Greenough, R., Peppard, J., & Tiwari, A. et al. (2007). State-of-the-art in product-service systems. *Journal of Engineering Manufacture*, *221*(10), 1543–1552. doi:10.1243/09544054JEM858

Baker, B. A., Manan, A., & Husband, T. M. (1997). Simulating maintenance work in an engineering firm: A case study. *Microelectronics and Reliability*, *16*(5), 571–581. doi:10.1016/0026-2714(77)90288-8

Berdinyazov, A., Camci, F., Sevkli, M., & Baskan, S. (2009) Economic analysis of maintenance policies for a system Diagnostics for Electric Machines. *Proceedings of theIEEE International Symposium onPower Electronics and Drives (SDEMPED '09).*

Bloom, N. (2006). *Reliability Centered Maintenance. Implementation made simple.* New York: McGraw-Hill Inc.

Campbell, J. D., & Jardine, A. (2001). *Maintenance Excellence*. New York: Marcel Dekker.

Cohen, M. A., Agrawal, N., & Agrawal, V. (2006). Winning in the Aftermarket. *Harvard Business Review*, *84*(5), 129–138. PMID:16770900

Crespo, A. (2007). *The Maintenance Management Framework: Models and Methods for Complex Systems Maintenance*. Springer Series in Reliability Engineering.

Crespo, A., Moreu de León, P., Sola, A., & Gómez, J. F. (2015). Criticality Analysis for Maintenance Purposes: A Study for Complex In-service Engineering Assets. *Quality and Reliability Engineering*, *02*. doi:10.1002/qre.1769

Duffuaa, S. O. (2000). Mathematical models in maintenance planning and scheduling. In M. Ben-Daya, S. O. Duffuaa, & A. Raouf (Eds.), *Maintenance, Modelling and Optimization*. Boston: Kluwer Academic Publishers. doi:10.1007/978-1-4615-4329-9_2

Gilabert, E., Fernandez, S., Arnaiz, A., & Konde, E. (2015) Simulation of predictive maintenance strategies for cost-effectiveness analysis, *Proceedings of the Institution of Mechanical Engineers, Part B: Journal of Engineering Manufacture*. doi:10.1177/0954405415578594

Huiskonen, J. (2001). Maintenance spare parts logistics: Special characteristics and strategic choices. *International Journal of Production Economics*, *71*(1-3), 125–133. doi:10.1016/S0925-5273(00)00112-2

Jones, R. (1995). *Risk-based management: A Reliability-Centered Approach* (1st ed.). Houston, Texas: Gulf Publishing Company.

Kaplan, R. S., & Norton, D. P. (1992). The Balanced Scored – measures that drive performance. *Harvard Business Review*, *70*(1), 71–79. PMID:10119714

Marshall Institute. (1999) *Maintenance Effectiveness Survey*. Retrieved from http://www.marshallinstitute.com/inc/eng/consulting/areas/customized_reliability_solutions/body/measurvey.pdf

Mobley, K. (1999). Root Cause Failure Analysis. Butterworth-Heinemann, Reed Elsevier Group.

Moubray, J. (1991). *RCM II: Reliability-Centered Maintenance*. New York, USA: Industrial Press Inc.

Osterwalder, A., & Pigneur, Y. (2010). *Business model generation: a handbook for visionaries, game changers, and challengers*. John Wiley & Sons.

Parra, C., Crespo, A., Kristjanpoller, F., & Viveros, P. (2012). Stochastic model of reliability for use in the evaluation of the economic impact of a failure using life cycle cost analysis. Case studies on the rail freight and oil industries. *Journal of Risk and Reliability*, *226*(4), 392–405.

Parra, C., & Omaña, C. (2001). Técnica de auditoría de la gestión del mantenimiento para el sector Refinación. *Proceedings of theVII Congreso de Ingeniería de Mantenimiento de Petróleos de Venezuela*, Caracas.

Patton, M. Q. (2001). *Qualitative Research & Evaluation Methods* (3rd ed.). SAGE Publications, Inc.

Persson, F., & Saccani, N. (2009). Managing the after-sales logistic network–a simulation study. *Production Planning and Control*, *20*(2), 125–134. doi:10.1080/09537280802707530

Ramani, S., & Krishnan Kutty, K. V. (1985). Management of multi-item, multi-group inventories with multiple criteria under service level constraints. *Engineering Costs and Production Economics*, *9*(1), 59–64. doi:10.1016/0167-188X(85)90010-2

Rausand, M., & Høyland, A. (2004). *System reliability theory: models, statistical methods, and applications* (Vol. 396). John Wiley & Sons.

Saaty, T. L. (1990). How to make a decision: The analytic hierarchy process. *European Journal of Operational Research*, *48*(1), 9–26. doi:10.1016/0377-2217(90)90057-I

Sagarna, I., Uribetxebarria, J., & Castellano, E. (2015, 19-20 November). Optimization of offshore wind parks maintenance service strategies using agent based simulation techniques.*Proceedings of the 4th International Conference on Business Servitization (ICBS)*, Madrid, Spain.

Shenoy, D., & Bhadury, B. (1998). *Maintenance resources management: Adapting MRP*. London: Taylor and Francis.

SPRI. (2015) *Business Development Basque Agency: Machinery and Equipment Sector*. Retrieved from http://www.spri.eus/en/invest-in-the-basque-country/significant_sectors/machinery_equipment_sector

Tersine, R. J. (1994). *Principles of Inventory and Materials Management* (4th ed.). Englewood Cliffs, NJ: Prentice-Hall.

Van Ostaeyen, J. (2014). *Analysis of the business potential of product-service systems for investment goods* [PhD thesis]. KU Leuven, Faculty of Engineering Science.

Vandermerwe, S., & Rada, J. (1988). Servitization of Business: Adding Value by Adding Services. *European Management Journal*, *6*(4), 314–324. doi:10.1016/0263-2373(88)90033-3

Wilson, R. H. (1934). A Scientific Routine for Stock Control. *Harvard Business Review*, *13*, 116–128.

Woodhouse, J. (1993). *Managing industrial risk*. London: Chapman Hill.

Woodward, D. G. (1997). Life cycle costing theory. Information, acquisition and application. *International Journal of Project Management*, *15*(6), 335–344. doi:10.1016/S0263-7863(96)00089-0

Zio, E., & Compare, M. (2013). Evaluating maintenance policies by quantitative modelling and analysis. *Reliability Engineering and System Safety, Elsevier*, *109*, 53–65. doi:10.1016/j.ress.2012.08.002

KEY TERMS AND DEFINITIONS

After-Sales Maintenance: All those activities (e.g., verifications, measurements, replacements, adjustments and repairs, etc.) offered by a manufacturer to ensure that, once sold and installed in a customer's production process, the asset made (e.g., machinery) performs its function in a due and proper manner.

Computerized Maintenance Management Information System (CMMS): Software package that maintains a computer database of information about an organization's maintenance operations.

Condition Based Maintenance (CBM): Maintenance that is performed after some indicators show that equipment is going to fail or that equipment performance is deteriorating.

Cyber-Physical System (CPS): System of collaborating computational elements controlling physical entities, e.g., sensor-based communication-enabled autonomous systems.

Industry 4.0: Term embracing a number of contemporary automation, data exchange and manufacturing technologies that allow; interoperability between humans, CPSs and smart factories, virtualization of physical processes, decentralization and autonomous CPS decision making processes, real-time capabilities, internet of services, modularity, etc.

Industrial Maintenance: All those activities (e.g., verifications, measurements, replacements, adjustments and repairs, etc.) needed to ensure an industrial installation or asset continues to perform its function in a due and proper manner.

Mean Down Time (MDT): Is the average time that a system is non-operational, including all the downtime associated with repair, corrective and preventive maintenance, self-imposed downtime, and any logistics or administrative delays.

Mean Time Between Failures (MTBF): Describes the expected time between two failures for a repairable system.

Mean Time To Repair (MTTR): It represents the average time required to repair a failed component or device. It is a basic measure of the maintainability of repairable items.

Product Service System (PSS): Mix of both products and services, in comparison to the traditional focus on just physical products, capable of jointly fulfilling a user's need.

ENDNOTES

1 Fagor automation cnc, fagor automation motores, ikusi

2 Fagor industrial, tifell, azkoyen

3 ABC compressors, urola, alstom, fagor arrasate presses and stamping systems, fagor arrasate blanking systems, lanbi, lazpiur, loramendi, caf, gh, lonxi, ulma handling

4 Ulma forklift

5 Ondoan

Chapter 8
Compatibility Welding Parameters with the Results Obtained in Testing of Fracture Mechanics in HSLA Steel

Rafael González-Palma
University of Cádiz, Spain

Carlos López-Escobar
Independent Researcher, Spain

María Carmen Carnero
University of Castilla – La Mancha, Spain & University of Lisbon, Portugal

David Almorza
University of Cádiz, Spain

Pedro Mayorga
EnerOcean S.L., Spain

ABSTRACT

Many investigations led to show that no crack begins to propagate to an increase of stress intensity factor. The life of the components of a structure containing premature cracks, can be governed by the degree of subcritical crack propagation. Thus, knowledge of crack propagation to determine the fatigue of the structure is necessary. One problem of steels of high resilience is their low toughness in the HAZ, when they are welded with a high heat input. In this work we have studied nine specimens that have been welded under a submerged arc welding process controlling the welding parameters and checking in the HAZ of such specimens, critical tensions at the ends of the cracks, the critical cracks lengths and stress intensity factors. It is intended to check that the parameters that indicate the values of fracture mechanics in the HAZ, after heat cycle to which the steel has undergone, under a process with a maximum heat input of 2.327kJ /mm, they are still valid, with the welding parameters applied. It is checked a correlation between the theoretical and experimental values.

DOI: 10.4018/978-1-5225-0651-5.ch008

BACKGROUND

Great deal investigations of crack propagation by fatigue led to show that no crack propagation begins until an increase of stress intensity factor, ΔK_{th}, threshold as discussed depends on several known factors.

Initial investigations that were conducted by Frost (1960) showed a low degree of slowdown in growth due to fatigue at low stress, the data provided offered the presumption that below certain values, the crack does not spread. The existence of this threshold, was predicted by Clintock (1963) works under elastic–plastic analysis.

Later Lindner (1965) and Paris (1970) showed that the propagation of fatigue crack can be analyzed in the context of linear elastic fracture mechanics, and the threshold ΔK_{th}, can be determined.

In the 60s, Elber (1971) demonstrated that fatigue cracks interfere each other via a locking mechanism. It was Schmidt (1973) who later it showed that the locking mechanism has a significant effect on the threshold behavior. The information provided led to indicate that there are four locking cracks.

Factors that may inGCHuence the threshold crack propagation by fatigue are: yield strength, grain size and other microstructural elements as residual stresses, CTOD mode, Young's modulus, and environmental temperature. The effects of most of these factors have on ΔK_{th}, they can be explained by its relationship with the locking mechanisms of the crack.

Researchers as Kitagawa and Takahashi (1976), El Haddad, Dowling, Topper and Smith (1980), found a constant value of ΔK_{th} for crack lengths greater than 0.5 mm.

For our steel EMZ 450, belonging to ferritic-pearlitic according to Barsom (1974):

$$\Delta K_{th} = 7(1 - 0.85R)\ \mathrm{MN/m^{3/2}} = 6.405\mathrm{MN/m^{3/2}} = 202.5\ \mathrm{N/mm^{3/2}} \quad (1)$$

Life of a structural element fatigue can be considered from three different states:

1. Initiation of fatigue crack.
2. Crack Propagation.
3. Fracture.

The life of the components of a structure containing premature cracks, can be governed by the degree of subcritical crack propagation. To this end, the many destructive and non-destructive testing can collaborate to establish any cracks before commissioning. Thus, knowledge of crack propagation through the concepts of fracture mechanics calculation to determine the fatigue of the structure is necessary.

Most testing the growth of fatigue cracks are performed on samples subjected to load fluctuations at constant amplitude. The lengths of the cracks to the cycles that are undergoing represented graphically as measured in Figure 1.

Another observable effect on the specimens, is relating the initial size of the crack with life or number of cycles, noting that the higher the initial crack length, fewer life cycles Figure 2.

The curves plotted in Figure 1 show that the greater the increase in the cyclic load fluctuation less is the life of the specimen by fatigue.

Thus, by varying the cyclic load fluctuation and initial crack size, different curves can be drawn $a =$ f (N). (Crack length according to the cycles).

However, the curves can be summarized in a single, when expressed in terms of speed of crack growth, depending on the amplitude of the stress intensity factor which it has no geometrical dependence and

Figure 1. Effect of cycle-stress range on crack growth

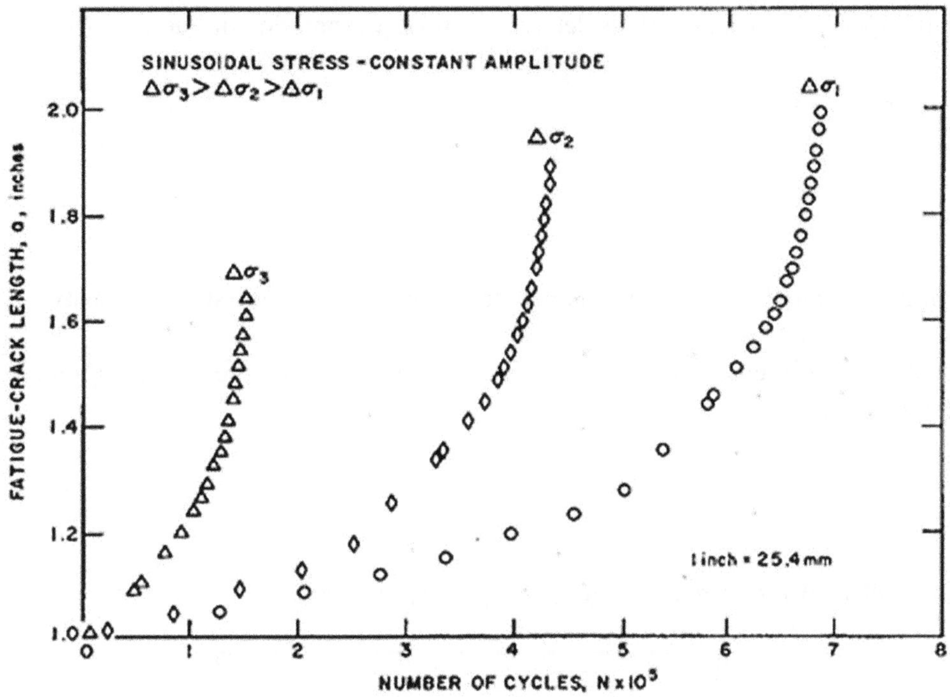

Figure 2. Effect of initial crack length on crack growth

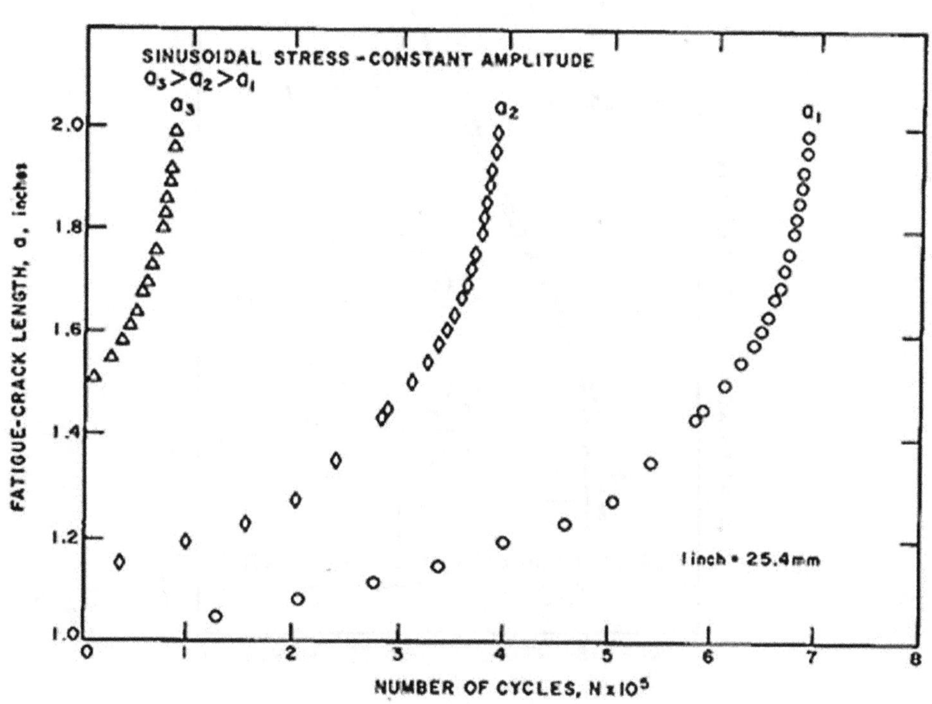

summarized in a single parameter the ($\Delta\sigma$) and crack length (a); being, $\Delta K = Y(\Delta\sigma)\sqrt{\pi a}$ normal curve representing the speed of crack growth, is determined by the expression of Paris (1970):

$$\frac{\mathrm{d}a}{\mathrm{d}N} = A\left(\Delta K\right)^{m}$$

where:

a = crack length, N = number of cycles, ΔK = range or amplitude of the stress intensity factor.

"A" and "m" are constants depending on the material. This curve is represented in Figure 3, it consists of three different parts:

Figure 3. Stress intensity factor range- crack growth rate per cycle

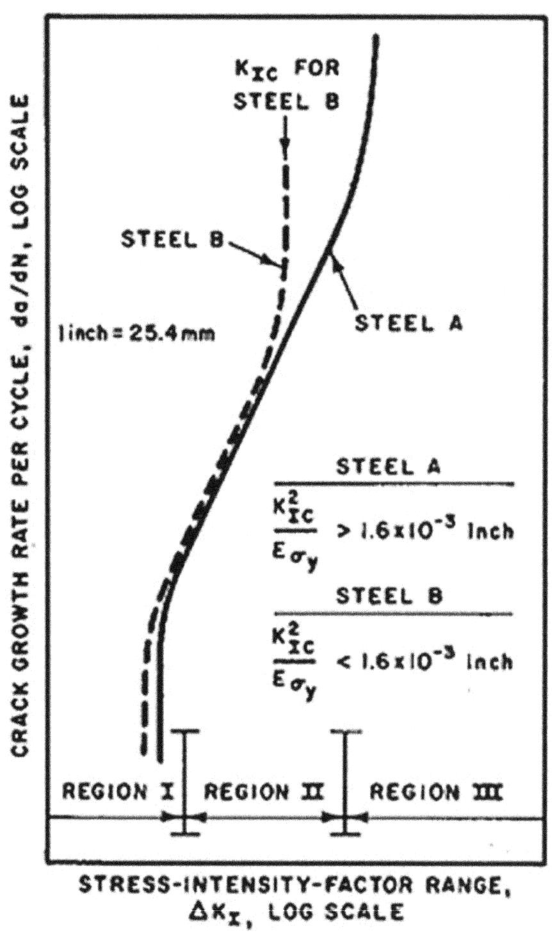

- **Region I:** Displays the threshold of fatigue fracture in which the crack does not spread under cyclic stress fluctuation (Frost, 1960; Paris, Bucci, Wessel, Clark, & Mager, 1972).
- **Region II:** Sample propagating fatigue crack above the fluctuation threshold stress intensity factor ΔK_{th}, in this region.

$$\frac{\mathrm{d}a}{\mathrm{d}N} = A\left(\Delta K\right)^{m}, \text{(Barsom, 1971)} \tag{2}$$

- **Region III:** Region where the growth cycle fatigue is greater than in the region II above. The data which the authors Barsom, (1974), Harrison (1970) and Silverio et al.(2003) present in this region show that the growth of a fatigue crack is increased from zero stress to the load, namely $\Delta K = K_{max}$.

This increase occurs at a constant value of displacement of the end of the crack δ_T and a value corresponding to the stress intensity factor, K_T, which can be determined by the following expression:

$$\delta_T = \frac{K_T^{\ 2}}{E\sigma_{ys}} = 1.6 \times 10^{-3}\text{inches } \left(0.04064 \text{ mm}\right) \tag{3}$$

Being:

- E = Young's Modulus
- K_T = The range of the stress intensity factor, corresponding to the beginning of accelerated growth of the fatigue crack.
- σ_{ys} = Yield stress that is closer to the yield point, and that can be taken as the average of the yield point and yield stress.

Acceleration the growth rate of fatigue crack, it is determined by the transition between the region II and the region III. This seems to be caused by a combination of a ductile tearing mechanism, on a cyclic mechanism subcritical crack extension, which leaves a fatigue striations on the fracture surface. Crack tearing occurs when the crack reaches a critical value. Thus the fatigue rate of the transition from region II to region III depends K_{max} and stress ratio R.

PURPOSE OF TEST

450 EMZ steel is known and proven by tests supplied by the mill, which at the beginning of the crack and during propagation obey the design parameters (σ_{cr}, a_{cr}, K_{IC}, da/dN =A (ΔK), etc,).

However, after the thermal cycle in a welding process, material suffers and the mechanism of the microstructure of the base metal can change, being able to modify the aforementioned parameters in such a way and to such an extent that if not properly controlled, could lead to a catastrophic failure of the structure.

In experimental work carried out here, we intend to ensure that these parameters governing fracture mechanism, remain valid under the rules compatible with the design of the base material and that as far as possible a correlation is established between welding parameters and those obtained in fracture tests, so that if the results of the tests are not satisfactory, we could apply the appropriate solution for welding in order to that parameters governing the fracture tests are acceptable. Also emphasis will be placed on the behavior of fatigue fracture after a welding cycle to verify the validity of its parameters.

PREPARATION OF THE SAMPLES PRIOR TO TESTING

Cross sections to welding were obtained in order to get some micrographs that allowed us analyze metallographically the grain size in different parts of the HAZ. These sections were obtained at the beginning and end of the welded panel.

Before being observed by the optical microscope specimens were polished and attacked with a solution of nital at 2%, which allowed us to identify different zones along the fusion line by extending at least five increases in the microscope optical.

The grain size evaluation was carried out by the method of HEYN in accordance with the ASTM (112-63) standard. Grain sizes were classified in the heat affected zone (HAZ) in the following categories, (Leiva, Donoso, Muehlich, & Labbe, 2004):

- Coarse area, called GCHAZ (grain course heat affected zone) and according to ASTM are those grains having size less than 5, i.e., over 65 µm.
- GCHAZ as above but with grain size of ASTM 5-7, that is, between 64 and 32 µm.
- Fine-grained zone, FGHAZ (fine grain heat affected zone) and whose grain size upper of ASTM 7, i.e., less than 32 µm.
- Intercritical zone it is so called as the peak temperature is between AC1 <Tp <AC3 and grain size between 9 ASTM (15µm) and 11 ASTM (7 µm).
- Subcritical area where Tp <AC3 and size of grain above 11 ASTM and below 7 ASTM.
- Base metal 12 ASTM grain size, i.e., 12 (5µm).

The different zones at different points thickness were marked and measured according to ASTM 112 and percentages of fine grain size measured in the entire thickness and in 2/3 of the thickness (central 2/3 = 100%), this last condition is in accordance with the EEMUA (158) standard requirements.

Macrograph coupon initially obtained (welded panel), and its grain size corresponds to Figure 4.

Of said coupon, nine samples were extracted in the preferred geometry *B* x 2*B* as Methods for Crack Opening Displacement C.O.D. (Testing) (1.979). In Figure 5 measured percentages of the various ASTM grain sizes and CTOD values for the specimen 1 are shown.

Mapping of fatigue fracture of the specimen 1 in the HAZ conditions AW

The nominal dimensions were:

- B= thickness =60 mm
- W = width = 120mm = 2B
- L = length = 630 mm

Figure 4. Micrograph of the specimen section (left). Schematic representation of grain size and its percentage of the HAZ (right).

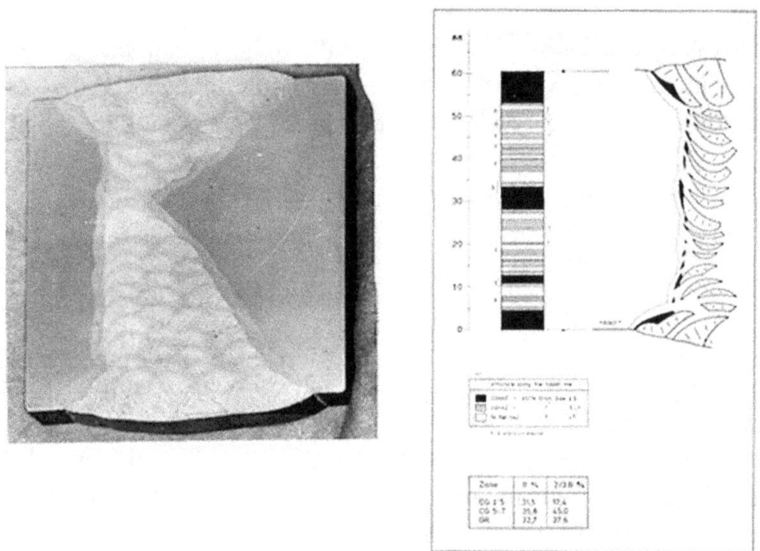

The cross sections were taken from the ends of the coupon of each of the samples were polished and attacked with a solution of nital 5% to allow the location of the notch on both sides and parallel zone to the fusion line, i.e., coarse grains area in the HAZ.

Given the difficulty of drawing a straight line completely along the area where the successive welding beads have not refined the grain, as GCHAZ, this line where is the notch includes IRGCHAZ (Intercritical reheated grain coarse heat affected zone) and SC (subcritical zone).

Figure 5. Microestructure size ASTM

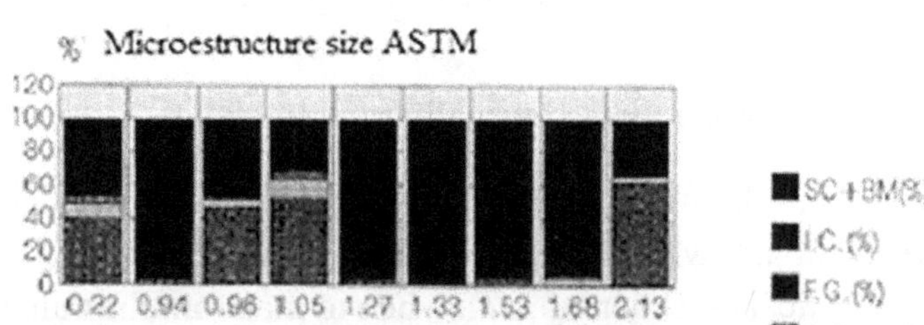

C.T.O.D.(mm)	0.22	0.94	0.96	1.05	1.27	1.33	1.53	1.68	2.13
SC+BM(%)	0	16.3	0	0	6,1	0	4.9	9.4	0
I.C.(%)	17	21.6	4	0	5,1	9.4	5,4	17.5	3.3
F.G.(%)	30.5	58.1	43.4	31,4	84,9	90.6	85,1	68	30.9
G.C.7-9(%)	4.7	4	1.3	5,4	3,8	0	4.6	2.9	0
G.C.<7(%)	7.8	0	5	12	0	0	0	4.2	3.8
W.M.(%)	40	0	46.3	51,3	0	0	0	0	61.9

C.T.O.D.(mm)

In all cases the machined notch was placed at the center of both regions. The onset of crack consisted of a machined notch of 2 mm wide and 50 mm deep finished in a V at 45 ° to facilitate the onset of fatigue crack.

Since the panel was tested in conditions "as welded" without heat treatment, a lateral pre-compression of the specimen 0.5% of its thickness was applied to ensure compliance as BS 5762 and better observation of stable growth, this compression was applied before the precracking by fatigue

PRECRACKING BY FATIGUE

Introduction

One problem that design of these steels of high resilience have shown is their low toughness in the HAZ, when they are welded with a high HI (heat input).

In this work we have studied nine specimens that have been welded under a SAW (submerged arc welding) process controlling the welding parameters and checking in the HAZ of such specimens, critical tensions (σ_{cr})at the ends of the cracks, the critical cracks lengths (a_{cr}) and stress intensity factors (K_{IC}). (Silverio, Moura, & Guedes, 2003).

It is intended to check that the parameters that indicate the values of fracture mechanics: $\left(\sigma_{cr}, a_{cr}, K_{IC}\right), da / dN, da / dN = A\left(\Delta K\right)^{m}$, etc.,) (Kujawski, and Sadananda, 2011) in the HAZ, after heat cycle to which the steel has undergone, under a process with a maximum HI of 2.327 kJ / mm, they are still valid, with the welding parameters applied.

It is checked a correlation between the theoretical values and those obtained experimentally.

Experimental Work

After the machined notch 50 mm deep it was proceeded to a lateral precompression, continuing the fatigue testing of specimens to ambient laboratory temperature in resonance pulsator 200 KN Schenck, with a ratio $R = \frac{P_{min}}{P_{max}} = 0.1$, being the span between supports four times the width of the test piece (*W*). The crack propagation (a) was followed visually stopping when $\frac{a}{W} \approx 0.5$.

The experimental values of the stress intensity factor based on the number of cycles, for each of the nine samples are shown in Table 1, where (*W*) is the width of the specimen and (*a*) the length of the crack.

Is represented by (a_f) critical crack length, i.e., when to maintain the value of the constant "P" load the crack extends to breakage and (a_{cr}) length corresponding to the first part of the crack extension, i.e., crack growth fatigue. The difference between values (a_f) and (a_{cr}) determines the stable growth (δ). In the last column *N* x1000 are represented number of cycles required to reach the crack length (*a*). By (K_{fmax}) is represented the maximum stress intensity factor, during the vibration test corresponding to the crack formation (a_{cr}), all values represented in Table 1 have been taken from experimental CTOD tests. These values were taken in the HAZ, and we will compare them with the base metal values, such as steel 450 EMZ should have under the point of view of design. If we start from the specimen with three-point

Table 1. Fracture propagation at"N" cycles

Specimen	Heat Input	Crack Depth		Relation Crack Specimen Width	Stress Intensity Factor K_{fmax}	Cycles
	kJ/mm	a_{cr} (mm)	a_f (mm)	a/W	N/mm $^{3/2}$	Nx1000
1	2.298	61.1	62.4	0.509	2320	74
2	2.298	60.9	64.5	0.506	2270	67
3	2.298	60.5	61.1	0.502	2250	69
4	2.298	60.2	61.2	0.499	2230	71
5	2.298	61.5	62.9	0.510	2320	76
6	2.298	58.5	59.5	0.488	2150	81
7	2.327	59.8	60.0	0.491	2090	67
8	2.327	60.5	61.2	0.503	2180	63
9	2.327	59.5	59.6	0.487	2070	62

bending and supporting on Elices-Calafat (1992) curves $(\frac{a}{W}, \frac{K_I}{K_0})$ Figure 6, the following empirical function is obtained:

$$\frac{K_I}{K_0} = 1.09 - 1.73\left(\frac{a}{W}\right) + 8.2\left(\frac{a}{W}\right)^2 - 14.2\left(\frac{a}{W}\right)^3 + 14.6\left(\frac{a}{W}\right)^4 \quad (4)$$

where

Figure 6. Flection on three points

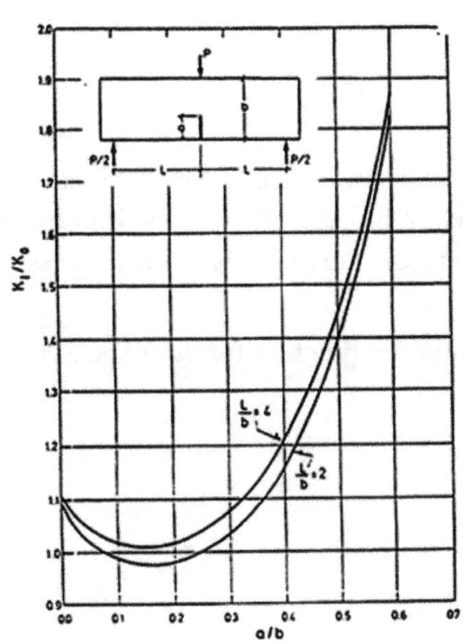

$K_0 = \sigma\sqrt{\pi a}$; and

$$\sigma = \frac{My}{I} = M\frac{W}{2}\frac{1}{\frac{W^3}{12}} = \frac{6M}{W^2}$$

Since the beam section is W×1, these values are substituted in:

$$K_0 = \sigma\sqrt{\pi a} = \frac{6M}{W^2}\sqrt{\pi a}\ ;$$

where:

$$M = \frac{P \times 2W}{2} = P \times W$$

From Table 1 it follows that the ratio $a / b = 0.5W / W$ is approximately 0.5, this value entered in the abscissas axis and for the curve $L / b = 2W / W = 2$ in Figure 6,

K_I / K_0 takes a value of 1.4, then:

$$K_I = 1.4K_0 = 1.4\sigma\sqrt{\pi a}$$

If we start from the stress intensity factor; and we represent in Table 2 the values of σ, by this expression, the following values are obtained for (a_{cr}):

$$\sigma = \frac{Kf_{\max}}{1.4\sqrt{\pi a_{cr}}} = \frac{Kf_{\max}}{2.481\sqrt{a_{cr}}} \Rightarrow a_{cr} = \left(\frac{Kf_{\max}}{2.481\sigma}\right)^2 \tag{5}$$

In Table 2 are represented the design stress values versus critical crack. Values obtained in this Table show that the design stress in the HAZ based on the values from the crack obtained in the fatigue tests are acceptable values because they are about 58% of the stress corresponding to yield strength $\sigma_{ys} = 483\ N / mm^2$.

From these values, we can obtain the function for each specimen $\sigma = f(a)$ that with these results obtained and the equation:

$$\sigma = \frac{Kf_{\max}}{1.4\sqrt{\pi}}a^{-1/2} \tag{6}$$

Table 2. Cracking stress/yield strength

Specimen	σ_{ys} N/mm²	K_{fmax} N/mm$^{3/2}$	a_{cr} (mm)	σ N/mm²	σ/σ_{ys} %
1	483	2320	11.1	280.67	58.1
2	483	2270	10.9	277.13	57.4
3	483	2250	10.5	279.87	57.9
4	483	2230	10.2	281.43	58.3
5	483	2320	11.5	275.75	57.1
6	483	2150	8.5	297.24	61.5
7	483	2090	9.8	269.10	55.7
8	483	2180	10.5	271.17	56.1
9	483	2070	9.5	270.70	56.0

are obtained for each of them the quantities shown in Table 3. From these values, those corresponding to specimen 1 have been represented in the Figure 7.

With these functions, for easier calculation, graphics have been obtained from which only those corresponding to specimen 1 have been represented.

This relation allows us to establish a correspondence between the crack size and nominal tension. That for the panel CTOD 01 AW that we are analysing in this work and from results obtained, are acceptable. Tension values are fluctuating between 269.10 N / mm² and 297.24 N / mm², as shown in Table 2. This means that these materials have acceptable fracture toughness in the HAZ.

Table 3. Relation between tension and crack size

Specimen	$\sigma = f(a)$ N/mm²
1	$935 \times a^{-1/2}$
2	$915 \times a^{-1/2}$
3	$907 \times a^{-1/2}$
4	$899 \times a^{-1/2}$
5	$935 \times a^{-1/2}$
6	$866 \times a^{-1/2}$
7	$842 \times a^{-1/2}$
8	$879 \times a^{-1/2}$
9	$834 \times a^{-1/2}$

Figure 7. Critical tension-critical crack

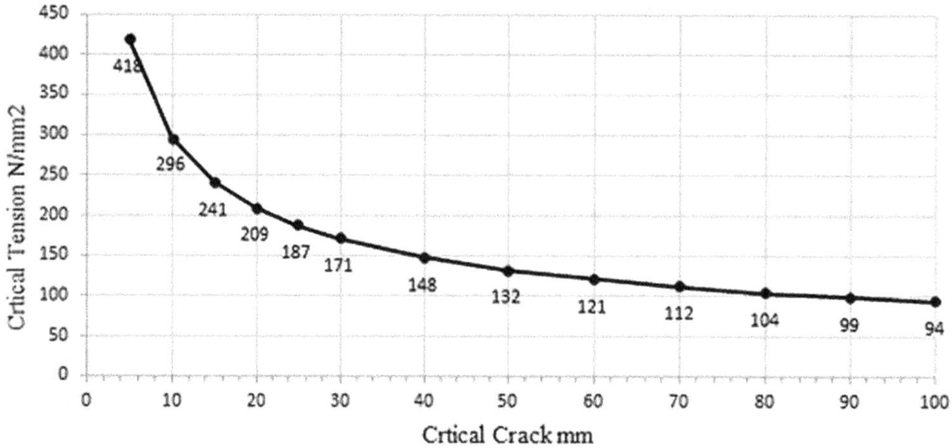

Crack Growth

The experimental work was carried out on a coupon of 450 EMZ HSLA steel 60 mm thick quenched and tempered steel referenced as 01AW CTOD (which is being studied in this paper), the coupon was tested in a submerged arc process in conditions "as welded" (without post-weld heat treatment). On the HAZ fracture mechanics test, CTOD were performed at a temperature of -10 °C. The tests were conducted conforms to Methods for Crack Opening Displacement C.O.D. (Testing) (1.979) and the publication of the EEMUA 158, Construction Specification for Fixed Off-Shore Structures in the North Sea (1986).

Samples were prepared and tested in order to obtain at least three valid data from them in accordance with said rule. The tests were conducted on preferred geometry test pieces($B \times 2B$) in which was made a notch through the entire thickness length (J) equal to the width of the specimen (W) Figure 8.

It was tried to carry out the Notch in the GCHAZ, next to FL (fusion line). This experimental work is based on the results of tests CTOD 450 EMZ steel, on which this paper consists, the results of nine samples were obtained according EEMUA, Construction Specification for Fixed Off-Shore Structures in the North Sea (1986), in the HAZ CTOD in trials of the panel 01 AW which is listed in Table 1.

As shown in the values specified in the Table 1, the crack length (a_{cr}) is between 58.5 mm and 61.5 mm. A tension intensity factor between $K_{f\,max} = 2320\ \mathrm{N / mm^{3/2}}$ and $K_{fmax} = 2070\mathrm{N / mm^{3/2}}$ and cycles between N= 81000 and N = 62000. So all the values move around similar results, we will take as example the specimen 1 which has the N value close to the mean. Considering that the notches made in the specimens, have the design shown in Figure 8, we begin by determining the threshold values of crack propagation.

Researchers like Harrison (1970), Clark (1974), Paris et al. (1972), and Cooper et al. (2004), among others, showed that for martensitic steels, bainitic and ferritic-pearlitic, the later belonging to the 450 hardened and tempered EMZ for the relationship:

$$R = \frac{\sigma_{\min}}{\sigma_{\max}} > 0.1$$

Figure 8. Sample dimensions

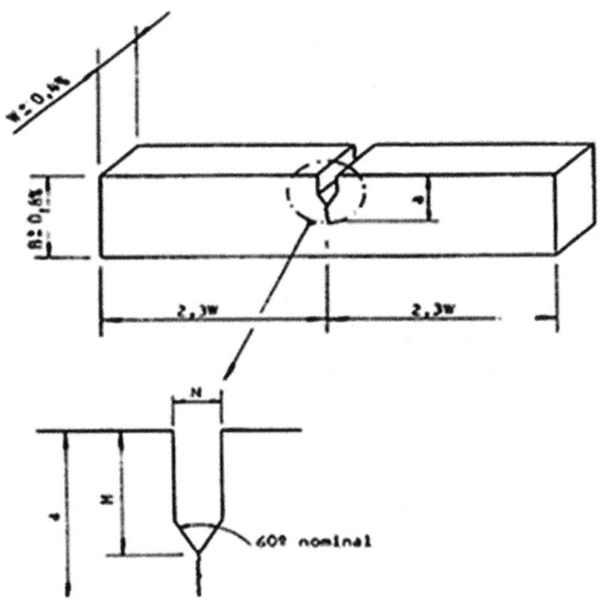

meet the following function that gives the amplitude of the amplitude of stress intensity factor threshold ΔK_{th} below which the crack does not propagate:

$$\Delta K_{th} = 7\ (1 - 0.85\ R)\ \mathrm{MN}\ \ /\ \mathrm{m}^{3/2}$$

When $R \leq 0.1$ our tests meet, ΔK_{th} have a constant value Equation 1) so, below these values, the crack does not spread.

Stable Growth by Fatigue

The procedure usually followed to analyze the behavior of the crack in the base metal, in weld metal and in heat affected zone is as follows, data for specimen 1:

1. It is estimated, the maximum size of the crack; in this chapter the results of testing9 specimens are summarize in which values of notches obtained mechanically and by precracked, have to meet again that 0.45 $< a / W <$0.55 in our case we started from value a / W = 0.509.
2. The critical crack size that would cause the failure by brittle fracture is considered. In our case the values of critical crack were obtained from experimental results, and are given in Table 1:

$a_{cr} = 11.1\text{mm}$, being:
$a_0 = 50\text{mm}$, $a_{cr} = 61.1\text{mm}$

Having a value of maximum stress intensity factor $Kf_{max} = 2.320 \text{ N/mm}^{3/2}$ where the value of P_f is obtained, starting from:

$K_f = \frac{Y\ P_f}{B\ \sqrt{W}}$, fulfilled according to Methods for Crack Opening Displacement C.O.D. (Testing) (1.979). Being:

$K_f < 0{,}63 x \sigma_{ys} x\ B^{1/2}$

and for the geometry of the test pieces: $B = 59$ mm and $W = 120$ mm a value of $P_f = 137\ 563$ N (specimen 1) is obtained for a geometrical factor Y = f (*a* / *W*) = 10.9; for an average value of yield stress of the base metal and weld metal; we can calculate the critical value $\Delta\sigma$ of crack propagation based on the results obtained:

$\Delta K = 1.4 \Delta\sigma \sqrt{\pi} \sqrt{a_m}$, being:

$\Delta K = 2088\ N\ mm^{3/2} = K_{fmax} - 0.1 K_{fmax}$;
for a $\Delta\ \sigma = 280.67 - 0.1 \times 280.67 = 252.60 \text{ Nmm}^{-2}$;

As average size of the crack specimen 1, we have: a_m=11,06mm

3. The next step is to determine an expression that meets the formula of Paris (1970), within the values for steels of more frequent use. We represent ferritic-pearlitic steels Structural Steel (BS4360.) Using the following expression:

$$\frac{da}{dN} = 5.01 x 10^{-12} \left(\Delta K_I\right)^{3,1} \text{ m/cycle, (Ewald's \& Wanhill 1991)} \quad (7)$$

where ΔK comes expressed in units of (MPa $m^{1/2}$ = MN $m^{-3/2}$), and crack length (*a*) in meters.
For the stress intensity factor:

$K = 1.4 \sigma \sqrt{\pi} \sqrt{a} \Rightarrow \Delta K = 1.4 \Delta\sigma \sqrt{\pi} \sqrt{a}$;

considering that the expression of Paris (1970) relates us the size of the crack with the cycles until the fatigue cracking, which occurs at a value that is obtained in this case experimentally, in Table 1 are given results for the nine specimens. To draw the curve that represent cycles fatigue-crack we have taken the values of the specimen 1.

$a = 61.1\text{mm} \quad ; K_{f_{\max}} = 2320\text{N} / \text{mm}^{3/2};$

- Bearing in mind that: $K_f < 0.63\, \sigma_{ys} B^{1/2}$;
  Starting from $P_f = 137563\text{KN}$
- And for geometric of the sample: $B = 59$ mm, $W = 120$mm
- And the relation of $a/W = 0.509$; we have:

$Y = 10.9$, and $K_f = \dfrac{YP_f}{BW^{1/2}}$ having in mind that yield strength is $\sigma_{ys} = 483$ N/mm^2.

4. Obtaining the increase stress intensity factor:

$$\Delta K = 1.4\, \Delta\sigma\sqrt{\pi a_m} = 626.8\sqrt{a_m}\ \text{N/mm}^{3/2} = 20.06\sqrt{a_m}\, MPa\sqrt{m} \tag{8}$$

5. Finally, the crack calculation depending on the number of cycles, using the equation of Paris (1973) as increments:

$$\Delta N = \frac{\Delta a}{A(\Delta K)^{3,1}} = \frac{\Delta a}{5.01 \times 10^{-12}(\Delta K)^{3,1}} \tag{9}$$

$$\Delta a = 0.6248\text{mm}$$

Being this Δa convenient value for 74000 cycles of our specimen 1.

And this way we obtain the following function, which is represented in Figure 9 (see Table 8.)

$$\Delta N = \frac{\Delta a}{5.01 \times 10^{-12}(\Delta K)^{3.1}}$$

From the results obtained in Table 3 curves of critical tension / critical crack are represented, (which we have plotted the values of the specimen 1 in Figure 7.), although the crack goes through the same material, different results are got, because the crack went through different zones of HAZ with different toughness values as if it the steel would have being made of different materials. The most toughness, as deduced from the values shown in Table 2 corresponds to the process where there is a greater margin of safety design stress with a percentage of the critical yield stress. For specimen 1 the average value corresponds to a stress ratio of 58.1%, and $K_f = 2320\text{Nmm}^{3/2}$. The expression of Paris (1973) gives the following values for ferritic – pearlitic steel:

$$\frac{da}{dN} = 5.01\text{x}10^{-12}(\Delta K)^{3,1}\text{m / cycle} \; ; being\ \sigma_{ys}$$

$$206.9\text{N/mm}^2 < \sigma_{ys} < 344.8\text{N/mm}^2$$

As we will see, our results are close to the values Ferritic - pearlitic (structural steel).

The differences found are not significant and they are due to changes in thermomechanical processes.

In any case, these different values depend on the different changes in the HAZ microstructure, however we have taken an average value of the nine test pieces if we would having taken values ΔK for each would have gotten slightly different results as obtained for the critical stress / critical crack, Figure 7.

The same author mentioned above Ewald's et al. (1991), consider using a single formula for ferritic-pearlitic, martensitic and austenitic stainless steels, this is due to the fact that the values obtained in the behavior of crack propagation are approximate.

The study of the speed of crack growth (da / dN) as a function of the amplitude of the stress intensity factor ΔK , conducted by Clark (1974) under submerged arc process on Steel ASTM A 533 Grade B Class 1, it shows that the growth rates of the crack in the weld and in the heat affected zone are similar to the base metal at room temperature.

METALLOGRAPHIC TEST

The results of the metallographic test are shown in Table 4. This Table links intercepted structures in a section of the sample taken by the end of the crack through the thickness and in the 2/3 thereof, as the standard relates mark and it is shown that in some cases the affected area is within the value of *B* / 6 that the standard EEMUA 158 (1986) Construction Specification for Fixed Off-Shore Structures in the North Sea (1986), does not consider.

Of The values given in Table 4, one can conclude that in the samples 4, 5 and 6 the front of a fatigue crack was found in the fine grain as shown in the mentioned Table, beyond than 0.6 mm from FL, in the three specimens no coarse region was intercepted in the two-thirds of the central area, due to that the crack was not found within 0.5 mm of the FL. Values from a viewpoint of the standard, should be considered not acceptable. The specimens 1, 2, 3 outperformed in this regard, with GGHAZ rates of 4.2%, 3.8% and 12%, see Table 5. The crack whose end was found in greater amounts in the area GCHAZ was

Figure 9. Crack transmission

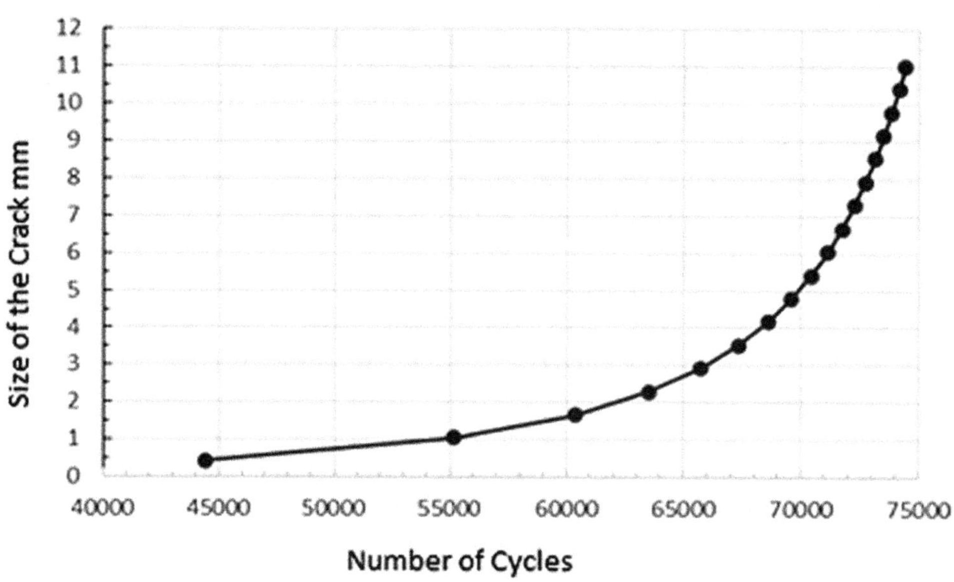

the specimen 3 with the highest percentage (12%) of ASTM <7 size, which accounted for the lower value of CTOD $(\delta) = 1.05$ mm.

We note that the sampling areas that cross the crack, those specimens that obtained a higher percentage of WM, GC (grain coarse) and I.C.(intercritical), the CTOD values obtained in the test were the lowest. Demonstrating the fragility of these three regions.

Also in the specimen 7, the end of the crack was found in the HAZ beyond than 0.6 mm from the fusion line, no part of the fatigue crack zone intersect the zone of grain coarse, that in accordance with the rule invalidates testing due to not be within 0.5 mm of the fusion line.

The specimens 8 and 9 showed that the crack went through at least 5% and 7.8% of grain coarse structure GCHAZ (i.e., within 0.5 mm from fusion line). The lower value corresponded to CTOD specimen 9, the crack is initiated in the fine grain FGHAZ very close to the GCHAZ.

We must highlight the extreme difficulty of getting the fusion line is completely straight and even perfectly normal to the surface of the plate, this helps the percentage found be of grain coarse where it crosses the crack are few, even zero. With regard to the previously noted, it is possible that the values obtained in the CTOD not represent the lower limit of the fracture toughness of the coarse-grained microstructure GCHAZ present in the testing.

Although the values deduced from Table 5, do not establish an inverse proportionality between CTOD and coarse amount found in the test, can be observed that the minimum values obtained in the CTOD test correspond to samples in which they found higher percentage of grain coarse in amounts of 12% and 7.8% GC size (ASTM) <7, with values δ = 1.05 and δ= 0.22 mm respectively (see Table 5.), the

Table 4. Review extract micrograph, fracture toughness of the specimens tested in CTOD

Specimen	CTOD value at -10 °C (mm)	Total Number of Tested Microstructures Throughout the Thickness, *B*, of the Specimen (*B* = 100%)						Total Number of Microstructures Tested With Two Thirds of the Thickness (2/3 *B* = 100%).					
		WM (%)	GC < 7 (%)	GC 7-9 (%)	FG (%)	IC (%)	SC + BM (%)	WM (%)	GC < 7 (%)	GC 7-9 (%)	FG (%)	IC (%)	SC + BM (%)
1	1.68 u	20.1	6.1	1.9	53.9	11.7	6.3	0.0	4.2	2.9	66.0	17.5	9.4
2	2.13 u	66.3	3.,9	0.0	27.5	2.2	0.03	61.9	3.8	0.0	30.9	3.3	0.0
3	1.05 u	59.9	12.,8	4.6	22.,8	0.0	0.0	51.3	12.0	5.4	31.4	0.0	0.0
4	1.33 u	16.4	3.8	0.0	68.1	11.7	0.0	0.0	0.0	0.0	90.6	9.4	0.0
5	1.53 m	17.1	3.6	3.1	69.4	3.6	3.2	0,0	0.0	4.6	85.1	5.4	4.9
6	1.27 u	18.4	3.2	2.6	66.3	5.5	4.1	0.0	0,0	3.8	84.9	5.1	6.1
7	0.94 u	12.7	4.0	2.7	43.7	20.5	16.4	0.0	0.0	4.0	58.1	21.6	16.3
8	0.96 u	44.0	6.2	1.2	46.0	2.7	0.0	46.3	5.0	1.3	43.4	4.0	0.0
9	0.22 u	37.0	7.8	4.0	32.3	19.0	0.0	40.0	7.8	4.7	30.5	17.0	0.0

Wm = Weld Metal; Gc < 7 = The grain size ASTM in HAZ lower 7

BM =Base Metal; Gc 7-9 = Grain Coarse size ASTM between 7 to 9 in HAZ.

FG = HAZ Fine Grain; IC = HAZ Intercritical. SC = HAZ subcriticaMaterial: Steel 450 EMZ; Identification of Welding: CA - CTOD-01-AW-1G-WM.Proportion of Diversion: 19 x 10^{-3} mm / sec .; Welding Process: Submerged Arc. CA - CTOD-01-AW-1G-WM.

Condition: Butt weld; Specimen geometry: B x 2B Specimen BS 5762: 1079.Location Notch: HAZ; Test temperature: -10 °C;Orientation Notch: Side by Side Thickness

latter value in the threshold of the objectionable. The last column of Table 5, informs us of where the fatigue crack had his initiation, these values were obtained by microscopic observation. It is noted that the crack initiation always occurs in areas adjacent to the GCHAZ, such as WM and FGHAZ, which in some cases can be assumed, that the aforementioned geometry of the FL, the actual initiation is in the grain coarse area. Table 6 informs us of the test results of fracture mechanics, corresponding to this test.

- Material: SteeL 450 EMZ.
- Welding Process: Submerged Arc. (SAW).
- Identification of the Weld: CA - CTOD-01-AW-1G-ZAC. Condition: Butt Weld.

EXPERIMENTAL STUDY OF FATIGUE CRACK.

The study of fatigue crack for testing CTOD fracture toughness was conducted in a push resonance Schenck 200 KN. The frequency of the electric motor driving the pushbutton, is adjusted by rotating an eccentric weight in order to obtain maximum load P_{max} which has been predetermined beforehand. This frequency, which is unknown but controlled depends inter alia on the resistance of the test piece. During fatigue process which the specimen is subjected, and due to the crack growth rate the frequency should be adjusted to maintain the P_{max} and ΔK (amplitude of the stress intensity factor).

Keep in mind that the stress intensity factor for fatigue fracture, according to British Standard BS5762 BS 5762, (González-Palma, Suárez-Bermejo, Vicario, & Muñoz, 2006; Suárez, Molleda, González, & Jiménez, 1994; Leiva, Donoso, Muehlich, & Labbe, 2004) is adjusted by the following equation:

Table 5. Extract of test results of fracture toughness on CTOD

Specimen	CTOD		Type	Relative Amount of GCHAZ Tested to Fatigue Failure		Point of the Microstructure of Fracture Initiation
Number	Individual (mm)	Average (mm)	(BS 5762)	GC < 7 (%)	GC 7-9 (%)	Zone
1	1.68	1.50	(u) Fracture	4.2	2.9	FGHAZ a 27.4 mm from the root surface
2	2.13		(u) Fracture	3.8	0.0	WM entre 31.7 y 38.5 mm from the root surface
3	1.05		(u) Fracture	12.0	5.4	GCHAZ a 42.5 mm from the root surface
4	1.33		(u) Fracture	0.0	0.0	FGHAZ a 35.5 mm from the root surface
5	1.53		(m) Máximum	0.0	4.6	no fracture
6	1.27		(u) Fracture	0.0	3.8	FGHAZ entre 29.2 from the root surface
7	0.94	0.71	(u) Fracture	0.0	4.0	FGHAZ a 30.0 mm from the root surface
8	0.96		(u) Fracture	5.0	1.3	FGHAZ a 24.5 mm from the root surface
9	0.22		(u) Fracture	7.8	4.0	FGHAZ (next to GCHAZ) a 23.5 mm from the root surface

GC < 7 = Grain Coarse in HAZ Grain Coarse size lower to ASTM 7.

Lower Surface = Bottom side in the respective mapping diagrams showing the distribution of the microstructures of the HAZ tested for fatigue failure. Material: Steel 450 EMZ; Identification of Welding: CA -CTOD-01AW-1G-WM. Proportion of Diversion: 19×10^{-3} mm / sec.; Welding Process: Submerged Arc. Condition: Butt Weld; Specimen geometry: B × 2B Specimen BS 5762: 1079.Location Notch: HAZ; Test Temperature: -10 °C; Orientation Notch: Side by Side Thickness.

$$K_f \leq 0.63\ \sigma_y \sqrt{B} = 0.63 \times 480 \times \sqrt{59} = 2322.77 N\ /\ \text{mm}^{3/2} > 2320\ \text{N/mm}^{3/2} \quad (10)$$

where:

σ_y = yield strength of the HAZ. B = thickness of the specimen.

From this stress intensity factor, and by the expression: $K_f = \frac{YP_f}{B\sqrt{W}}$ according to that rule, where P_f is the values of the maximum load to which the machine must be set for the fatigue test (**N**)

- B = thickness of the specimen (mm).
- W = width of the specimen (mm).

Y' depends on $a\ /\ W$, that is, the ratio of the length of the crack to specimen width. These values are expressed in Table 1 of BS 5762, and they are obtained through experimental polynomial formula, which Broek gives in his book of fracture mechanics by the following expression:

$$Y' = 2.9(a/W)^{1/2} - 4.6(a/W)^{3/2} + 21.8(a/W)^{5/2} - 37.6(a/W)^{7/2} + 38.7(a/W)^{9/2} \quad (11)$$

From which, the increase of the stress intensity factor by expression is obtained:

$$\Delta K = \frac{\Delta PSY'}{BW^{3/2}} \quad (12)$$

where: S = span between supports = 4W (as standard). So that:

Table 6. CTOD, data table fracture toughness

Specimen	B (mm)	W (mm)	a (mm)	$a_{y\,(mm)}$	P (KN)	$V_{p(mm)}$	(MPa) σ	(mm) (*) δ_e	(mm) (*) δ_p	(mm) (***) δ	Fracture Kind (**)	Valid (BS 5762)
1	59.0	120.0	61.1	62.4	358	5.76	483	0.18	1.50	1.68	u	Si
2	59.6	120.3	60.9	64.5	360	7.76	483	0.20	1.93	2.13	u	Si
3	59.3	120.5	60.5	61.1	358	3.28	483	0.16	0.89	1.05	u	Si
4	59.3	120.6	60.2	61.2	368	4.28	483	0.17	1.16	1.33	u	Si
5	59.1	120.5	61.5	62.9	352	5.24	483	0.18	1.36	1.53	m	Si
6	59.7	119.8	58.5	59.5	374	3.96	483	0.17	1.10	1.27	u	Si
7	61.5	121.7	59.8	60.0	376	2.76	456	0.16	0.78	0.94	u	Si
8	61.7	120.2	60.5	61.2	378	2.88	456	0.18	0.77	0.96	u	Si
9	61.2	122.2	59.5	59.6	304	0.40	456	0.10	0.11	0.22	u	Si

(*)Are the elastic and plastic components of CTOD fracture toughness respectively, while (***) is the CTOD fracture toughness.

(**): The symbol "u" is used when there is suggestive evidence that a pop-in or an unstable fracture was preceded by a ductile elongation, however "m" indicates the maximum force was reached.

$$\Delta K = \frac{\Delta P 4 W Y'}{B W^{3/2}} = \frac{\Delta P 4 Y'}{B\sqrt{W}}$$

substituting $4Y' = Y$, we obtain the value of the BS 5762 standard.

$$\Delta K = \frac{\Delta P Y}{B\sqrt{W}}$$

In our case, and following the results obtained for one of nine similar specimen comprising 01 AW CTOD test, which is corresponding to the authenticated as the specimen 1 and the results are around a mean value of the nine composing this test. We obtain a value of $Y' = 2.743$; where $a / W = 0.509$, and according to experimental results, this value $4Y' = 10.972$, agrees approximately with $Y = 10.93$, which is obtained from Table 1 of the BS, to $a / W = 0.509$.

However, throughout the test, no control of the point values of ΔP were taken, as well as neither of ΔK, as it would be impossible to detect the continued growth of the crack during it.

So, we only know the initial value of the depth of the crack, a_0, which corresponds to the length of the crack mechanically obtained by saw and the depth value, a, of the final crack obtained by fatigue. The fatigue fracture test, stops when a / W, is near of following values: $0.45 < a / W < 0.55$.

From $K_f = 2320$ N / mm$^{3/2}$ (obtained above), the P_{fmax} is obtained by the expression:

$$K_f = 2320 = \frac{Y P_{f\,max}}{B\sqrt{W}} \tag{13}$$

substituting the known values:

Y=10.9; K_f=2320N/mm$^{3/2}$; B=59mm, W=120mm

it is got to a P_{fmax} = 137 564 N, which ΔP is obtained by the following expression:

$\Delta P = P_{fmax}\, 0.9 = 123807$ N.

Putting the above function as their increments, we obtain:

$$\Delta K = \frac{\Delta P Y}{B\sqrt{W}}$$

Substituting the values listed, $\Delta K = 2088$ N / mm$^{3/2}$ is obtained. In this paper, we have presented the expression:

$$\Delta K = 1.4 \Delta\sigma \sqrt{\pi a_m} \tag{14}$$

in which:

a_m = mean value of the crack and Δσ= increasing tension applied

The following relationships must be taken into account:

$$\frac{P_{min}}{P_{max}} = 0.1\ ;\ \ \frac{\sigma_{min}}{\sigma_{max}} = 0.1\ ;\ \ \Delta\sigma = \sigma_{max} - \sigma_{min} = 0.9\sigma_{max}$$

From the above expressions we get:

$$\Delta\sigma = \frac{\Delta K}{1.4\sqrt{\pi a_m}} \tag{15}$$

resulting the following:

- ΔK=2088 N/mm$^{3/2}$
- a_m = 11 mm which corresponds to the depth of the fatigue crack, since a_0 = 50 mm and a= 61.1 mm for the specimen 1 which are considered in this calculation.
- a_0-a=11 mm; $\Delta\sigma \cong$ 253 N/mm^2; $\Delta\sigma$=0.9σ

in which σ represents tension of design for HAZ obtained from fatigue tests (see Table 2.).

Therefore being a_m = 11 mm, is obtained:

$$\begin{aligned}\Delta K &= 1.4\Delta\sigma\sqrt{\pi a_m} = 1.4 \times 252.6\sqrt{\pi a_m} = 627.8\ \sqrt{a_m}\,\mathrm{N/mm}^{3/2} \\ &= 20.06\sqrt{a_m}\ M\,Pa\ m^{1/2}\end{aligned} \tag{16}$$

where:

da/dN = growing of the crack per cycle of charge (m/cycle).

ΔK= Amplitude stress intensity factor.

$$\begin{aligned}&\frac{da}{dN} = 5.01 \times 10^{-12}(\Delta K)^{3.1} m\ /\ cycle\ ; \\ &206.9N\ /\ mm^2 < \sigma_{ys} < 344.8N\ /\ mm^2\end{aligned} \tag{17}$$

Writing the above formulas in the form of amplitude, we can calculate integrating the curve of cycle - crack growth, by the expression:

$$\Delta N = \frac{\Delta a}{5.01 \times 10^{-12} (\Delta K)^{3.1}}$$

Taking account that for ferritic-pearlitic steels with R = 0.1, the threshold value ΔK, researchers Harrison et al. (1970), Paris et al. (1972), Schweizer et al. (2011) and Kujawski (2013) can be estimated at:

ΔK_{th}=7(1-0.85R) N/mm$^{3/2}$

for R = 0.1 we obtain:

$\Delta \boldsymbol{K}_{th}$=6.405 MN/m$^{3/2}$=200 N mm$^{-3/2}$ <202.54 N/mm$^{3/2}$

The threshold value of the crack will be:

$$a_m = \left(\frac{\Delta K_{th}}{627.8} \right)^2 \approx 0.105 \text{mm}$$

Correspond to the size of the crack from which it begins its propagation. Based on these thresholds, we obtain the frequency curves - crack length to the specimen 1 and to a convenient value Δa = 0.625 mm is obtained:

$$\Delta N = \frac{0{,}625 \times 10^{-3}}{5.01 \times 10^{-12} \Delta K^{3.1}} \tag{18}$$

Starting from the experimental values given in the following Table 7.

Referencing the first specimen and expressions derived above, we can determine the values listed in Table 8 which have been calculated for the values of a_m to ΔN in cycles, summing them, we obtain the total cycles for the crack extends i.e. for 74427 cycles, as we can see practically it coincides with the experimental cycles for the specimen 1.

In this Table 8 is also shown the values of the speed of crack growth da / dN given in m / cycles theoretically determining the stretch of the formula corresponding to Paris (1970). The amplitude of the stress intensity factor and its logarithm allow us to check the nature of the ferritic- pearlitic steel of our specimens. All these values are plotted for the corresponding conclusions.

The first column of Table 8 was obtained starting from the crack threshold value a_0 = 0.105 mm which was obtained by Equation (16) and $a_m = a_0 + \Delta a / 2$, where $a_f = a_m + \Delta a / 2$ being Δa = 0.625 mm, which was mentioned above and is taken in an amount with a quantity acceptable enough for the eighteen values shown in Table 8, reaches a total amount of N and a_m to correspond to the experimental

Table 7. Crack length- stress intensity factor

Specimen Number	Crack Size a (mm)	Relation a/W	Stress Intensity Factor K(N/mm$^{3/2}$)	Cycles N
1	61.1	0.509	2320	74000
2	60.9	0.506	2270	67000
3	60.5	0.502	2250	69.000
4	60.2	0.499	2230	71000
5	61.5	0.510	2320	76000
6	58.5	0.488	2150	81000
7	59.8	0.491	2090	67000
8	60.5	0.503	2180	63000
9	59.5	0.487	2070	62000

results. The amplitude of the stress intensity factor ΔK was obtained from Equation (16) in MPa m$^{1/2}$. The quantity represented by ΔN is given by Equation (18) where Δa is expressed in meters. ΣN cycles total are obtained by adding to the column ΔN_1 successive values ΔN_i. The rate of crack growth in m / cycles is obtained from the Equation (17) in which ΔK values are expressed in MPa m$^{1/2}$ based in column 4 of Table 8.

Table 8. Stress intensity factor – number of cycles

a_0 mm	a_m mm	a_f mm	ΔK MPa m$^{1/2}$	Ln ΔK	ΔN ciclos	ΣN ciclos	da/dN m/ciclos
0.1050	0.417	0.729	12.954	2.561	44421	44421	$1.407x10^{-8}$
0.7290	1.041	1.354	20.467	3.019	10759	55180	$5.810x10^{-8}$
1.3540	1.666	1.979	25.892	3.254	5191	60371	$1.204x10^{-7}$
1.9790	2.291	2.604	30.363	3.413	3168	63539	$1.973x10^{-7}$
2.6040	2.916	3.220	34.255	3.534	2180	65719	$2.867x10^{-7}$
3.2290	3.541	3.853	37.748	3.631	1613	67332	$3.874x10^{-7}$
3.8540	4.166	4.478	40.944	3.712	1254	68586	$4.985x10^{-7}$
4.4780	4.790	5.080	43.903	3.782	1010	69596	$6.188x10^{-7}$
5.1030	5.415	5.728	46.680	3.843	835	70431	$7.484x10^{-7}$
5.7280	6.040	6.353	49.300	3.898	705	71136	$8.865x10^{-7}$
6.3530	6.665	6.977	51.788	3.947	605	71741	$1.033x10^{-6}$
6.9780	7.290	7.602	54.162	3.992	527	72268	$1.187x10^{-6}$
7.6030	7.915	8.227	56.436	4.033	464	72732	$1.348x10^{-6}$
8.2280	8.540	8.852	58.622	4.071	412	73144	$1.516x10^{-6}$
8.8520	9.164	9.541	60.726	4.106	369	73513	$1.692x10^{-6}$
9.4780	9.790	10.102	62.766	4.139	334	73847	$1.874x10^{-6}$
10.1020	10.414	10.726	64.735	4.170	303	74150	$2.062x10^{-6}$
10.7270	11.039	11.351	66.649	4.199	277	74427	$2.257x10^{-6}$

Although the threshold of crack formation in the samples was determined as 0.105mm, crack detection and the threshold value for the expression of Paris (1972) gives us a result that is slightly higher than determined and we can calculate as follows:

$$\frac{\mathrm{d}a}{\mathrm{d}N} = 5.01\times10^{-9}\left(\Delta K\right)^{3.1}\mathrm{mm/ciclo}$$

where from:

$$\int_0^N \mathrm{d}N = \int_{a_d}^{a_{cri}} \frac{\mathrm{d}a}{5.01\times10^{-9}\left(\Delta K\right)^{3.1}}$$; for a value of $a_{\text{crit.}} = 11.039$ mm,

$$\Delta K = 20.06\sqrt{a}\ \mathrm{MNm}^{-3/2},\ N = 74427 \text{ cycles.}$$

Is obtained an $\boldsymbol{a}_{\mathbf{d}} = 0.189$ mm, (Fatigue-Thresholds-Fundamentals and Engineering, 1982), which is the first detectable threshold of the crack.

Starting from the theoretical values in Table 8 we have been calculated following graphs which are shown:

Corresponding to the region III of the Paris (1973) equation graph is drawn representing $\Delta K = \mathrm{f}(\mathrm{d}a/\mathrm{d}N)$ Figure 11, to a ΔK value of 66.649 Mpa $\mathrm{m}^{1/2}$ where propagation occurs at 74427 cycles. This region has a higher incidence microstructure, medium tension and a small environmental impact surrounding the specimen. We see that the profile of the graph of Figure 12 corresponds to the values of the pearlitic-ferritic steels (Barsom, 1971), on being Figure 11 parable very laid with a regression line given by Barsom, (1977). Thus it is found that with the obtained experimental values of the stress intensity factor applied to the function $\frac{\mathrm{d}a}{\mathrm{d}N} = 5.01\times10^{-9}\left(\Delta K\right)^{3.1}\mathrm{mm/ciclo}$, the graph corresponds to the pearlitic-ferritic steels for the values obtained in the HAZ.

$$K = (\Delta K/\Delta\sigma)\sigma \tag{19}$$

Function is shown in Figure 10. This function represents us factor stress intensity depending on the fatigue stress of the material at the end of the crack as is indicated in Table 2 is 280.67 N / mm^2 corresponds to an acceptable value of 58.1% stress specimen 1 of the Table.

So check with the experimental values obtained stress intensity factor, applied to the function da/dN $=5.01\mathrm{x}10^{-12}\ (\Delta K)^{3.1}$ m/cycle, and drawn the curve of the crack propagation cycles obtained experimentally, that the behavior in the HAZ are satisfied from a viewpoint of fracture mechanics corresponds to that of the aforementioned steels.

In the Figure. 11 it checks that the values obtained from ($\mathrm{d}a/\mathrm{d}N$), ΔK corresponds to the ferritic-pearlitic steels according to Barsom (1971). This claim becomes more obvious when compared in Table 9 column 1 with column 5, being column 1 the values corresponding to our stress intensity factor when working in English units, wherein the intensity factor stress for pearlitic-ferritic steels is given by:

Figure 10. Stress intensity factor-fatigue tension

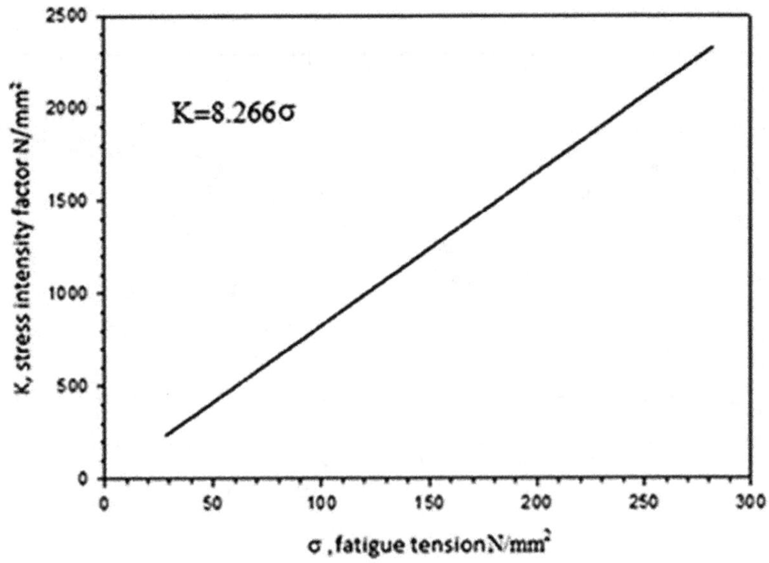

Figure 11. Amplitude of stress intensity factor - crack rate

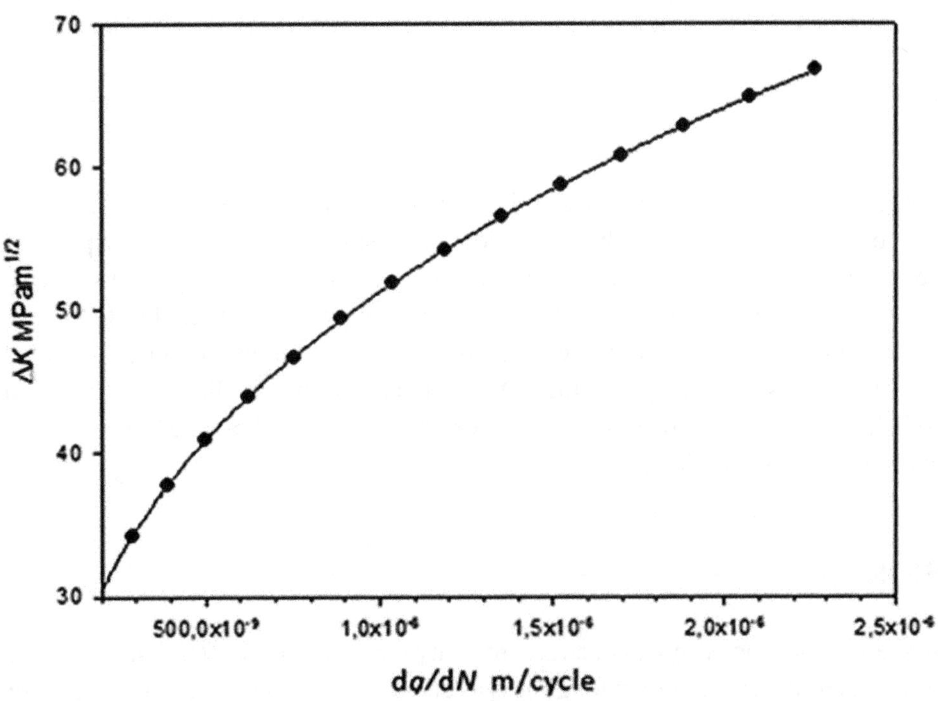

Figure 12. Stress intensity factor- crack growing

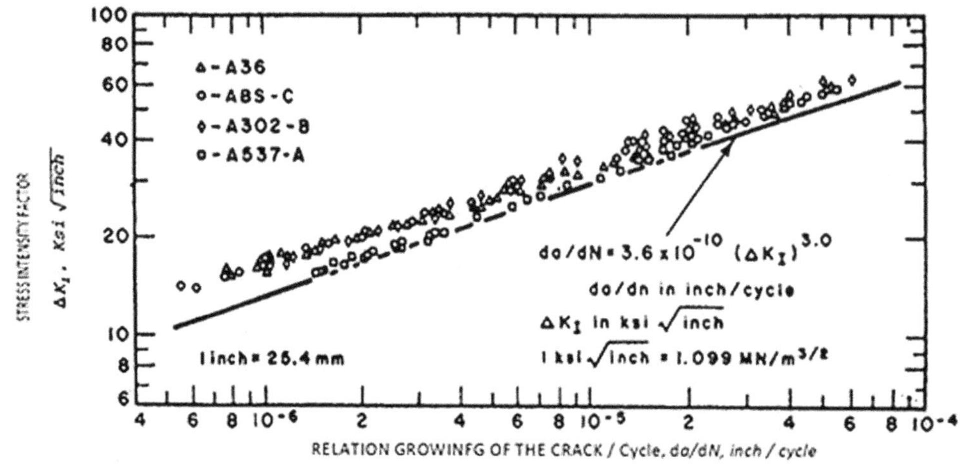

$$\Delta K = 1.4 \times \Delta\sigma\sqrt{\pi\ \mathrm{a}} = 1.4 \times 36\ .\ 625\ \ \sqrt{\pi\ \mathrm{a}} = 90\ .\ 9\sqrt{a}\ \ \mathrm{Ksi}\ \sqrt{\mathrm{inch}}$$

Being 36.625 the strain amplitude at the tip of the crack expressed in Ksi for sample 1.

The growth rate corresponding to the column 2 in inches per cycle for pearlitic-ferritic steels crack is given by:

$$\frac{\mathrm{d}a}{\mathrm{d}N} = 3.6 \times 10^{-10}(\Delta K)^3;$$

where ΔK is obtained from the column 1 Ksi $\sqrt{\text{inch}}$ the result is given in column 2 inches/ cycle.

Interestingly observe, the results of the amplitude factor stress intensity when expressed in English units coincides approximately with the results of column 5 as Barsom et al.(1977) so we conclude the nature ferritic-pearlitic of our steels based on the experimental results obtained in HAZ.

The crack growth for pearlitic-ferritic steels which have been given in column 2 of Table 9 are represented in Figure 12, expressed in English units. One can observe the parallelism of the results depicted in the graph of Figure 11 and Figure 12 especially for values above $\Delta K >$ 41 MPa $m^{1/2}$. Both represent a straight equation virtually identical.

CONCLUSION

1. We have seen that for the purposes of stress intensity factor, the microstructure of the HAZ behaves as different materials with different toughness as deduced from the values of fatigue stress obtained for the samples in Table 1, where it is observed that most tenacity corresponds to the specimens

Table 9. Comparison of experimental results with Barsom et al. (1977) results

ΔK	da/dN (pulg./ciclo)	ΔK (MPa $m^{1/2}$)	da/dN (m/ciclo)	ΔK Barsom
11.65	5.69x10^{-7}	12.954	1.407x10^{-8}	11.4
18.40	2.24x10^{-6}	20.467	5.810x10^{-8}	18.2
23.28	4.54x10^{-6}	25.892	1.204x10^{-7}	22.0
27.30	7.32x10^{-6}	30.363	1.973x10^{-7}	27.0
30,80	1.05x10^{-5}	34.255	2.867x10^{-7}	31.5
33.94	1.41x10^{-5}	37.748	3.874x10^{-7}	32.8
36.81	1.80x10^{-5}	40.944	4.985x10^{-7}	38.0
39.47	2.21x10^{-5}	43.903	6.188x10^{-7}	40.0
41.97	2.66x10^{-5}	46.680	7.484x10^{-7}	42.1
44.32	3.13x10^{-5}	49.300	8.865x10^{-7}	43.5
46.56	3.63x10^{-5}	51.788	1.033x10^{-6}	45.5
48.70	4.16x10^{-5}	54.162	1.187x10^{-6}	48.0
50.74	4.70x10^{-5}	56.436	1.348x10^{-6}	50.5
52.71	5.27x10^{-5}	58.622	1.516x10^{-6}	52.5
54.60	5.86x10^{-5}	60.726	1.692x10^{-6}	54.0
56.43	6.47x10^{-5}	62.766	1.874x10^{-6}	57.5
58.20	7.10x10^{-5}	64.735	2.062x10^{-6}	59.2
59.93	7.75x10^{-5}	66.649	2.257x10^{-6}	60.1

1 and 5; as we see in Table 7 gives the greatest life cycles with the specimen 6. It was also noted that these specimens experienced high stable growth at low coarse content.

2. Since the maximum stress is 297.24N / mm2. There is a wide margin of safety of 61.5% over the yield strength in the HAZ.

Graphs corresponding to the nine specimens for the experimental values of crack length to fracture stress although only shown the graph of the specimen 1 was drawn, being acceptable values and profiles Figure 7 graphs obtained, corresponding to the function:

$$\sigma = \frac{K_{f\max} \times a^{-\frac{1}{2}}}{1.4\sqrt{\pi}}$$

It has not been shown a correlation values in the specimens 1 and 5 of greater fracture toughness and microstructure percentages through the crack, which implies that there is a direct relationship or dependency on other factors.

3. Compliance is checked with the function of Paris (1973) for values of ferritic steels - pearlitic (structural steel)

$$\frac{da}{dN} = 5.01 \times 10^{-12} (\Delta K)^{3.1} m \; / \; ciclo \; ;$$
$$206.9N \; / \; mm^2 < \sigma_{ys} < 344.8N \; / \; mm^2$$

The value of the critical crack corresponding to the specimen 1 start in 0.105 mm determining that the first value of the inspectable crack of said sample according to the equation Paris (1970) corresponds to 0.189 mm.

4. Although not found a correlation between the grain size and the values of CTOD establish an inverse proportionality, it is noticed that the lower values obtained CTOD correspond to higher values of content of coarse sampled by the crack, They corresponded to those specimens that in the two series of tests obtained lower value of CTOD, ie CTOD = 0.22 mm, which obtained 7.8% GC <7. See Table 5.

Since the fusion line was not straight enough, could not find a higher percentage of crack in the 0.5 mm inside it. However, the areas adjacent to the coarse (WM and FG) showed the highest percentages of area where he crossed the crack.

5. CTOD values obtained were all accepted as the standard since all values were greater than 0.2 mm. In Table 4 CTOD values and percentages of grain are represented for each of the samples.
6. It was shown that the heat energy (HI) provided was adequate for this type of steel finding CTOD results according to standard.

REFERENCES

Barsom, J. M. (1971). Fatigue Crack Propagation in Steels of Various Yield Strengths. Transactions of the ASME. *Journal Engineering for Industry,* Series B, 93(4). doi:10.1115/1.3428061

Barsom, J. M. (1974). *Fatigue Behavior of Pressure-Vessel Steels. WRC Bulletin, 194.* New York: Welding Research Council.

Barsom, J. M., & Rolfe, S. T. (1977). *Fracture and Fatigue Control in Structures. Applications of Fracture Mechanics* (p. 07632). Englewood Cliffs, New Jersey: Prentice-Hall, Inc.

Broek, D. (1982). Elementary Engineering Fracture Mechanics. Hingham, MA, USA: Kluwer Academic Publishers. doi:10.1007/978-94-011-9055-8

Clarck, J. R. (1970). Fatigue Crack Growth Characteristics of Heavy Section. *ASTM A533 Grade B Class 1 Steel Weldments, ASME Paper N° 70-PVP 24.* American Society of Mechanical Engineers, New York.

Clark, J. R. (1974). How Fatigue Crack Initiation and Growth Properties Affect the Material Selection and Design Criteria. *Metals Engineering Quarterly*, Aug.

Clintock, M. C. (1963). On the Plasticity of the Growth of Time Fatigue Cracks. In Fracture of Solids (p. 65). New York: Willey-Interscience.

Construction Specification for Fixed Off-Shore Structures in the North Sea (1986). *The Engineering Equipment and Materials Users Association EEMUA*. Publication, *158*.

Cooper, R. J., Silva, H. F., & Trevisan, R. E. (2004). Influencia del precalentamiento en las propiedades de uniones soldadas de acero API 5L-X80 soldadas con alambre tubular autoprotegido. *Rev. Metal Madrid*, *40*(4), 280–270. doi:10.3989/revmetalm.2004.v40.i4.275

El Haddad, M. H., Dowling, N. E., Topper, T. H., & Smith, K. N. (1980). J Integral Application for Short Fatigue Cracks at Notches. *International Journal of Fracture*, *16*(1), 15–30. doi:10.1007/BF00042383

Elber, W. (1971). *The Significance of Crack Closure. Damage Tolerance Aircraft Structure. ASTM STP 486* (pp. 230–242). Philadelphia: American Society for Testting and Materials. doi:10.1520/STP26680S

Elices, M. (1992). Mecánica De La Fractura Aplicada A Sólidos Elásticos Bidimensionales. *Departamento de Ciencias de Materiales E.T. Superior de Ingenieros de Caminos de Madrid,* January.

Ewald's, H. L., & Wanhill, R. J. H. (1991). *Fracture Mechanics*. London: Edward Arnold.

Bäcklund, J., Blom, A.F., & Beevers, C.J. (1982). Fatigue-Thresholds-Fundamentals and Engineering (Vol 1 & 2). UK: ApplicationsEngineering Materials Advisory Services Ltd.

Frost. (1960). Notch Effects and the Critical Alternating Stress Required to Propagate a Crack in an Aluminium Alloy Subjected to Fatigue Loading. *Journal of the Mechanical Engineering Society, 2*(2), 109-119.

González-Palma, R., Suárez-Bermejo, J. C., & Vicario, F. J., & Aurelio Muñoz. (2006). Procesos metalúrgicos de aceros microaleados de alta resistencia 450 EMZ tipo II en la zona afectada térmicamente debido al efecto multipasada en soldaduras por arco sumergido. *Rev. Metal. Madrid*, *42*, 256–269.

Harrison, J. D. (1970). An Analysis of Data on non Propagating Fatigue Cracks on a Fracture Mechanics Basis, *British. Welding Journal*, *2*(3).

Kitagawa, & Takahashi, S. H. (1976). Applicability of Fracture Mechanics to Very Small Cracks or the Cracks in the Early Stages. *Proceedings of the Second International Conference on Mechanical Behaviour of Materials*, Metal Park, Ohio, USA. American Society for Metals.

Kujawski, D. (2013). Environment and Stress Interaction in Corrosion Fatigue and Stress Corrosion Cracking. *Metallurgical and Materials Transactions. A, Physical Metallurgy and Materials Science*, *44A*(3), 1340–1352. doi:10.1007/s11661-012-1507-8

Kujawski, D., & Sadananda, K. (2011). Effect of Crack-Tip Stresses on SCC Behavior. *Metallurgical and Materials Transactions. A, Physical Metallurgy and Materials Science*, *42A*(2), 377–382. doi:10.1007/s11661-010-0373-5

Leiva, R., Donoso, J. R., Muehlich, U., & Labbe, F. (2004). Un enfoque de dos parámetros para el análisis del efecto del cordón de soldadura sobre el constreñimiento. *Rev. Metal Madrid*, *40*, 280–287.

Lindner, B. M. (1965). Extremely Slow Crack Growth Bates Aluminium Alloy 7075 T6 [M.S Thesis]. Lehigh University, Bethlehem, PA.

Matsuoka, S., Tanaka, H., Homma, N., & Murakami, Y. (2011). Influence of hydrogen and frequency on fatigue crack growth behavior of Cr-Mo steel. *International Journal of Fracture*, *168*(1), 101–112. doi:10.1007/s10704-010-9560-z

Methods for Crack Opening Displacement C.O.D. (Testing). (1979). Methods for Crack Opening Displacement C.O.D. (Testing) DO: 19, (1.972). Methods for Plane Strain Fracture Toughness K_{ic} Testing. DD3. (1971). *British Standard BS5762*.

Paris, P. C. (1970). Testing for Very Slow Growth of Fatigue Cracks. *MTS Close Loop Magazine*, 2(5).

Paris, P. C., Bucci, R. J., Wessel, E. T., Clark, W. G., & Mager, T. R. (1972). *Extensive Study of Low Fatigue Crack Growth Rates in A533 and A508 Steels, ASTM STP513*. Philadelphia: American Society for Testing and Materials.

Schmidt, R. A., & Paris, P. C. (1973). Threshold for Fatigue Crack Propagation and Effect of Load Ratio and Frequency. In Progress in Flow Growth and Fracture (p. 79). Philadelphia: ASTM.

Schweizer, C., Seifert, T., Nieweg, B., Von Hartrott, P., & Riedel, H. (2011). Mechanisms and modelling of fatigue crack growth under combined low and high cycle fatigue loading. *International International Journal of Fatigue*, *33*(2), 194–202. doi:10.1016/j.ijfatigue.2010.08.008

Silverio, R. C., Moura, T., & Guedes, P. (2003). Evaluation of cold crack susceptibility on HSLA steel welded joints. *Rev. Metal. Madrid*, *39*(2), 83–89. doi:10.3989/revmetalm.2003.v39.i2.318

Sree, P. C. R., & Kujawski, D. (2014). A Deviatoric Stress-Strain Function for Fatigue Life Analysis. *Advanced Materials Research*, *891-892*, 789–802. doi:10.4028/www.scientific.net/AMR.891-892.797

Standard method for estimating the average grain size of metals, ASTM: E 112-63. (n. d.).

Suárez, J. C., Molleda, F., González, R., & Jiménez, R. (1994). Correlation of modified crack tip opening distance with heat input to the heat affected zone of high-strength low-alloy steels. *Theoretical and Applied Fracture Mechanics*, *21*(1), 17–22. doi:10.1016/0167-8442(94)90004-3

KEY TERMS AND DEFINITIONS:

AC1: Region of peak temperature at 720°C.

AC3: Region of peak temperature at 900°C.

AW: Aswelded conditions, metal without post-weld heat treatment.

FGHAZ: Fine grain, it is region within the heat affeced zone of the metal that is being welded, where the temperature reached is between 900°C and 1100°C.

FL: Fusion line. It is the line between the welding and heat affected zone.

GCHAZ: Grain coarse heat affected zone, it is region within the heat affeced zone of the metal that is being welded, where the temperature reached is between 1100°C and 1450°C. It is a fragile zone.

HEYN: It is a method to measure the grain size in the heat affected zone of steel after welding, observed by microscope, grain size in accordance with the astm (112-63) standard.

HI: Heat input, heat energy, it depends of electrical and voltage of the welding arc and the speed of the tourch. in the hsla steel is limited to 2.6 kJ/mm.

HSLA: High strength low alloy steel, they are microalloyed steels with high yield point. they have low content of carbon compensated with microalloys.

IRGCHAZ: Intercritical grain coarse zone, it is region within the heat affeced zone of the metal that is being welded, where the temperature reached is between 720°C and 900°C. Reheated by passes of welding.

SAW: Submerged Arc Welding. Automatic welding process.

SC: It is region within the heat affeced zone of the metal that is being welded, where the temperature reached is lower than 720°C.

Chapter 9
Model of a Performance Measurement System for Maintenance Management

José Contreras
INGECON, Venezuela

Carlos Parra
University of Seville, Spain

Adolfo Crespo Márquez
University of Seville, Spain

Vicente González-Prida
University of Seville, Spain

Fredy Kristjanpoller
Universidad Técnica Federico Santa María, Chile

Pablo Viveros
Universidad Técnica Federico Santa María, Chile

ABSTRACT

This chapter book will analyze different systems of indicators by which maintenance management is evaluated. The models analyzed are: 1) The Balanced Scorecard. 2) The Alsyouf Model. 3) The Maintenance Scorecard Model. 4) The Metrics for Maintenance of SMRP (Society of Maintenance and Reliability Professionals). 5) The EN-15341-Maintenance Key Performance Indicators (European Standard). This paper presents a model for indicators based on various hierarchical levels and different functions and processes taking place in a maintenance department. With this model a comprehensive assessment of the most important aspects at all levels of the organization is achieved and shows the relationship between the various indicators to understand the overall performance of maintenance management and so align departmental objectives with the strategic objectives. Also is developed the C-KPI-M model, which is the definition of 18 chains of key indicators for viewing the cause and effect of the maintenance KPI that contribute to achieving the maximum Economic Value Added (EVA).

DOI: 10.4018/978-1-5225-0651-5.ch009

INTRODUCTION

In many companies the maintenance departments perform management control based on a list of indicators that, in principle, are necessary and important to verify the level of performance resulting from the maintenance management. The problem with using indicators in isolation, without knowing the cause and effect indicators for different functions and hierarchical levels, you do not have a comprehensive view of both operational performance and financial management system. Another problem of not having a system of performance measurement is that those responsible for carrying the indicators, are often unaware of the applicability and the importance of calculating certain causing discouragement in the performance of their duties indicators. Therefore, it is necessary that a system of indicators for maintenance management where everyone involved understands the importance and impact of the function it performs in the overall results of the company is formalized.

The model of a system for maintenance management indicators proposed has its foundation in the basic organizational structure, which in our view, should have all maintenance department and its relationship to the highest hierarchical level or senior management of any industrial organization. A functional analysis is performed to define indicators tactical-operational level and model of Balanced Scorecard is used to define indicators in the strategic and corporate levels.

SYSTEMS FOR MAINTENANCE MANAGEMENT INDICATORS

To measure the performance of maintenance management, a set of indicators to measure different aspects of the management process is used, the most common reflect the performance of the equipment and the costs of maintenance. These two types of indicators are clearly lagging indicators and indicate the status after the action has taken place. For this reason, you should design a system of indicators that includes preventive actions to avoid undesirable results.

Indicators should be formulated for the different levels of the organizational structure and for each level, indicators have specific purposes for specific users.

Users at the highest management level relate to the overall administrative performance, while those in the functional levels are related to the physical condition of the assets. The use of multiple indicators of performance levels of systems and subsystems helps solve problems. If a corporate indicator shows a problem, then lower level indicators should define and clarify the cause of the weakness that has caused this problem. The relationship of the indicators at different hierarchical levels with business goals is vital to the success of a program of physical asset management.

It is shown below the basics of five models of systems management indicators that are widely used to measure the performance of maintenance management and constitute the conceptual basis of the model presented here.

The systems in question are:

1. The Balanced Scorecard (BSC)
2. The Alsyouf model (BSC adapted)
3. The Maintenance Scorecard (MSC)
4. The system of indicators of SMRP
5. The system of indicators of EFNMS

1. The Balanced Scorecard (BSC)

The *Balanced Scorecard (BSC)* is a strategy performance tool - a semi-standard structured report, supported by design methods and automation tools, that can be used by managers to keep track of the execution of activities by the staff within their control and to monitor the consequences arising from these actions.

The critical characteristics that define a balanced scorecard are:

- Its focus on the strategic agenda of the organization concerned,
- The selection of a small number of data items to monitor,
- A mix of financial and non-financial data items.

Balanced scorecard is an example of a closed-loop controller or cybernetic control applied to the management of the implementation of a strategy Closed-loop or cybernetic control is where actual performance is measured, the measured value is compared to an expected value and based on the difference between the two corrective interventions are made as required. Such control requires three things to be effective:

- A choice of data to measure,
- The setting of an expected value for the data,
- The ability to make a corrective intervention.

Within the strategy management context, all three of these characteristic closed-loop control elements need to be derived from the organisation's strategy and also need to reflect the ability of the observer to both monitor performance and subsequently intervene - both of which may be constrained. Initially, Balanced Scorecard emerged as a performance management system, over a period of time it has come to be known as a strategy management system, with its ultimate aim being the achievement of long term financial performance. Balanced scorecard is seen as a strategic management system enabling business leaders to meet the challenge of strategy execution.

Two of the ideas that underpin modern balanced scorecard designs concern facilitating the creation of such a control - through making it easier to select which data to observe, and ensuring that the choice of data is consistent with the ability of the observer to intervene.

The BSC was proposed by Kaplan and Norton (1992) and the pioneering application in the field of maintenance is Tsang (1998). It is considered a model that translates the vision, mission and strategy of an organization into a set of numerical targets and performance indicators grouped and link all levels and functions of that organization.

It is a system of performance management that can be used in any organization, large or small, to align the vision and mission with customer requirements, daily chores, managing business strategies, monitoring improvements in operational efficiency, create organizational capacity, communicating progress to all staff. The BSC brings together financial and non-financial indicators to measure performance. It is a vehicle that reflects the mission of a business unit strategy and a set of objective and quantifiable measures built around four perspectives:

1. Financial (points of view of the investor).
2. Client (performance attributes valued by customers).

3. Internal processes (the means short and long term to achieve the financial and customer objectives).
4. Development (Learning) and growth (ability to improve and create value).

The BSC is often used to group key performance indicators and show the different faces of the maintenance function. Performance indicators in maintenance are obtained from the perspectives of the BSC and applied from the divisions to employees. Maintenance objectives are related to key success factors and the result will be the key performance indicators for maintenance that will contribute to achieving the objectives. In "The Balanced Scorecard applied to Maintenance Management" (Contreras, 2014b) may get a detailed explanation of the application of this technique to measure maintenance performance.

2. The Alsyouf Model (Adapted BSC)

Alsyouf discusses the general framework of business processes and then introduces an adapted BSC model which can be used to measure maintenance performance. Discuss the impact of the maintenance function on the elements of business performance according to the BSC approach.

- Although the BSC has been adopted by the industry as a basis for strategic management systems, many authors have criticized the following:
- Failure to identify performance measures as a two-way process, as it focuses only on performance measurement from the top down.
- It is unable to answer what the competition is doing.
- It is rigid and limited to customers, ignores other parts of the company as employees, suppliers, alliance partners.
- No highlights the contribution of employees and suppliers in achieving the objectives of the company.
- Not identify the role of the community in defining the environment within which the company works.

Imad Alsyouf (2006), in his work "Measuring Maintenance Performance Using a Balanced Scorecard Approach" proposes a system of indicators for maintenance management based on the BSC but expanding perspectives on which indicators are developed as follows: Financial, Customers, Society, Production, Maintenance, Innovation and Growth. Figure 1 shows the structure of Alsyouf model that consists of three components:

- **Part 1:** Downstream organisation (corporation, partnership, clients)
- **Part 2:** Operations (Production Support Functions)
- **Part 3:** Upstream organisation (HR, suppliers, consultants)

For each of the proposed management perspectives indicators that reflect the performance of maintenance management and its influence on the operational and financial performance of the corporation are defined.

Figure 1. Adapted BSC model structure
(Alsyouf, 2006)

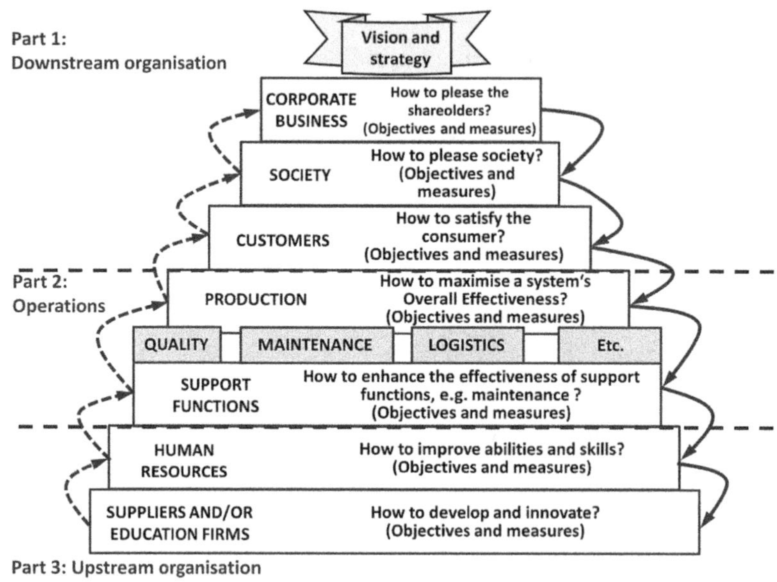

3. The Maintenance Scorecard (MSC)

The MSC is a comprehensive approach used to develop and implement strategy in the area of asset management. It also serves to identify strategic improvement initiatives, along with the areas that they are focused on, early in the process.

As a methodology based in the measurement of performance, the MSC is built around a use of management indicators, or metrics, to lead the development and implementation of strategy. It also recognizes that sustainable competitive advantages come from the implementation of strategy, not merely its development.

While the balanced scorecard has been in existence for over a decade at the time of writing, the level of take-up within the areas of asset management has been very low, despite the growing strategic importance of the area.

This model is proposed by Daryl Mather (2005) who defines it as a tool to implement and communicate corporate strategy throughout the company. The MSC provides a framework to help companies understand how to make asset management a source of competitive advantage by maintaining a balance between profitability and other responsibilities. The structure of the MSC is based on dividing the organization into three hierarchical levels: Corporate, Strategic and Functional (Figure 2), defining for each level, strategic requirements based on 6 prospects, to achieve maximum competitiveness of the organization.

At the corporate level, the competitive advantages consist of the set of unique skills, competencies and capabilities of an organization that allows to compete in the markets involved. At the strategic level, there are the strategic advantages that are unique set of skills, competencies and capabilities of an organization supporting the competitive advantages of the company. Finally there is the functional level comprising the strategic assets defined as the set of skills, competencies and capabilities of an organization that are required to achieve strategic advantages.

Figure 2. Maintenance scorecard model structure
(Mather, 2005)

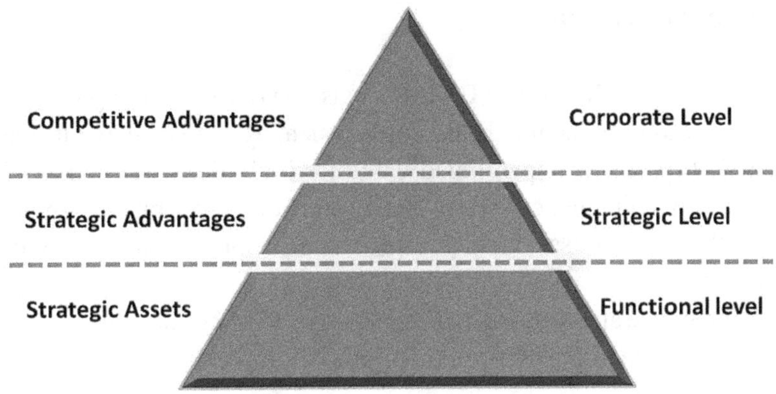

After determining the strategic requirements for each hierarchical level of the organization, the MSC raises the definition of performance indicators to meet such requirements based on the perspectives shown in Figure 3.

For example, at the corporate level should develop competitive advantages from each of the perspectives. From the perspective of productivity, how can the management of assets to the ability to produce more?

This approach may include:

- How can we increase the production time of the machines by reducing time to repair?
- How can we increase the production capacity through the most reliable machinery?
- How can we increase the production time by reducing time lost due to administrative processes?

Figure 3. The "Maintenance Scorecard" model
(Mather, 2005)

The process described above must be developed for all perspectives and at all levels.

4. The SMRP Indicators System

The Society for Maintenance & Reliability Professionals (SMRP) is a nonprofit professional society formed by practitioners to advance the reliability and physical asset management industry and to create leaders in the field. SMRP provides unparalleled value for individual practitioners looking to expand their knowledge and skills in maintenance and reliability and build more business connections with other practitioners. SMRP is the premier asset management resource for companies looking to improve their processes and procedures.

The Society for Maintenance and Reliability Professionals (SMRP) developed a system comprising a total of 67 indicators divided into five groups called pillars of knowledge for the profession "Maintenance and Reliability". These pillars are:

1. Business and Management
2. Manufacturing Process Reliability
3. Equipment Reliability
4. Organization and Leadership
5. Work Management

On the website of the SMRP (http://library.smrp.org/metrics/) there is the detailed information about indicators that has been developed for each of the pillars of knowledge. Below is a brief description of each.

Pillar 1: Business and Management (4 Indicators)

It is the first pillar of the profession, describe the skills used to translate business objectives of an organization in appropriate maintenance and reliability objectives that support and contribute to business results.

Pillar 2: Manufacturing Process Reliability (6 Indicators)

This pillar relates to the contribution of the activities of maintenance and reliability to improve the manufacturing process of the organization.

Pillar 3: Equipment Reliability (9 Indicators)

This pillar establishes requirements to determine the current capabilities of the assets and improvement strategies in relation to its reliability, availability, maintainability and criticality.

There are two kinds of activities that apply to this set of skills:

- Activities used to assess the current capabilities of the equipment and processes in terms of reliability, availability, maintainability and criticality.
- Activities used to select and apply the most appropriate maintenance practices, so that equipment and processes continue delivering their expected capabilities of the safest way and at the lowest cost.

Pillar 4: Organization and Leadership (4 Indicators)

This pillar describes the processes to ensure that reliability and maintenance personnel have the necessary expertise and is the most qualified to achieve the goals of the organization.

Pillar 5: Work Management (44 Indicators)

This pillar of the profession focuses on the skills used for maintenance work and reliability is performed correctly. It includes planning and programming, quality assurance of the activities of maintenance and inventory management.

The Figure 4 shows the structure of the indicator system management SMRP.

5. The EFNMS Indicators System

The EFNMS, the European Federation of National Maintenance Societies, was established in 1970. The EFNMS transformed to a formal non-profit organization ("Vereniging zonder Winstoogmerk") according to Belgian law created on January 18, 2003, in Amsterdam.

The EFNMS objectives are the following: the improvement of maintenance for the benefit of the peoples of Europe.

By the term 'maintenance' is meant: the combination of all, technical, administrative and managerial actions during the lifecycle of an item intended to retain or restore it to a state in which it can perform its required function.

Maintenance is of utmost importance for trade and commerce, for the environment, and for general health and safety. In order to pursue its goals, the EFNMS shall be an umbrella organization for the non-profit National Maintenance Societies in Europe.

In order that it may realize its objectives, the EFNMS shall, amongst other activities:

Figure 4. Structure of SMRP maintenance performance measuring system

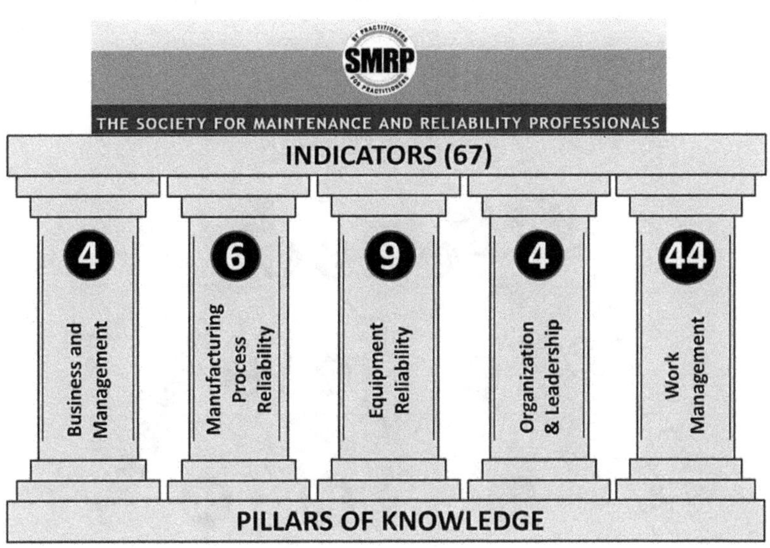

- Co-ordinate maintenance matters between European National Maintenance Societies and establish contacts with Maintenance Societies outside of Europe.
- Study and introduce good maintenance management and promote and initiate maintenance techniques and processes.
- Promote the awarding of prizes and distinctions relating to maintenance.
- Provide direction in the development of maintenance.
- Initiate and promote the publication of the results of scientific and practical work in the maintenance field.
- Draft maintenance guidelines.
- Organize bi-annual international conferences and conventions about maintenance.
- Represent "maintenance", as developed and implemented in Europe, and support it in contacts with official national, supra national, and international institutions.

The European Federation of National Maintenance Societies (EFNMS) developed the "EN 15341-Maintenance Key Performance Indicators", which aims to describe a system to manage key indicators to measure maintenance performance in the framework of the factors that influence it, such as the economic, technical and organizational.

The system includes a total of 71 indicators of three types, called: economic, technical and organizational. They are distributed into three levels equivalent to three hierarchical levels of the organizational structure. For level 1 are defined 19 indicators, of which 6 are economical, 5 technical and 8 organizational. Level 2 has 12 indicators, of which 8 are economical, 2 technical and 2 are organizational. Finally, Level 3 contains 40 indicators, 10 economical, 14 technical and 16 organizational.

The economic indicators are 24, all directly related to maintenance costs. Technical indicators are 21, related to the availability, reliability and maintainability. Organizational indicators are 26, related to the use of resources and time spent for the execution of maintenance activities.

Figure 5 shows the structure of the system of indicators EFNMS where the number of indicators by type and level shown is observed.

Figure 5. Structure of EFNMS maintenance performance measuring system

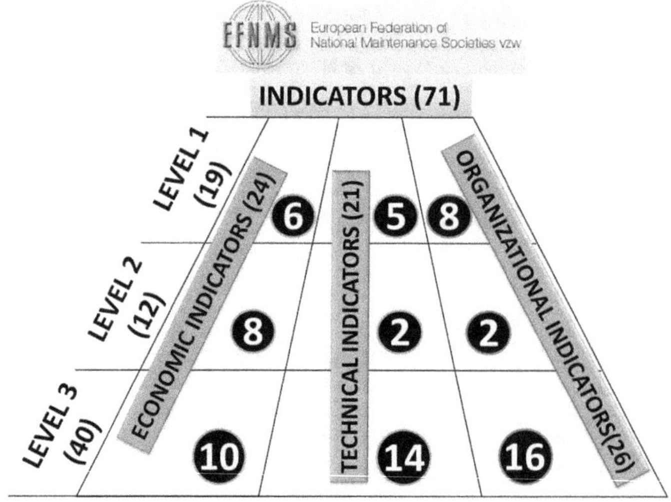

PROPOSED MODEL: HIERARCHICAL-FUNCTIONAL MODEL OF A SYSTEM FOR MAINTENANCE MANAGEMENT INDICATORS

The detailed study and analysis of each of the systems described above, gives the conceptual basis and understand very clearly the importance of adopting a system of indicators rather than a list of indicators for monitoring maintenance management. Each has its strengths, as have already been described, but none related the indicators to the functions and hierarchical levels simultaneously. This weakness is what allows us to propose a model that is based on the definition of functional indicators for tactical-operational level and strategic measures for corporate or senior management levels.

The elements of the proposed model are the following:

1. **The Hierarchical Levels:** The maintenance department is depicted in a three levels organizational structure:
   a. **Corporate Level**: Represented by the plant management or operations of an organization.
   b. **Strategic Level:** Represented by the maintenance department.
   c. **Functional Level**: Represented by the functions developed in the maintenance department.
2. **The Functions of the Maintenance Department:** The basic functional structure of the entire maintenance department should consist of at least the following three areas:
   a. Planning and Scheduling:
      i. **Process:** Identification, Planning, Scheduling
   b. Execution:
      i. **Processes:** Execution, Completion
   c. Maintenance Engineering:
      i. **Process:** Root Cause Analysis and Criticality Analysis, Failure Modes and Effects Analysis, Maintenance Strategies, Equipment Performance.

Management Indicators

As already indicated, the model proposes indicators for three hierarchical levels, of which, level 2 and 3 correspond directly to the maintenance function (strategic level and functional level). The use of level 1 (corporate level) is useful to visualize the importance of maintenance management in the overall results of the company.

To define indicators of corporate and strategic levels, we use the Balanced Scorecard model that proposes the use of four perspectives to measure the performance of management. For its part, the functional level indicators are defined for the processes taking place in each of the functional areas of the maintenance department.

Figure 6 shows the structure of the proposed model, where we can observe that the indicators at higher levels are "Lagging Indicators", also known as indicators of "delay". At the functional level, indicators are mostly "Leading Indicators", also known as indicators of "progress".

The detailed model is shown in Figure 7, in which you can see the number of indicators for each hierarchical level. For level 1 (corporate level) are defined 24 indicators, corresponding 6 to each of the perspectives of Balanced Scorecard. Strictly speaking, these first 24 indicators, are not directly part of a system of indicators for maintenance, but their inclusion is very important to visualize and understand

Figure 6. Structure of hierarchical-functional model of a performance measurement system for maintenance management

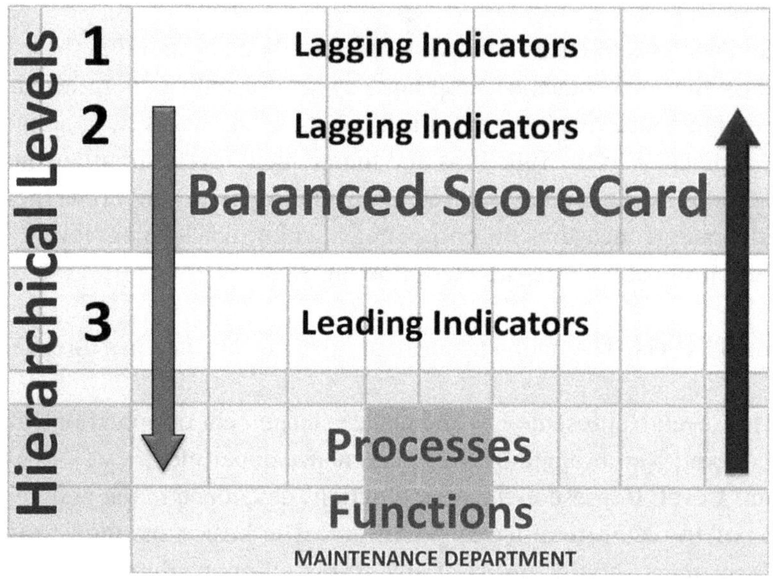

the importance of maintenance for business strategy. For level 2 (strategic level), are defined 20 indicators, corresponding 5 to each of the perspectives of the Balanced Scorecard and show the requirements to be met by the maintenance department to help achieve the strategic objectives. For level 3 (functional level), we propose a total of 38 indicators, distributed into the processes to be developed by the basic functions of a maintenance department. Importantly, this model offers a total of 58 indicators and no 82, as it seems, as it only the indicators proposed for levels 2 and 3 are directly part of the proposed system of indicators.

Finally, in Figure 8 there are the specific indicators proposed for each hierarchical level, function-processes and perspectives of BSC.

Advantages of the Proposed System

The use of the Hierarchical-Functional Model for defining indicators for maintenance management has the following advantages:

1. Define indicators for each hierarchical level of the organizational structure of the maintenance department.
2. Define indicators for each function and each process is carried out in the maintenance department.
3. Each person in the maintenance department clearly understand what is your responsibility within the department and has clearly defined its objectives and specific targets to be achieved.
4. The cause-effect relationship between the various indicators of the processes is clearly seen by the natural sequence in which they are developed.
5. The importance of lagging indicators is clearly understood to achieve the objectives and achieve the goals expressed in the lagging indicators.

Figure 7. Detail of hierarchical-functional model of a performance measurement system for maintenance management

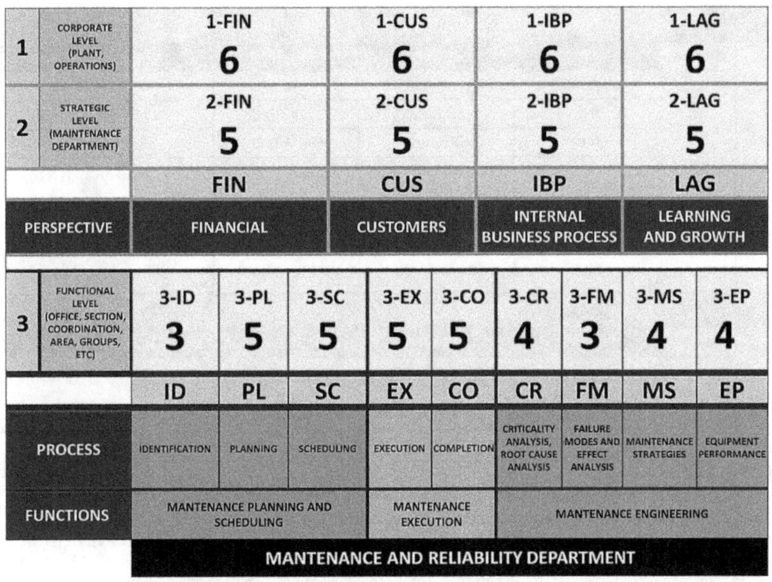

CHAINS OF KEY PERFORMANCE INDICATORS FOR MAINTENANCE

As we all know, a chain is defined as a series of links interwined to form a single body that has some utility. Under this concept, we make the analogy of maintenance indicators with the links of a chain to form the C-KPI-M.

The C-KPI-M are developed from the analysis of cause and effect that an indicator has on another. Under this criterion, there are two types of indicators to measure performance. The indicators related with causes are called "Leading Indicators" and indicators related with the effects are called "Lagging Indicators".

A proper translation for "Leading Indicators" can be "advance indicators", these are indicators that measure performance before the result begins to follow a particular pattern or trend. They can be used to predict changes and trends. Meanwhile, the "Lagging Indicators" can be translated as "delay indicators" or "result Indicators", which measure the performance after the result begins to follow a particular pattern or trend. These indicators confirm long-term trends but does not predict these trends.

In a Maintenance Performance Measurement System, each component is a process that can be measured using leading and lagging indicators. This depends on the place or part of the process is used the indicator. There is a cause-effect relationship between leading and lagging indicators. The action being measured cause a resulting action or effect which in turn is also being measured. This means that given a measure, this could be both a lagging indicator for a preliminary cause of the chain and a leading indicator for the next effect. Exist a serie of causes-effects in the chain until at last the lagging indicators is reached.

Figure 9 shows that an indicator may be leading and lagging at the same time, depending on the application of the measure. For example, the schedule compliance of preventive maintenance (PM) is used to measure how much scheduled work has been completed. Seen in the context of the implementa-

Figure 8. Specific indicators of hierarchical-functional model of a performance measurement system for maintenance management

OPERATIONS MANAGEMENT – PLANT MANAGEMENT					
	PERSPECTIVE	FINANCIAL	CUSTOMERS	INTERNAL PROCESS	LEARNING AND GROWTH
CORPORATE LEVEL (PLANT, OPERATIONS) 1	INDICATORS	PRODUCTION COSTS	SATISFACTION LEVEL	OEE	NEW PRODUCTS QUANTITY
		EBIT	TIME DELIVERIES	QUALITY	NEW PRODUCTS SELLS
		EBITDA	CLAIMS QUANTITY	PRODUCTIVITY	TRAINING INVESTMENT
		ROI	MEAN TIME TO PROCESS ORDERS	PROUCTION LEVEL	STRATEGIC HABILITIES LEARNED
		ROE	RETURNED ORDERS	ENVIRONMENTAL PERFORMANCE	R &D INVESTMENT
		ROCE	NEW CUSTOMERS	CAPACITY UTILIZATION	SUGGESTIONS AND CONTRIBUTIONS OF EMPLOYEES

MANTENANCE DEPARTMENT AND RELIABILITY					
	PERSPECTIVE	FINANCIAL	CUSTOMERS	INTERNAL PROCESS	LEARNING AND GROWTH
STRATEGIC LEVEL (MAINTEN. DEPT) 2	INDICATORS	LABOR COSTS	SATISFACTION LEVEL OF OPERATIONS	BACKLOG	TRAINING HOURS PER EMPLOYEE
		MATERIALS COSTS	CLAIMS QUANTITY	AVAILABILITY	NUMBER OF TRAINED STAFF
		SUPPORT ACTIVITIES COSTS	WORK ORDERS SERVED ON TIME	LABOR UTILIZATION LEVEL	CMMS UTILIZATION LEVEL
		COSTS OF HIRING	SPEED OF ATTENTION	INVENTORY SERVICE LEVEL	NUMBER OF PEOPLE TRAINED IN MAINTENANCE MANAGEMENT
		UNAVAILABILITY COSTS	EXECUTED WORK QUALITY	SHUTDOWNS PERFORMANCE	NUMBER OF PEOPLE TRAINED IN OPERATIONAL RELIABILITY

MANTENANCE DEPARTMENT AND RELIABILITY										
	FUNCTION	MAINTENANCE PLANNING AND SCHEDULING			MAINTENANCE EXECUTION		MAINTENANCE ENGINEERING			
	PROCESS	IDENTIFI-CATION	PLANNING	SCHEDULING	EXECUTION	COMPLETION	CRIT. AN. & RCA	FMEA	MAINTEN STRATEG.	EQUIPM. PERFORM.
FUNCTION LEVEL (OFFICE, AREA, SECTION, GROUP) 3	INDICATORS	PROACTIVE WORK IDENTIFIED	PLANNING VARIATION INDEX	SCHEDULING COMPLIANCE	BACKLOG BY SPECIALITY	WRENCH TIME	SPENT TIME FOR CRITICALIT Y ANALYSIS	SPENT TIME FOR FMEA	% PREVENTIVE MAINTENAN	MTBF
		% ATTENDED IDENTIFIED WORK	PLANNING VARIATION INDEX	SCHEDULING INTENSITY	MTTR	FORMAL CLOSURE OF WORK ORDERS	SPENT TIME FOR RCA	SYSTEMS COVERED BY FMEA	% PREDICTIVE MAINTENAN	MTTF
		COMPLIANCE WITH PROCEDURES TO IDENTIFY AND COMMUNIC. NEW WORKS	PLANNING INTENSITY	RESPONSE OF SCHEDULING	COMPLETION OF WORK ON SCHEDULE	VARIATION OF PLANNED TIME VS REAL TIME	SYSTEMS COVERED BY CRITICALIT Y ANALYSIS	MAINTEN. PLANS ORIGINATED FROM FMEA	% PLANNED CORRECTIVE MAINTENAN	TOTAL OPERATING TIME
			RESPONSE OF PLANNING	SCHEDULED TIME PRECISION	EXECUTED WORK QUALITY	VARIATION OF PLANNED COSTS VS REAL COSTS	SYSTEMS COVERED BY RCA		% CORRECTIVE MAINTEN. FOR EMERGENC.	MEAN OPERATING TIME
			PLANNING QUALITY	SCHEDULED LABOR UTILIZATION	SCHEDULED LABOR UTILIZATION LEVEL	IMPROVEMENT IDEAS				

Figure 9. C-KPI-M basis (SMRP)

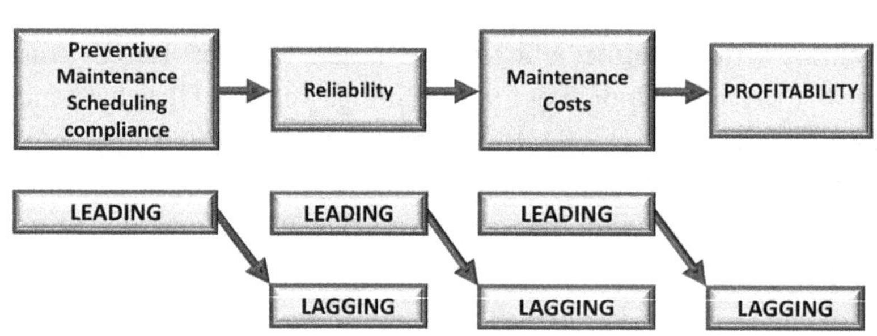

tion of the work, the result of what has been accomplished MP is a lagging indicator. However, when this indicator is seen as an indicator of the reliability of the equipment, it is a leading indicator of the process of operational reliability. The higher the compliance of preventive maintenance schedules, it is likely to lead to increased equipment reliability. Similarly, improved reliability will lead to a reduction in maintenance costs, which is a lagging indicator of the overall maintenance process. When a leading indicator is analyzed, it is very important to recognize the lagging indicator that is being affected.

The C-KPI-M to Reach the Maximum Economic Value Added

This model of C-KPI-M is based on the use of the financial indicator Economic Value Added (EVA) serving as a reference for analyzing the impact of maintenance management on the financial results of the company. EVA is a financial indicator that incorporates the calculation of the cost of equity and so investors can evaluate the performance of the company in terms of wealth creation, above the coverage of basic needs for subsistence. It provides a measure of the profitability of a company as the result of the net profit after taxes less the applicable charge for the opportunity cost of all capital that is invested in the company.

How Is EVA Calculated?

The EVA is calculated by subtracting the cost of capital employed to Earnings Before Interest and Taxes (EBIT). The cost of capital employed is calculated by multiplying the value of the Operating Assets (OA) by the Weighted Average Cost of Capital (WACC). The mathematical expression is: EVA = EBIT - (OA x WACC).

Analyzing the above expression, we can determine that the EVA can be increased if profits (EBIT) is increased and the operating assets (OA) is decreased. Now, the only way to increase profits is increasing revenues and reducing costs. Meanwhile, the operating assets can only be reduced if the value of the assets used to produce the benefits, decreases.

Let us see, from the strict point of view of maintenance management, how to increase the EVA.

Increased Revenue

Sales revenue of a company can be increased by increasing the availability and this in turn increases with increasing reliability, reduces Mean Time to Repair (MTTR) and if managed properly shutdowns, turnarounds and other maintenance projects. Increased reliability is synonymous with decreasing the frequency of failures, the MTTR reduction mainly will be achieved as long as the work is executed with proper planning and better performance in the shutdowns and turnarounds will be achieved with the formal implementation of project management methodology for all maintenance projects which will result in the execution of them in the shortest time possible.

When is analyzed as decrease the frequency of failures, the highlight aspects are: The Root Cause Analysis (RCA) and the planned maintenance resulting of predictive maintenance inspections and preventive maintenance that routinely performs all maintenance departments. Decisions related to the increase in maintenance activities based on the condition (predictive) and based on time intervals (preventive) must be the result of the correct application of the methodology of Reliability Centered Maintenance (RCM).

The decrease in the Mean Time to Repair (MTTR) is primarily achieved by developing fully the cycle of maintenance work management comprising the following processes: identification, planning, scheduling, execution and completion.

The best performance during the execution of maintenance activities framed as a project (shutdowns, turarounds) will only be achieved if all the activities are developed under the methodology of project management.

Reduced Costs

To reduce costs, we consider three main elements. Two of them have already been analyzed as key players to achieve the objective of increasing the availability, it is the cycle of maintenance work management and management of plant shutdowns. Just as a good performance in these aspects contribute to increase the availability, simultaneously it produces a lower total cost of maintenance management. The third key element for reducing the total cost is inventory management, which unfortunately has not been given the importance, because management inefficient inventory leads to high costs and the worst is that they are not clearly visible, which are considered most often as hidden costs.

The overall objective of inventory management for maintenance is to minimize investment to keep stored materials, that is, it seeks to have the minimum total value of the inventory, which basically depends on order point and order quantity. These parameters must be estimated correctly to optimize inventory levels maintained by the company. Another factor that negatively affects the pursuit of the objective of minimizing costs is the level of obsolescence that exists in the store. Obsolete materials for all they serve is to increase the value of inventory.

Decrease in Operating Assets

As we know, the Operating Assets (OA) are equipment, machinery and inventories maintained by the company to meet the requirements of operations.

Maintenance management contributes to the decrease in Operating Assets maintaining the minimum inventory value possible without affecting the continuity of operations and full use of the machinery and equipment available to counter the need to acquire more assets that increase in value.

It is obvious that the higher value of inventory is the greatest value of the OA in the same way that increases costs, as was previously discussed.

In case of utilization factor is very important to recognize the impact of maximizing the equipment available and this is done from the point of view of maintenance, reaching maximum availability, something that was also previously analyzed.

Global Map of the C-KPI-M

From the above analysis, we have developed the global map of the C-KPI-M shown in Figure 10. In this map there are 18 roads or routes, from the point of view of maintenance management, leading to achieve maximum Economic Value Added (EVA). In these ways we have called "Chains of Key Performance Indicators for Maintenance" or "C-KPI-M", which graphically illustrate the causal relationship from the leading indicators obtained in the tactical-operational levels to the lagging indicators obtained at

Figure 10. Global map of C-KPI-M

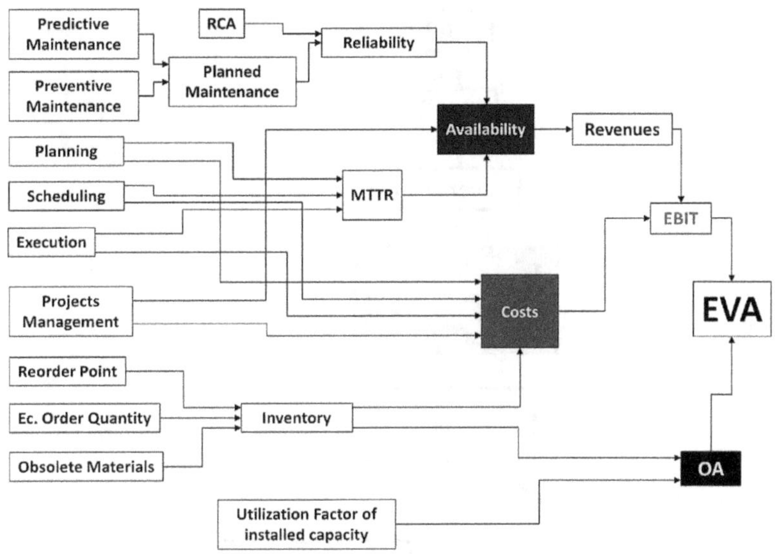

the management-strategic level of the maintenance department and the organization. The C-KPI-M is a model that integrates technical and administrative indicators for maintenance to the objective of achieving the highest Economic Value Added (EVA).

Looking at the map, you can think about the possibility of generating more than 18 chains, as there is a greater number of relationships than shown, but what is proposed is to consider the most important influence of a factor over another. Under this premise arise 18 chains which can be classified as follow: Three chains from reliability, 8 chains from work management, 6 chains from inventory and 1 chain from utilization factor.

Reliability chains owe their name to the fact that since the implementation of activities arising from the application of reliability centered maintenance methodology is that most assets achieves the highest reliability and thus greater availability.

The work management Chain show the impact on the average repair time and costs if properly develop the cycle of maintenance work management (identification, planning, scheduling, execution and completion), as well as the formal application the methodology of project management to plant shutdowns and other maintenance projects.

Inventory chains show the influence of the proper management of inventories on decreased costs and operating assets. Finally, the utilization factor that directly indicates that a higher level of utilization of assets less will be the need to acquire new assets.

Figures 11, 12, and 13 show the chains according to the classification developed.

C-KPI-M Indicators Formulas

Table 1 shows the formulas of the C-KPI-M indicators.

Figure 11. 3 reliability chains

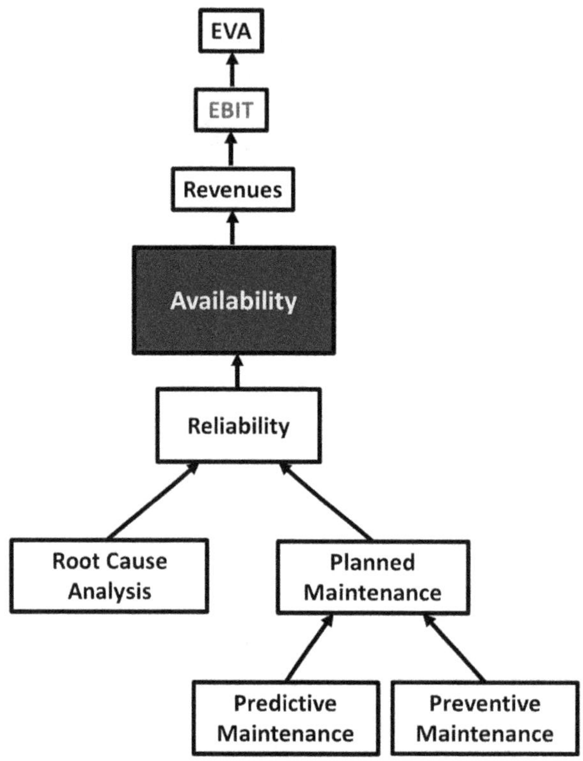

Figure 12. 8 work management chains

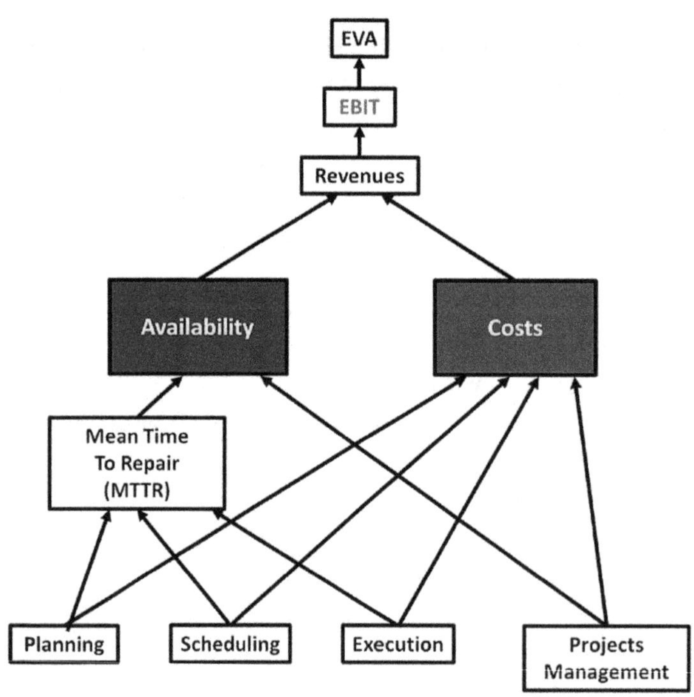

Figure 13. 6 inventory chains and 1 utilization factor chain

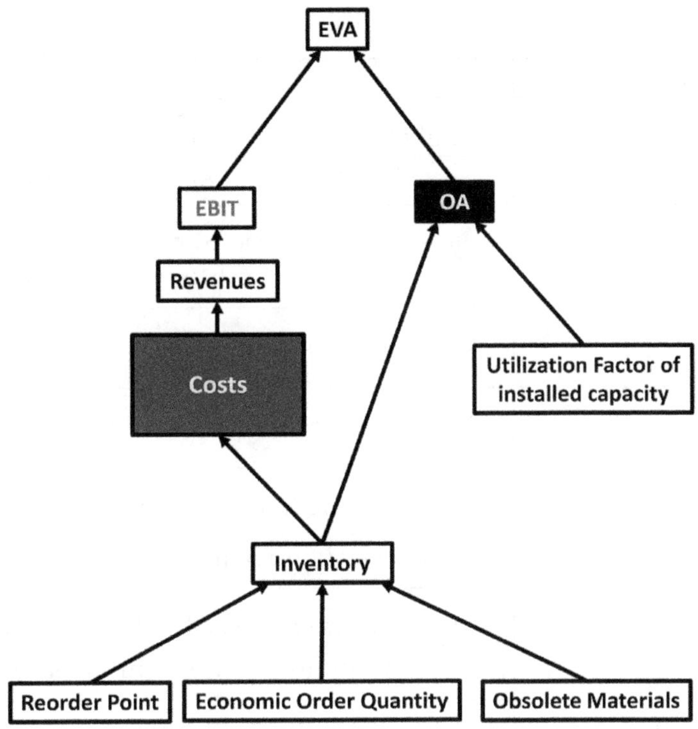

Table 1. Formulas of the C-KPI-M indicators

Indicator	Formula
Root Cause Analysis	$\frac{\text{Number of systems cover by a Root Cause Analysis}}{\text{Total number of systems}}$
Predictive Maintenance	$\frac{\text{Man-hours of Predictive Maintenance}}{\text{Total Man-hours of Maintenance}}$
Preventive Maintenance	$\frac{\text{Man-hours of Preventative Maintenance}}{\text{Total Man-hours of Maintenance}}$
Planned Maintenance	$\frac{\text{Man-hours of Planned and Scheduled Maintenance}}{\text{Total Man-hours of Maintenance}}$
Planning	$\frac{\text{Number of WO closed with actual cost between } \pm \text{ 20\% of the planned costs}}{\text{Total Number of WO executed}}$
Scheduling	$\frac{\text{Number of WO executed}}{\text{Total Number of WO scheduled}}$

Continued on next page

Table 1 continued

Indicator	Formula
Execution	$\frac{\text{Number of WO executed that needs rework}}{\text{Total Number of WO executed}}$
Project Management	$\frac{\text{Number of Projects closed with actual time} \pm 15\% \text{ of the planned time}}{\text{Total Number of Projects executed}}$
Reliability	Mean Time to Failure (MTTF)
MTTR	Mean Time to Repair (MTTR)
Availability	$\frac{\text{Operating Time (Uptime)}}{\text{Uptime + Maintenance Downtime}}$
Reorder Point	$\frac{\text{Number of items with valid Reorder Point}}{\text{Total number of items in inventory}}$
Economic Order Quantity	$\frac{\text{Number of items with valid Economic Order Quantity}}{\text{Total number of items in inventory}}$
Obsolete Materials	$\frac{\text{Value of Obsolete Materials Inventory}}{\text{Total Value of Inventory}}$
Inventory	$\frac{\text{Total Value of MRO Inventory}}{\text{Total Value of Operating Assets}}$
Costs	$\frac{\text{Total Maintenance Cost}}{\text{Total Quantity of Units Produced}}$
Utilization Factor	$\frac{\text{Operating Time (Uptime)}}{\text{Total Time Available}}$

CONCLUSION

In this article different systems of indicators by which management evaluates maintenance are analyzed. All analyzed systems offer many advantages when used according to the model they are based, but not directly present the relationship between the various indicators. A common problem is that many companies use a list of indicators on which decisions are based on the higher levels.

This paper presents a model for developing a system of indicators for maintenance management based on the definition of indicators for the various hierarchical levels and different functions and processes taking place in a maintenance department. With this model a comprehensive assessment of the most important aspects at all levels of the organization is achieved and shows the relationship between the various indicators to understand the overall performance of maintenance management and so align departmental objectives with the strategic objectives.

The C-KPI-M model define the sequence of key indicators in the management of maintenance directly contribute to achieve the maximum EVA. This model is useful when you want to implement quickly and effectively a system of performance measurement in keeping in line with the financial objectives of the organization.

REFERENCES

Alsyouf, I. (2006). Measuring maintenance performance using a balanced scorecard approach. *Journal of Quality in Maintenance Engineering*, *12*(2), 133–149. doi:10.1108/13552510610667165

Amaratunga, D., Haigh, R., Sarshar, M., & Baldry, D. (2002). Application of the Balanced Score-Card concepts to develop a conceptual framework to measure facilities management performance within NHS facilities. *International Journal of Health Care Quality Assurance*, *15*(4), 141–151. doi:10.1108/09526860210431805

Arora, R. (2002). Implementing KM: A balanced score card approach. *Journal of Knowledge Management*, *6*(3), 240–249. doi:10.1108/13673270210434340

Contreras, J. (2014a). Gestión del mantenimiento. Una lista de indicadores o un sistema de indicadores de gestión. Retrieved from https://app.box.com/s/tnp0pw8kjo6s3o2vzek0

Contreras, J. (2014b). El Balanced Scorecard aplicado a la gestión del mantenimiento. Retrieved from https://app.box.com/s/syfy3mklpt8nca7iq2k9

De Groote, P. (1995). Maintenance performance analysis: A practical approach. *Journal of Quality in Maintenance Engineering*, *1*(2), 4–24. doi:10.1108/13552519510089556

European Federation of National Maintenance Societies. (2007). *EN-15341 - Maintenance Key Performance Indicators.*

Kaplan, R. S., & Norton, D. P. (1992). Using the Balanced Scorecard as a Strategic Management System. *Harvard Business Review*, *74*(1), 75–85.

Kumar, U. (2006). Development and implementation of maintenance performance measurement system: *Issues and challenges. WCEAM Paper 127.*

Mäkeläinen, E. (1998). Economic Value Added, EVA. Retrieved from http://www.evanomics.com

Manjarres, M. (2013). Valor economico agregado, ebitda y flujo de caja ¿qué es el valor económico agregado? Retrieved from http://www.ebanque-pdf.com/fr_valor.html

Mather, D. (2005). *The Maintenance Scorecard – Creating Strategic Advantage*. New York: Industrial Press.

Muralidharan, R. (2004). A framework for designing strategy content controls. *International Journal of Productivity and Performance Management*, *53*(7), 590–601. doi:10.1108/17410400410561213

Parida, A., & Kumar, U. (2006). Maintenance performance measurement (MPM): Issues and challenges. *Journal of Quality in Maintenance Engineering*, *12*(3), 239–251. doi:10.1108/13552510610685084

Society for Maintenance and Reliability Professionals. (2014, December 5). Retrieved from http://library.smrp.org/metrics/

Tsang, A. H. C. (1998). A strategic approach to managing maintenance performance. *Journal of Quality in Maintenance Engineering*, *4*(2), 87–94. doi:10.1108/13552519810213581

Tsang, A. H. C., Jardine, A. K. S., & Kolodny, H. (1999). Measuring maintenance performance: A holistic approach. *International Journal of Operations & Production Management*, *19*(7), 691–715. doi:10.1108/01443579910271674

Visser, J., & Pretorious, M. (2002). The development of a performance measurement system for maintenance. Retrieved from http://sajie.journals.ac.za

Chapter 10
AI and Statistical Technologies for Manufacturing and Maintenance Strategies improvement:
Health Monitoring for Electromechanical Actuators

Susana Ferrerio Del Río
IK4-Tekniker, Spain

Iñaki Bravo-Imaz
IK4-Tekniker, Spain

Santiago Fernández
IK4-Tekniker, Spain

Egoitz Konde
IK4-Tekniker, Spain

Aitor Arnaiz Irigaray
IK4-TEKNIKER, Spain

ABSTRACT

The development and the implementation of advanced actuation systems has increased in recent years, as many factors are driving the migration from hydraulic actuators to electromechanical actuators (EMAs) in aeronautics. But not only do we have to consider the right design to customize the system from the requirements oriented to the final application, also additional functions that can provide the system with additional value, to make it more competitive in this market. This is the case of the Health Monitoring (HM) systems. The development, implementation and integration of predictive algorithms into the environment of the EMA provide the system with an additional functionality, from which it is possible to detect failures at an early stage in order to avoid catastrophic accidents and improve maintenance activities. This work shows how to develop HM algorithms based on AI and Statistical technologies to detect and predict early stages of failure in a gearbox, which can directly affect to the transmission of power in EMAs.

DOI: 10.4018/978-1-5225-0651-5.ch010

1 INTRODUCTION

Nowadays, large aeronautical multinationals, aircraft Original Equipment Manufacturers (OEMs) and their suppliers are immersed in the concept called "More Electric Aircraft", which will enable the aircraft industry to improve significantly in terms of reduction in aircraft weight, fuel consumption, total life cycle costs, maintainability and reliability. This trend is accelerating in recent years and these companies are investing heavily towards this concept of more electric aircraft, in which traditional hydraulic and pneumatic systems are being replaced by electrically-driven systems.

In this context, the development and the implementation of advanced actuation systems has increased, as many factors are driving the migration from hydraulic actuators to electromechanical actuators (EMAs). EMAs provide significant advantages in complex applications because they increase the controllability of the system (shortening changeover times), provide re-configurability, maintain functionality during faults (improving the accuracy and reliability), and make it possible to carry out advanced diagnostics and prognostics for a more intelligent maintenance, furthering the increment of the aircraft availability with the long-term planning for maintenance activities. Overall, the main objective for the companies in the last few years is to save operating costs, fuel burn and environmental impact.

Due to these necessities and regarding the development of EMAs, many issues must be taken into account. Not only do we have to consider the right design to customize the system from the requirements oriented to the final application (i.e., cargo door, landing gear..), but also additional functions that can provide the system with additional value, to make it more competitive in the market. This is the case of the Health Monitoring (HM) systems. The development, implementation and integration of health monitoring algorithms into the environment of the EMA provide the system with an additional functionality, from which it is possible to detect failures at an early stage in order to avoid catastrophic accidents. Nevertheless, it is not an easy and simple task because a lot of subsystems or components are responsible for causing the fault of the EMA. A HM system must use data that already exist within the EMA and its developed algorithms, addressing all the parts of the system (i.e., electronics, sensors, motor and mechanical parts) in order to detect or predict the fault before affecting the actuator output in a critical way. In order to develop a HM system for the EMA, it is necessary to identify and define the architecture and the parameters that can be monitored from the system, and to develop the algorithms for anomaly detection and diagnostics to be implemented and integrated into the EMA finally.

Gearboxes are a common component in numerous actuation systems and EMAs in particular. They provide speed and torque conversion from the rotating power source to another mechanism, such as a screw. Direct drive mechanisms exist to take power from the source without any reduction. Nonetheless gearboxes add flexibility in design and application.

Component failures impact overall effectiveness, by means of reducing availability, reliability and safety, increasing cost of maintenance and repairs, and impacting performance. The role of gearboxes in the transmission of power makes them important in EMAs. Transmission of power in EMAs can be, therefore, affected by gear faults such as backslash, wear, or even broken tooth. As a consequence, health monitoring and assessment of this component is desirable.

A typical approach to gear health assessment focuses on direct analysis, such as threshold crossing, of sensor measurements and characteristic feature values. However, the use of statistical and computational techniques is becoming more common in order to handle uncertainty and complex decisions, and also because of their capabilities for flexible modeling and self-learning abilities.

In Al-Atat, Siegel, and Lee, (2011) studied broken tooth in gear, shafts and bearing defects in a gearbox. Regime segmentation from test rig signals was approached, plus an extensive set of features was extracted from vibration signals. In addition to this, a diagnostics and health assessment method was proposed based on a similarity (distance) approach and probability of defect calculation. Bechhoefer and He (2012) describe a process for data driven prognostics and, as an example, use a gear fault run to failure test. Their approach fuses features into a health indicator through statistical process. In (Chen, Yang, & Hu, 2012), a technical framework for embedded diagnostics and prognostics of complex systems is proposed and illustrated on a gearbox case study. The use of advanced data fusion and predictive techniques is envisaged. (Cuc, 2002) illustrates the applicability of neural network-based and neuro-fuzzy methods for helicopter gearbox health monitoring. (Lei & Zuo, 2009) studied gear crack level identification using a classification algorithm (weighted K nearest neighbor). (Li, lin Zhang, Tian, shan Mi, sheng Liu, & quan Ren, 2011) performs hybrid fault diagnosis of a gearbox using four classifiers, namely K nearest neighbor, nearest mean, linear discriminant and least-square support vector machine. (Saravanan & Ramachandran, 2009) and (Saravanan & Ramachandran, 2010) studied gear box fault diagnosis using decision tree classification and artificial neural network. Artificial neural network was also chosen by (Wu, Saxena, Patrick, & Vachtsevanos, 2005) for fault diagnosis of helicopter planetary gears. And more recently, a neural network ensemble classifier was used by (Yuan, Lan, & Gan, 2013) for gearbox fault location. In (Lu & Qiao, 2014a) and (Lu & Qiao, 2014b) employed multiclass SVM classifier for diagnosis of gearbox faults.

This work shows how to develop health monitoring algorithms to detect and predict early stages of failure in a gearbox, taking into account vibration signals, which can be obtained from the EMA by means of triaxial accelerometer in on-ground testing.

Our work presents experimental testing and data collection of healthy and faulty gears (chipped/broken tooth, wear, abrasion) and extracts an extensive set of features with two pre-processing techniques (TSA, Windowing). Moreover, an emphasis is put on data mining techniques for diagnostics and health assessment. A number of technologies are introduced and applied for diagnostics and health assessment of gears from vibration sensor signals.

The paper is organized as follows. In Section 2 a description of the activities related with the experimental tasks is given: the failure modes considered for the health assessment of the gearbox, the test rig description, the experimental testing definition and execution and the first impression of the signals. Then, Section 3 presents a brief introduction of artificial intelligence technologies utilized in the analysis and their ability to predict an early health assessment of the gearbox. And finally, Section 4 introduces the conclusions and future work.

2 DATA ACQUISITION AND EXPERIMENTATION

As previously mentioned in the introduction section, there are a lot of components or subsystems that need to be monitored in order to avoid faults in the EMA (i.e., motor, power electronics, bearings, screw, gearbox..). Each and every one of them has different failure modes and propagation, which are directly related to some failures at system level. These failure modes must be detected in their early stages to take actions and fix them before affecting the EMA performance, which leads to a fault.

This section presents the main failure modes identified and defined for the gears (which form the gearbox). Some of these failure modes have been simulated and tested in a test rig in order to study and establish the way to detect them from vibration signals, and then to be able to assess an index of the state of the health of the component. A more detailed description of these failures, an overview of the test rig used for the experimentation and data acquisition, and the summary of the experimental testing and some initial impressions from the data are given below.

2.1 Failure Modes

Table 1 presents a description and a summary for the most common failure modes for the gearbox: pitting, adhesion, scuffing, abrasion, plastic deformation, ductile fracture and corrosion. It defines each failure, the way the failure spreads, how it could affect at system level and some recommended task to be performed to prevent the fault in the system.

Notice that some of the failures are complex to be simulated in a test rig, either because the complexity of the phenomenon or the number of operating cycles needed for it to be reproduced. Moreover, almost all the failures at component level produce the same type of failures at system level; the EMA has some difficulties to perform its function because of a mechanical breakage, the required torque is not well transmitted (which implies problems establishing system accelerations) or there are oscillations and vibrations (possibly causing fluctuating loads that can cause fatigue problems in other components, and higher motor consumption and torque that may increase the temperature of the motor).

Only ductile fracture is related with other types of failures at system level, as it could generate undetected misalignment, increasing resistance to move the actuator, friction, and system lock as result. Also the actions, which are executed in order to prevent the fault at EMA level, are similar to all of the failure modes and depend on the severity of the failure and the final application of the EMA. These actions could be component repair, maintenance or replacement.

2.2 Test Rig Description

Gearbox Prognostics Simulator (GPS) has been specifically designed to simulate industrial gearboxes for diagnostics and prognostics research (see Figure 1).

This experimental system consists of a two-stage parallel shaft test gearbox with rolling or sleeve bearings, which can be configured with a gear ratio from 1 to 6. The gearbox can be submitted to a torque large enough to induce wear and damage in the gears. All elements of the GPS have been designed to maximize the number of gearbox configurations to investigate gearbox dynamics and acoustic behaviour, health monitoring, and vibration based diagnostic and prognostics techniques. It is robust enough to handle heavy loads and spacious enough for easy gear placement, setup, and installation of monitoring devices.

The test rig consists mainly of two confronted motors, a reduction gear box for the load motor and the monitored gearbox. One acts as a drive motor and the other as the load motor. The drive motor, which is commanded by the control, provides the speed. And the load control supplies the axial load applied to the gearbox. The motors of the test rig are two 10 Hp three-phase asynchronous motors. The configuration provides a range of speeds (maximum axel speed is 3.600rpm) and loads (radial load from 0 to 1500N, and axial load from 3.5 to 52 Nm) for the test of the different spur gears. Moreover, radial loadings can be applied to study damage signature or propagation in gears and/or bearings: the axial load is applied via a 10 HP variable frequency AC drives with a programmable, user-defined speed profiles. With the programmable load drive, load fluctuations can be applied to simulate real life loading conditions.

Table 1. Description of the failure modes (gearbox)

Name	Short Description	Propagation
Pitting (contact fatigue)	The pitting is a surface fatigue failure which occurs when the endurance limit of the material is exceeded; a failure of this nature depends on surface contact stress and number of stress cycles. A fatigue crack initiates either at the surface of the gear tooth or at a small depth below the surface. The crack usually propagates for a short distance in a direction roughly parallel to the tooth surface. When the cracks grow they cause a piece of surface material break forming a pit with sharp edges.	• Not a proper alignment across the full face width of the gear mesh. • Excessively high contact stresses. • Surface overload conditions (high load, speed). • Adhesion, abrasion. • Often a bending fatigue crack will originate from a pit, causing a premature tooth breakage failure. • The gear does not transmit the necessary torque and it does not rotate to the required velocity.
Adhesion	Layers of metal are removed because two metals rub together with sufficient force to cause removal of material from the less wear resistant surface. Adhesion is the transfer of material from the surface of one tooth to another due to welding and tearing. It is confined to oxide layers on the tooth surface. It is categorized as "Mild", "Moderate" or "Severe=Scuffing":	• The loads are so high that the lubricating film breaks. • Inadequate lubrication. • The gears are no longer fit for reliable service. • The gear does not transmit the necessary torque and it does not rotate to the required velocity.
Abrasion	Layers of metal are removed, or "worn away", more or less uniformly from the contacting surfaces of the gear teeth. This failure results from the transport of abrasive particles in the lubricant film, which can be trapped between the gear teeth. The abrasive particles from the environment or other types of faults such as pitting. The abrasion is caused by contaminants in the lubricant such as sand, scale, rust, machining chips, grinding dust, weld splatter and wear debris (ex. Pits).	• The heat of the mesh, which results in only marginal lubrication. • The heat of the mesh and the bulk temperature of the rotating gears combine to break down the lubrication film. Excessive temperature, surface load, or surface speed, all of which generate excessive heat. • Foreign material in the lubrication system. • High contact temperature and pressure and marginal lubrication. • Misalignment under operation, with more loads being carried out on the teeth than the lubrication film can support. • Pitting. • Loss of tooth geometry. Premature tooth breakage failure. • The gear does not transmit the necessary torque and it does not rotate to the required velocity.
Scuffing	An adhesive type wear occurring at normal temperatures where smooth burnished appearing radial striations are observed in the direction of sliding on the tooth surfaces. It is severe adhesion or scuffing transfer metal from the surface of one tooth to that of another. Typically it occurs in the addendum or dedendum in bands along the direction of sliding. Load concentrations can cause localized scuffing. Surfaces have a rough or matte texture that, under magnification, appear to be torn and plastically deformed. Scuffing ranges from mild to severe	• It can appear where tooth pressures are high and oil films are in the boundary regime and where speeds are slow enough that high contact temperatures do not occur. • Loss of tooth geometry. Premature tooth breakage failure. • The gear does not transmit the necessary torque and it does not rotate to the required velocity.
Plastic deformation	It is a surface deformation resulting from the yielding of the surface and subsurface material, and is usually associated with the softer gear materials, although it often occurs in heavily loaded case-hardened and through-hardened gears as well. It is caused by high contact stresses and the rolling and sliding action of the mesh.	• High contact stress. • Slow-speed operation (it does not build up adequate elasto-hydrodynamic film thickness). • Repeated cycles. • Inadequate lubricating film. • Loss of tooth geometry. Premature tooth breakage failure. • The gear does not transmit the necessary torque and it does not rotate to the required velocity.

Continued on next page

Table 1. Continued

Name	Short Description	Propagation
Ductile fracture	It is a failure case by breakage of a whole tooth or a substantial portion of a tooth. This can result from overload or by cycling stressing of the gear tooth beyond the endurance limit of the material.	• Excessive loads. • High stress concentration. • From fatigue, pitting, abrasion wear. • Stress risers from holes in the web. • Web vibrations. • The gear does not transmit the necessary torque and it does not rotate to the required velocity.
Corrosion	Corrosion is the gradual destruction of material, usually metals, by chemical reaction with its environment. It means electrochemical oxidation of metals in reaction with an oxidant such as oxygen. It can be caused by lubricant contamination by chemicals or water, absence or deficiency of anticorrosive additive, mixed of oil brands, or by the use of an inappropriate lubricant for bronze gears.	• Lubricant contamination. • Absence of anticorrosive additives. • Inappropriate lubrication system. • The required torque is not being transmitted and the component is not rotating at the needed speed. • It may causes loss of tooth geometry.

Its design enables changing the gears and bearing components quickly and easily, as seen in Figure 1. The gearbox is adaptable to install either rolling element bearings or oil-impregnated sleeve bearings with either spur or helical gears. The modular design makes the introduction of faulty bearing and/or faulty gears an easy task.

The gearbox is designed to accommodate different types of sensors easily. Accelerometers can be installed on the bearing housing to measure the vibrations in all three directions. A torque meter enables precise load measurement. The input and output shafts can be fitted with an encoder or a tachometer to measure the transmission error or for time synchronize averaging.

The monitored gear box is composed of three shafts with different gears and bearings. In the next diagram (Figure 2) you can see which gear is tested.

The gears tested are spur gears with a module of 1,5875, a pressure angle of 14.5 °, and a β of 0 °. In Figure 3 a blueprint of the tested gear can be seen.

Figure 1. GSP test bench and gearbox

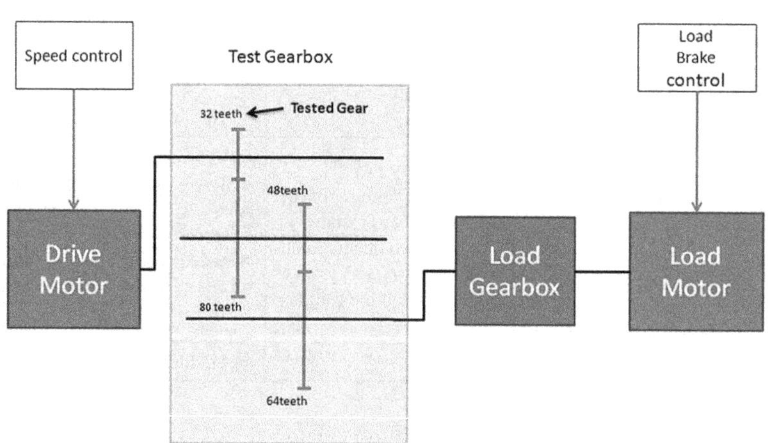

Figure 2. Diagram of the GPS configuration

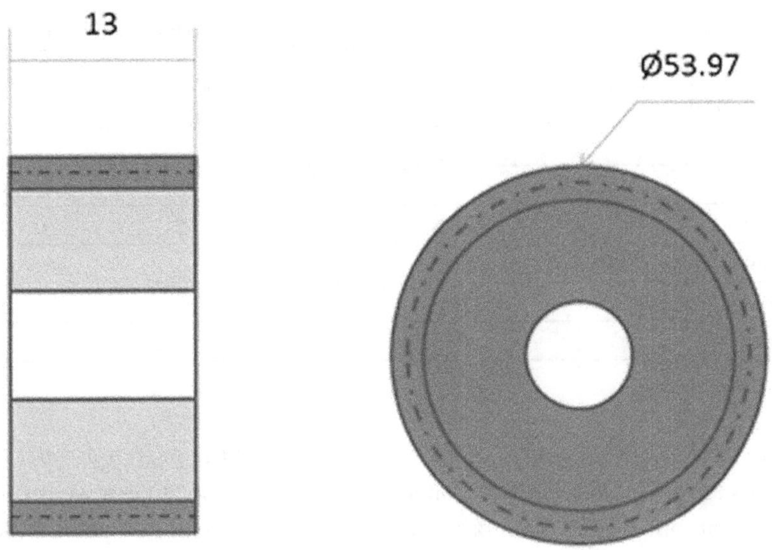

Figure 3. Blueprint of the tested gear

2.3 Experimental Testing

Initially, eight gears were tested in the test rig, two of them were healthy gears, without damage, whereas the rest were modified artificially to induce or simulate the failure. Table 2 presents the failure modes considered in the experimentation, it shows the identification for the gear tested in the test rig, a short description of it and the way in which it was obtained.

From the set of the most common failure modes for the gearbox, ductile fracture has been considered as the most serious one, as it can cause misalignments, and an increment of the friction that can lead to system lock, impending the proper functioning of the system (i.e., the extension movement of the landing gear during take-off is not completed successfully and the drone impacts). This failure has been simulated testing a gear with chipped tooth and another one with a missing tooth. Abrasion has been

Table 2. Simulation of failures

Gear ID	Short Description	Failure Generation
0003G	Chipped tooth	Artificially, breaking the gear
0005G	Missing tooth	
0007G	Faulty	Wear out through a number of operating cycles
0010G	Faulty	
0011G	Healthy	None
0012G	Abrasion level 1 37.5%	Artificially, by rubbing the surface of the gear
0013G	Abrasion level 2 68.75%	
0014G	Healthy	None

simulated as well but artificially, causing damage to the surface of the gear at different levels of severity. The rest of the failures are difficult to be simulated independently because a wear out process not only leads to one phenomenon, but it is also a mixture of them. In this case, two gears (0007G and 00010G) have been tested under heavy operating conditions until some damage effects appeared, according to the expert assessment of tribologists. The gear 0007G and 0010G has signs of fatigue (pitting) and areas of reddish rust. As the torque was set as low as possible for the tests, further damage was not generated while testing. The next groups of pictures (Figure 3-10) show the type of damage of each of the gears to be tested next.

The test rig has been utilized to test these gears and acquire the vibration signal from a triaxial accelerometer (channel AX, AY and AZ). The tests have been performed increasing the speed rate and load at different levels, although only the test obtained at maximum speed and no load in the load motor test condition has been considered, i.e. 3.5 Nm and 1500rpm, for the development of the algorithms for an early prediction of failures. Mainly for two reasons:

Figure 4. ID 0003G

Figure 5. ID 0005G

Figure 6. ID 0012G

Figure 7. ID 0013G

Figure 8. ID 0007G

Figure 9. ID 0010G

Figure 10. ID 0011G

- Initial testing of the algorithms with the signals acquired at different levels of load and speed shows a clearer detection of these failures with higher speeds, but not so obvious regarding the increase of the load.
- The algorithms will be used in an on-ground pre-flight check, from an extension/retraction movement commanded to the EMA. And the quickest and easiest way to do it is not subject the system to stresses or loads.

The data have been acquired initially in different operating conditions (load/speed) as shown in Table 3. The profile for the test is set for 15 seconds and repeated 15 times for every test in order to acquire many repetitions for each operating condition. However, as argued above, only the data saved from the test performed at 3.5 Nm and 1500rpm have been used for the analysis.

3 HEALTH ASSESSMENT THROUGH VIBRATION AND DATA MINING ANALYSIS

A methodology based on data-driven analysis method, similar to that shown in the following Figure 11, has been followed to predict early stages of failures in the gearbox of the EMA.

The methodology starts obtaining data from the accelerometer. At present, accelerometers are not considered when the EMA is designed, however they provide additional measurements that could be considered for the future generations, especially because they could provide vital information for condition monitoring and failure prediction. Spectral features can be obtained from vibration signals by means of their analysis. These features can be used as failure indicators, and machine learning classifiers applied to learn from them will be able to determine the most relevant frequencies and create prediction rules to detect the failure before it happens.

This methodology follows a 'Data Mining' approach, a multidisciplinary field easy to put into practice, which combines several techniques in order to extract knowledge from a databank, in this case, knowledge from the set of data obtained from the experimentation of the gears with the test rig. The most typical 'Data Mining' process consists of some general phases explained below:

- An initial phase to select and prepare the dataset (data acquisition from sensors and pre-processing techniques application);
- The next phase for feature extraction and selection whose objective is to get the most significant features from which to identify the failure to be predicted.
- And finally, develop a model and interpret it for knowledge extraction.

Table 3. Design of experiments to establish operating conditions to be evaluated

Condition Matrix		**Speed (rpm)**			
		250	**500**	**1000**	**1500**
Axial load (%)	0	X	X	X	X
	40	X	X	X	X
	80	X	X	X	X
	100	X	X	X	X

Figure 11. ID 0014G

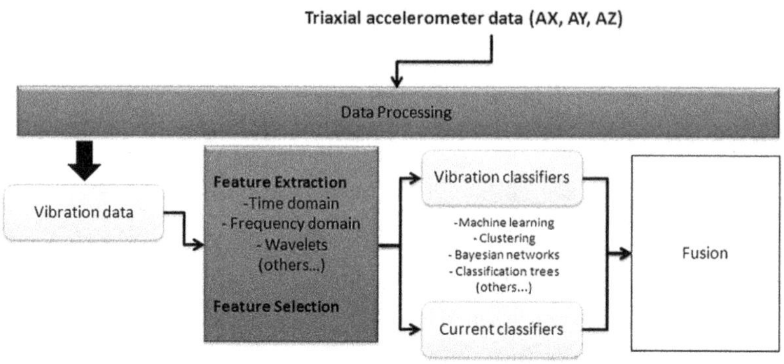

One of the main characteristics of this process is that the output of any of the phases may turn back to the previous phases and that some iteration could be necessary to extract high-quality knowledge. Usually, during the whole process several techniques and methods should be explored to find the most useful ones and finally select them in order to develop a good model to predict early stages of the failure. Furthermore, the other characteristic of the process is that it is interactive because the experts on the issue help during data preparation and evaluation. This evaluation not only needs to have well-defined training and validation stages to decide which model obtains better performance and accuracy, but also it needs to make sense.

3.1 Signal Pre-Processing and Feature Extraction

The aim of the work performed in this section is to process raw data and get the most relevant frequencies to study the failure.

Figure 12 gives an overview about the technologies used for the signal pre-processing. Statistical and spectral features have been extracted and used to detect the wear out of the gear. But before calculating some features, the algorithm of Fast Fourier Transform (FFT) has been computed because it is powerful mathematical tool to view the signal in another domain from which difficult problems become simple

Figure 12. Methodology development plan for early failure prediction

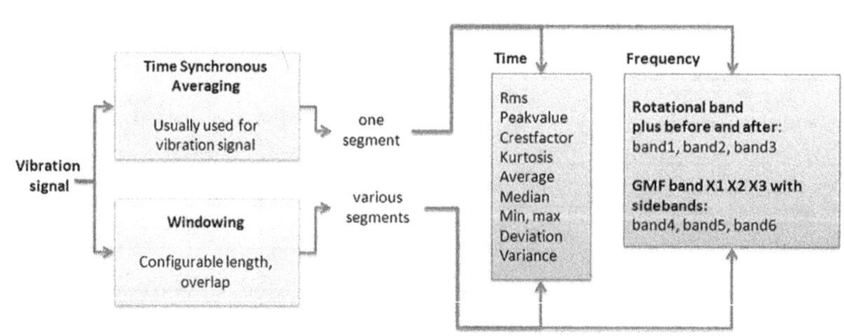

to analyze. FFT gives what frequencies are present in the signal and in what proportions (Schmid, 2012). Fast Fourier Transform has been used combined with two signal processing techniques in order to significantly enhance the availability to detect early stages of the failure in the gears: Time Synchronous Averaging (TSA) (Braun, 2011) and (Al-Atat et al., 2011)) and Windowing (Harris, 1978). The first technique requires the tachometer signal to synchronize the sampling of the vibration signal with the rotational frequency of the shaft. The samples of 15 seconds have been divided into segments of one rotational period and the average has been calculated for enhancing non-synchronous frequencies. Concerning the Windowing processing technique, the basis is quite different; it divides each sample of 15 seconds into equal segments of 1 second long. Each segment is used later as an independent sample.

After these processing techniques, some methodologies have been applied to the signal in the time domain and frequency domain (see Table 4). In the time domain, there are a number of statistical parameters that have been used in the literature. Common features are the root-mean-squared (RMS) amplitude, peak value, crest factor, kurtosis, etc. that attempt to capture unusual behavior and/or impacts associated to early degradation stages and failures. In the frequency domain, theoretical frequencies can be calculated for rotating gears, where gear mesh frequencies and sidebands associated to these frequencies or harmonics can be indicative of failures. Whilst these features are the most common in the literature, other features or transformations are proposed by other authors but not used in this work owing to the good results obtained with the features mentioned before. Among them, there are studies employing the Wavelet transform, which offers capabilities for multi-resolution analysis, as opposed to the FFT time-frequency resolution trade-off.

3.2 Clustering Analysis to set the Severity of the Damage

At this stage of the analysis vibration signals have been processing and the features extracted. There is a data set consisting of a set of time-domain and frequency-domain features plus the category of the signal ('Failure' or 'Not Failure') depending on the gear that has been tested. This set of data is well prepared to apply supervised machine learning classifiers. However, the aim of this study is to go a step further and be able not only to predict if there is a failure or not from the vibration signal, but also to be able to establish a health index in order to establish a level of severity for the failure, i.e., 3 groups {little, moderate and severe}. With this objective in mind, clustering analysis have been done, using k-Means and EM algorithm. These algorithms are non-supervised algorithms which group the set of pre-processed signals (set of features for each signal) in such a way that the set of objects that are in the same group or cluster are more similar among them than those in the other group. They recognize different groups of samples with similar characteristics. The mathematical basis of these algorithms is not explained in this work but there is more information about the algorithms available in (Dasarathy, 1990) and (Mitchell, 1977).

Table 4. Common used methodologies applied for feature extraction

Time-Domain	Frequency-Domain	Others
▪ Root-mean-squared amplitude ▪ Peak value ▪ Crest factor ▪ Kurtosis ▪ Descriptive statistics	▪ Characteristics frequencies and sidebands ▪ Mesh frequency	▪ Wavelets

After applying these algorithms, the result is explained below. EM and k-Means algorithms group the set of data in the same three clusters as presented in Table 5: Cluster 0 {0011G, 0014G and part of the signals recorded from 0013G}, Cluster 1 {0012G and part of the signals recorded from 0013G} and Cluster 2 {0003G, 0005G, 0007G and 0010G}.

Notice that the gears 0003G, 0005G, 0007G and 00010G are grouped in the same cluster. This seems reasonable because they are the gears whose failures are more severe and evident. Gear 0011G and 0014G are grouped as well in the same cluster but it seems they do not have such good quality when the algorithms group together with part of the samples of the gear 0013G (in which the abrasion effect was simulated artificially). A closer look at the gears 0011G and 0014G reveals that the finishing of castings does not have very good quality; they have some visible traces of the machining process. And this is the reason why the samples of these gears are grouped in the same cluster as some of the samples (the first tests/repetitions) of the gear 0013G. The last cluster groups the samples from the gear 0012G and the rest of the samples (last repetitions) of the gear 0013G. It could seem strange that the samples obtained from the gear subjected to the more severe abrasion effect is grouped by the algorithms together with the samples of the most moderate abrasion, and even for some samples that it seems to have less damage. However, it has a physical explanation as the abrasion failure was simulated artificially and manually, and for the gear 0012G the simulation was performed in a uniform manner and it is well distributed in the piece, while for the gear 0013G the effect was simulated more pronounced but not uniformed, which makes it less harmful and quite similar to that produced for the gear 00012G.

Hereafter, the objective of the work is to predict stages of failure for the gearbox based on this data set (computed features from the vibration signal, plus the label which identifies the category of the failure) by applying machine learning algorithms to it. The algorithms will be able to detect the stage of the failure, not a specific failure (i.e., chipped tooth, abrasion,..), but they will be able to detect from vibration signals if there is any damage or wear out in general and its scale of severity {little, moderate, severe}.

3.3 Machine Learning Techniques to Predict the Damage

Machine Learning is a scientific discipline closely related to data mining that allows computers to learn automatically from data and to improve with experience, recognizing complex patterns hidden in the data and making intelligent decisions based on it. It is a subfield of artificial intelligence that develops

Table 5. EM and k-means clustering

Gear ID	Short Description	Cluster Number	Health Assessment
0003G	Chipped tooth	Cluster 2	Severe damage
0005G	Missing tooth	Cluster 2	Severe damage
0007G	Faulty	Cluster 2	Severe damage
0010G	Faulty	Cluster 2	Severe damage
0011G	Healthy	Cluster 0	Little damage
0012G	Abrasion level 1 37.5%	Cluster 1	Moderate damage
0013G	Abrasion level 2 68.75%	Cluster 0 / Cluster 1	Little damage / Moderate damage
0014G	Healthy	Cluster 0	Little damage

algorithms to discover knowledge from specific data and experience based on statistical and computational principles.

The field of Machine Learning integrates many distinct approaches such as probability theory, logic, optimization, search, statistics, reinforcement learning and control theory. The developed methods are the basis of many applications, ranging from vision to language processing, forecasting, pattern recognition, data mining, expert systems, robotics, etc. Within it, several types of algorithms are distinguished depending on their functionality and application: Supervised Learning and Unsupervised Learning. Supervised Learning gives the possibility of learning from a categorized set of data, resulting in a predictive model able to represent and generalize the behaviour contained in the data. Once the model is created, it will be able to classify and categorize new samples of the problem that it is trying to solve. On the contrary, Unsupervised Learning consists of samples where the class is unknown and will be analyzed by the models in order to recognize different groups of samples with similar characteristics. The creation of these groups allows the extraction of information from the available data set, making certain characteristics that hide information visible.

Unsupervised Learning has been used when applying clustering algorithms in order to establish the levels of the severity. Now, as the level of damage is the category to predict from the features, some supervised algorithms have been applied (Figure 13): Bayesian methods (BayesNet, NaiveBayes), Logistic regression (Logistic, SimpleLogistic), Neural Networks (MLP, SOM), Instance-based-learning (IB), Classification trees (J48, ID3, RandomTree) and Induction Rules (JRip, Ridor). These algorithms have been executed from Weka software, which provides a collection of machine learning algorithms written in Java. More information of these algorithms can be found in numerous sources of information such as (Witten, Frank, & Hall, 2011) and (Hall et al., 2013).

4 EXPERIMENTAL RESULTS

Table 6 shows the results obtained by some of the machine learning algorithms for the prediction of the stage or severity of the failure ('little', 'moderate', 'severe') after the data pre-processing and feature extraction.

In order to estimate the real error rate of the prediction, the accuracy of each model was calculated as the average of the ten rates, obtained from the execution of the ten times the 10-fold cross validation (a method commonly used).

Figure 13. Pre-processing and feature extraction methods

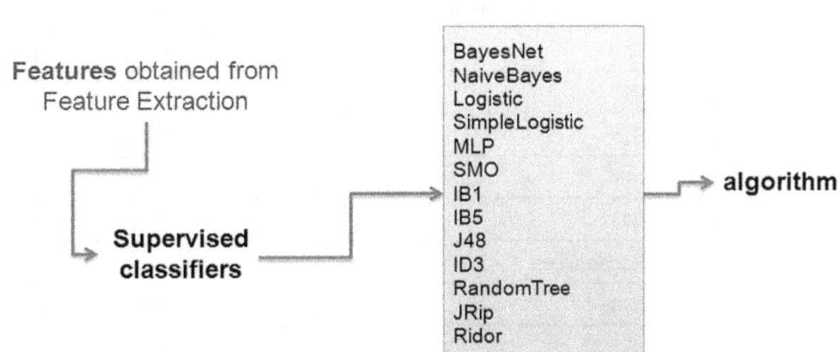

Figure 14. Set of algorithms applied to predict the stage of the failure

Table 6. Results of machine learning algorithms after different pre-processing techniques

Pre-Processing Technique	Algorithm	% of Estimated Success Rate (Accuracy)	Standard Deviation
TSA	BayesNet	93.42	6.52
	NaiveBayes	88.33	9.48
	Logistic	94.50	6.18
	SimpleLogistic	95.50	5.73
	MultilayerPerceptron	96.08	5.22
	SMO	91.58	7.63
	IB1	97.50	4.02
	IB5	96.33	5.21
	J48	90.17	9.94
	ID3	91.50	7.77
	RandomTree	88.08	10.41
	JRip	88.58	8.99
	Ridor	87.58	10.15
Windowing	BayesNet	99.00	2.97
	NaiveBayes	100.00	0.00
	Logistic	100.00	0.00
	SimpleLogistic	99.58	1.82
	MultilayerPerceptron	100.00	0.00
	SMO	100.00	0.00
	IB1	100.00	0.00
	IB5	100.00	0.00
	J48	96.58	5.81
	ID3	97.50	4.81
	RandomTree	98.00	4.12
	JRip	96.58	5.31
	Ridor	96.50	5.19

It can be seen that the results provided by the algorithms when applying TSA are not much better than when Windowing is used, there are even some algorithms in which the percentage of estimated success rate (or the accuracy) seems worse and for almost all of them the deviation is higher, which implies that numerical stability is worse and the algorithms are less robust against this pre-processing technique.

Bearing in mind Windowing pre-processing technique, all the algorithms from machine learning provide a good accuracy, however some of them seem to be more robust, reaching an estimated success rate very close to 100%. A 100% of estimated success rate does not imply that the algorithm never fails, as the values have been rounded up to no more than two decimals places; it means that the accuracy of the algorithm is very good and close to 100%, with a minimum error.

Once the results have been obtained and taking into account the final application, which is the use of the algorithm to predict the stage of failure of the gear in an EMA, in order to diagnose at EMA level, it is necessary to select and implement any of the algorithms as a function together with other functions developed for the rest of the most critical components in the system. Although some of them are more complex, difficult and time consuming to be implemented in C or Ansi C code, it is possible if the function is required to be locally implemented in a simulated environment similar to the EMA's. The next task that should be performed is to manufacture a test bench for the whole EMA, able to simulate the real work environment and operating conditions for the system, and try these techniques in order to finally validate the knowledge derived from this research.

5 CONCLUSION AND FUTURE WORK

A lot of the works performed regarding the state detection and prediction of failures are centered on the signal pre-processing and analysis in order to determine the most relevant features or frequency bands that could prevent the failure and set thresholds in an intuitive and visual way. However, applying machine learning allows the extraction of knowledge that can remain hidden in data, which is not possible to be quantified without using this type of techniques, and which can help to identify the failure and even identify the level of damage (its severity) as we have also seen in this work, together with the expert tribologists who have helped us to validate and interpret some of the results.

The type of research presented in this paper is necessary for knowledge extraction to be adapted, transferred and validated later with data obtained from the test benches used at system level experimentation, which simulate a real environment. This process can be complex to obtain when only the requirements of the process, the design and the manufacturing of these systems can take years. Nevertheless, these tasks are part of the technological advances which have been necessary in the last few years. There is a long way to go, not only for the research and development of technologies applied to early failure detection and prediction for some of the components of the EMA and their transference to the real environment of the system, but also to combine all of them and achieve a global intelligent diagnosis and prognosis at EMA level, which covers the every failure mode of the most critical components or subsystems.

As future work, we are working on replicating the process used in this research but this time to evaluate the accuracy of the current signal as a possible useful signal in order to predict the failure. The current signal is a signal that can be acquired from the EMA without extra sensors, which could allow forecasting in real time. Also as future work, an accelerated life testing will be raised in the GPS at gearbox level, wearing out a gear in good conditions until a replacement is needed. This process implies a lot of hours of experimentation and many operation cycles, and it will be carried out following a standard procedure utilized by the tribologists in other equipment from which it is not possible to extract data today. The objective to carry out this long test is to extract the features explained in this work to analyze and monitor their evolution measured operation cycles, and then, based on this, to be able to perform a prediction and forecasting to determine the remaining useful life (number of operation cycles) of the component before it needs to be replaced, since the wear out would not allow this function to be performed.

ACKNOWLEDGMENT

The authors gratefully acknowledge the support of the European Commission Seventh Framework program from Research and Technological Development. This article summarizes the work performed as part of FP7 project ACTUATION 2015 'Modular Electro Mechanical Actuators for ACARE 2020 Aircraft and Helicopters'.

REFERENCES

Al-Atat, H., Siegel, D., & Lee, J. (2011). A Systematic Methodology for Gearbox Health Assessment and Fault Classification. *International Journal of Prognostics and Health Management*.

Bechhoefer, E., & He, D. (2012). A Process for Data Driven Prognostics. *Proceedings of MFPT 2012 Conference*.

Braun, S. (2011). The synchronous (time domain) average revisited. *Mechanical Systems and Signal Processing*, *25*(4), 1087–1102. doi:10.1016/j.ymssp.2010.07.016

Chen, Z. S., Yang, Y., & Hu, Z. (2012). A Technical Framework and Roadmap of Embedded Diagnostics and Prognostics for Complex Mechanical Systems in Prognostics and Health Management Systems. *IEEE Transactions on Reliability*, *61*(2), 314–322. doi:10.1109/TR.2012.2196171

Cuc, A. I. (2002). Vibration-Based Techniques for Damage Detection and Health Monitoring of Mechanical Systems. University of South Carolina.

Dasarathy, B. (1990). *Nearest neighbor: Pattern classification techniques (nn norms: Nn pattern classification techniques)*. IEEE Computer Society.

Hall, M., Frank, E., Holmes, G., Pfahringer, B., Reutemann, P., & Witten, I. H. (2013). Weka 3: Data mining and open source machine learning software in java. Retrieved from http://www.cs.waikato.ac.nz/ml/weka/

Harris, F. J. (1978). On the use of windows for harmonic analysis with the discrete fourier transform. *Proceedings of the IEEE*, *66*(1), 51–83. doi:10.1109/PROC.1978.10837

Lei, Y., & Zuo, M. J. (2009). Gear crack level identification based on weighted K nearest neighbor classification algorithm. *Mechanical Systems and Signal Processing*, *23*(5), 1535–1547. doi:10.1016/j.ymssp.2009.01.009

Li, B., Zhang, P.-l, Tian, H., Mi, S.-s, Liu, D.-s, & Ren, G.-q. (2011). A new feature extraction and selection scheme for hybrid fault diagnosis of gearbox. *Expert Systems with Applications*, *38*(8), 10000–10009. doi:10.1016/j.eswa.2011.02.008

Lu, D., & Qiao, W. (2014a). A GA-SVM hybrid classifier for multiclass fault identification of drivetrain gearboxes. *Proceedings of the IEEE Energy Conversion Congress and Exposition*, Pittsburgh, PA, USA (pp. 3894-3900). doi:10.1109/ECCE.2014.6953930

Lu, D., & Qiao, W. (2014b). Fault diagnosis for drivetrain gearboxes using PSO-optimized multiclass SVM classifier. *Proceedings of the IEEE PES General Meeting Conference & Exposition*, National Harbor, MD, USA. doi:10.1109/PESGM.2014.6939892

Mitchell, T. (1997). *Machine Learning*. McGrawHill International.

Saravanan, N., & Ramachandran, K. (2009). Fault diagnosis of spur bevel gear box using discrete wavelet features and Decision Tree classification. *Expert Systems with Applications*, *36*(5), 9564–9573. doi:10.1016/j.eswa.2008.07.089

Saravanan, N., & Ramachandran, K. I. (2010). Incipient gear box fault diagnosis using discrete wavelet transform (DWT) for feature extraction and classification using artificial neural network (ANN). *Expert Systems with Applications*, *37*(6), 4168–4181. doi:10.1016/j.eswa.2009.11.006

Schmid, H. (2012). *How to use the FFT and Matlabs pwelch function for signal and noise simulations and measurements (Tech. Report)*. University of Applied Sciences of Northwestern Switzerland.

Witten, I., Frank, E., & Hall, M. (2011). *Data mining. Practical Machine Learning Tools and Technques*. San Francisco: Morgan Kaufmann.

Wu, B., Saxena, A., Patrick, R., & Vachtsevanos, G. (2005). Vibration monitoring for fault diagnosis of helicopter planetary gears.*Proceedings of the 16th IFAC World Congress*.

Yuan, G., Lan, Z., & Gan, M. (2013). Composite fault location of gearbox based on wavepack and BP neural network ensemble.*Proceedings of the 32nd Chinese Control Conference* (pp. 3795-3799).

Chapter 11
Sensor-Based Decision Making in Uncertain Context

Eric Villeneuve
Université de Toulouse, France

Cedrik Beler
Université de Toulouse, France

François Pérès
Université de Toulouse, France

Vicente González-Prida
University of Seville, Spain

ABSTRACT

Decision makers, whether human or computer, using sensor networks to instrument information collecting necessary for decision, often face difficulties in assessing confidence granted to signals transmitted and received in the network. Several organizational (network architecture or nature, distance between sensors ...), internal (sensor reliability or accuracy ...) or external (impact of environment ...) factors can lead to measurement errors (false alarm, non-detection by misinterpretation of the analyzed signals, false-negative ...). A system-embedded intelligence is then necessary, to compare the information received for the purpose of decision aiding based on margin of errors converted in confidence intervals. In this chapter, the authors present four complementary approaches to quantify the interpretation of signals exchanged in a network of sensors in the presence of uncertainty.

INTRODUCTION

Automation and complexity of technological systems are experiencing a continuous growth. Therefore, decision making related to these systems is generally based on sensor networks which are becoming more complex and more heterogeneous. These sensors allow the collection of information concerning the system state by redundant measurements of the same variable or by a variety of measures of different variables. These multiplicity and heterogeneity of measures make the analysis of this information for decision-making complex and can lead to ambiguities or inaccuracies in the measurement (Daniel & Lauffenburger, 2014; Frikha & Moalla, 2014). Moreover, the confidence about the various sensors can be variable depending on their nature or their environment of use.

DOI: 10.4018/978-1-5225-0651-5.ch011

Therefore, there is a real need for mechanisms to assist the decision based on multiple sensor inputs. These mechanisms have to formalize the collected knowledge in the light of the uncertainty inherent to the system. They also enable the knowledge fusion from multiple and/or heterogeneous sensors taking into account the physical structure of the system (e.g. the spatial position or the immediate environment of the sensor) (Chen, Jiang, Li, Gao, Xu & Ding, 2013; Flammini, Marrone, Mazzocca, Nardone & Vittorini, 2015; Mkrtchyan, Podofillini & Dang, 2015; Savic, Wymeersch & Zazo, 2013; Shang, Yang, Huang & Lyu, 2014). In addition, these mechanisms should take into account the quality and the quantity of the capitalized knowledge to provide the decision-maker (whether human or computer based) information concerning the knowledge uncertainty to make its decision easier and wiser.

In this context, this paper aims to present four complementary approaches to quantify the interpretation of signals exchanged in a network of sensors in the presence of uncertainty. The following section introduces concepts regarding the uncertain knowledge formalization. Then the four approaches are detailed and illustrated through case studies. Finally, the last section provides a formal outlook of these approaches compared to each other to identify strengths, weaknesses and the conditions of use of each approach.

INTEREST AND BACKGROUND ON MULTI SENSOR FUSION

Multi-sensor fusion can provide more accurate and reliable information than the information given by each sensor individually. In addition, data from multiple heterogeneous sensors have different degrees of uncertainty and confidence. Among the techniques of multi-sensor fusion, Bayesian methods and Theories of evidence such as the theory of Dempster-Shafer (DS), are commonly used to treat the degrees of uncertainty in the merge process. The research carried out takes place within this context. The multi-sensor fusion refers to the combination of sensory data from multiple sensors to provide more accurate and reliable information. The potential benefits of the multi-sensor fusion are redundancy and complementarity of the acquired information. The fusion of redundant information can reduce the overall uncertainty and thus helps to provide more accurate target information. Multiple sensors providing redundant information may also be used to increase reliability in the case of error or failure of the sensors. Additional information from multiple sensors provide environmental characteristics that would be impossible to collect using only each isolated sensor information.

Within the field of science and engineering, data imperfection requires the use of tools to define mechanisms for reasoning with partial knowledge and uncertain information. In the work of Dubois and Prade (2009) several types of imperfections are discerned:

- Incompleteness and inaccuracy are used to qualify the status of a data. It is said to be incomplete if it is impossible for the source to provide information regarding all or part of the aspects of a problem. Inaccuracy is considered to be a form of incompleteness as when the source provides an imprecise data, the resulting information is necessarily incomplete.
- Uncertainty applies when the source is unable to distinguish the veracity of a piece of information (that is to say whether the information is true or false). It therefore characterizes the extent of information compliance compared to reality. It is possible to distinguish two kinds of uncertainty. Random uncertainty is induced by the variability of an entity in a population and is the outcome

of random experiments. This type of uncertainty cannot be reduced since it is the result of chance. Epistemic uncertainty is due to the lack of knowledge and therefore relates to the concept of incompleteness.

- Ambiguity represents the fact that a same information can have several interpretations. It is therefore linked to the formalism of information representation which is not always clear and shared by all the stakeholders. This type of imperfection is very common and is often a source of misinterpretation. It can be avoided when formalizing the representation.
- Granularity of information characterizes the difficulties that appear when two very close values have to be distinguished (Livi & Sadeghian, 2015).

A decision requires integrating these imperfections to justify the actions that will be undertaken (Klir, 1995). In the field of sensor networks, decisions must be taken at every moment with respect to signals emitted individually by each sensor or regarding the signals received and compiled by a centralized database. Taking into account imperfect data allows the decision maker to legitimate its choices since he will have at its disposal additional knowledge associating the data with an interval of confidence and a margin of error.

The sensor information merge process generally involves four main steps namely, modelling, estimation, combination and decision. Modelling consists in formalizing the method to be applied. This step is based on the knowledge of experts. In the context of merging data multi sources, each source provides an information on a potential decision. This information can be modelled by a probability distribution, a function, a formula according to the chosen formalism. The estimation stage comes next. It consists in quantifying the information sources according to their natures, for example, in the form of probabilities, if it is the chosen formalism. The aim of combination, is to consolidate information from different sources, using an operator according to the formalism previously chosen. The result will be of the same nature and shall be interpreted in the same way as the merged information. The decision is the heart of the merging process. The goal is to make a decision more developed that the one which could have been taken by considering each of the sources separately. The criterion is chosen on the basis of the formalism. Generally, it is the minimization or maximization of a function from the combination.

Several methods of fusion have been proposed in the literature. There are, generally, two main categories of fusion divided between those based on physical models and those based on the use of artificial intelligence techniques and automatic classification which can be based on several theories such as the theory of probabilities, evidence theory, theory of possibilities and fuzzy subsets. One can also refer to neuron networks and Support Vector Machines as possible methodologies to be used for merging the information provided by sensor networks.

As mentioned so far, there are many formalisms used to represent uncertain knowledge. The best known and most used formalism is the probability theory. Probability theory is a mathematical tool for representing random and uncertain phenomena. It is the oldest uncertain representation theory and the most represented in the literature. It is very well endowed in terms of mathematical tools to model a system as well as to develop learning mechanisms, including the framework of Bayesian networks. Bayesian Networks (Jensen & Nielsen, 2007) are graphic models designed to formalize knowledge with the purpose of reasoning about a problem. Bayes theorem is central in the mechanism of inference in Bayesian Networks. It makes the link between a series of hypotheses, characterized by probabilities of occurrence, and a series of observations representing the actual state of the system.

However, uncertainty modeling is complex because of the limitations set by the probabilistic mechanisms. If the probability theory can be used to quantify the randomness of the information (variability), it does not allow an easy integration of epistemic information (incompleteness, imprecision). Indeed, the symmetry or indifference principle (equiprobability) does not allow to make a difference between the representation of a random situation and a partially described context (Dubois & Prade, 2009). Within this framework, belief functions can be of a great help. Based on the work of Dempster (1967) generalizing the theory of probabilities by using probability intervals, the belief function theory (or evidence theory) was developed by Shafer (1976) to provide a general framework for the representation of uncertainties. This combined approach is commonly referred to as the Dempster-Shafer Theory (DST). It is used to represent information based on the belief or the state of knowledge of an entity (person, sensor…) and provides a very rich mathematical framework for:

- The characterization of partial knowledge (including total ignorance),
- The fusion of information from various homogeneous or heterogeneous sources,
- The modelling of uncertainty (random and epistemic) related to the state of a system,
- The assessment of the degree of confidence associated with a result.

The theory of belief functions is particularly well suited to the representation of different forms of uncertainty. It allows the modelling of problems where the lack of information prevents the reasonable use of probability theory. Belief function techniques are used in many fields (data analysis, classification, decision, diagnosis, multi-sensor perception, image processing ...) for the very varied tasks such as pattern recognition, likelihood analysis or information merging.

Deep Belief Network (DBN) (Bengio, Lamblin, Popovici & Larochelle, 2006; Hinton & Salakhutdinov, 2006) is based on a multilayered architecture forming a network with undirected connections between layers. DBN can be seen as a recent evolution of neural networks that have shown impressive performance improvements in machine learning and classification problems. They avoid typical problems of training large neural networks (Hochreiter, Bengio, Frasconi, Schmidhuber & Elvezia, 2001; Bengio et al., 2006) and has been recently named one of the Breakthrough Technologies of 2013 (Turell, 2013). Regarding sensor fusion tasks, they provide a simple and powerful solution and this is what will be reviewed in this paper.

DECISION-MAKING METHODS

Several methods, using the formalization tools presented above can be used to facilitate decision-making based on multi-sensor knowledge. The four proposals that follow allow to meet needs that differ depending on the quantity and quality of available knowledge. Each approach is illustrated by a case study to highlight its strengths and weaknesses.

Bayesian Networks

Principles

A Bayesian network is defined by an oriented acyclic graphic model (Directed Acyclic Graph or DAG) noted G=(V,E) where $V = \{V_1, V_2, \cdots, V_i, \cdots, V_n\}$ is a set of nodes, and E is a set of arcs. Each node of the graph G is a variable characterizing a knowledge. Every oriented arc represents a direct dependency link (generally, a causal link). Thus, an arc from a variable $C(V_i)$ (ascendant or parent) to a variable V_i (descendant or child) implies that a knowledge on $C(V_i)$ has an influence on V_i, and conversely a knowledge on V_i may affect the knowledge about $C(V_i)$ (Jensen & Nielsen, 2007).

Figure 1 introduces an example of a Bayesian network which represents a sensor system. This system uses two sensors, S1 and S2, to detect if an object is present in the studied area.

To each node belonging to the set V of graph G is associated a conditional probabilities table aiming at quantifying the effect of the ascending nodes on another one. When the node has no parent (root node), the probabilities are no more conditional but priors (Charniak, 1991; Verma & Pearl, 1991; Jensen & Nielsen, 2007; Verron, 2007). In the previous example (Figure 1), nodes S1 and S2 are root nodes and are defined by prior probabilities. The node D symbolises a decision using conditional probabilities to merge knowledge from the two sensors. Once the probabilities have been defined, it is possible to take into account the probability of false-positive or false-negative according to the reliability of the sensors. Thanks to this model, several uses are possible. Firstly, if the variable S1 is true and S2 is false (e.g. only sensor S1 detects something), node D provides to the decision-maker the probability that an object is really in the area. Secondly, if an object is really in the area (variable D is true), it is possible to know the probabilities that it has been detected for each sensor.

Figure 1. Example of Bayesian network

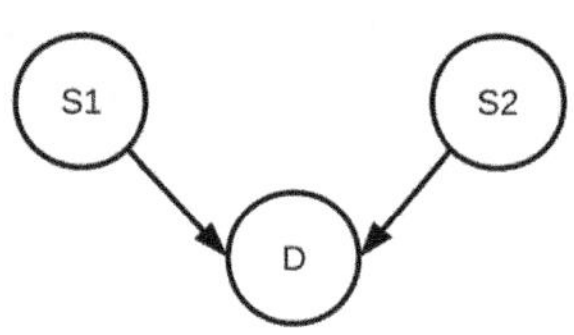

Variable (V)	Description	V=Yes	V=No
S1	An object is detected by sensor 1	P(S1=Yes)	P(S1=No)
S2	An object is detected by sensor 2	P(S2=Yes)	P(S2=No)

Variable	Description	S1	S2	D=Yes	D=No
D	There is an object in the area	Yes	Yes	P(D=Yes|S1=Yes and S2=Yes)	P(D=No|S1=Yes and S2=Yes)
		Yes	No	P(D=Yes|S1=Yes and S2=No)	P(D=No|S1=Yes and S2=No)
		No	Yes	P(D=Yes|S1=No and S2=Yes)	P(D=No|S1=No and S2=Yes)
		No	No	P(D=Yes|S1=No and S2=No)	P(D=No|S1=No and S2=No)

The joint probability of the nodes of the graph is given by the following equation (Naim, Wuillemin, Leray, Pourret & Becker, 2011):

$$P\left(V_1, V_2, \ldots, V_n\right) = \prod_{i=1}^{n} P\left(V_i \middle| C\left(V_i\right)\right)$$

where $C(V_i)$ corresponds to the set of parents of V in graph G given that each node V can take a certain number of values called modalities.

Application

The length of the cables in complex systems increased significantly to meet the requirements of new technologies (flow rate, error rate, real time, etc.). Whether in the distribution of energy, aeronautics, automobile or even telecommunications, this increase resulted in a complexity of wiring networks.

Aeronautics field is the best example to illustrate this complexity. In the "more electric" aircraft, electric power current replaces pneumatic, hydraulic and mechanical energies for the control of some vital functions. The advantage is twofold: on the one hand, electrical systems are less complex, less expensive and more reliable than the hydraulic and mechanical mechanisms, on the other hand, electrical systems allow lowering the weight of an aeroplane and consequently, the fuel consumption. This technological breakthrough gave rise however to a much more complex electric network reaching 10000 cables and 40000 connectors for largest aircrafts.

For several years, wires have been considered as permanently operational over the life of the system they belong to (Verron, 2007). This simplistic hypothesis is today no longer topical. It is known now that a wire is aging and may exhibit signs of failure. These failures can result in more or less serious abnormalities (loss of electrical signal, distortion of information, power surges, system dysfunctions, smoke, fire, etc.) whose consequences can unfortunately be dramatic if the wires are part of critical systems for the safety of people, equipment and the environment. As an illustration the crashes of TWA 800 flight (in 1996) and Swissair 111 (1998) as well as the accident of Apollo XIII (1970) (Naim et al., 2011) turned out to be initiated by wire failures.

The establishment of a diagnosis system for the detection and localization of faulty wires is therefore necessary. Within this framework, reflectometry techniques appear to be relevant. These techniques are based on the same principles as a radar. A wave is injected into a wire, and when facing a discontinuity, part of its energy is reflected to the injection point. The reflected signal analysis allows interpreting information about this discontinuity to detect a possible defect (Auzanneau, 2010).

Note that reflectometry is not limited to the diagnosis of transmission lines, but can be applied to more complex wiring network topologies. Figure 2 shows the reflectogram corresponding to network Y shown in Figure 2.

The first peak in the negative direction corresponds to the junction. Then, the second peak corresponds to the open circuit at the end of the B line. It is located 35 m from the point of injection. Similarly, the third peak corresponds to the open circuit at the end of line B positioned at 45 m from the point of injection. Finally, other peaks relate to the test signal round trips in the network until its mitigation. The simulation window is limited to 80 m in this case.

Figure 2. Multiple paths in a Y shape network

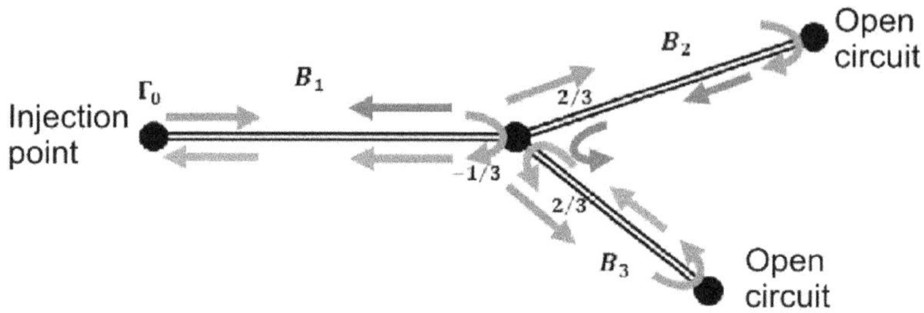

Reflectometry methods have proven their effectiveness in detecting defects in wiring networks, even when facing complex topologies (Y networks, CAN, bus etc.). However, in the case of a complex network, the number of branches can result in ambiguities related to the localization of defect (s).

Figure 3 shows an example of a complex network consisting of 5 branches B_i, i ∈ {1, 2, ···, 5} with the same features. Branches B_2, B_4, B_5 are terminated with open circuits. An open circuit is observed on B branch at a distance of 25m of the reflectometre at the end of branch B.

Following injection of a signal in R1, the corresponding reflectogram will actually detect a defect at a distance of 25 m from the reflectometre but it will be impossible to know if the corresponding error is located on branch B1 or B2. An ambiguity associated with fault location appears here. This uncertainty worsens if other defects exist on the network. In addition, the test signal will be attenuated due to the multiple reflections encountered along its path. This makes uneasy the detection of a fault at a significant

Figure 3. Reflectogram of a Y shape network with an open circuit at the end of lines B_2 and B_3

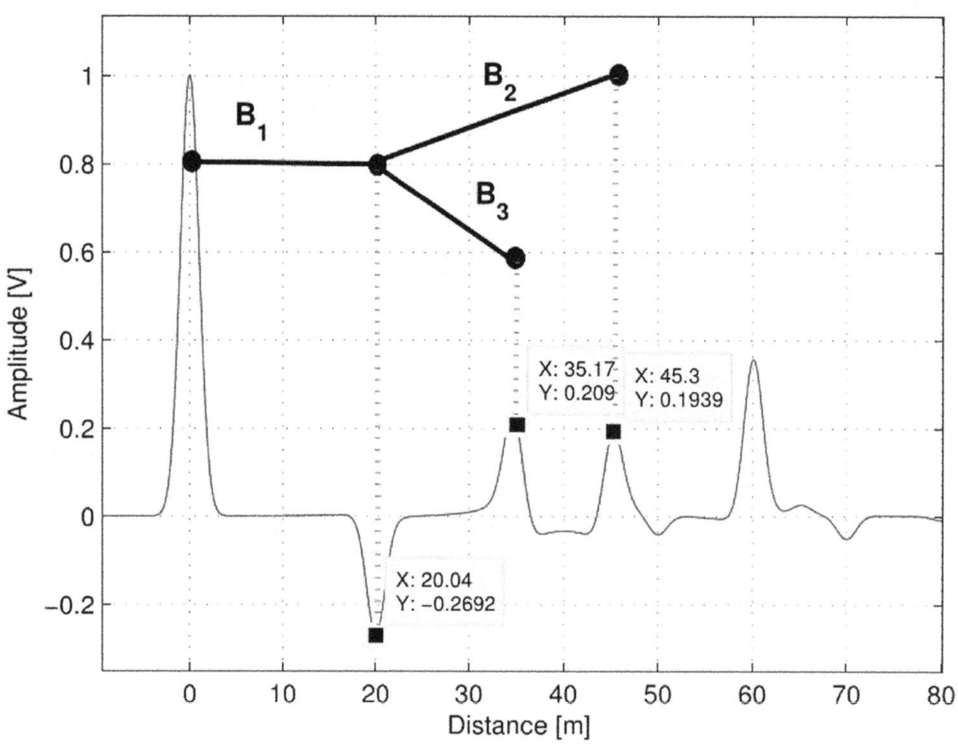

distance from the point of injection if not impossible. In the case of a wiring network affected by several flaws, only the first defect (closest to the source of injection) can be detected and localized. Others can be interpreted as secondary echoes (Figure 3).

The model of Bayesian network proposed for an analysis carried out on the basis of a single reflectometre was developed in the works of Ben Hassen et al. (2012, 2013a, 2013b, 2013c, 2014, 2015). It is built on the idea that a wire involves seven classes of variables that may explain the presence of a defect: failure characteristics, reflectometre features, procedure of reflectometry, disturbance effects, wire characteristics and thermal stress. Each class contains variables of interest that may or may not be related by causal links. Each variable of interest contains different states and a table of conditional probabilities or prior probabilities according to the presence or not of ascendants.

This network is used to calculate the probability of detection of the defect on each branch supervised by a reflectometer. For every reflectometer, a conditional probability of detection of the defect on the branch B_m of the network is calculated as follows:

$$P\left(X_D | X_{Bt}, X_{Tm}, X_{Ar}, X_{At}, X_C\right) = \frac{P\left(X_{Dd}; X_{Bt}, X_{Tm} X_{Ar}, X_{At}, X_C\right)}{P\left(X_{Bt}, X_{Tm} X_{Ar}, X_{At}, X_C\right)}$$

To solve the problem of a possible ambiguity of location of a defect in a complex network, a new diagnosis strategy is proposed. It consists in dividing the Bayesian network into multiple Bayesian subnets where each branch B_m, monitored by a reflectometre R_j is modelled by a sub Bayesian network, with $m \in \{1, 2, \ldots, N_b\}$ and $j \in \{1, 2, \ldots, N_r\}$ where N_b and N_r represent the number of branches and reflectometres, respectively. Each Bayesian subnet allows the calculation of the conditional probability associated with the presence of a defect on the B_i branch according to the reflectometre R_j.

The results corresponding to this local diagnosis lead to an information vector characterizing the probability of detection on each branch B_m by all the reflectometer R_j:

$$\begin{array}{c} \\ R_1 \\ R_2 \\ \vdots \\ R_{N_r} \end{array} \begin{array}{c} \begin{array}{cccc} B_1 & B_2 & \cdots & B_{N_b} \end{array} \\ \begin{pmatrix} P_{B_1}^{R_1} & P_{B_2}^{R_1} & \cdots & P_{B_{N_b}}^{R_1} \\ P_{B_1}^{R_2} & P_{B_2}^{R_2} & \cdots & P_{B_{N_b}}^{R_2} \\ \vdots & \vdots & \ddots & \vdots \\ P_{B_1}^{R_{N_r}} & P_{B_2}^{R_{N_r}} & \cdots & P_{B_{N_b}}^{R_{N_r}} \end{pmatrix} \end{array}$$

Then, a fusion of data collected from all reflectometres on all branches is made to take the decision on the location of the defect on the network. At this stage, the presence of internal (non-reliable reflectometre) or external factors branch under thermal stress) may influence the reflectometry measures and degrade consequently the quality of the diagnosis through uncertain results leading to incorrect decision on the defect location. The idea is then to merge the information previously obtained.

With the aim of carrying out a global diagnosis, we use the fusion of information to facilitate the decision-making on the location of the defect by combining, in a first step, the information provided by the N_r reflectometres taken individually. The probabilistic information from the Bayesian subnet (local diagnosis) is used to assess the global probability P through the mean of the data collected on each branch as follows:

$$P^{Bm} = \frac{1}{Nr}\sum_{j=1}^{Nr} P_{Rj}^{Bm}$$

The probability P compiles the information gathered from different sources and will be used for decision making. The branch representing the highest probability will be considered as the defective one. The authors use the maximum to take decision d_k such that:

$$P\left(d_k\right) = max_{j\in\{1,2,\ldots,Nb\}} P^{Bj}$$

Transferable Belief Model

Principles

The Transferable Belief Model (TBM) is an evolution, proposed by Smets and Kennes (1994), of the theory of belief functions originally developed by Dempster (1967) and Shafer (1976). This approach is a subjectivist interpretation of belief functions which introduces two levels of reasoning. The credal level allows the static representation (knowledge formalization) and dynamic manipulation (fusion, revision and reasoning with formalized knowledge) of belief states. The pignistic level intends to facilitate decision making by transforming (irreversibly) the belief functions in probability measures. It then becomes possible to implement risk or bet concepts without altering the credal level.

Modeling a problem using belief functions, requires to determine the value of the variable ω, which represents the system states. The Frame of Discernment (FoD) represents all possible n values (or hypothesis) for the variable ω and is denoted by Ω.

$$\Omega = \left\{\omega_1, \omega_2, \ldots, \omega_i, \ldots, \omega_n\right\} = \bigcup_{i=1}^{i=n} \left\{\omega_i\right\}$$

The definition of the belief mass function, denoted m^{Ω}, allows the translation of an observation (or advice) provided by an agent on the "power set" of Ω (denoted 2^{Ω}). The power set corresponds to all the subsets that can be formed from the assumptions and the unions of assumptions of Ω. Consequently, the belief mass function is defined as:

$$\begin{array}{lcl} m^{\Omega} : 2^{\Omega} & \rightarrow & \left[0,1\right] \\ A & \mapsto & m^{\Omega}\left(A\right) \end{array}$$

$$avec\, 2^{\Omega} = \left\{\varnothing, \left\{\omega_1\right\}, \left\{\omega_2\right\}, \left\{\omega_1, \omega_2\right\}, \left\{\omega_3\right\}, \left\{\omega_1, \omega_3\right\}, \left\{\omega_2, \omega_3\right\}, \ldots, \left\{\omega_1, \ldots, \omega_n\right\}\right\}$$

Notation $m_s^\Omega(A)$ stands for the identification of a source S which provides information on proposal A. The basic belief assignment (bba) is the set of belief masses on the proposition $A \subseteq \Omega$ that satisfies:

$$\sum_{A\subseteq\Omega} m_S^\Omega(A) = 1$$

A bba can be transformed to highlight information and to improve the dynamic aspects of TBM, including the fusion rules. The credibility measure or belief (denoted $bel^\Omega(A)$) symbolizes the minimal belief in a hypothesis A. It is the part of belief specifically attributed to A (without the part corresponding to the empty set). The plausibility measure (denoted $pl^\Omega(A)$) represents the maximal belief in a hypothesis A. It is the sum of all the masses for A. Both measures are defined as follows:

$$bel^\Omega(A) = \sum_{\varnothing \neq B \subseteq A} m^\Omega(B), \forall A \subseteq \Omega$$

$$pl(A) = \sum_{B \cap A \neq \varnothing} m^\Omega(B), \forall A \subseteq \Omega$$

Two other measures can be implemented to facilitate the operations related to the dynamic part of TBM (in particular information fusion): the implicability (denoted $b^\Omega(A)$) and communality (denoted $q^\Omega(A)$). Both measures have no real physical interpretation but they are tools to facilitate the calculation.

$$b^\Omega(A) = \sum_{B \subseteq A} m^\Omega(B) = bel^\Omega(A) + m^\Omega(\varnothing), \forall A \subseteq \Omega$$

$$q^\Omega(A) = \sum_{B \supseteq A} m^\Omega(B), \forall A \subseteq \Omega$$

All these measures (bel, pl, b and q) are equivalent representations of a bba.

The transformation from credal level to pignistic level is called pignistic transformation (Smets, 2005). This non-reversible transformation aims to reduce the bba in a probability distribution so as to be compatible with decision theory. Hence, the mass of proposal A is distributed using equiprobability on the singletons of the frame of discernment, Ω, i.e., on the hypotheses related to A. Therefore, this probability distribution, denoted BetP, can be obtained as follows:

$$\begin{aligned} BetP\{m^\Omega\} : \quad \Omega \rightarrow & \quad [0,1] \\ \omega_i \mapsto & \quad BetP\{m^\Omega\}(\omega_i) \end{aligned}$$

With regard to decision-making itself, it consists in selecting the singleton ω_i of the frame of discernment Ω having the highest pignistic probability to maximize the chances that the hypothesis symbolized by ω_i represents the actual state of the system (or the smallest, if for instance, the goal is to minimize the probability of occurrence of an event). It is therefore to maximize the expected utility in order to rationalize the decision. From a mathematical standpoint, this decision can be expressed by:

$$\omega = \begin{vmatrix} argmin \\ argmax \end{vmatrix} BetP\left\{m^{\Omega}\right\}\left(\omega_i\right)$$

The dynamic aspect of the TBM is mainly produced by the fusion of information. The two fusion rules (commonly called rules of combination) most commonly used are the conjunctive and disjunctive rules of combination. The use of these rules is conditioned by attributes characterizing the information sources that have to be merged:

- A source is said to be reliable (Smets, 1993) if the user believes it is possible to trust the veracity of the information transmitted.
- Two sources are said distincts (Smets, 1993), if both sources have nothing in common. For example, two sensors are considered as distinct sources if they have no common information (different technologies, different brands, no direct communication between the sensors ...). This concept is fairly close to the independence concept in the probabilistic framework (Yaghlane, Smets & Mellouli, 2001).

The Conjunctive Rule of Combination (CRC), introduced by Shafer (1976), allows the fusion of bba from reliable and distinct sources. Noted $\cap^*$, this rule transfers the mass on the subsets with lower cardinality (specialization effect (Smets, 2002)) and is defined by:

$$q_{1\cap^* 2} = \left(q_1 \cap^* q_2\right)\left(A\right) = q_1\left(A\right) \times q_2\left(A\right), \forall A \subseteq \Omega$$

The Disjunctive Rule of Combination (DRC), introduced by Smets (1993), allows the fusion of bba from distinct sources including at least one reliable source. Noted $\cup^*$, this rule transfers the mass on the supersets with higher cardinality (generalization effect (Smets, 2002)) and is defined by:

$$b_{1\cup^* 2} = \left(b_1 \cup^* b_2\right)\left(A\right) = b_1\left(A\right) \times b_2\left(A\right), \forall A \subseteq \Omega$$

There are many other rules of combination that allows, for example, the fusion of non-distinct sources. Interested readers can refer to the works of Dempster (1967), Dubois and Prade (1988) or Denoeux (2008).

Application

An example to illustrate the concept of belief functions applied to sensor networks can be found in the works of Senouci et al. (2012, 2014). The author of the work deals with the area of Wireless Sensor Networks (WSN) based especially on infrared or ultrasonic techniques. These networks are made up of smaller equipment whose number varies from hundreds to hundreds of thousands. They can be deployed on one or more locations to observe a particular phenomenon and their mutual collaboration enables the triggering of alerts or the gathering of information on a supervised phenomenon.

The topology of the network is determined by the number and the position of sensors deployed in the area to be studied (called Region of Interest - RoI) and it allows to characterize the attributes of the network (coverage, connectivity, cost or lifetime). The network architecture is mainly dependent on the reliability in transmission of the information from sensors which determines the quality of the network coverage. The use of belief functions in this context aims to analyze the imperfection management problem by modeling the uncertainty in the data collection process.

The problem is modeled from two states to determine if a space point that belongs to the RoI (p ∈ RoI) is covered (θ_1) or uncovered (θ_2). The set $\Theta = \{\theta_1, \theta_2\}$ is therefore the FoD of the problem. Let s be a sensor, R_s, be its sensing range and R_u be a distance ($0 \leq R_u \leq R_s$) as illustrated in Figure 4.

The information concerning the coverage of a space point p ∈ RoI by a sensor s is modeled by a bba $b_{s/p}$. The complementary information 1 - $b_{s/p}$ is assigned to the whole FoD because it encodes the sensor ignorance. Therefore, the bba that represent the information resulting from the sensor s about the space point p ∈ RoI can be expressed by:

$$m_{s/p}\left(\left\{\theta_1\right\}\right) = b_{s/p}, b_{s/p} \in \left[0,1\right]$$
$$m_{s/p}\left(\Theta\right) = 1 - b_{s/p}$$
$$m_{s/p}\left(\varnothing\right) = 0$$

Figure 4. Complex network with open circuit on line B_3 and the associated Bayesian network

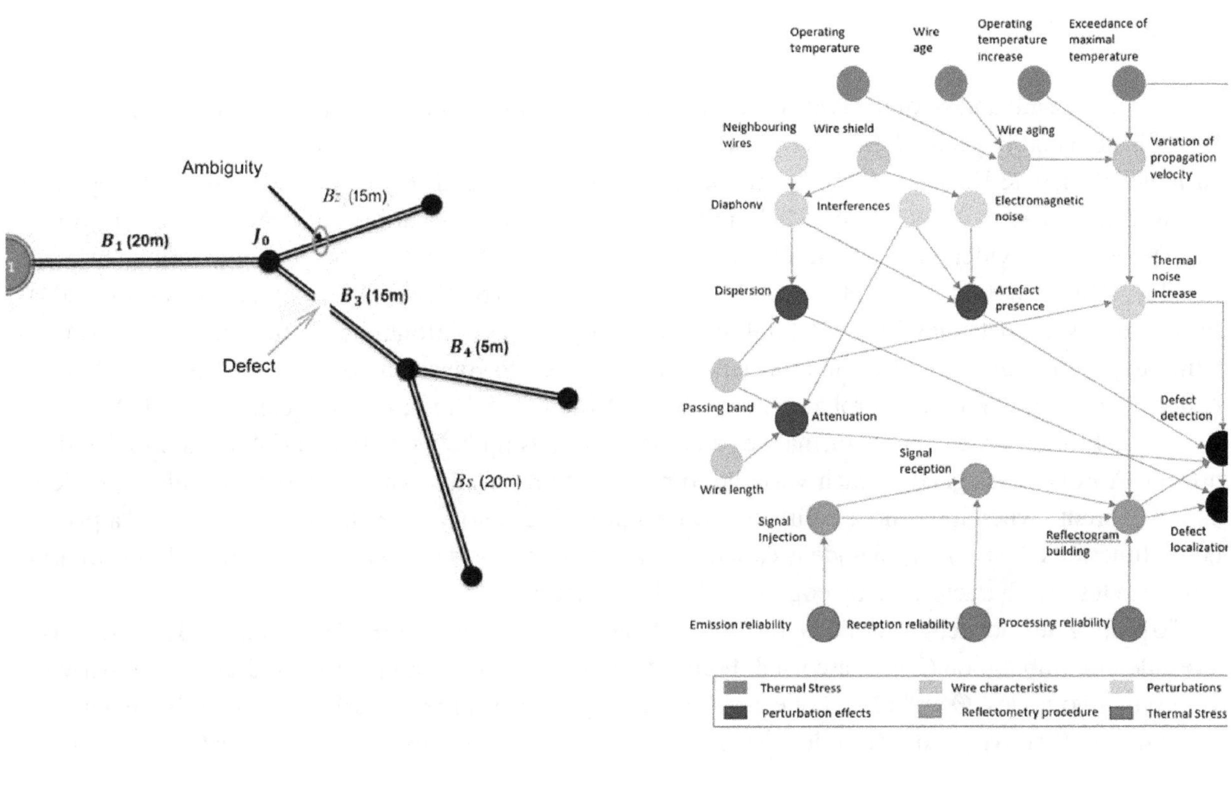

Decision making is done using the pignistic transformation of $m_{s/p}$ (denoted by $BetP_{s/p}$) to determine whether the point p ∈ RoI is covered by the sensor s. The chosen hypothesis $\hat{\theta}$ is the one corresponding to the state with the highest pignistic probability:

$$\hat{\theta} = \underset{i=0,1}{\operatorname{argmax}}\, BetP_{s/p}\left(\left\{\theta_i\right\}\right)$$

A space point p is considered as covered by a sensor s if:

$$\hat{\theta} = \theta_1$$
$$BetP_{s/p}\left(\left\{\theta_1\right\}\right) \geq Th_p$$

The threshold (Th_p) value is an application-specific user-specified parameter. It allows the variation of the model sensitivity depending on the applications. For example, an application where a false-negative is considered as dangerous, the value of Th_p will be close to 1. One current application of WSNs is target/event detection; the sensors collaborate to reach a consensus decision to determine if a target is present in the RoI. A TBM-based approach is considered for reaching this consensus: one of the sensors (called a fusion center: it can be a sink, a cluster-head or any sensor), gathers the evidence from the other sensors, combines the evidence and decides whether an event has occurred or not.

Directed Evidential Networks

Principles

In order to take advantage of the two approaches mentioned so far, it is possible to combine Transferable Belief Model and Bayesian Networks. Basic approaches enable this integration but are limited to binary variables (Smets & Kennes, 1994). Evidential networks, introduced by Smets (1993) and developed by Xu and Smets (1994), use conditional belief functions for belief propagation in acyclic oriented networks. Each node of the evidential network represents a random variable that is associated with a finite number of values. Directed Evidential Networks (DEN) were introduced by Yaghlane, Smets and Mellouli (2004). In the same way that "classic" evidential networks, they use conditional belief functions to spread the belief in graph without circuit. This formalism is very close to Bayesian networks but uses conditional beliefs instead of conditional probability functions (Villeneuve, Beler, Peres & Geneste, 2012).

In this formalism, each arcs formalizes a causal relationship between two variables (symbolized by nodes). A FoD is assigned to each variable to represent all possible values of the variable. A node is called "Parent" when it has no arc directed toward him. These nodes are characterized by an "a priori" belief functions. Conversely, a node is called "Child" when there is at least one arc directed toward him. These nodes are characterized by conditional belief functions.

To propagate knowledge in a DEN, the Generalized Bayes Theorem (GBT) (Smets, 1993) and disjunctive rule of combination (DRC) are used. Figure 5 represents a very simple directed evidential network with only two nodes, Θ and Ω and one arc representing the causal relationship between the two nodes. From the perspective of the variables, which are represented by nodes, the edge indicates that node Ω,

Figure 5. Sensor sensing range representation

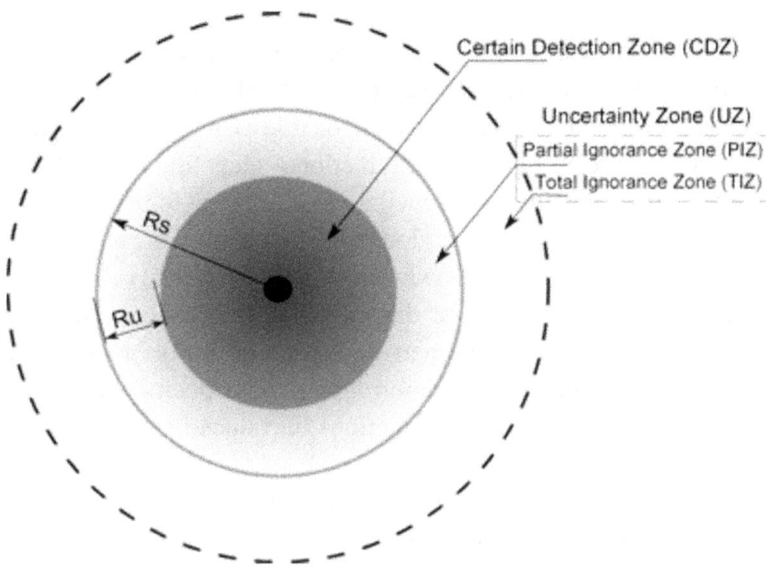

which represents the observation space, is conditioned by node Θ, standing for the hypothesis space. Therefore, the plausibility distribution on Ω is $m^{\Omega}[\theta]$.

In Figure 5, in addition to the structure of the network, two belief distributions are set. The Θ node is characterized by a priori belief mass distribution, noted m_0^{Ω} and the Ω node is characterized by a conditional belief mass distribution $m^{\Omega}[\theta]$. There are several ways to propagate knowledge in a directed evidential network depending on the node receiving the current knowledge expressed in the form of a new belief mass distribution (Smets, 1993).

The knowledge is spread in the direction of the arc if the Θ node receives a new distribution of masses m^{Θ}, the Ω node is then updated taking into account this new information. This type of spread, called forward propagation enables the calculation of the distribution of plausibility pl^{Ω} using the equation of the Generalized Bayes Theorem based on an a priori knowledge. This equation uses the disjunctive rule of combination to determine the plausibility $pl^{\Omega}[\theta_i]$ for each subset $\theta_i \in \Theta$ and evaluate pl^{Ω} by combining the conditional plausibilities $pl^{\Omega}[\theta_i]$ using the Conjunctive Rule of Combination (CRC). The knowledge can also be propagated in the opposite direction of the arc if the Ω node receives a new distribution of masses m^{Ω}. The information contained in the Θ node is therefore updated to take account of this new information. This type of spread, called backward propagation enables the calculation of the plausibility distribution pl^{Θ} by using the following equation based on the Generalized Bayes Theorem to determine plausibilities $pl^{\Theta}[\omega_i]$:

$$pl^{\Theta}\left(\theta\right) = \sum_{\omega \subseteq \&} m^{\&}\left(\omega\right) \times \left(1 - \prod_{\omega_i \subseteq \omega}\left(1 - pl^{\Theta}\left[\omega_i\right]\left(\theta\right)\right)\right), \forall \theta \subseteq \Theta$$

Application

A case study using evidential network applied to sensor networks is being processed. It is related to Advanced Sensor Technologies for Nondestructive Testing (NDT) in the field of aeronautics Structural Health Monitoring (SHM). Sensors based on ultrasonic or Eddy Current techniques are arranged on an aircraft structure to measure the level of damage. Functional recycling of products or just testing the integrity of an aeronautical structure (fuselage or wing of an airplane for example) requires a diagnosis of their current state. The main goal is to detect, locate and identify defects, and, subsequently, follow their evolution. The use of NDT techniques allows the decision-maker to call on the compliance of the parts analyzed by quantification of the level of damage. The discussed example aims to develop an informational interface based on an evidential networks overlaid on the NDT sensor network for the interpretation of the signals received and the decision support. The position and number of sensors lead to a level of uncertainty characterized by belief functions introduced and propagated in the evidential network (Figure 6).

As a possible illustration of the interest of using evidential networks and with respect to the sensor architecture as well as the emitted and received signals, one can use here this technique to:

- Predict the presence of a defect in a forward propagation of the marginal belief functions knowing the signal values
- Predict the detection of a defect in a backward propagation of the conditional belief function knowing the existence of a proven deterioration or failure.

The authors observe also that upon receiving the information where a particular defect has been detected by any one of several distinct sensors, one can identify the source as belonging to this set. Therefore, one can assign the belief that the signal is present to the set of possibilities. The use of a Bayesian network would have forced to distribute this belief over all the individual nodes (sensors). Evidential reasoning avoids so the need for assumptions on missing data. When beliefs on a sensor is later required they are under constrained resulting from the disjunction and an interval representation is needed to capture the real constraint which enable the explicit representation of uncertain and certain knowledge.

Figure 6. Basic directed evidential network

Deep Belief Networks

Principles

Deep Belief Network (DBN) is based on a multilayered architecture forming a network with undirected connections between layers. They are a kind of deep learning architectures, subsuming convolutional networks (LeCun, Bottou, Bengio & Haffner, 1998), deep autoencoders (Hinton & Salakhutdinov, 2006).

Hinton showed that stacking Restricted Boltzmann Machine (RBM) was forming a DBN and it was proven that the learning process was very fast compared to alternatives techniques (Hinton, Osindero & The, 2006) (most of the works on DBN use RBM for the hidden layers). In RBM, as shown in Figure 7, the lateral connections in the visible and hidden layer are not considered. The consequence is that random variables encoded in hidden units are conditionally independent given the state of the visible units.

Each layer (i.e. a RBM) when trained individually using a very fast unsupervised learning algorithm (i.e. contrastive divergence) reveals to be very efficient. Once each layer is trained, the whole network can be adjusted (fine tune of the weight) by using back-propagation. DBN have similarities with belief network as they are graphical models in which the underlying (generated by learning) model contains many layers of hidden causal variables. It allows to extract a deep hierarchical representation of large training data. The strategy is to learn simpler concepts first (lower layers for low-level features) and build on them to learn more abstract concepts (upper layers for the more abstract concepts) (Bengio et al., 2006).

Figure 7. Instrumented aircraft monitored by sensor networks

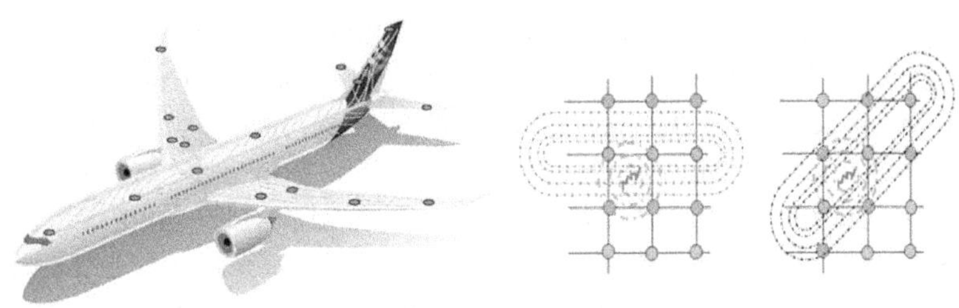

Figure 8. Boltzmann and Restricted Boltzmann Machine according to O'Connor, Neil, Liu, Delbruck & Pfeiffer (2013)

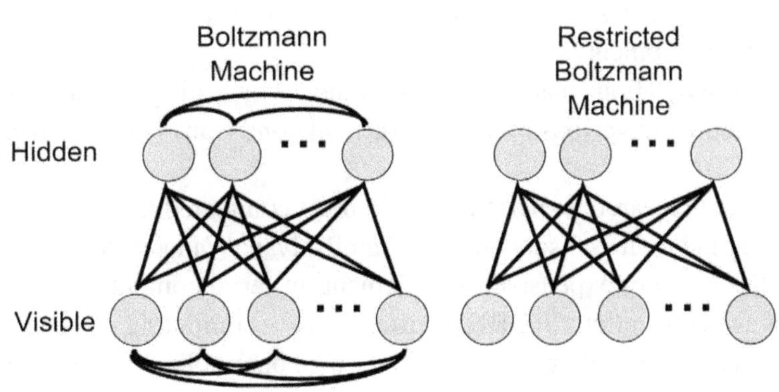

Using unsupervised learning, each RBM is trained to encode in its weight matrix a probability distribution that predicts the activity of the visible layer from the activity of the hidden layer. By stacking such models, and letting each layer predict the activity of the layer below, higher RBMs learn increasingly abstract representations of sensory inputs. This is why, although DBN has recently excelled among machine learning approaches, it provides a simple and powerful solution for sensor fusion tasks.

The data required for pre-training does not have to be labeled, and can thus make use of massive data that can be found today. Indeed, the emergence of Big Data explains this success as (pre)training can take advantage of giant databases of images, text, sounds or videos available from the Internet. An additional interesting and attractive feature is that the performance of deep networks typically improves with network size. Actually, there is new hope of achieving brain-like artificial intelligence simply by scaling up the computational resources (O'Connor et al., 2013). However, DBNs are very expensive to implement on serial computers because of their inherent need for feedback and parallel update of large numbers of units (O'Connor et al., 2013).

Application

Most applications of DBN are targeted at machine learning, artificial intelligence tasks such as voice and visual recognition. But given DBN intrinsic qualities regarding multi-sensor fusion tasks, several works exist in this area. Two researches on that topic are presented below.

In the first research, DBN were used for multi-sensor health diagnostic of industrial systems (Tamilselvan & Wang, 2013). A novel multi sensor health monitoring and failure diagnosis methodology was implemented for Kansas industries to manufacture reliable and failure preventive aircrafts. It has been applied to the diagnosis of the health status of an aircraft engine and of an electric power transformer. Aircraft health state detection is a key factor for prevention of performance degradation at different stages of damage, so aircraft reliability depends on continuous monitoring of current system health status. Due to the nature of observed data and the available knowledge, health diagnostic methods are often a combination of statistical inference and machine learning. The presented research proposes the use of DBN for structural health monitoring applications of aircraft and develops multi-sensor health diagnosis method. DBN works based on Restricted Boltzmann machine and it learns layer by layer considering priors and network posteriors. Enhanced diagnostic system can be structured in three steps:

- Definition of the health conditions and the corresponding sensor data for training and testing of the DBN model (collection of data and preprocessing);
- Development of the DBN classifier model to diagnose the system health according to the training models (classifier development);
- Validation of the classifications made by the DBN model by testing its behavior with existing sensor datasets extracted from various real-time practical conditions (validation).

In the second presented research, experiences have been conducted to improve the performance of DBN in a work that combine multi-sensors outputs (Tamilselvan, Wang & Wang, 2012). Indeed, and as it was said earlier, DBNs are very expensive to implement on serial computers. Despite their potential, the number of neurons and connections in DBN requires massive computing power, time, and energy, and thus makes their use in real-time applications infeasible (e.g., on mobile devices or autonomous robots). In their work (Tamilselvan et al., 2012) evaluate the spiking DBNs. They additionally present an event-based DBN architecture that can associate visual and auditory inputs, and combine multiple uncertain cues from different sensory modalities in a near-optimal way. One hypothesized role is in multisensory integration, and as generative Bayesian models, DBNs are very well suited to perform such tasks, e.g., by combining visual and auditory cues for improved recognition (Hinton et al., 2006). Among several experiments, they created a task in which visual stimuli from a silicon retina and auditory stimuli from a silicon cochlea were associated with each other in real-time. During training, the presentation of a pure tone was always paired with the presentation of an image of a handwritten digit. The network thus had to learn to associate the two sensory domains. The idea was to resolve the ambiguity of one sensory stream through information from the other stream.

In the first presented research, results of the application of the DBN classification have shown that the diagnosis was better than when classification standard algorithms, such as neural back-propagation network (BNN), self-organizing maps (SOM) or self-organizing maps (SOM), are used (Hinton et al., 2006). In the second presented work, results of the task where the two information streams were associated have better result. When considering only one source, the network alternate between several solutions whereas when combining both inputs (mixing at 50%), there is only one consistent interpretation.

COMPARATIVE ANALYSIS

Each of the approaches described above has advantages and disadvantages that make its application more or less relevant depending on the studied problem. Therefore, this section aims at comparing the different formalisms according to five criteria. These criteria are used to characterize the possibilities offered by each formalism based on the modeling needs of the studied problem, the quality of available data, the amount of usable data or the needed resources for the implementation of the problem. The different chosen criteria are:

- Amount of data needed by the formalism: if the chosen formalism is based on statistical (or probabilistic) principles, the required amount of data will be important to get a relevant and reliable result.
- Ability to model causal links: if the problem requires the implementation of a deductive reasoning based on the modeling of a chain of events (this type of reasoning can usually be expressed by sentences like "*If A happens then B happens*"), it is necessary to use a formalism that allows to model these causal links. The formalisms used to model such links are usually based on directed graphs (such as Bayesian networks or Directed Evidential Networks for example).
- Computational complexity of the modeling: this criterion allows to characterize the number of mathematical operations required to infer the problem using the formalism studied and, therefore, the computing power that is required to obtain the results in a reasonable time. So this criterion helps to have an idea about the hardware requirements to implement the problem.
- Involvement of humans during the uses of the formalism: this criterion defines whether studied formalism requires human intervention. This intervention can occur at three levels of the process:
    - The human formalizes the problem: the human intervention is often required for this phase, however, some formalisms can automatically build a model from a data set. For example, there are machine learning mechanisms based on Bayesian networks that allow to build the network structure but also to evaluate the variables.

- The human is a source of information in the problem: some formalisms allow to use a human expert as an information source (e.g., to assess the quality of information transmitted by a sensor). The ability to model human opinion and quality of this modeling are important criteria to compare formalisms.
- The human supervises the model: most formalisms provide the opportunity for the human to interact with the model to supervise its operation. This supervision often results in the ability to modify the settings in order to refine results.

- Integration of formalism in an embedded system: this criterion is closely linked to the criterion expressing the computational complexity. Required computing power varies a lot according to formalism choice and embedded applications often have limited resources. In addition, depending on the application, it is sometimes necessary to produce a result in a very short time to react in real time to change in the problem. The choice of modeling formalism is critical for this kind of applications.

Table 1 shows the comparison of the different formalisms presented in this chapter based on the five criteria that have just been introduced.

CONCLUSION

The aim of this chapter was to present four formalisms to consider the notion of uncertainty in a sensor network. The underlying idea is to be able to assess the measurement error and the corresponding risk to reduce the oversizing of the monitoring architectures and better define the level of confidence placed in the information received from the network. Bayesian networks, Transferable Belief Model (TBM),

Table 1. Comparison of formalisms

	Bayesian Networks	Transferable Belief Model (TBM)	Directed Evidential Networks (DEN)	Deep Belief Networks (DBN)
Amount of data	• High (frequentist approach) • Low (subjectivist approach)	Low	Low	High
Causality modeling	Yes	No	Yes	No (can only manage statistical correlation)
Computational complexity	Low to mean (depending on the complexity of the network)	Mean	High to very high (depending on the complexity of the network)	High
Human intervention	• To formalize the model: possible but not mandatory • As an information source: possible but not mandatory • To supervise the model: possible but not mandatory	• To formalize the model: Yes • As an information source: possible but not mandatory • To supervise the model: possible but not mandatory	• To formalize the model: Yes • As an information source: possible but not mandatory • To supervise the model: possible but not mandatory	• To formalize the model: No • As an information source: No • To supervise the model: possible but not mandatory
Embedded systems	Can easily be embedded	Can be embedded	Difficult to embed	Difficult to embed

Directed Evidential Networks (DEN) and Deep Belief Networks (DBN) have been presented. Each formalism has been illustrated with an application in the field of sensor networks. Each of these formalisms have advantages and disadvantages that make their use more or less relevant depending on the field of application. The comparative analysis in the last section aims to help choose the most suitable method for each problem instrumented by a sensor network. This analysis compares the different formalisms described in this chapter according to different criteria of quality, quantity and/or type of data collected by the sensor network. It also considers the resources necessary to implement each method in order to reduce and/or understand measurement or interpretation errors related to uncertainty in a sensor network.

From a methodological point of view, whereas Bayesian Networks and Transferable Belief Model are quite common techniques that are today sound and commonly used in the context of sensors networks, Directed Evidential Networks and Deep Belief Networks have not been applied much in such contexts. Therefore, the two last methods need to be more applied so as to be considered sound and computationally tractable.

From an application point of view, perspectives of such techniques are very large especially since the advent of Internet of Things. The most discriminant criteria likely to be used for the selection of a merging method are the amount of available data and the necessity to model causal relationships as shown in Table 1. Let just introduce two possible applications that could be tackled in an next future. One is about PHM (Prognostic Health Management) where lots of sensors are set up on machine tools so as to anticipate failure and problems. Another is about distance health monitoring and fall detection of elder people when staying at home. Indeed, technology enables today to equip someone and his house with several networked sensors. In both applications, analysis of their sensor outputs can be done through the four presented models still depending on the amount of available data, the possibility to have an underlying causal model, ... One interesting contribution will be to compare the usage of the four techniques to the same application context so as to underline setup differences (number of sensors, data quantity, measured phenomenon…) and of course the relevance of the results.

REFERENCES

Auzanneau, F. (2010). Diagnostic Filaire: Détection, localisation et caractérisation de défauts dans des réseaux filaires complexes. In Recherche et Innovation des Techniques de l'Ingénieur (pp. 1–11).

Ben Hassen, W., Auzanneau, F., Incarbone, L., Pérès, F. & Tchangani, A. P. (2013a). OMTDR using BER estimation for ambiguities cancellation in ramified networks diagnosis. In *Intelligent Sensors, Sensor Networks and Information Processing* (pp. 414-419).

Ben Hassen, W., Auzanneau, F., Incarbone, L., Pérès, F., & Tchangani, A. P. (2013b). On-line diagnosis using Orthogonal Multi-Tone Time Domain Reflectometry in a lossy cable.*Proceedings of IEEE International Conference on Systems, Signals and Devices*. doi:10.1109/SSD.2013.6564144

Ben Hassen, W., Auzanneau, F., Incarbone, L., Pérès, F., & Tchangani, A. P. (2015). Distributed Sensor Fusion for Wire Fault Location Using Sensor Clustering Strategy. *International Journal of Distributed Sensor Networks*, *2015*, 1–17. doi:10.1155/2015/538643

Ben Hassen, W., Auzanneau, F., Pérès, F., & Tchangani, A. P. (2012). A distributed diagnosis strategy using bayesian network for complex wiring networks.*Proceedings of IFAC Workshop on Advanced Maintenance Engineering, Services and Technology*. doi:10.3182/20121122-2-ES-4026.00031

Ben Hassen, W., Auzanneau, F., Pérès, F., & Tchangani, A. P. (2013c). Diagnosis sensor fusion for wire fault location in CAN bus systems. *Proceedings of IEEE Sensors, 2013*, 1–4. PMID:25126153

Ben Hassen, W., Auzanneau, F., Pérès, F., & Tchangani, A. P. (2014). Ambiguity Cancellation for Wire Fault Location based on Cable Life Profile. *Proceedings of IFAC World Congress*, Cape Town, South Africa. doi:10.3182/20140824-6-ZA-1003.02411

Bengio, Y., Lamblin, P., Popovici, D., & Larochelle, H. (2006). Advances in Neural Information Processing Systems: Vol. 19. *Greedy layer-wise training of deep networks.*

Charniak, E. (1991). Bayesian Networks without Tears. *AI Magazine*, *12*(4), 50–63.

Chen, W., Jiang, X., Li, X., Gao, J., Xu, X., & Ding, S. (2013). Wireless Sensor Network nodes correlation method in coal mine tunnel based on Bayesian decision. *Measurement*, *46-8*(8), 2335–2340. doi:10.1016/j.measurement.2013.04.018

Daniel, J., & Lauffenburger, J. P. (2014). Fusing navigation and vision information with the Transferable Belief Model: Application to an intelligent speed limit assistant. *Information Fusion*, *18*, 62–77. doi:10.1016/j.inffus.2013.05.013

Dempster, A. (1967). Upper and lower probabilities induced by a multivalued mapping. *Annals of Mathematical Statistics*, *38*(2), 325–329. doi:10.1214/aoms/1177698950

Denoeux, T. (2008). Conjunctive and disjunctive combination of belief functions induced by non-distinct bodies of evidence. *Artificial Intelligence*, *172*(2-3), 234–264. doi:10.1016/j.artint.2007.05.008

Dubois, D., & Prade, H. (1988). Representation and combination of uncertainty with belief functions and possibility measures. *Computational Intelligence*, *4*(3), 244–264. doi:10.1111/j.1467-8640.1988.tb00279.x

Dubois, D., & Prade, H. (2009). Formal representation of uncertainty. In D. Bouyssou, D. Dubois, M. Pirlot, & H. Prade (Eds.), *Decision-Making Process*. UK and Wiley. doi:10.1002/9780470611876.ch3

Flammini, F., Marrone, S., Mazzocca, N., Nardone, R. & Vittorini, V. (2015). Using Bayesian Networks to evaluate the trustworthiness of '2 out of 3' decision fusion mechanisms in multi-sensor applications, *IFAC-PapersOnLine*, 48-21, 682-687.

Frikha, A. & Moalla, H. (2014). Analytic hierarchy process for multi-sensor data fusion based on belief function theory. *European Journal of Operational Research*, 241-1, 133-147.

Hinton, G. E., Osindero, S., & Teh, Y. (2006). A fast learning algorithm for deep belief nets. *Neural Computation*, *18*(7), 1527–1554. doi:10.1162/neco.2006.18.7.1527 PMID:16764513

Hinton, G. E., & Salakhutdinov, R. R. (2006). Reducing the dimensionality of data with neural networks. *Science*, *313*(5786), 504–507. doi:10.1126/science.1127647 PMID:16873662

Hochreiter, S., Bengio, Y., Frasconi, P., Schmidhuber, J., & Elvezia, C. (2001). Gradient flow in recurrent nets: the difficulty of learning long-term dependencies. In S. C. Kremer & J. F. Kolen (Eds.), *A Field Guide to Dynamical Recurrent Neural Networks* (pp. 237–244). New York, NY: IEEE Press.

Jensen, F., & Nielsen, T. (2007). *Bayesian networks and decision graphs*. Springer. doi:10.1007/978-0-387-68282-2

Klir, G.J. (1995). Principles of uncertainty: What are they? Why do we need them? *Fuzzy Sets and Systems*, 74-1, 15-31.

LeCun, Y. L., Bottou, L., Bengio, Y., & Haffner, P. (1998). Gradient-based learning applied to document recognition. *Proceedings of the IEEE, 86*(11), 2278–2324. doi:10.1109/5.726791

Livi, L. & Sadeghian, A. (2015) Data granulation by the principles of uncertainty. *Pattern Recognition Letters*, 67-2, 113-121.

Mkrtchyan, L., Podofillini, L., & Dang, V. N. (2015). Bayesian belief networks for human reliability analysis: A review of applications and gaps. *Reliability Engineering & System Safety, 139*, 1–16. doi:10.1016/j.ress.2015.02.006

Naim, P., Wuillemin, P. H., Leray, P., Pourret, O., & Becker, A. (2011). *Réseaux bayésiens*. Algorithmes. Eyrolles.

O'Connor, P., Neil, D., Liu, S.-C., Delbruck, T., & Pfeiffer, M. (2013). Real-time classification and sensor fusion with a spiking deep belief network. *Frontiers in Neuroscience, 7*, 178. PMID:24115919

Savic, V., Wymeersch, H., & Zazo, S. (2013). Belief consensus algorithms for fast distributed target tracking in wireless sensor networks. *Signal Processing, 95*, 149–160. doi:10.1016/j.sigpro.2013.09.005

Senouci, M. R. (2014). *On the use of the belief functions theory in the deployment and control of wireless sensor networks* [PhD thesis]. Paris-Est University.

Senouci, M. R., Mellouk, A., Oukhellou, L. & Aissani, A. (2012). Using the Belief Functions Theory to Deploy Static Wireless Sensor Networks. In *Belief Functions: Theory and Applications* (pp. 425-432). Springer Berlin Heidelberg.

Shafer, G. (1976). *A Mathematical Theory of Evidence*. Princeton University Press.

Shang, C., Yang, F., Huang, D., & Lyu, W. (2014). Data-driven soft sensor development based on deep learning technique. *Journal of Process Control, 24*(3), 223–233.

Smets, P. (1993). Belief functions: The disjunctive rule of combination and the Generalized Bayesian Theorem. *International Journal of Approximate Reasoning, 9*(1), 1–35. doi:10.1016/0888-613X(93)90005-X

Smets, P. (2002). The application of the matrix calculus to belief functions. *International Journal of Approximate Reasoning, 31*(1-2), 1–30. doi:10.1016/S0888-613X(02)00066-X

Smets, P. (2005). Decision making in the TBM: The necessity of the pignistic transformation. *International Journal of Approximate Reasoning, 38*(2), 133–147. doi:10.1016/j.ijar.2004.05.003

Smets, P., & Kennes, R. (1994). The transferable belief model. *Artificial Intelligence*, *66*(2), 191–234. doi:10.1016/0004-3702(94)90026-4

Tamilselvan, P., & Wang, P. (2013). Failure diagnosis using deep belief learning based health state classification. *Reliability Engineering & System Safety*, *115*, 124–135. doi:10.1016/j.ress.2013.02.022

Tamilselvan, P., Wang, Y., & Wang, P. (2012). Deep Belief Network based state classification for structural health diagnosis. *Proceedings of theIEEE Aerospace Conference*, Big Sky, Montana, USA (pp. 1-11). doi:10.1109/AERO.2012.6187366

Turell, J. (2013). 10 breakthrough technologies 2013: Deep learning. *MIT Technology Review*. Retrieved from.http://www.technologyreview.com/featuredstory/513696/deep-learning/

Verma, T., & Pearl, J. (1991). A Theory of Inferred Causation. *Proceedings of theSecond International Conference on the Principles of Knowledge Representation and Reasoning*, Cambridge, Massachusetts, USA.

Verron, S. (2007). *Diagnostic et surveillance des processus complexes par réseaux bayésiens* [PhD thesis]. Institut des Sciences et Techniques de l'Ingénieur d'Angers.

Villeneuve, E., Beler, C., Peres, F., & Geneste, L. (2012). Hybridization of statistical and cognitive experience feedbacks to perform risk assessment. *Proceedings of theIEEE International Conference on Industrial Engineering and Engineering Management*, Hong-Kong, China.

Xu, H., & Smets, P. (1994). Evidential reasoning with conditional belief functions.*Proceedings of the 10th Conference on Uncertainty in Artificial Intelligence* (pp. 598–605).

Yaghlane, B. B., Smets, P., & Mellouli, K. (2001). Belief function independence: I. The marginal case. *International Journal of Approximate Reasoning*, *29*(1), 47–70. doi:10.1016/S0888-613X(01)00055-X

Yaghlane, B. B., Smets, P., & Mellouli, K. (2004). Directed Evidential Networks with Conditional Belief Functions.Symbolic and Quantitative Approaches to Reasoning with Uncertainty, *LNCS* (Vol. *2711*, pp. 291–305). doi:10.1007/978-3-540-45062-7_24

KEY TERMS AND DEFINITIONS

Bayesian Network: Probabilistic and graphical model that represents random variables and causal relations between these variables as an acyclic oriented graph.

Decision: Act of an agent (human or computer) who makes a choice between several solutions according to one or several objectives.

Deep Belief Network: Deep Belief Networks is a multilayered architecture forming a network with undirected connections between layers. They are often built as hierarchies of recurrently connected simpler probabilistic graphical models, so called Restricted Boltzmann Machines (RBMs).

Directed Evidential Network: Graphical model similar to Bayesian network using belief functions instead of probabilities to formalize the knowledge included in the problem variables.

Rule of Combination: In the theory of belief functions, a rule of combination is a mathematical tool that allows the fusion of information provided by multiple sources.

Theory of Belief Functions: Modeling and reasoning framework with uncertain knowledge that generalizes, among others, the probabilistic framework. This generalization provides a more comprehensive view of uncertainty and, especially, allows to model the human uncertainty.

Uncertainty: State of an agent when it is unable to distinguish the veracity of a piece of information (that is to say whether the information is true or false).

Section 2

Chapter 12
Asset Life Cycle Plans:
Twelve Steps to Assist Strategic Decision-Making in Asset Life Cycle Management

R. J. (Richard) Ruitenburg
University of Twente, The Netherlands & Liander N.V., The Netherlands

A. J. J. (Jan) Braaksma
University of Twente, The Netherlands

L. A. M. (Leo) van Dongen
University of Twente, The Netherlands

ABSTRACT

Effective management of physical assets should deliver maximum business value. Therefore, Asset Management standards such as PAS 55 and ISO 55000 ask for a life cycle approach. However, most existing methods focus only on the short term of the asset's life or the estimation of its remaining life. These methods do not consider alignment to changing corporate objectives in a variable context, nor do they adopt a multidisciplinary perspective. This chapter argues that, to create maximum value, Asset Management should be a multidisciplinary and strategic practice that considers the complete life cycle of the asset: Asset Life Cycle Management. A practical twelve-step approach is presented to develop an Asset Life Cycle Plan (ALCP) in which expert sessions are used to identify the main lifetime impacts that influence the creation of business value from the use of the asset. The steps are illustrated with an example from practice. The chapter concludes that the ALCP supports asset managers in making long-term strategic decisions in a timely and effective manner.

INTRODUCTION

Physical assets are indispensable for society. Oftentimes, the assets are not noticed until a seemingly minor technical failure disrupts daily life. At these moments, it becomes clear that these assets fulfil vital functions in our lives. Examples include bridges and roads for transportation, the electricity grid to keep the lights on, and water treatment plants for drinking water. Next to their societal value, the

DOI: 10.4018/978-1-5225-0651-5.ch012

financial (replacement) value of these assets is substantial as well. An estimate of this value in the Netherlands lies around 400 billion (milliard) euros (Veenman & Besselink, 2010), which is around 50% of the Dutch yearly GDP. This clearly shows the large financial value represented by physical assets in a modern Western country. Additionally, the costs of the maintenance needed to keep these assets fit for use are considerable. In the Netherlands, the maintenance costs amount to 30 to 35 billion euros per year (NVDO, 2015). Because such high financial and social values are at stake, the effective management of these assets is important for asset owners and society alike.

The fact that the end of the expected functional lives of many assets is approaching, increases the importance of effective Asset Management. This applies to Dutch infrastructure assets (de Leeuw & Pries, 2014; Jongepier, 2007; Wetzer & Bouwman, 2007). The same applies to Western Europe and the USA (Allan, 2005; Brown & Humphrey, 2005). Assets reaching the end of their useful life tend to need more intensive maintenance, and modernization or life extension may be worthwhile. On the other hand, timely disposal may be necessary to prevent excessive costs, or risks for health, safety, and the environment (Jongen, 2012; Rouse & Chiu, 2009). Additionally, these ageing assets often have to fulfil different functions than those for which they were designed, often decades ago. The assets have to comply with new regulations, satisfy changing market demands, and may be needed to produce higher quantities than anticipated (Al-Turki, 2011).

It is in this complex context that the asset manager operates. The management of ageing assets in a changeable context asks for strategic decisions, next to the daily work of maintenance and tactical planning. Strategic decisions directly impact corporate objectives by asking questions such as: How long can the operation of asset X be continued? What modifications need to be made to asset Y at the next overhaul? How can we convince management that the purchase of asset Z is worthwhile, even though its price is higher than all other alternatives? How can we keep the operation and maintenance of the assets aligned with changing corporate objectives? and What external developments may have an effect on the business value created with the operation of the assets?

Such a strategic focus on the complete life cycle of the asset is also required by recent Asset Management standards such as PAS 55 (IAM & BSI, 2008) and ISO 55000 (ISO, 2014). These require a life cycle approach. However, these standards do not provide much guidance as to how such a life cycle approach should be implemented. The same applies to well-known maintenance concepts, such as Reliability Centred Maintenance (RCM), which focus primarily on the short term (Coetzee, 1999; Murthy, Atrens, & Eccleston, 2002). In other words, there are many strategic Asset Management decisions to be made, but there is very little guidance available for asset managers who need to make these decisions. Therefore, the aim of this chapter is to present a tool to assist strategic decision-making in Asset Management.

The next section will present a theoretical background on Asset Management. It will conclude that Asset Management should be a strategic and multidisciplinary practice, focusing on the complete lifetime of an asset. Such an approach to Asset Management will be termed "Asset Life Cycle Management" (ALCM). Current maintenance approaches often just focus on the short term and do not adopt a multidisciplinary approach. These approaches do not address changes in the environment in which the asset operates, changes in regulation, or new corporate objectives. Hence, the concept of "lifetime impacts" will be introduced in the next section. These lifetime impacts are a key ingredient for an Asset Life Cycle Plan (ALCP), which is a tool developed to support strategic decision-making in Asset Management. An ALCP does this by presenting a holistic view of the current and expected performance, costs of the assets, and the main challenges that may affect performance over an asset's lifetime. An extensive and practical twelve-step approach will be presented to develop such an ALCP, illustrated with an ALCP on distribution transformers from a Dutch network operator. The chapter concludes that the ALCP aids asset managers in making strategic decisions in a timely and effective manner. In the last section, the implications for further research and for practitioners will be presented.

BACKGROUND ON ASSET MANAGEMENT

It is worthwhile to investigate the concept of Asset Management in some more detail before presenting the tool that assists strategic decision-making in Asset Management. Pudney (2010) proposes an elaborate definition of Asset Management based on a study of Asset Management literature. He defines Asset Management as "an organisation's coordinated multidisciplinary practice that applies human, equipment and financial resources to physical assets over their whole life cycle to achieve defined asset performance and cost objectives at acceptable levels of risk whilst taking account of the relevant governance, geo-political, economic, social, demographic and technological regimes" (Pudney, 2010, p. 8). From this definition, five characteristics of Asset Management stand out. Asset Management (AM):

1. Is a multidisciplinary practice;
2. Takes the whole life cycle of a physical asset into account;
3. Achieves certain objectives;
4. Considers the limits of risk and relevant regimes; and
5. Determines the allocation of resources (Ruitenburg, Braaksma, & van Dongen, 2015).

Each of these five characteristics will be discussed in more detail.

Asset Management Is a Multidisciplinary Practice

Many different disciplines should be combined in Asset Management. This applies to technical disciplines such as engineering and maintenance. Additionally, a financial perspective on the assets should be included in Asset Management. However, there are additional disciplines that play a crucial role, such as data collection and analysis, planning and scheduling, operational excellence, rules and regulations, IT, human resources, and innovation (Smit, 2014). To create maximum value from the exploitation of an asset, all these disciplines may play a role. One should understand degradation processes that deteriorate the condition of the asset. One should know what maintenance actions may be performed to counter this deterioration. Based on the costs of the possible interventions, a choice has to be made for the most cost-effective solution. This intervention should comply with existing regulations and fit with the capabilities of the engineers and operators employed by the company. The innovation department may have sought out new technologies that can be implemented while making the intervention. And of course, the implementation of the intervention needs to be scheduled, taking the production planning into account to ensure a good fit with the ongoing operations.

In spite of the importance of the involvement of many disciplines in Asset Management, Haffejee and Brent (2008) conclude, "[t]he economic, environmental, social and technical dimensions of Asset Management are not explicitly depicted." Rather, approaches are often limited to the technical aspects of the asset (Frangopol, Saydam, & Kim, 2012; Garg & Deshmukh, 2006) or focus mainly on the financial performance of the asset (Asiedu & Gu, 1998; Márquez, Márquez, Fernández, Campos, & González-Prida Díaz, 2012). To conclude, multidisciplinary approaches to Asset Management are scarce.

Asset Management Takes the Complete Life Cycle of the Asset into Account

Asset Management should consider the asset's complete life cycle. Different reasons for this are given in the literature. One of these is that most of the life cycle costs are already fixed in the design of the asset; hence, it is important to take the complete life cycle of the asset into account from the moment of design onwards (Schuman & Brent, 2005; L.A.M. van Dongen, 2011). In the design, a large part of the future performance of an asset is already fixed. One may be able to change the asset's throughput, operation speed, or safety during its service life, but often at considerably larger costs than if it had been part of the original design. L.A.M. van Dongen (2011) illustrates this with a figure (refer to Figure 1). The figure shows the influence the asset manager may exert over the performance of an asset during its life cycle (in this case the life cycle of a train). Future performance requirements should be considered as early as possible to prevent unnecessary high refurbishment costs later in the asset's life. In other words, the complete life cycle of the asset should be taken into account.

A second reason to consider the complete life cycle of the asset is the increasing market turbulence in many sectors. This requires the asset manager to make the right strategic decisions to ensure that the asset will remain valuable to the company even in a changed market (Komonen, Kortelainen, & Räikkönen, 2012). Additionally, many external developments (e.g. new regulations) may have an impact on the asset (Al-Turki, 2011) and should be taken into account as early as possible in the asset's life. However, the same authors stressing the importance of considering the asset's complete life cycle also state that a focus on the complete life cycle is often lacking. This also applies to well-known Asset Management methods, such as Reliability Centred Maintenance (RCM), which primarily focuses on the short term (Coetzee, 1999; Murthy et al., 2002).

Figure 1. Overview of a typical life cycle at NedTrain
(adapted from L.A.M. van Dongen, 2011)

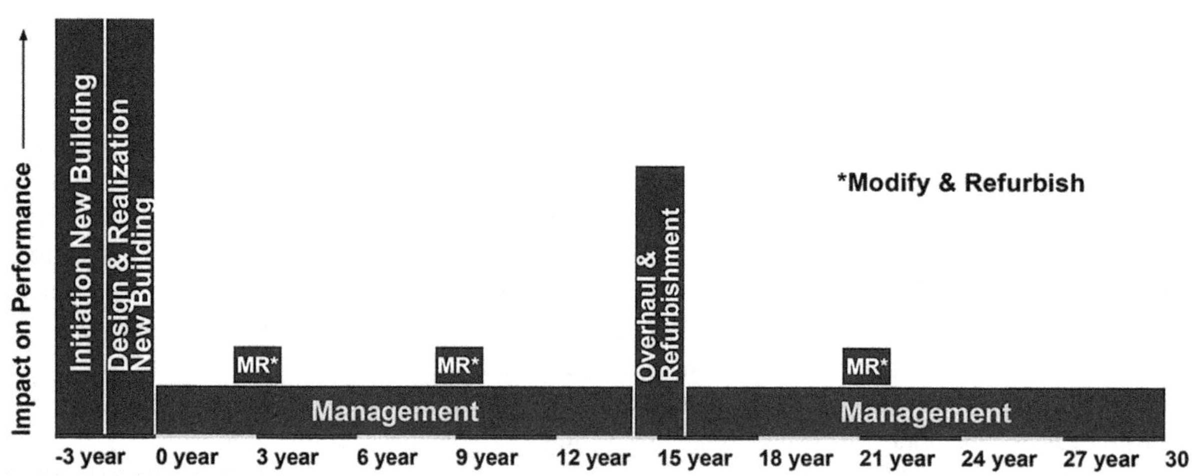

Asset Management Aims to Achieve Specific Objectives

The third characteristic of Asset Management shows that Asset Management aims to reach certain objectives. These objectives should relate to the objectives of the asset owner and thus ultimately to the corporate strategy. The asset is not valuable to the company in its own right (apart from its potential resale or scrap value). The asset is valuable only if it fulfils a function that fits the strategic goals of the company, goals that may change over time. Combining this with the previous characteristic shows that the management of assets should not be considered a stand-alone process carried out in a vacuum. Rather, the operation and maintenance of assets should continuously be aligned to changes in corporate objectives and the operating environment (Komonen et al., 2012).

The importance of Asset Management with respect to the corporate goals is also shown by the PAS 55 definition of Asset Management: Asset Management consists of "systematic and coordinated activities and practices through which an organization optimally and sustainably manages its assets and asset systems, their associated performance, risks and expenditures over their life cycles for the purpose of achieving its organizational strategic plan" (IAM & BSI, 2008, p. 2). This definition shows that Asset Management is inextricably connected to the achievement of the "organizational strategic plan" and therefore the corporate strategy. As assets are necessary means to reach the corporate objectives, the management of these assets is a strategic concern to a company. For example, the assets used by a chemical company facilitate the chemical processes that create the added value for the company. For a utility company, the infrastructure assets allow the company to deliver electricity, gas, or water to its customers. For an airline, the aircraft are indispensable to transport travellers to destinations all over the world. To summarize, since companies are highly dependent on the assets they operate to achieve their corporate objectives, the strategic aspects of maintenance and Asset Management should be emphasized. However, the connection between Asset Management and the corporate strategy is often lacking (Komonen et al., 2012).

Asset Management Considers the Relevant Risks and Regimes

Asset Management should take relevant risks and regimes into account. Risks can be failures or safety risks, the type of risks most maintenance methods focus on (e.g. RCM or Condition-Based Maintenance [CBM]). However, risks may also be related to topics as diverse as production errors, reputation, sustainability, or cyber security. Asset Management should also consider regimes, which relate to the external developments mentioned earlier: new regulations, changes in the labour market, social pressure to increase sustainability, or a lobby to improve working conditions.

The concept of risks emphasizes the inherent uncertainty related to Asset Management. Asset Management should consider the asset's complete life cycle, but our knowledge of the future is incomplete by definition. Additionally, often only limited information on the asset is available, and—as Asset Management is a multidisciplinary practice—this information may be dispersed over different people, departments, and software systems. In addition, the quality of data (e.g. failure data or condition data) may be unreliable (Braaksma, 2012). Nevertheless, decisions have to be made about the assets and the allocation of resources to the assets.

Asset Management Determines the Allocation of Resources

Ultimately, Asset Management is all about the allocation of resources. Where should money be invested? Where may a reduction of working hours be achieved? What maintenance tasks are necessary, and which may be postponed? Which activities do we do ourselves, and which can be carried out by service providers in a more cost-effective way? The allocation of resources applies to both the daily operations (e.g. production schedules, spare part availability) and the strategic management of the assets (e.g. the budget reserved for an overhaul or the purchase of a new asset). Many methods and tools exist that may be used to effectively allocate resources, especially for operational and tactical planning. The field of maintenance management addresses these topics (for two good reviews of this topic, refer to Frangopol, Saydam, & Kim, 2012; Garg & Deshmukh, 2006).

Asset Life Cycle Management (ALCM)

To conclude, Asset Management should be a multidisciplinary practice focusing on the achievement of changeable corporate objectives over the complete life cycle of the asset and taking into account the relevant external developments and risks. Such a strategic, multidisciplinary, and long-term approach to Asset Management will be termed Asset Life Cycle Management (ALCM) throughout the remainder of this chapter.

LIFETIME IMPACTS

Asset Management ultimately focuses on the realization of business value from the use of physical assets over their complete lifetime. For the short term, maintenance concepts such as Reliability Centred Maintenance (Moubray, 1997; Smith & Hinchcliffe, 2003), Total Productive Maintenance (Wireman, 2004), and many others exist (Waeyenbergh & Pintelon, 2002). These often produce an overview of maintenance tasks to be carried out in a certain time period, and these same tasks may be repeated over and over again. Additionally, there are many tools available to continuously improve the performance of assets, such as Pareto analysis or root cause analysis (e.g. Ishikawa diagrams) (for an overview, see Zaal, 2013).

Looking at the remainder of the asset's life, many approaches exist that focus on the estimation of the Remaining Useful Life (RUL) of the asset (e.g. Si, Wang, Hua, & Zhou, 2011). These tools and methods aim to estimate the point in time when the asset will cease to perform its function because of technical failures, for example, after a specified amount of operation hours.

Maintenance concepts and RUL estimation cover the short term and the end of the asset's life. What is lacking is a method that covers the period in between, a method that assists the asset manager in ensuring the asset remains valuable to the company in a changeable context (see Figure 2). After all, repeating the same maintenance tasks during the asset's lifetime may overlook internal or external changes to the company that may have an impact on the asset. These changes may present opportunities or threats that have an impact on the realization of value from the use of the asset. Therefore, they should be taken into account to ensure a maximum contribution of the asset to the business objectives. Additionally, both the maintenance concepts and RUL estimation primarily take a technical perspective on the asset, rather than the multidisciplinary approach that is important in Asset Management.

Figure 2. The gap in asset management between maintenance concepts—covering the short term and sometimes repeated periodically—and the outcome of RUL estimations giving the expected end-of-life of the asset

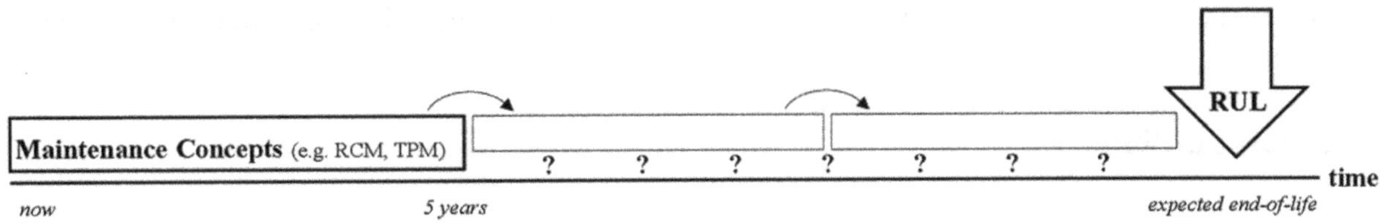

To fill this gap in the asset's life cycle, the authors have proposed the concept of "lifetime impacts" in earlier publications (Ruitenburg, Braaksma, & van Dongen, 2014; Ruitenburg & Braaksma, in press). Lifetime impacts can be defined as "probable (technical and non-technical) events or trends that may have a positive or negative influence on the value creation with the asset in the intermediate or long term" (Ruitenburg & Braaksma, in press).

The concept of lifetime impacts is explained most easily using an analogy of the Titanic. For a moment, imagine yourself as the asset manager responsible for the safe and reliable operation of the Titanic. From a purely maintenance perspective, you might carry out a sophisticated FMEA-analysis to create a maintenance program to keep this massive, valuable ship in optimum condition. Additionally, an RUL-estimation could give an idea of the projected end-of-life moment of the vessel, based on assumptions of operation hours, operation conditions, and knowledge of degradation mechanisms. However, it is well-known that neither the maintenance program nor the RUL-estimations would have prevented the Titanic from sinking. It was an iceberg that brought this massive, high-tech asset to a premature end, resulting in massive financial and reputational losses, not to mention the human tragedies caused by this event.

One of the many intriguing stories surrounding the Titanic disaster is about the lookout in the crow's nest (Salkeld, 2007). It is said that he did not have the key to open the locker with the binoculars. In the official inquiry, the lookout was asked about the consequences of the lack of binoculars. He stated that, with binoculars, "We could have seen it [the iceberg] a bit sooner." When asked, "How much sooner," he replied, "Well, enough to get out of the way."

Just as a single iceberg sank the Titanic, there are many proverbial icebergs that may impact the creation of business value with an asset (see Figure 3), for example, a large increase in the costs of spare parts or the ban of a certain toxic substance used in the asset. Only through timely identification of such icebergs can timely measures be taken to circumvent them. Knowing the remaining useful lifetime will only become relevant if the asset does not find a premature end-of-life due to the consequences of a lifetime impact.

Figure 3. Lifetime impacts[1]

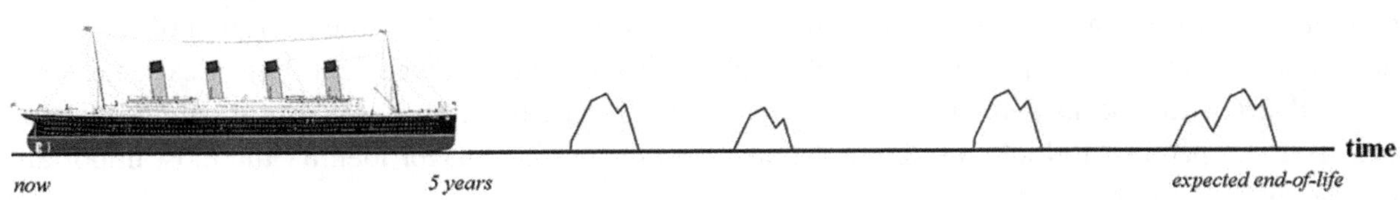

For example, a plant in the process industry only produces business value as long as it complies with regulations on safety, emissions, hazardous materials, etc. If a change in regulation, which would be a lifetime impact, makes the operation of the plant illegal, the plant may need to be shut down. In such a case, the plant will be worthless to the asset owner, irrespective of its technical condition and remaining useful life.

Five Perspectives

As ALCM is a multidisciplinary practice, the lifetime impacts that are relevant for an ALCP should be multidisciplinary as well. Based on an industry project of Dutch asset owners (R. van Dongen, 2011) and a first application of the concept of lifetime impacts in practice (Ruitenburg & Braaksma, in press), five different perspectives have been identified. Using the Titanic analogy, there are five different types of binoculars to look into the future in order to identify the relevant opportunities and threats.

1. **Technical Perspective:** Technical lifetime impacts concern the ability of the asset to comply with the technical specifications set regarding its technical condition and performance. Reliability, availability, production quality, remaining useful lifetimes, and bathtub curves are all related to this perspective.
2. **Economic Perspective:** The economic perspective is concerned with the financial aspects of the asset and its operation. Does the asset still deliver financial value to its user or owner? Maintenance costs, operating costs, costs of spare parts, life cycle costing, and total cost of ownership are relevant concepts here.
3. **Compliance Perspective:** The compliance perspective asks if the company is still allowed to operate the asset. This perspective looks at all kinds of regulation (e.g. safety, environmental, worker conditions, or hazardous materials). Additionally, the social acceptance of the asset and its operation is relevant here since a decreasing acceptance may precede official regulation.
4. **Commercial Perspective:** The commercial perspective looks at the extent to which the asset still fulfils the demands of the market. An asset producing Discmans or mp3 players may be in great condition from a technical perspective, but it probably no longer fulfils the demands of the market.
5. **Organizational Perspective:** Finally, the organizational perspective concerns the ability and willingness of the company to operate the asset. There may be a lack of specialized knowledge limiting the ability of the company to maintain the asset in the desired condition, or the organization may aim for an increased standardization of its assets. This may make certain asset types redundant for the company, even though from the other perspectives there may be no reason to decommission these assets.

These five perspectives, abbreviated by the acronym TECCO, together give a multidisciplinary insight into the main lifetime impacts regarding the future of the asset. The perspectives are not intended to be mutually exclusive. For example, an incentive towards a more sustainable operation of the assets may come both from social pressure (Compliance) and from higher management (Organizational), or a certain degradation mechanism (Technical) may result in higher maintenance costs (Economic). Using these five perspectives to look into the future will help the asset manager identify the most important lifetime impacts for the asset.

Asset Life Cycle Plans

The identification of lifetime impacts is a crucial step to ensure continuous value creation with an asset over its complete lifetime. However, knowing these impacts in itself does not necessarily result in the desired performance of the asset, nor in an alignment of the operation with the asset and the corporate objectives. Therefore, in response to the required "life cycle approach" by ISO 55000, this paper proposes the concept of the Asset Life Cycle Plan (ALCP). In their earlier research, the authors carried out a case study on Asset Life Cycle Management (ALCM) using these ALCPs at a rolling-stock maintenance company (Ruitenburg et al., 2015). In this company, the ALCP was used as a means to align corporate objectives with the management of trains, as well as to communicate the expected performance and the associated costs of the train to the relevant stakeholders (idem). In this way, the ALCP can be a means to assist ALCM practices in a company and to support strategic Asset Management decisions.

Figure 4 shows the main structure of the ALCP. This structure has been developed based on the means-end analysis (Holmström, Ketokivi, & Hameri, 2009), the Pyramid Principle for structuring texts (Minto, 2008), and the ALCPs in the case study on ALCM carried out by the authors (Ruitenburg et al., 2015).

The ALCP starts with the strategic objectives the company has for the asset. These are compared with the current and expected performance and costs of the asset. If the performance does not comply with the objectives, interventions may be needed to bring the performance into accordance with the goals. Lifetime impacts may have an impact on either the strategic objectives (e.g. goals for the reduction of CO_2 emissions) or the asset's future performance and costs (e.g. higher costs due to higher spare part prices). In the ALCP, the main focus lies on the value realized from the exploitation of the asset. This involves the continuous alignment of the asset's performance with the corporate objectives in a changeable context.

This general structure has been translated into the following chapter structure for an ALCP:

0. **Management Summary:** States the main message of the ALCP, aiming at higher management. This should answer the question, "What needs to be done to keep the asset (population) valuable to the company in the long term?"
1. **Introduction:** Gives a brief introduction to the ALCP, including its purpose, aim, the scope of the ALCP, and an outline of the document.

Figure 4. The structure of the ALCP

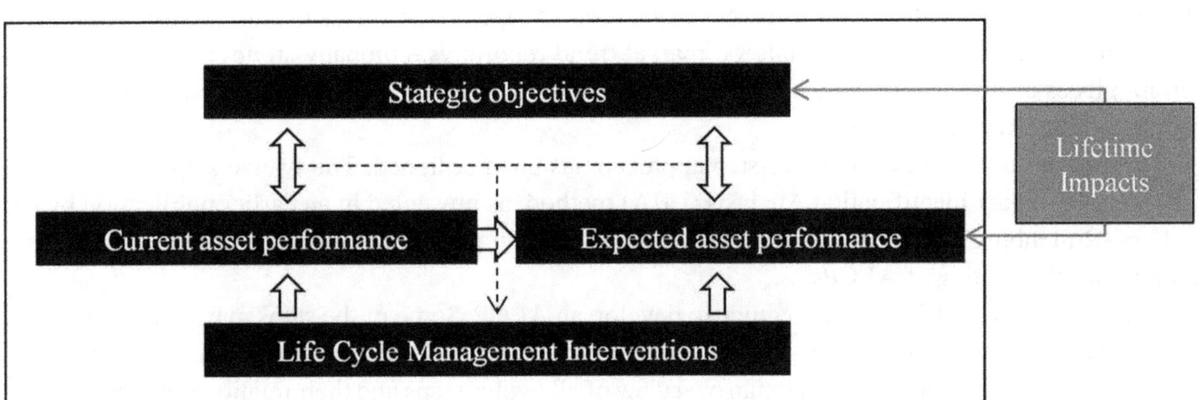

2. **Strategic Objectives and Function:** Lists the main corporate goals relevant for the asset and the main strategic developments and decisions that are relevant, for example, the aim to enter new markets. Additionally, it states how the asset contributes to these goals.
3. **Current Situation and Performance:** Gives an overview of the main characteristics of the asset (population), including its age, main components, replacement value, and a brief summary of the current maintenance policies. It lists the current performance of the asset related to the strategic objectives and relevant goals for the asset. Since ALCM is a multidisciplinary practice, the performance should be measured from different perspectives as well, for example, production output, availability, reliability, maintenance costs, and safety.
4. **Lifetime Impacts:** This chapter lists the main lifetime impacts that affect the value the asset will have for the company in the future. Lifetime impacts are listed for all five TECCO perspectives. Attention is paid to both positive and negative lifetime impacts.
5. **Expected Performance and Costs:** Based on the current performance and the effects of the lifetime impacts identified, an overview of the expected performance on the asset objectives and the costs of the asset is given. This may be by means of KPIs or other quantitative measures, but can be qualitative as well.
6. **Life Cycle Management:** Based on the analysis thus far, this chapter recommends actions be taken to keep the asset valuable to the company. This may include additional maintenance, design changes, overhauls, modernization projects, preventive replacement, and actions to collect information needed to manage the asset more effectively in the future.
7. **List of Actions:** Gives an overview of the main actions resulting from the ALCP, including projected start and end dates and the responsible person or department.

WRITING AN ALCP: STEP-BY-STEP INSTRUCTIONS

Writing an ALCP is not just a matter of writing a number of chapters according to a specific structure. It is not so much the structure that is important, but rather the process of creating an ALCP. One of the main challenges of writing an ALCP is to bring all the relevant knowledge, information, and expertise on the asset together in one document. As previously stated, Asset Management is a multidisciplinary practice. Many companies are used to organizing these disciplines in separate functional silos. This information is often scattered over different databases (e.g. failure data vs. maintenance budgets vs. outsourced maintenance), departments (e.g. regulation vs. operations vs. financial vs. purchasing), and documents (e.g. maintenance manuals vs. market trend reports vs. company strategies). Additionally, certain pieces of information may not be available in explicit form, but are rather stored in the heads of people working with the assets.

To structure this process, a twelve-step approach has been designed. The twelve steps are based on the Lifetime Impact Identification Analysis (LIIA) method, as presented in an earlier publication by the authors (Ruitenburg et al., 2014; Ruitenburg & Braaksma, in press), and on the experience of the first author with the writing of ALCPs.

The twelve steps can be used as a guide to develop an ALCP. Some of the steps may be carried out in tandem, but the order of the steps has been designed in such a way that each step produces the preconditions for the next steps (see Figure 5 for an overview of all twelve steps and their relation with the ALCP).

Figure 5. The twelve steps towards the ALCP, including its relation to the ALCP document. The three deliverables in the process are indicated in bold.

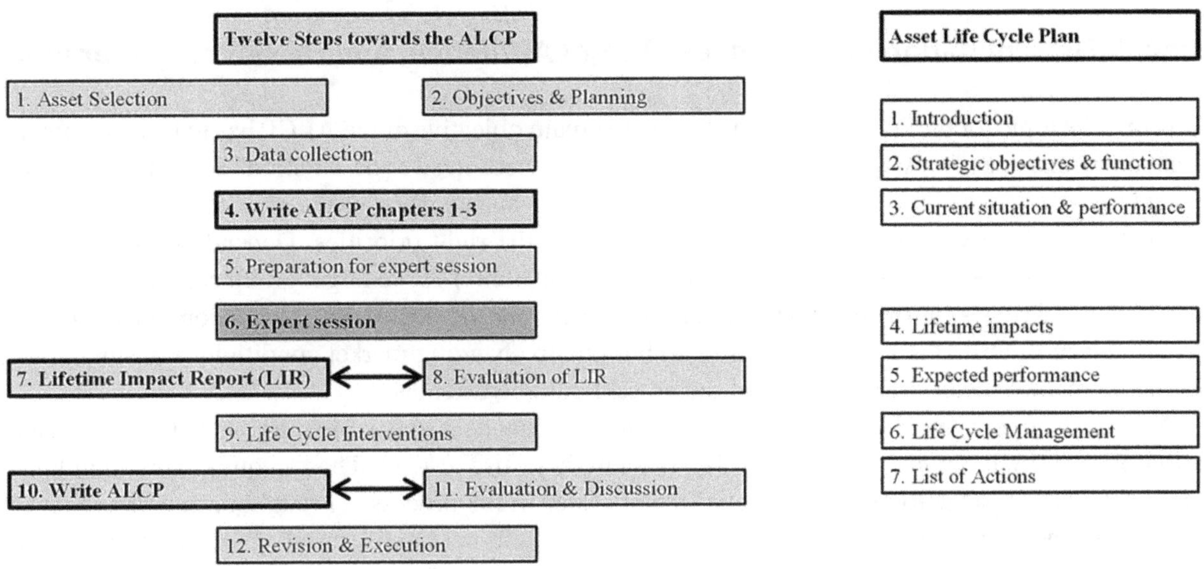

The description here is aimed primarily at the person responsible for the creation of the ALCP. Oftentimes this will be the asset manager. Not only should the asset manager take the lead over writing the ALCP, but they should also act as a director of the execution of the twelve steps that will be presented.

To illustrate the twelve steps, a case is presented from Liander, a Dutch network operator applying ALCPs in their Asset Life Cycle Management practices. Please refer to the illustration presented in Appendix 1.

Step 1: Asset Selection

The first step is the selection of the asset (population) as the topic of the ALCP. This may seem obvious, but it requires some thought. Depending on the availability of information and the depth of the analysis, the whole process of writing an ALCP may take one fully employed asset manager a couple of weeks. Therefore, only those assets should be selected that deserve such an investment of time. For example, assets with questionable performance or assets on the verge of an important life cycle event, like a modernization or overhaul, or maybe assets with a high criticality for the company or those representing a very large financial value. Additionally, a very dynamic environment may increase the desire for an ALCP to investigate the consequences for the asset.

Additionally, the level of detail of the analysis has to be determined. What are the system boundaries that will limit the analysis? Will the analysis cover the asset as a whole, or will it take the main functional components as the object of analysis? When a population of similar assets is taken as the subject, scoping decisions have to be made. What assets are similar enough to be treated as one population, and which are too different? It is important to reason from the function of the asset and the value it produces rather than the technical characteristics. Some assets may be different from a technical point of view but can be treated as similar in an ALCP because they have the same function and need to fulfil the same objectives. Certainly, when differences between subpopulations are relevant, these should be reported, but only if necessary for the main message of the ALCP.

Step 2: Determination of Objectives, Target Audience, and Preliminary Planning

Now that the topic for the ALCP has been chosen, the main objective of the ALCP has to be determined. What questions does the ALCP have to answer, and what needs need to be satisfied? This clearly relates to the target audience of the document. This may be higher management or the asset manager for whom the ALCP can be a working document and a way to set the right priorities. Depending on the target audience, a short and preliminary answer to the main question(s) of the target group can be given. This may help the writer(s) of the document set the right focus for the document. Such a core message may be, "In the short term, the asset performs well, but due to changing market conditions a business case should be made to determine if the life of the asset should be extended."

Next to the content of the ALCP, the target audience may have special requirements for the presentation of the ALCP, for example, format, layout, length, or use of data. These requirements should be made clear at this stage to prevent potential rework later in the process. Throughout the remainder of these steps, the target audience and the company at large will not be discussed again. However, this is clearly important to effectively meet the demands of the target audience.

Additionally, this is the moment to set preliminary planning. Input is needed from other departments and people, and these dependencies should be managed to make the process of writing an ALCP as efficient as possible. Depending on the organization and the availability of information within the organization, the collection of information from different departments may take several weeks to months.

Step 3: Data Collection

Now that the goal of the ALCP has been set, the time has come to collect all the information that is needed to answer the main question(s) of the target audience. The information should cover at least three different topics:

1. **The Corporate Objectives Relevant for the Asset:** What are the business values the asset should contribute to (e.g. safety, reliability, costs, and customer satisfaction)? And what strategic decisions have been made that are relevant for the asset? An example would be the decision to enter new markets to increase quality or quantity of production, or maybe to become a front-runner in sustainability. These objectives set the stage for the remainder of the ALCP. The asset is only valuable insofar as it contributes to these objectives. Therefore, it is of crucial importance to align the management of the asset with these goals.

Next, these (often relatively broad and intangible) objectives should be translated to specific goals for the asset. What does the asset have to achieve? This can be done by means of KPIs or other quantitative measures. Alternatively, this can be qualitative descriptions of performance, such as a continuation of the contract with the main customers or a happy workforce. However, it may still be advisable to make these objectives measurable in some way since this will make it easier to determine if a goal has been reached.

2. **Static Asset Data:** The asset should be described in a way that gives the target audience a better background on the asset. As the ALCP may be used in departments not directly involved with Asset Management, for example, in the strategy department or the purchasing department, some of the most important facts and figures may be needed: at least function, age, (replacement) value, population size, projected date of decommission (if any), and a brief overview of the current maintenance policy.
3. **Dynamic Asset Performance Data:** Next to the static data, the current performance of the asset should be reported and related to the objectives relevant for the asset. The performance should be reported in a multidisciplinary way, covering, for example, the five TECCO perspectives. At least the production output of the asset, its reliability/availability, safety, and cost performance should be covered. Ideally, the performance should be given over a period of time (e.g. the previous five years) to give an indication of the trends visible in the performance. This will also allow a better prognosis of future performance.

It is important to note that this information will probably not be available within one department; hence, there is a need to cooperate with the relevant people to bring all the information together. This may very well cross functional boundaries within, and maybe even across, the organization. While this may be a cumbersome and time-consuming process, it is crucial to establish a multidisciplinary Asset Management. Additionally, it is a way to bring the relevant stakeholders together and to get them involved in the process of writing an ALCP. As a result, the ALCP can be positioned as a "shared understanding" of the performance of the asset and its future.

Additionally, one should realize that—especially when writing the first ALCP—not all relevant information may be available for the ALCP. This is an excellent opportunity to identify the relevant information gaps (see also Haffejee & Brent, 2008) and to discuss which information may be necessary for an effective management of the assets. Depending on the costs of collecting this information, new information may be collected for a future ALCP.

Step 4: Writing the First Chapters of the ALCP (Chapters 1-3)

With the work done thus far, the first three chapters of the ALCP can be written. The reason to start writing the ALCP is threefold:

1. It can serve as a first tangible milestone in the process of crafting an ALCP;
2. It allows a first discussion with the stakeholders on this fundamental part of the ALCP; and
3. It is an important input for the next steps, where the main lifetime impacts relevant to the asset will be identified using one or more expert sessions.

Please refer to the illustration presented in Appendix 2.

Step 5: Preparation for the Lifetime Impact Identification Expert Session[2]

As stated before, it is crucial in Asset Management to look ahead and to notice proverbial icebergs in the future. The identification of these lifetime impacts will be done in Step 6 by means of an expert session. The current step prepares for this session. The preparation consists of content-related and practical preparation.

Content-Related Preparation

The main goal of content-related preparation is to collect the relevant information to identify the lifetime impacts. A large part of this information was already collected in Step 3 and reported in Step 4. But there is some additional information that may be relevant, since the information from Steps 3 and 4 predominantly describes the current and past situations while lifetime impacts relate to the future. Therefore, information should be collected regarding the future of the asset. This may include scenarios of market developments, analyses of failure data (e.g. bathtub curves), Remaining Useful Life (RUL) estimations of components, cost prognoses, or lists of upcoming changes in regulation. It may be useful to think about the information that may be available within the company on future developments from the five TECCO perspectives. The goal is not to get a full overview of every bit of information within the company but instead to facilitate an efficient and effective expert session. There is a trade-off to be made here: more time spent on the preparation of the expert session will reduce the time spent in the session.

Practical Preparation

Practical preparation mainly concerns the organization of the expert session(s). This starts with the selection of the experts to be invited. It is advisable to invite experts from the five different TECCO perspectives and to keep the group size limited to between four and eight persons. Since many experts have to be able to join the meeting, setting a date for the meeting should be done several weeks in advance. Additionally, the goal of the meeting should be explained, as should the uses of an ALCP. Some of the experts may have already been involved by providing information in Step 3; others may be new to the concept of an ALCP. Still others may have little knowledge of the asset (e.g. a financial or regulation expert). A level playing field should be created to allow them to join the discussion in the expert session. This can be achieved by distributing the first chapters of the ALCP, as written in Step 4. The experts may be asked to prepare themselves for the meeting, for example, by preparing a short statement to start the discussion on the TECCO perspective they represent.

Other than the experts, it is important to think of two other roles. The first is the chair or facilitator of the expert session. This person should guide the expert session, keep the discussions focused, keep track of the time, and may need to ask relevant questions that are overlooked in the session. These may, for example, be topics that were raised in the content-related preparation. It is recommended that a list of topics that need to be discussed be prepared for the facilitator. Additionally, the facilitator should make sure that sufficient attention is paid to both positive and negative lifetime impacts. It is not necessary for the facilitator to be an expert on the asset, but the facilitator must have the right chairman skills to keep the discussion efficient. The second role is the secretary or note-taker, who records the discussion.

Furthermore, the structure of the session should be prepared. This may include a time schedule and a presentation that guides the process and gives the necessary information inputs and a background on ALCM and lifetime impacts.

Step 6: The Lifetime Impact Identification Expert Session

The expert session lies at the heart of the process to write an ALCP. In the expert session, experts are brought together to share their knowledge and expertise. This resembles the expert sessions as used in the Failure Mode and Effect Analysis (FMEA) to develop maintenance plans in RCM (Moubray, 1997). Regarding the ALCP, the expert session is the means by which all the information collected in the previous steps is processed, combined, enriched with the tacit knowledge of the experts, and put to use in the identification of lifetime impacts. This section will present the main parts of the expert session. Additionally, two important topics related to the expert session—time and uncertainty—will be discussed at the end of the section.

Introduction

The expert session starts with an introduction. In the introduction, the experts should briefly introduce themselves since some may be from very different departments. Additionally, the ALCP, lifetime impacts, and the different TECCO perspectives are explained, for example, using the Titanic analogy.

Evaluation of the Current Maintenance Program

After the introduction, the expert session proceeds with an analysis of the current maintenance program because some lifetime impacts may be hidden in the current maintenance program. For example, the maintenance program may prescribe certain maintenance actions that are not carried out in practice or the work quality may be poor. Alternatively, a critical evaluation of the current maintenance policy may reveal opportunities for improvement, for example, by adopting new innovative techniques or by lengthening the maintenance intervals. After a critical discussion of the current maintenance program and its execution, the current maintenance program will be out of scope for the remainder of the expert session.

The Identification of Lifetime Impacts

After this evaluation, the time has come to look into the future and to identify the relevant lifetime impacts. This is done by means of the five TECCO perspectives, which are used as binoculars to look into the future from a certain perspective. Each perspective is discussed separately to ensure that explicit attention is paid to each of the perspectives. The goal is to identify the main lifetime impacts for each perspective, negative as well as positive. The discussion begins with an introductory statement from the expert representing the relevant TECCO perspective. Then, the other experts react to this statement while the facilitator facilitates the discussion in order to get a good overview of all the relevant positive and negative lifetime impacts. The focus should be on the likelihood of the impact, its effect, and the certainty of the expert(s) on the impact. The amount of discussion on a lifetime impact may be a useful indicator of the level of certainty: the more disagreement among the experts, the lower the certainty.

Each discussion ends with a brief summary by the facilitator or the note-taker to verify that the most important impacts have been identified. This may trigger additional discussion because it may take a moment to realize that something important has not yet been discussed.

If there are specific future scenarios available within the company (e.g. describing different demands from the market), these may be used to elicit additional lifetime impacts. In this case, after the discussion of the TECCO perspectives, each scenario is discussed separately. Some scenarios may give rise to additional lifetime impacts while others may make certain lifetime impacts that were already identified more important while reducing the importance of others. Additionally, one may make use of brainstorming techniques to elicit the lifetime impacts. For example, the expert session may start off with a brainstorm using sticky notes where each expert can individually write down the lifetime impacts considered relevant.

Closure

At the end of the session—with an advised length of no longer than four hours, including one or two breaks—the session is closed. The closure should include an outlook on the remainder of the process. If a second expert session is needed, this can be planned. In addition, the next steps in the process should be explained, including the possibility to give feedback on the Lifetime Impact Report (see Step 8).

Two Notes on the Topics of Time and Uncertainty

The expert session may raise an important topic of discussion: the time horizon used in the identification of lifetime impacts. Are all lifetime impacts relevant, or does one limit the discussion to impacts within a certain timeframe (e.g. five or ten years)? This depends on the needs of the target audience of the ALCP: Do they want an overview of the lifetime impacts to expect during the next five years, or would they rather know all impacts to expect during the asset's remaining life? Either way, it may be advisable to pay attention to the complete life cycle of the asset. Some high-impact lifetime impacts may have an influence on short-term interventions. For example, in a modification of the asset, future changes in regulation should be taken into account. Additionally, the time horizon may depend on the flexibility of the organization compared to the changeability of its context. If the context is highly changeable while the organization needs a long time to adapt the asset or its operation to these new circumstances, there may be a need to look further into the future, and in more detail. If the organization is very flexible and the context is stable, the time horizon for a detailed analysis may be shorter. Irrespective of the time horizon chosen, the facilitator needs to ensure that the discussion covers the complete time horizon rather than just covering the lifetime impacts nearby.

Additionally, the topic of uncertainty may require some attention here. Some lifetime impacts are known to happen in the future (e.g. if a supplier has announced it will stop the production of spare parts); others are uncertain (e.g. changing regulations). For some, the effect on the asset's performance may be clear; for others, this is entirely unsure. It may be advisable to elicit all reasonably likely lifetime impacts. However, it may not be necessary to think of suitable interventions for all of these because some may not be worth the effort. Additionally, the interventions may depend on the certainty of the lifetime impact: those that are certain may need direct action while those that are uncertain may need to be monitored or to be studied in more detail.

Step 7: Writing the Lifetime Impact Report (LIR)

A second important deliverable in the process is the Lifetime Impact Report (LIR). The LIR lists all the lifetime impacts identified during the expert session. Additionally, it relates these to the performance of the asset. This gives a first idea of the future performance of the asset. Some impacts may, for example, only relate to the maintenance costs of the asset (e.g. higher price of spare parts) while others may relate to many performance measures (e.g. technical degradation of a specific component may result in downtime, lower reliability, and safety risks). Additionally, some lifetime impacts may be very hard to directly relate to the performance but may be very important to register, for example, the realization that it becomes harder to find skilled technicians, which in the long term may reduce the maintenance capacity of the organization.

It is important to list all of these lifetime impacts, including the less likely or seemingly unimportant ones. These may not all find their way into the final ALCP, as the effect of some impacts on the value creation with the asset may be negligible. However, the asset manager may want to monitor these by revisiting the list periodically to see if new developments make one of these more important. In this way, the LIR can be viewed as a priority list for the asset manager: it lists all the potentially important lifetime impacts on the asset. Some may be considered very important, others may seem irrelevant, but depending on the future developments, all may require the attention of the asset manager at some point in time.

There are four other uses of the LIR that are worth mentioning. The first function is to get feedback from the experts in the expert session. Second, the LIR may be used as an input when decisions need to be made regarding life cycle events, such as overhauls or modernization. The LIR can then be used as an overview of everything that may be relevant and ensures that no important issues are overlooked in the decision-making process. Third, the LIR can be used as an input to the expert session for other ALCPs. Some impacts may not be very relevant for one asset but may have a large impact on others. In this way, the LIR facilitates information sharing. A final use of documenting these insights is that important expertise is made available on paper, which makes the company less vulnerable if knowledgeable employees leave.

Please refer to the illustration presented in Appendix 3.

Step 8: Evaluation of the Lifetime Impact Report (LIR)

Now that the LIR has been written, summarizing the lifetime impacts and their impacts on the asset performance, it can be discussed with the experts that were present in the session. One could organize a special evaluation session for this purpose, have short meetings with each of the experts individually, or just send them the LIR and ask for their feedback. Irrespective of how this is done, it is crucial to ask for the input of the experts again. They may think of additional impacts, may criticize the summary, or they may comment on the effects on the asset performance as described in the LIR. In addition, they may give different priorities to the lifetime impacts listed.

Additionally, the LIR may be used to involve other experts in the process, for example, those that could not be present at the expert meeting or those with a very specific expertise that may shed light on specific lifetime impacts.

By taking the time for an evaluation, the resulting ALCP will be more reliable. Additionally, the ALCP will become more of a shared understanding of the future of the asset, rather than just a view of the asset manager writing the ALCP.

Step 9: Think of Appropriate Life Cycle Management Interventions

After rewriting the LIR based on the feedback of the experts, a joint understanding of the main lifetime impacts relevant for the asset evolved. Based on this understanding, it may be necessary to take actions to prevent negative lifetime impacts from happening (or to reduce their effects) or to grasp the full potential of positive lifetime impacts. In other words, there may be a need for life cycle management interventions. These interventions may range from changes in the maintenance program, starting pilot projects to test out new innovations, or the collection of additional information on particular lifetime impacts to design changes, overhauls, modernization projects, or even the preventive replacement of particular assets. Additionally, some lifetime impacts may be very uncertain and/or only materialize over time, and monitoring their progress may be sufficient.

Whatever the specific solution, it may be important to give a brief indication of the cost of the solution, its benefits, and the moment in time when actions will have to be taken. Still, for the asset manager, some actions may have a high priority and need to be listed.

Important to note here is that the ALCP is not meant to be a decision-making document. Rather, it is a document that gives an overview of the asset, the main future opportunities and threats, and the recommended actions to be taken. In most organizations, a specific decision-making process will exist, which needs to be followed to make large and high-impact decisions. Nevertheless, just giving an overview will probably not satisfy the needs of the target audience. Hence, it is important to think of the main actions that are necessary to produce maximum business value with the assets in the long term.

Step 10: Writing the ALCP

The tenth step is the actual writing of the ALCP. The first three chapters have already been written in Step 4. The information for Chapters 4 and 5 can be taken from the LIR. The information for the remainder of the ALCP was collected in Step 9. The chapter on life cycle management should list the main actions to be taken and their main consequences. Additionally, it gives an outlook regarding the future of the asset, for example, by means of a description of the situation in 10 or 20 years. In this way, the target audience can be prepared for decisions that may need to be made in the future, as well as their consequences. In this way, the ALCP is not just a document that gives an overview of the future of the asset. It is also a communication device to manage expectations and to create a shared understanding of the future of the asset.

Please refer to the illustration presented in Appendix 4.

Step 11: Discussion and Evaluation with Target Audience and Other Relevant Stakeholders

The ALCP itself is just a piece of paper, irrespective of the huge amount of effort, time, and expertise it has taken to produce. It has to be put to use. This happens by means of a presentation of the ALCP to the target audience and other relevant stakeholders and a discussion and evaluation of the main results. If the target audience is higher management, they may be especially interested in the financial implications of the ALCP and the main decisions that need to be made. The ALCP may help create a joint understanding of the future of the asset and facilitate decision-making on strategic issues relevant to the asset.

If the discussion and evaluation of the ALCP give rise to changes in the ALCP, these need to be made. Then the ALCP can be accepted and authorised by the target group and/or the management. This makes the ALCP not only a joint understanding of the asset's future but also the official view of the organization.

Step 12: Determine Revision Date

It is important to realize that the ALCP is not a one-time exercise. Rather, it is a way of doing Asset Life Cycle Management. This implies that every ALCP will need to be kept up to date. Depending on the internal and external dynamics relevant to the asset, this may need attention every one to five years. Updating an existing ALCP will of course take less time because only the changes to the previous ALCP will have to be processed. However, it is of crucial importance to carry out a new expert session because new lifetime impacts may appear on the horizon.

The revision date should be set already to make sure a revision will be carried out in time. Still, it may be necessary to do some updates before the revision date, for example, if a large life cycle event happens or a large decision needs to be made.

This is also the moment to start the execution of the actions listed in the final chapters of the ALCP. Their results can be reported in the next ALCP, and their impact on the performance can be evaluated. In this way, a loop is created, from objectives, current and expected performance, and the development and execution of lifetime interventions to the evaluation of the new performance. This loop results in a growing understanding of the asset and a more effective management of the asset.

Please refer to the illustration presented in Appendix 5.

CONCLUSION

Asset Management should aim to create maximum business value from the use of physical assets. To do so, it is crucial to take a multidisciplinary approach towards Asset Management, to be aligned with the corporate objectives, and to focus on the complete lifetime of the asset. Asset Life Cycle Management (ALCM) has been used to describe this "ideal" way of doing Asset Management.

To assist strategic Asset Management decision-making, an integrated insight into the remaining lifetime of the asset is necessary. This insight can be developed by the use of an Asset Life Cycle Plan (ALCP). An ALCP starts with the strategic objectives a company has for the asset in order to let it create business value. The current and expected performance of these objectives is reported to show how the asset delivers business value. Additionally, future impacts and trends on the value creation with the asset are identified, and their effect on the performance is estimated. Finally, the ALCP gives an overview of the Asset Life Cycle Management interventions that may be useful to secure or increase the asset's value for the company.

Current maintenance concepts and tools primarily focus on the short term of the asset's life and often do not take a multidisciplinary perspective; hence, these tools do not assist in the identification of future impacts and trends with an impact on the asset. To support strategic decision-making in Asset Management, the identification of these "lifetime impacts," as they have been called in this paper, is vital. Therefore, a twelve-step approach has been presented to identify lifetime impacts from a multidisciplinary perspective. Five different perspectives are used for the identification: Technical, Economic, Compliance, Commercial, and Organizational (TECCO) lifetime impacts result. The lifetime impacts from these perspectives are identified by making use of the knowledge and information available within the company, primarily by bringing experts together in an expert session. In such a session, a joint understanding of the asset and its future will develop, and the relevant lifetime impacts can be identified.

By following the practical twelve-step approach presented in this paper, an ALCP can be written. The approach is illustrated with an ALCP written by Liander, the largest Dutch network operator. This illustration shows how an ALCP supports asset managers in making strategic decisions in a timely and effective manner.

DISCUSSION AND IMPLICATIONS

This final section discusses the conclusions from this chapter as well as the implications for further research and for practitioners. A first topic of discussion concerns the prioritization of lifetime impacts. During the expert session, many lifetime impacts will be identified. Which of these are the most important and most relevant then becomes the question. In RCM, the failure modes identified in an expert session are prioritized based on likelihood and effect size, resulting in a Risk Priority Number (RPN) (Moubray, 1997). A similar approach may be applicable to lifetime impacts.

Second, further research may also address the question of how to mitigate negative lifetime impacts and to exploit the full potential of lifetime impacts. The twelve-step approach presented in this chapter does not offer much guidance on this topic. The identification of lifetime impacts will only deliver value if the right interventions are made in a response to the lifetime impacts identified.

A third topic to discuss is the generalizability of the ALCP and the steps presented in this chapter. While developing the method, the authors aimed for a general applicability in Asset Management. The focus on the creation of business value allows users to adapt the method to a specific industry. However, thus far the ALCP has only been applied in populations of similar assets in the energy sector.

A last issue concerns the title of this book: *Optimum Decision-Making in Asset Management*. The authors are of the opinion that optimum decisions do not exist in long-term Asset Management. Obviously, some decisions are better than others, and some may be plainly wrong. But as one never knows what the future holds, making optimum decisions on the long term is a myth. Rather than optimal decisions, good and timely decisions are needed. Therefore, this chapter has set out to assist asset managers with the difficult and uncertain decisions they often need to make.

IMPLICATIONS FOR PRACTITIONERS

Apart from the twelve practical steps to develop an ALCP, this chapter has four important implications for practitioners in Asset Management. Firstly, practitioners should focus on the complete lifetime of their asset(s) to produce maximum business value for their companies. Since decisions made now may have important implications for the remainder of the lifetime of the asset, maximum value can only be created by taking the full lifetime into account. It is crucial that higher management supports this approach to Asset Management. Because many of the benefits of the ALCP will only be visible in the long term, the long-term support of management is needed. Such lengthened support will allow the asset manager to keep working with the ALCPs until tangible results can be seen in a higher value created with the use of the asset.

Secondly, this chapter has shown that Asset Management should be a multidisciplinary practice. One may compare the disciplines represented in one's Asset Management practices with the TECCO perspectives to see if all important disciplines are covered.

As a third implication, practitioners in Asset Management may ask themselves if their Asset Management practices have closed the loop of objectives, performance, interventions, expected performance, and new performance figures. Closing this loop will facilitate a process of continuous improvement.

Finally, this chapter has stressed the importance of the knowledge of experts. Often, asset managers face limited data availability and/or data of a doubtful quality. Additionally, the future is uncertain by definition. Nevertheless, decisions need to be made. The use of the knowledge of experts, added to existing information and data, may allow for better decisions to be made.

ACKNOWLEDGMENT

The authors gratefully acknowledge Liander N.V. for funding this research project. Additionally, the first author thanks all the individual Asset Management experts from Liander who cooperated in the development of the ALCP and provided valuable feedback from their daily experience.

REFERENCES

Al-Turki, U. (2011). A framework for strategic planning in maintenance. *Journal of Quality in Maintenance Engineering*, *17*(2), 150–162. doi:10.1108/13552511111134583

Allan, D. M. (2005). Asset management challenges in ageing power systems. *Proceedings of 2005 International Symposium on Electrical Insulating Materials* (ISEIM '05) (pp. 265–268). ISEIM. doi:10.1109/ISEIM.2005.193395

Asiedu, Y., & Gu, P. (1998). Product life cycle cost analysis: State of the art review. *International Journal of Production Research*, *36*(4), 883–904. doi:10.1080/002075498193444

Braaksma, A. J. J. (2012). *Asset information for FMEA-based maintenance*. Groningen, The Netherlands: Faculty of Economics and Business. University of Groningen.

Brown, R. E., & Humphrey, B. G. (2005). Asset management for transmission and distribution. *IEEE Power and Energy Magazine*, *3*(3), 39–45. doi:10.1109/MPAE.2005.1436499

de Leeuw, J., & Pries, F. (2014, June4). Deltaplan voor infrastructuur nodig: Verouderde infrastructuur in Nederland vereist gezamenlijke aanpak op verschillende niveaus. In *Het Financieele Dagblad* (p. 10). Opinie & Dialoog.

Frangopol, D. M., Saydam, D., & Kim, S. (2012). Maintenance, management, life-cycle design and performance of structures and infrastructures: A brief review. *Structure and Infrastructure Engineering*, *8*(1), 1–25. doi:10.1080/15732479.2011.628962

Garg, A., & Deshmukh, S. G. (2006). Maintenance management: Literature review and directions. *Journal of Quality in Maintenance Engineering*, *12*(3), 205–238. doi:10.1108/13552510610685075

Haffejee, M., & Brent, A. C. (2008). Evaluation of an integrated asset life-cycle management (ALCM) model and assessment of practices in the water utility sector. *Water S.A.*, *34*(2), 285–290.

Holmström, J., Ketokivi, M., & Hameri, A. (2009). Bridging practice and theory: A design science approach. *Decision Sciences*, *40*(1), 65–87. doi:10.1111/j.1540-5915.2008.00221.x

IAM, & BSI. (2008). *Asset management - Publicly available standard*. IAM & BSI.

ISO. (2014). *ISO 55000 - Asset management*. Geneva, Switzerland: International Standards Organization.

Jongen, R. A. (2012). *Statistical lifetime management for energy network components*. Delft, The Netherlands: Technische Universiteit Delft.

Jongepier, A. G. (2007). *Ageing assets - Consume, prolong or replace. Leonardo Energy*. Arnhem, The Netherlands: KEMA Consulting.

Komonen, K., Kortelainen, H., & Räikkönen, M. (2012). Corporate asset management for industrial companies: An integrated business-driven approach. In T. van der Lei, P. Herder, & Y. Wijnia (Eds.), *Asset management: The state of the art in Europe from a life cycle perspective* (pp. 47–63). Dordrecht, The Netherlands: Springer Netherlands. doi:10.1007/978-94-007-2724-3_4

Márquez, A. C., Márquez, C. P., Fernández, J. F. G., Campos, M. L., & González-Prida Díaz, V. (2012). Life cycle cost analysis. In *Asset management: The state of the art in Europe from a life cycle perspective* (pp. 81–99). Dordrecht, The Netherlands: Springer Netherlands. doi:10.1007/978-94-007-2724-3_6

Minto, B. (2008). *The pyramid principle - Logical writing, thinking and problem solving* (3rd ed.). Pearson Education Limited.

Moubray, J. (1997). *Reliability centered maintenance* (2nd ed.). Oxford, UK: Butterworth-Heinemann.

NVDO. (2015). *Wie zijn wij? – NVDO* [Who are we?]. Retrieved from http://www.nvdo.nl/wie-zijn-wij/

Pudney, S. (2010). *Asset renewal decision modelling with application to the water utility industry* [Ph.D. Thesis]. Faculty of Built Environment and Engineering. Queensland University of Technology, Queensland, Australia.

Rouse, P., & Chiu, T. (2009). Towards optimal life cycle management in a road maintenance setting using DEA. *European Journal of Operational Research*, *196*(2), 672–681. doi:10.1016/j.ejor.2008.02.041

Ruitenburg, R. J., & Braaksma, A. J. J. (in press). The multidisciplinary lifetime impact identification analysis to support strategic asset life cycle management. *CIRP Journal of Manufacturing Science and Technology*.

Ruitenburg, R. J., Braaksma, A. J. J., & van Dongen, L. A. M. (2014). A multidisciplinary, expert-based approach for the identification of lifetime impacts in asset life cycle management. *Procedia CIRP*, *22*, 204–212. doi:10.1016/j.procir.2014.07.007

Ruitenburg, R. J., Braaksma, A. J. J., & van Dongen, L. A. M. (2015). Towards a model for effective asset life cycle management control – A case study in rolling stock maintenance. *Proceedings of the 22nd EurOMA Conference* (pp. 1–10). Neuchatel, Switzerland: EurOMA.

Salkeld, L. (2007). Is this the man who sank the Titanic by walking off with vital locker key? *Daily Mail*. Retrieved from http://www.dailymail.co.uk/news/article-478269/Is-man-sank-Titanic-walking-vital-locker-key.html

Schuman, C. A., & Brent, A. C. (2005). Asset life cycle management: Towards improving physical asset performance in the process industry. *International Journal of Operations & Production Management, 25*(5/6), 566–579. doi:10.1108/01443570510599728

Si, X. S., Wang, W., Hua, C. H., & Zhou, D. H. (2011). Remaining useful life estimation – A review on the statistical data driven approaches. *European Journal of Operational Research, 213*(1), 1–14. doi:10.1016/j.ejor.2010.11.018

Smit, K. (2014). *Maintenance engineering and management*. Delft, The Netherlands: DAP Delft Academic Press.

Smith, A. M., & Hinchcliffe, G. R. (2003). *RCM--Gateway to world class maintenance*. Burlington, MA: Butterworth-Heinemann.

RMS Titanic. (2015, September 29). *Wikipedia*. Retrieved from https://en.wikipedia.org/wiki/RMS_Titanic

van Dongen, L. A. M. (2011). *Maintenance engineering: Maintaining links*. Enschede, The Netherlands: Faculty of Engineering Technology, University of Twente.

van Dongen, R. (2011). *Referentiemodel voor levensduurverlenging van technische assets* [Reference model for life extension of physical assets]. VITALE Mainnovation.

Veenman, A., & Besselink, L. (2010). *Asset owners & maintenance*. Breda, The Netherlands: Academic Press.

Waeyenbergh, G., & Pintelon, L. (2002). A framework for maintenance concept development. *International Journal of Production Economics, 77*(3), 299–313. doi:10.1016/S0925-5273(01)00156-6

Wetzer, J. M., & Bouwman, S. (2007). *Managing risks in ageing networks: A case study. Leonardo Energy - Electricity briefing paper*. Arnhem, The Netherlands: KEMA Consulting.

Wireman, T. (2004). *Total productive maintenance*. Industrial Press.

Zaal, T. M. E. (2013). *Profit-driven maintenance for physical assets* (2nd ed.). Maj Engineering Publishing.

KEY TERMS AND DEFINITIONS

Asset Life Cycle Management: An approach to Asset Management with explicit emphasis on the multidisciplinary and strategic approach to the performance of an asset and the business value created with the use of the asset throughout its complete life cycle.

Asset Management: An organization's structured and organized practice to create maximum value with its assets.

Life Cycle: The complete lifespan of a physical asset from design and construction to decommissioning and recycling.

Lifetime Impact: A probable (technical and non-technical) event or trend that may have a positive or negative influence on the value creation with the asset in the intermediate or long term.

Long Term: Refers to the longest time horizon applied in an organization. Depending on the organization, the long term may commence from 3 up to 15 years into the future (after the short and intermediate term).

Multidisciplinary: A practice in which different fields and specialities are deliberately brought together.

Physical Asset: Any physical object having an expected lifetime of at least 10 years and representing a significant value to its owner, both in terms of money and the function it fulfils.

Strategic: Related to the corporate strategy.

Strategic Alignment: In agreement with the (corporate) strategy; the goals of the Asset Management department are tuned to the higher (corporate) strategy.

TECCO: Acronym to describe five important perspectives that need to be covered in order to make Asset Management multidisciplinary. TECCO is an acronym for Technical, Economic, Compliance, Commercial, and Organizational. TECCO is intimately linked to the identification of lifetime impacts.

ENDNOTES

1 The picture of the Titanic is created by Boris Lux, available at Wikipedia (RMS Titanic, 2015).

2 Earlier work of the author (Ruitenburg et al., 2014; Ruitenburg & Braaksma, in press) has introduced the Lifetime Impact Identification Analysis (LIIA), a method consisting of several steps. The twelve steps as presented in this chapter include and extend the steps of the LIIA.

APPENDIX 1

Illustration: An Introduction to the Use of ALCPs at Liander

Liander is the largest Dutch network operator, responsible for the safe and reliable distribution of electricity and gas to its 3.1 million customers. As shown earlier, the ageing of assets is a well-known issue in Western Europe. Liander experiences the same problem, as large parts of its networks were constructed in the 1960s and 1970s. Because many of the assets in the network have expected lifetimes between 40 and 60 years, large-scale replacement may be needed in the years to come. The current changes in the production and use of energy—often called the energy transition—offer an additional challenge. Distributed electricity production, for example, by means of solar and wind power, is on the rise, while new applications such as electric vehicles may cause a large increase in demand. This results in new demands on the existing electricity grid. To support the strategic decisions Liander has to make, the Asset Management department decided to start working with ALCPs.

One of the ALCPs covers the distribution transformers Liander operates. Its main question was, Is there reason to expect a large increase in the replacement rate of the distribution transformers? Since the population of distribution transformers consists of 28,000 assets, representing a replacement value of 200 million euros, this clearly is a strategic topic that has the interest of higher management. The twelve steps will be illustrated using both the process of developing this ALCP as well as the structure of the final document. The goal is not to show every detail of the ALCP, but rather to give the reader an idea of the scope and logic of such a document. For reasons of confidentiality, the numbers as well as some of the lifetime impacts presented in this example are fictive.

APPENDIX 2

Illustration: The First Three Chapters of the ALCP on Distribution Transformers

Chapter 1: Introduction

This ALCP describes the future of the population of distribution transformers for the coming 20 years. The main question it will answer is, Is there reason to expect a large increase in the replacement rate of the distribution transformers? The scope of this ALCP is limited to the 28,000 distribution transformers in the grid. The large majority of these assets transforms electricity from medium voltage (10 kV) to low voltage (400 V). Figure 6 shows a picture of a distribution transformer.

Chapter 2: Strategic Objective and Function

Liander aims to create maximum business value from the exploitation of its assets by achieving the required performance (i.e. availability, reliability, safety) at minimum costs and at an acceptable risk. The performance on each of the relevant KPIs should minimally remain constant in the next five years of operation.

Figure 6. A distribution transformer. In the interior the medium and low voltage coils are visible.

The main strategic developments relevant to the distribution transformers are energy transition and sustainability. The energy transition may cause temporal reversals of the electricity flow in residential areas with a high number of photovoltaic power cells or wind turbine generators. Since this is a new phenomenon, it is unknown if this will negatively influence the remaining service life of the asset. The corporate ambitions on sustainability aim for a strong reduction in CO_2 emissions. Because distribution transformers cause a significant part of the current CO_2 emissions through the energy losses inherent to the transformation process, the sustainability ambition may have implications for the distribution transformers. Additionally, Liander has decided to apply 20 kV on its medium voltage grids. This transition may take decades, but irrespective of its planning, which is currently unknown, the management of the distribution transformers should be aligned to its implementation since the 10 kV transformers will be obsolete in a 20 kV grid.

Chapter 3: Current Situation and Performance

Figure 7 shows the age of the distribution transformers of Liander (28,000 in total). The average age is 30 years, compared to an expected life of at least 40 years. Based on the current experience, this may be an underestimation of the life expectancy, which experts agree lies around 60 years. Currently, every year 600 transformers are replaced based on capacity (quality of the transformers is not yet a reason for preventive replacement). A preventive maintenance policy based on periodic inspections is used.

The current performance of the population is good. The reliability of the transformers is very high (99.98% reliability). Their contribution to the yearly outage per customer is less than 1%. Regarding safety, no incidents have been reported involving the distribution transformers. From a sustainability point of view, the energy losses of the transformers are substantially higher for older transformers.

Figure 7. Example of age distribution of distribution transformers (based on fictive numbers)

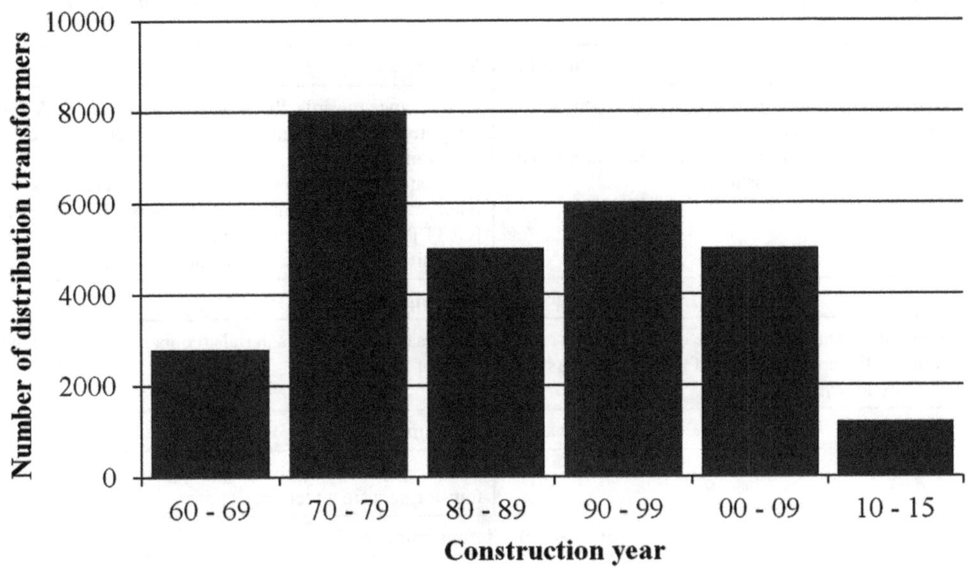

APPENDIX 3

Illustration: The Distribution Transformer Expert Session and the LIR

The Expert Session

The expert session was attended by eight experts from the departments of Asset Management, Innovation, Grid Planning, Purchasing, and Maintenance Execution. The session took four hours, which was barely enough time to discuss the five TECCO perspectives at sufficient length. Lifetime impacts were elicited in two ways: first, by means of an individual brainstorm using Post-it® notes, which were collected on five large sheets of paper (one per perspective). Second, each perspective was discussed in a plenary discussion. This yielded a total of 124 lifetime impacts. After filtering out double impacts and combining closely related ones, 34 unique lifetime impacts remained. The filtered-out lifetime impacts were retained on a list for safe keeping.

The LIR (Input for ALCP Chapters 4 and 5)

These 34 lifetime impacts were reported in the Lifetime Impact Report. Table 1 shows examples of the lifetime impacts reported in the LIR on the distribution transformers.

Based on the current performance of the transformers and the lifetime impacts identified, the performance is expected to remain at the current levels for the next five years. An increase in the amount of failures is not expected, even though the population is ageing. However, in the long term, three lifetime impacts may increase the need for replacement:

Table 1. Examples of lifetime impacts related to the distribution transformers for each of the five TECCO perspectives. Positive impacts (+) are listed on the left, negative impacts (-) on the right

Technical Lifetime Impacts	
+ Real-time load measurement may yield additional insights into the remaining lifetime of the assets. + The limited availability of condition data limits the insights into the remaining useful life of the transformers.	- The change towards 20 kV grids will render 10 kV transformers obsolete; the impact depends completely on the speed of this transition. - The energy transition may give rise to a need for power management in distribution transformers (On-Load Tap-Changers [OLTC]). Because modification of existing transformers is expensive and difficult, replacement may be necessary.
Economic Lifetime Impacts	
+ The use of aluminium coils may reduce the costs of new transformers (but this will increase the size of the transformers, limiting their suitability for existing stations).	- Increasing prices of materials (copper, aluminium) increases the price of transformers.
Compliance Lifetime Impacts	
+ None.	- New regulations on power quality and energy losses in transformers are under development.
Commercial Lifetime Impacts	
+ None.	- The energy transition may give rise to new demands, for example, power management in distribution transformers. - The use of DC (Direct Current) instead of AC (Alternating Current) in the grid would make transformers redundant.
Organizational Lifetime Impacts	
+ The use of bio oil in the transformers may increase their recyclability, fitting the organizational ambitions on sustainability. + Innovative amorphous transformers may reduce energy losses in new transformers.	- The technicians in the transformer workshop will all retire in the coming years, giving rise to potential capacity and knowledge problems. - Liander pays increasing attention to sustainability and energy losses, which may lead to the decision to preventively replace a substantial part of the population.

1. The energy transition may lead to new demands on the transformers (e.g. power management, additional monitoring), which may result in the modification or replacement of transformers in areas with a high local production of solar power.
2. The transition to 20 kV grids will make the existing 10 kV transformers obsolete.
3. The corporate ambition to reduce CO_2 emissions may lead to a replacement of the oldest transformers (which are still reliable but less energy efficient).

APPENDIX 4

Illustration: The Remaining Chapters of the ALCP on Distribution Transformers

Chapter 6: Life Cycle Management

Currently, there is no need for large changes in the current management of the distribution transformers. Innovations such as bio oil, OLTC, and amorphous transformers are being monitored by the innovation department. A business case will be developed to investigate if replacing old transformers for new energy-efficient ones would be profitable. If so, this may have an impact on the older transformers in the population. This particularly applies to the transformers built before 1980. According to the fictive age structure given in Figure 7, this amounts to a total of 10,800 units.

The strategic decision of Liander to change existing 10 kV grids to 20 kV potentially has a large impact on the existing population. The size of the impact depends on the speed of this transition. To make this transition as cost-efficient as possible, the existing population of distribution transformers should be taken into account when planning this transition. Additionally, the new demands on the transformers caused by the energy transition need to be monitored closely since these may need to be modified or replaced in the future.

APPENDIX 5

Illustration: Evaluation and Discussion of the ALCP on Distribution Transformers

Evaluation

The ALCP has been evaluated with the asset manager responsible for the distribution transformers. He indicated that the ALCP is a foundation for further improvement. It gives an overview of the main topics relevant to the asset for its complete lifetime. This makes it different than existing maintenance policy documents, which often have a more limited focus both in time and in perspective. This makes it easier to identify the main priorities in the management of the transformers. Additionally, it is a communication device to share the expected performance of the asset with other stakeholders or departments.

The ALCP was presented to the management of the Asset Management department. It was well received. The strategic focus of the document and the discussion of the future replacement needs were stated to be strong characteristics.

Discussion

The application at Liander shows the potential value of working with ALCPs. The ALCP on distribution transformers allowed the company to get an integrated understanding of the asset's future from the different TECCO perspectives. Based on such an understanding, well-informed strategic decisions can be made. The experts joining the expert session regarded this session as useful and interesting. They especially valued the fact that experts from different departments were present, allowing an integrated and multidisciplinary view of the transformer's future to emerge.

By means of an ALCP, Asset Management may become proactive rather than reactive. It looks ahead, presents the main challenges and opportunities that lie ahead, and comes up with suitable solutions that can be implemented in time. This also implies that the chances are small that new problems will suddenly appear and ask for immediate and often less cost-effective solutions.

Chapter 13
Towards Informed Maintenance Decision Making:
Guiding the Application of Advanced Maintenance Analyses

W. W. (Wieger) Tiddens
University of Twente, The Netherlands & Netherlands Defence Academy, The Netherlands

A. J. J.(Jan) Braaksma
University of Twente, The Netherlands

T. (Tiedo) Tinga
University of Twente, The Netherlands & Netherlands Defence Academy, The Netherlands

ABSTRACT

Advanced maintenance techniques (AMTs) are practices that can help practitioners to make better-informed maintenance decisions, such as ensuring just-in-time maintenance, corporate business planning or lifetime extension of physical assets. These techniques take the current, but preferably also the future, state of assets into account. Although there is much literature on the development of specific techniques, reports on their adoption and use show that only few companies have successfully applied these methods. Guidance is needed on their selection and application. In this chapter, a typology of companies–based on an ongoing multiple-case study–that apply these maintenance techniques is proposed. This typology identifies typical difficulties practitioners experience in applying AMTs. Finally, a four-step procedure is offered with the aim of helping practitioners to overcome the discussed difficulties in the application of AMTs.

DOI: 10.4018/978-1-5225-0651-5.ch013

INTRODUCTION

In challenging economic times, the costs of maintenance–which form around 25% of the total operating costs of a typical manufacturing company (Cross, 1988; Komonen, 2002)–can make or break a business. Current progress in advanced maintenance techniques (AMTs) enable companies to improve maintenance decisions, for example overhaul or lifetime extension of assets, and thereby decrease maintenance costs. Research literature describes diffused enhancement on developing prognostic techniques, a type of AMTs, that can be used to make more informed maintenance decisions. AMTs are practices that help decision makers to support maintenance decision making by taking the current, but preferably also the future, state of assets into account. Practitioners rarely know for certain when their equipment will fail. Therefore, knowledge about the current and future state can help in determining whether the revenues that are generated by the asset will compensate the risk of its failure. In other words, operators need to define if the operation can be continued or if the remaining lifetime can be thrown away by replacing parts (Haddad et al., 2014).

AMTs can help practitioners to detect imminent failures, make a health assessment of the system and predict the remaining useful life (Lee et al., 2014), or ascertain the probability a machine operates without failures or faults up to a certain time given the current machine condition and operational profile. The latter might be important in a nuclear power plant, for example (Jardine et al., 2006). More exact information about the failure behavior of assets and the time it will (probably) fail enables decision makers to make more effective (doing the right things) and efficient (doing things right) maintenance decisions.

By supporting maintenance decisions, these techniques help to:

1. **Reduce Life Cycle Costs:** Fewer unnecessary repairs and replacements (not throwing remaining useful life away); and
2. **Reduce Business and Safety Risks:** Fewer (safety) incidents and less unplanned downtime (Bo et al., 2010).

Conservative maintenance intervals (conducting more maintenance than strictly required) can create inefficient–costly–maintenance processes as these are set by asset owners to prevent their critical equipment from failing unexpectedly (Tinga, 2010). However, these high costs and safety issues caused by unexpected failures do not legitimize infinite preventive expenditures when other options, such as more advanced practices to predict imminent failures, are available.

Problem Statement

Grubic et al. (2011) show that around ten per cent of UK manufacturers have adopted prognostic techniques. However, as promising as these techniques are, research conducted by the authors (Tiddens et al., 2015) shows that companies experience several difficulties in the application of these AMTs. This struggle with the effective application of these techniques is also shown in the study of Grubic et al. (2011), who show that many practitioners that have applied AMTs experience a gap between the potential and realized benefits. To overcome technical problems in applying prognostic techniques, many scholars in the field have proposed models and algorithms to predict future failures and assess the current state of equipment. However, as also other authors have pointed out (Garg & Deshmukh, 2006; Kerkhof et al., 2014; Veldman et al., 2011), most relevant research excludes organizational and managerial aspects, and is purely focused on the technical aspects of prognostics, including the development of accurate sensors, algorithms, and models. As also Grubic et al. (2011) suggest, research in this area should, therefore, embrace both the technological and business aspects of diagnostics and prognostics.

Many practitioners experience difficulties in:

1. The selection of the correct AMT to apply and
2. In the route they have to take towards successful maintenance decision making (Tiddens et al., 2015).

AMTs can be difficult to implement due to frequent change of the design of assets, a change in the use of technologies, changing business policies, and changing management executives (Jardine et al., 2006).

Furthermore, as also Kerkhof et al. (2014) note, the difficulties associated with successful and systematic organization of a condition-based maintenance (CBM, a maintenance technique) program within an organizational context are often underestimated. Moreover, there is a lack of communication between theory developers and practitioners in the area of reliability and maintenance (Jardine et al., 2006). Widespread adoption of these technologies needs an in-depth evaluation of their use within companies (Benedettini et al., 2009). This gives academics an overview of typical problems that companies experience while applying these worthwhile techniques.

Research Approach and Overview of the Chapter

This chapter illustrates the application of advanced maintenance techniques, in order to support companies and practitioners to step towards better-informed maintenance decision making. Moreover, this work provides scholars with insights about real-life applications of AMTs.

The chapter starts with an analysis of various classifications of AMTs. This overview gives decision makers a basis for the application and the selection of these techniques. After this, three stereotypical companies with different maturity levels in the usage of AMTs will be introduced, in order to highlight typical problems enterprises experience in the application of prognostic techniques. In addition, since company and asset fleet differences influence this application process, the chapter will elaborate on how organizational aspects could impede this process. The subsequent section of the chapter will describe a four-step-framework that guides the application of AMTs. Finally, building on existing theory and supported by multiple case studies, solutions will be offered to support decision making for maintenance by using AMTs and enable the implementation of these technologies.

BACKGROUND

Maintenance techniques help practitioners in enabling better-informed maintenance decision making. These techniques are used to provide information about the current and preferably also the future state of an asset. The supportability of the assets can be improved when maintenance decisions can already be made in an early phase of the lifetime of a physical asset. For example, knowing spare part requirements, maintenance planning, and when the asset should be faced out.

Traditionally, chosen conservative maintenance intervals rely on experience-based predictions (Singpurwalla, 1995; Van Noortwijk, 2009; Zio et al., 2012). Even large manufacturing companies do not always measure indicators of equipment degradation (Bousdekis et al., 2015), or are not able to utilize the benefits offered by these sensors that measure expressions of degradation (i.e., temperature, vibration). These (static) experience-based predictions are often limited in prognosticating reliability and future behavior, especially when the asset has a dynamic behavior. When opting for the appropriate maintenance policy, a run-to-failure or a time-based maintenance policy will only be satisfactory in a few cases. In many industrial cases, both the costs of maintenance and the costs of failure are high, and condition-based maintenance (CBM may, therefore, be a better choice (Jardine et al., 2006).

Simple static (not updated) lifetime predictions are based on, for example, the number of running hours, loading cycles, or current mileage. However, this does not offer exact knowledge, since, in many situations, a complete database with records of components with various damage conditions, which ran to failure, is too expensive or is not available. Moreover, such a database would only give insight in problems that occurred before and the gathered probability distribution function of equipment failures would only be applicable to known and tested (operational) situations. Furthermore, static predictions will not offer correct forecasts when systems are loaded dynamically (e.g., a military aircraft performing different types of missions and operating in different regions with differing operational conditions). A dynamic load-based maintenance concept (maintenance based on the stresses applied to the system) would be more appropriate in these situations (Tinga, 2010). Quantitative data is preferred for such a concept, but the load factors (the impacts of usage types) can also be quantified based on expert knowledge.

Nowadays, on-condition maintenance (conducting maintenance at just the right moment) is becoming more viable when the developments in prognostic technologies are combined with developments in information technology (Dibsdale, 2011). On-condition maintenance uses the actual condition of the equipment to determine the required maintenance, instead of using a prediction for an average component under historical average conditions or a general trending of a prognostic parameter to a certain threshold. CBM, for example, is a maintenance program that only recommends maintenance actions when abnormal behavior is detected in the condition monitoring information of a physical asset (Jardine et al., 2006): maintenance is only conducted when it is required. CBM thereby enables decision making for both corrective and preventive maintenance actions. In the case of corrective maintenance, CBM informs the decision maker about occurring failures (detection) and classifies these failures based on their root cause and effect (diagnosis). For proactive maintenance actions, CBM uses prognostics results that inform about the future condition of the asset (Bousdekis et al., 2015).

Strategic Maintenance Management

Although maintenance has not always be seen as a strategic unit in the organization, it should be considered for all decisions that involve major investments in the future, such as corporate business planning, capacity planning, or acquiring facilities (Al-Turki, 2011).

Traditional maintenance decisions only consider repair/replace options or maintenance planning. However, the available maintenance capacity and the capability to maintain (new) items in the future influence company's business successes dramatically. Strategic maintenance management approaches should "be managed strategically, as a vital core business activity; and it needs to be based on quantitative models integrating maintenance with other business decisions such as production" (Murthy et al., 2002, p. 290).

Prognostic techniques can support informed maintenance decision making by describing the current, and preferably also the future, state of equipment. This is often required since human-made decisions are frequently not sufficiently reliable or accurate when dealing with complex systems with interrelating failure modes (Sikorska et al., 2011). Moreover, studies show that model-based decision making is superior to human-based decision making (Druzdel & Flynn, 2002).

Decisions that are taken on the basis of the information of the prognostic technologies do not only regard maintenance, but also include logistic and supply issues (designing the logistical and supply function), planning options (when systems can be used), inventory (how many spares, when, where), (improving) system design, safety (avoiding catastrophic failures), reliability and other strategic management concerns, such as future expansion, quality management, and corporate business planning.

Grubic et al. (2011) found that only a low percentage of companies adopted such decision support systems. They suggested that the captured, transferred, and analyzed data is often not readily available and integrated with a company's operational system. When decision support systems are effectively applied, they can provide three types of recommendations: changes in maintenance schedule or strategy, the optimal time of applying a pre-defined action, and pairs of optimal actions and the optimal time for their implementation (Bousdekis et al., 2015). Improvements in the system architecture and information flow can help to make these recommendations earlier and more accurate. This provides more reaction time for the decision maker. This improved knowledge about the state of the system and the remaining capabilities is not only of importance to the asset owner, but is also essential for service providers or original equipment manufacturers (OEMs) when offering performance or availability-based contracts and services, for example.

Advanced Maintenance Techniques

AMTs can provide three types of technical results: detection, diagnostics, and prognostics. Thereby, AMTs offer a form of decision support to the decision maker, enabling more effective (doing the right things) and efficient (doing things right) decisions.

Using detection techniques in equipment such as aircraft, vehicles, and machines, as for example build-in tests in systems, helps to detect anomalies in the system and provide (early) warning signals. On the basis of these binary (0 or 1) signals, users can make decisions about further operation of the system, such as continuing to use the system, conducting maintenance, or shutting the system down immediately.

Diagnostics provides more information; it will identify and quantify the occurred damage (Sikorska et al., 2011). Moreover, diagnostics aims to provide information about the root causes of the failures (Mathur et al., 2001). The final technical product of the maintenance techniques, the prognostics, aims to provide information about the future state of the system. This includes the remaining useful life (RUL) and the fault free period of the system, given a certain operating profile (Jardine et al., 2006).

Using the technical results of the AMTs for strategic business decisions can improve their potential. One can think of phasing in or out new and current systems, sourcing and investment strategies, and product mix changes. As for the latter, a detailed analysis could bring up, for example, that certain output in the product mix cause higher wear and tear of machinery. This could make these products less profitable than they were supposed to be before.

Making Informed Maintenance Decisions

AMTs allow for more informed–technical or business–maintenance decisions by providing exact information about the future predicted behavior of assets.

Typically, the technical outputs of these techniques can support maintenance decision making by giving:

1. An estimated time to failure (or remaining useful life); and
2. An associated confidence limit (Sikorska et al., 2011).

These outputs of the maintenance techniques are critical to make informed decisions for planning, maintenance scheduling, and spare parts provision (Jardine et al., 2006). If a system has failed, these decisions depend on knowing when it will possibly fail (again) and how it has failed (Haddad et al., 2014).

In a broader perspective, the results of AMTs can support typical business decisions that consider three characteristic problems engineering plants face:

1. Pre-scheduling and pre-planning of maintenance (when can systems be used? When should maintenance be conducted?);
2. Inventory control (how many spares? When? Where?); and
3. Safety risk control (avoiding catastrophic failures) (Tse & Atherton, 1999).

Furthermore, typical business decisions in the form of corporate business planning, capacity planning, or acquiring facilities can be supported (Al-Turki, 2011). Finally, these techniques can facilitate improved decisions about logistic and supply challenges (designing the logistical and supply function), changing usage type (what type of usage causes the highest system loads), and improving (future) system design (knowing the weakest parts of the system).

The core value–and often one of the main difficulties–of maintenance decision making is the determination of the (monetary) values of all options; the decision maker has to make good trade-offs between the options. Within the decisions that are directly focused on the actual maintenance, the value of wait-to-maintain, the maximum time to be taken before conducting maintenance to prevent severe failures from occurring (Haddad et al., 2014), is essential. The determination of this value is often difficult for three reasons. Firstly, when dealing with complex systems with interrelating failure modes, human decisions are often not sufficiently reliable or accurate (Sikorska et al., 2011). Secondly, multi-component dependencies have to be included in these predictions (Zio, 2009). Finally, system deterioration, and therefore the maintenance needs of complex moving assets, can vary dramatically when the system is operated under highly variable working conditions (Tinga, 2010). In order to overcome these difficulties, it is important to ensure that the correct AMT is applied.

Selecting Advanced Maintenance Techniques

The need to conduct AMTs differs per type of company and industry, and depends on factors such as profit margins, safety standards, and inherent risks. The type of analysis that would be applicable to the situation under consideration depends on the (type of) usage of the asset, the number of similar assets, and the (potentially) available data.

A vast amount of literature describes many specific maintenance techniques. However, many of these methods for prognostics and health management are introduced and applied to solve specific problems without the argumentation of why and how these methods have been selected (Lee et al., 2014).

The selection of the correct AMT to apply is therefore not a trivial task. Different prognostics models are not equally well suited for different business needs and not all models are equally well proven as others, and therefore they imply different business risks (Sikorska et al., 2011). This means that an owner of a nuclear power plant who has the same problem as the owner of a wind turbine might have to select a different type of maintenance technique since the business risks and needs differ.

The selection of AMTs requires prior consideration of the available input data and the required outcomes. Both diagnostics and prognostics heavily rely on system knowledge and input from systems. The knowledge and data could be incomplete or imprecise, especially the data about (actual) failures. Since many preventive maintenance programs ward off components from actual failure, failure data could be limited.

On the other hand, the required outcomes determine the required quality level and accuracy of the maintenance techniques. The qualification and quantification of uncertainties are important factors to be considered in the decision making process. Uncertainties can lead the prediction to being too early, leading to false positives (no fault found), or too late, leading to unplanned downtime. Uncertainty of the prediction can be caused by (i) uncertainty in the model, (ii) uncertainty in the measurement and forecast due to environmental and operational loads, and (iii) uncertainties with the characteristic parameters caused by the production process (Bo et al., 2010).

Typology of Advanced Maintenance Techniques

The basis of the categorization of the AMTs is the type of information used in the AMT. The AMTs use a mix of three elements: data, knowledge, and physics of failure, as shown in *Figure 1*. The balance between these elements differs per type of analysis as the position of the ellipse in the three triangles of Figure 1 illustrates:

1. **Data - Based Predictions:** Aim to identify anomalies in the system by identifying changes in the data characteristics (Jianhui et al., 2003). This type of predictions are therefore highly dependent on the quantity and quality of the input data (Jianhui et al., 2003). The strength of data models is the ability to transform high-dimensional noisy data into lower dimensional information (Jianhui et al., 2003).
2. **Knowledge-Based Predictions:** Use experience from subject matter experts and form rules to apply that knowledge (Sikorska et al., 2011). These models however, require a high degree of exactness and completeness to be useful (Biagetti & Sciubba, 2004).
3. **Physics-of-Failure-Based:** Predictions quantitatively characterize the behavior of a failure mode using physical laws (Sikorska et al., 2011). These predictions are especially useful for predicting the

Figure 1. Elements of a typical maintenance analysis; the oval shape shows that the balance of the three elements can differ per type of analysis

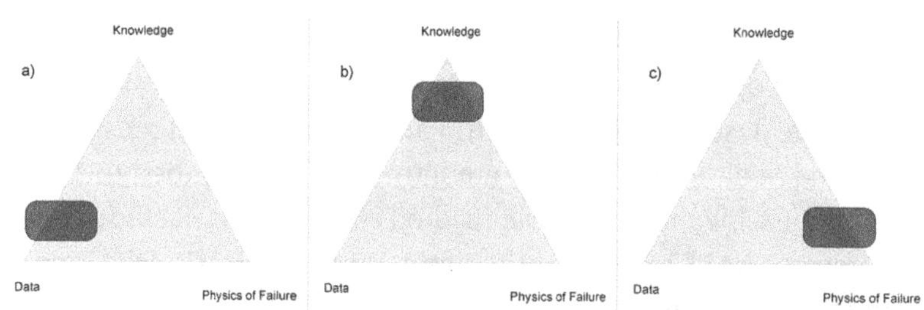

response of assets to new loading conditions or system configurations. However, these predictions require the availability of an accurate mathematical model (Jianhui et al., 2003) and are typically more computationally intensive than data-based models (Farrar & Lieven, 2007).

The vast majority of maintenance techniques are composed of a mix of at least two of these three elements. The basic form of the classification given below was first proposed by Coble and Hines (2008), and has been extended by Dibsdale (2015) with the fifth category, the mechanism-based predictions, and thereafter by Tiddens et al. (2015) with the first category, the experience-based predictions. The classification now consists of five maturity levels of prognostics:

1. **Experience-Based Predictions:** Based on knowledge, previous experience, and OEM knowledge, and are supported by little or scattered data. The prediction results in the estimated lifetime for an average component under historical average conditions.
2. **Reliability Statistics:** Based on historical (failure) records of comparable equipment without considering component specific (usage) differences. The prediction results in the estimated average lifetime of a component operating under historically average conditions.
3. **Stressor-Based Predictions:** Based on a combination of historical records and stressor data (for example temperature or humidity) to include environmental and operational variances. The prediction results in the estimated lifetime of an average component in a specific environment.
4. **Degradation-Based Predictions:** Based on the extrapolation of the general path of a prognostic parameter to a failure threshold. The prognostic parameter indicates signs of incipient failure, is inferred from sensor readings (for example temperature or vibration), and includes the current state of degradation. The prognosis therefore results in the estimation of the lifetime of a specific system in a specific environment.
5. **Mechanism-Based Predictions:** Based on the direct sensing of the critical failure mechanisms of individual components. The prognostic parameter is calculated with a physical model of the degradation mechanism. The inputs for this model are the sensed variations of loads or usage. The prediction results in the estimation of the lifetime of a specific system in specified conditions.

The decision maker can select the correct technique to apply on the basis of the described outcomes of the various techniques given in the classification above. This already structures the choice of the technique. The next sections will discuss the organizational difficulties for the application and implementation of AMTs.

DIFFICULTIES IN APPLYING ADVANCED MAINTENANCE TECHNIQUES

In the previous sections, the general characteristics of AMTs have been described. To specifically study how companies use these techniques for maintenance decision making, several case studies have been conducted. The findings of this case study are used in the upcoming sections to present a company typology of several types of AMT users. Based on the literature, the steps that have to be taken towards informed maintenance decision making will be described. The classification of the different types of AMT users will be employed to reflect on these steps.

Case Study

To be able to reveal and identify the central issues companies experience while implementing and applying prognostic technologies, a multiple case study has been carried out among thirteen departments within nine companies that employ diverse types of AMTs. Interviews have been conducted with relevant staff from their maintenance departments (i.e., maintenance and reliability engineers, and maintenance managers). The case study includes companies from different industries in the Netherlands: railway, process, nuclear, maritime, defense & aerospace, and energy. The results show preliminary findings and are part of ongoing work. The analysis of the case study was presented in a previous work (Tiddens et al., 2015).

The case studies highlighted a number of issues that hindered the successful usage of the maintenance techniques. These issues vary from organizational aspects to the technological features of the techniques. The company typology sheds a further light on the issues and challenges these different kinds of companies experience. However, this typology, which differentiates three types of AMT users, cannot be regarded as an exact black and white description of the diverse types of companies conducting AMTs; intermediate forms do exist and no form is better per se.

THE DESIGN OF COMPANY TYPOLOGIES

The company typologies, based on the aforementioned case studies, describe three types of companies that use more or less AMTs. This section will describe the typologies and will highlight their recurrent difficulties and challenges. Many companies have evolved to a certain typology, *Experiencers, Improvers,* or *Predictors*, due to their history. Four central themes that play a vital role in the application and implementation of prognostic techniques differentiate the three company typologies. These themes were deducted using cross-case analysis between the thirteen case studies. Every theme is more or less developed in the companies; however, the companies that have focused more on the adoption of AMTs have progressed further in the improvement of these themes and are therefore often more successful in the application of AMTs. The four themes are:

1. **The Need for Just-in-Time Maintenance:** That is the importance of just-in-time maintenance in the organization, the need for improvement in knowing the future equipment performance and upcoming failures, and the extent to which these techniques are the basis for decision making processes for both technical as management practices.
2. **Commitment:** That is the willingness of top management to invest in the development of more AMTs and, secondly, the capabilities and competencies of the workforce to conduct these AMTs.
3. **Information System Capabilities:** That is the effectivity and efficiency of information and data collection and storage systems, and the options to use these to support maintenance techniques.
4. **Maturity of the Maintenance Processes:** That is the processes that are in place to incorporate the usage of maintenance techniques in the daily practice, such as continuous improvement loops. It also means that there are processes in place to support the workforce in developing these techniques.

Type A: The Experiencers

Experiencers base their predictions on previous practice from both explicit and tacit knowledge within and outside the company, and build-up history with comparable equipment. Operator knowledge and technical expertise are the core input of the maintenance techniques. Many *Experiencers* can be found in industries such as the process or maritime industry. Given the fact that in these industries equipment and machines are typically one-of-a-kind products and have a long lifetime, the amount of build-up history data of unique systems is often low.

Experiencers use the extensive knowledge of their equipment to improve current maintenance plans by techniques such as failure mode and effect analyses (FMEA) and reliability centered maintenance (RCM). However, these evaluations are often a one-time action; the processes to continuously improve them are not in place. Since the analyses are mainly knowledge-based, the relevant concepts are often not evidence-based. Large scatter in the predictions (an inaccurate determination of the moment of failure) is therefore foreseeable. Preventive maintenance is sometimes caught up by the daily practices of equipment failures, which require all attention of the maintenance departments. Production and operation is often the focus of top management. Maintenance, however, seems not to be regarded as a strategic area; the urgency of radically improving in this area is not communicated within the company. As a result, there is not a clear innovation and investment strategy for maintenance and the processes which are–sometimes–in place to improve maintenance are not implemented well. Since there is not a clear urgency of radical improvement, risk taking is not encouraged; consequently, together with the large scatter in the predictions, this seems to lead to conservative maintenance intervals. Finally, the lack of communicated urgency is also reflected in the workforce's build-up skills, and little progress is made in the development of AMTs. Some *Experiencers* have adopted simple condition-based maintenance practices as only replacing equipment after–visual–inspections or periodical monitoring. More advanced practices are often advised by external companies and OEMs or triggered by major disruptions in the operational process.

Type B: The Improvers

Improvers have established a solid basis of experience-based predictions and have become able to gradually improve them by using more AMTs. *Improvers* undergo a strong urgency to ameliorate maintenance predictions; therefore, they are seeking to find and implement new techniques and technologies. Although maintenance concepts are primarily FMECA- and RCM-based, small improvement projects are defined to test new maintenance practices and improvement cycles are started to update the established predictions. This type of companies starts to see maintenance as a strategic improvement area. Although *Improvers* experience many difficulties in the development and application of maintenance techniques, local successes help to create a sense of urgency within the company and to create top management's support. Improvers find it difficult to create the business case for new projects, collect data, select the applicable technique, select the components to apply prognostics to, and communicate the achieved successes throughout the company

Improvers seem to develop more mature techniques with a trial and error approach, which is started by collecting data on many different parameters. For the companies it is very difficult to determine beforehand the relevant methods and parameters, and they therefore experience a long and costly implementation process. Thus, some projects are only conducted as technology demonstrators where (pre-)

installed sensors are utilized for the initial projects, often organized together with external knowledge partners or research institutes. The companies' workforce is being trained in developing and implementing maintenance techniques. Many *Improvers* experience difficulties with good data collection, as data is not always of good quality or not logged in accessible systems. This hinders the effective application of maintenance techniques.

Type C: The Predictors

Predictors have become able to have insight in actual and expected performance of equipment. For *Predictors*, real time performance data is key. They keep track of their critical assets, preferably in real time, and are continuously updating predictions about failures and lifetime. These companies consider maintenance as a strategic part of the company and conduct maintenance proactively. Strategic business decisions can use inputs based on maintenance predictions. The company strongly communicates the urgency of maintenance and local successes achieved with the application of AMTs. *Predictors* aim to invest business case wise in the development of AMTs, which means that a rigid business case (that expresses the costs and benefits) is constructed before implementing the AMT. These companies often start with experience-based methods, which are enriched with data from historical failure records to create evidence-based maintenance concepts. Different methods are utilized and feedback loops are established to assess the effectivity and efficiency of the techniques. Although *Predictors* use many mature techniques, also lower mature techniques like FMECA are used for less critical equipment. These techniques are often more evidence-based than those of the *Experiencers*. Top management acts like a project champion for these improvement trajectories and is therefore willing to invest in large improvement projects like the monitoring of all performance and condition data of equipment. *Predictors* have established this sense of urgency by communicating the successes of (small) improvement projects, which are used to roll out the tested techniques. Although *Predictors* are able to predict the moment of failure of equipment, they experience difficulties in extending the lifetime of equipment and in extending maintenance intervals.

Relation between the Company Typologies and the Main Themes

The company typologies are differentiated by the four main themes that play a role in the implementation of AMTs. Table 1 summarizes the role of these themes in the company typologies as discussed above.

THE ROUTES TOWARDS INFORMED MAINTENANCE DECISION MAKING

As shown in the previous section, the route to informed maintenance decision making contains many challenges for companies.

Two facets of implementation issues, technical and organizational, will be discussed to enable effective implementation of the technical sides of maintenance techniques. Greitzer et al. (2001) proposed a six-step process to implement prognostic technologies. Greitzer et al. emphasized that companies should start with a failure mode, effect, and criticality analysis of the target system to identify the top faults and failures in terms of frequency, costs and other maintenance issues. After this step, necessary sensors should be identified and designed (which requires a detailed analysis of the failure mechanisms and the technical expertise of sensor application). Parallel to this, a cost-benefit analysis has to be conducted to

Table 1. Company typologies and their relation with the influencing factors

	The Need for Just-In-Time Maintenance	Commitment	Information System Capabilities	Maturity of the Maintenance Processes
Type A: *Experiencers*	Experience-based maintenance is set in stone, little perceived need for improvement (perhaps due to e.g. outsourcing).	Top management is not convinced of benefits. Capabilities to develop techniques lack within the workforce.	Little to no data storage. Often poor input quality.	Plans have been made. Continuous improvement loops not in place. New techniques based on old experiences.
Type B: *Improvers*	Perceived need for improvement. Small projects defined.	Capabilities to develop new techniques are developed. Organization starts to see advantage of new techniques.	Historical data is collected. However, the company struggles with quality of historical data.	Continuous improvement loops are set up. Predictions with more mature techniques carried out for critical equipment.
Type C: *Predictors*	High sense of urgency experienced to improve decision making.	Top management acts as project champion for new projects. Workforce is capable to conduct advanced techniques.	Data is collected and centrally accessible. Sometimes in real-time.	Company invests business-case wise in the development of mature maintenance techniques.

determine the benefits of the application of the sensor technology. After the prognostic algorithms have been developed in a prototype system, the benefits of the sensors have to be assessed before continuing to develop onboard sensor capabilities.

Many companies, even the *Experiencers,* have already conducted the first step, the FMEA. Therefore, a good consecutive step to implement AMTs that are more advanced is to create more evidence-based maintenance concepts. One of the techniques suitable for this is CBM, which uses condition monitoring to continually or periodically measure signs of incipient failure of equipment. Equipment will only be repaired or replaced when a certain threshold has been reached. Condition monitoring techniques can only be used if failures can be detected before occurring and signs of incipient failure can be described by one or more parameters. According to Adams (2007), the application of CBM techniques is useful when there are variable or unknown operational and environmental loads. Moreover, these techniques are beneficial when the designs are under or over conservative, many components interact in the system, unanticipated changes can occur in the life cycle, or damage can occur in hard-to-reach areas.

Organizational issues are the other crucial part of the implementation of new techniques. This is especially important for the *Experiencers.* Top management's support, for example, is not always guaranteed as maintenance is seldom seen as a strategic function (Al-Turki, 2011). In business environments, Kotter (1995) his multi-step process is often used as a good way to overcome organizational issues. Kotter's process should be driven by high-quality leadership and create power and motivation to overwhelm reactionaries. It focusses on establishing a sense of urgency to discuss potential crises and on creating vision within the company. Moreover, Kotter states that this vision has to be communicated and others within the company have to be empowered to act on it. Finally, short-term wins have been created by starting up small projects which are visible within the company.

The Route Towards Informed Maintenance Decision Making

To enable successful application of advanced maintenance techniques, four steps are proposed, as described in Figure 2. The steps are illustrated by an example of a case study on the use of AMTs for a

Figure 2. The steps towards informed maintenance decision making applied to a practical situation
The four steps are based on Jardine et al. (2006), Coble and Hines (2008) and Dibsdale (2015)

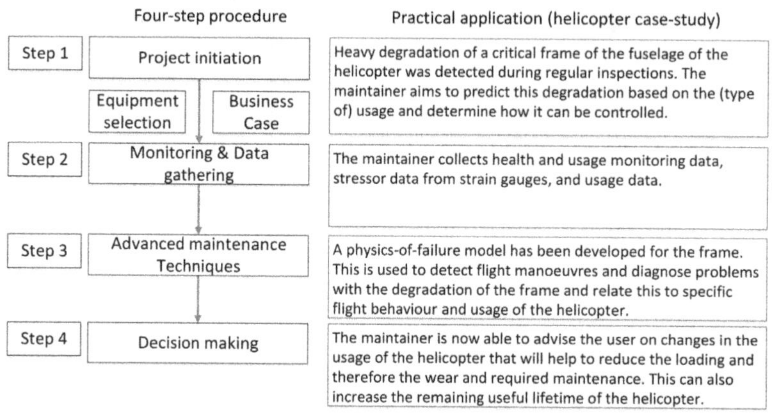

helicopter. The framework starts with the project initiation. In this step, Kotter (1995) and Greitzer et al. (2001) principles are included. In order to establish a sense of urgency, the right selection components and the validation of this selection is incorporated by the development of a business case. Moreover, the top management can share the vision by defining a strong project start and communicating the benefit of such projects. The continuing steps focus on the collection of data and the maintenance techniques. According to Jardine et al. (2006), a typical condition-based maintenance process consists of three steps: data acquisition, data processing, and maintenance decision making. This latter step is the ultimate goal of the start of the project. To apply the right types of maintenance techniques, the principles described in Adams (2007) handbook have to be followed. The proposed framework describes the technical approach related to the consecutive steps that should be taken when companies are considering implementing certain types of AMTs. Moreover, organizational principles are included to achieve successful informed maintenance decision making.

Step 1: Project Initiation

First, a deliberation should be made about the how and why to start an advanced maintenance analysis. The urge to do such an analysis can be driven by a technology push (e.g., a new technology is developed and offers perspectives to improve maintenance decision making) or by decision pull, where the need has arisen to improve a decision for maintenance. While the former is characterized by the development of a new technology, which could potentially be used for measurements or analyses for example, the latter, decision pull, can be divided into proactive and reactive improvements. In a proactive decision pull the need has arisen to improve–strategic–decision making; reactive decisions are often led by major process interruptions or incidents related to safety, downtime or costs.

In this first step, two actions have to be taken: the selection of equipment and the development of a business case. Selecting critical components is a task typically carried out with an FMECA analysis. This analysis can be input for *Figure 3*, which shows a selection chart to identify the components that are suitable to apply prognostic techniques. In Figure 3, the identified failures are plotted against their failure frequency and failure effect (e.g. downtime or costs). Four main categories of failures are differentiated by drawing the bold lines in the figure. These lines are based on the company's demands on production and maintenance. The chart has similarities to a risk matrix used in FMECA studies. The lines differentiating the four areas should not be exactly horizontal or exactly vertical since a trade-off between downtime and failure frequency will result in the specific acceptance of risk. Indeed, risk is company dependent–risk is defined as: probability of occurrence multiplied by the effect of failure. The four areas are:

A1: These failures cause major disruptions to the process and should therefore be designed out during design.
A2: These failures occur relatively often, but do not cause major disruptions; therefore, it is advisable to put these parts in stock to reduce downtime.
A3: These failures do not occur very often and do not cause major disruptions, regular maintenance will therefore suffice.
A4: These failures rarely occur; however, they have a critical impact on the process; predictive maintenance is an advisable strategy for these components.

The second action, creating the business case, is an important factor to demonstrate the added value of the project and to create a sense of urgency for the project. However, investing in AMTs should also be viewed as a strategic improvement decision. Therefore, a company should invest in the start of multiple projects and only continue with those projects that show real added value. This is especially true for *Improvers* and *Predictors* who can become learning organizations when this strategy is implemented. The business cases can be constructed by showing the investment costs and balance against the value of less downtime, fewer incidents, and other benefits that the prediction offers to the company. For *Experiencers* it is important to create a sense of urgency to improve the current maintenance concepts. The value

Figure 3. Four Quadrant Chart for the selection of critical components
The figure has been adapted from Lee et al. (2014) who defined a chart with four strict quadrants. However, components with extreme high failure frequency (top performance killers) should not be tolerated and components which can cause extreme high downtime or costs should be modified.

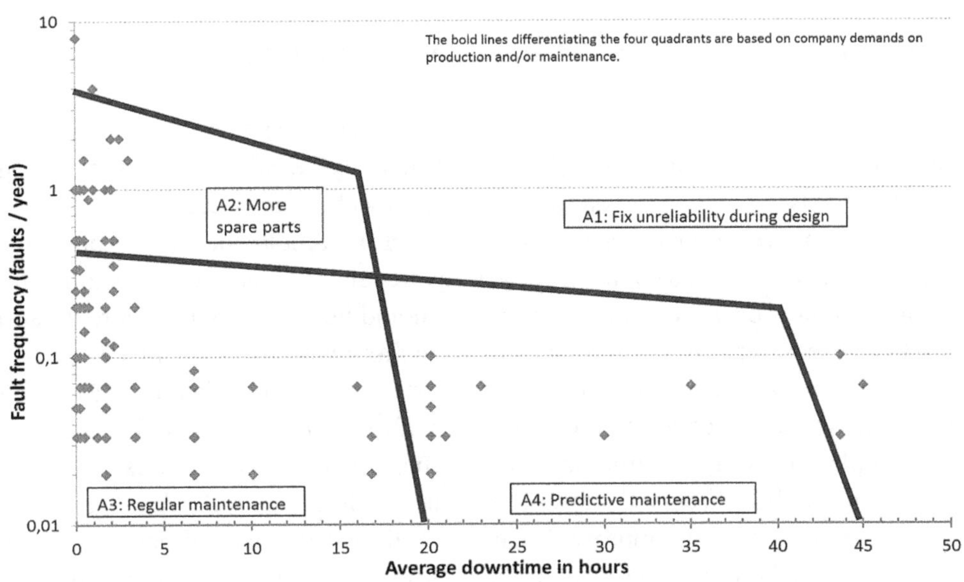

of this improvement trajectory to the company has to be determined and communicated throughout the organization as a first step of the innovation and investment strategy.

To further improve the maintenance concept with more mature maintenance techniques, the enterprise should invest in both the workforce's knowledge on these techniques and in creating small projects where these techniques can be conducted for critical equipment. These small projects can be used to share local successes and convince the workforce and the top management of the urgency and the added value of more AMTs.

Step 2: Monitoring and Data Gathering

Step 2 focuses on the monitoring process and the collection of (available) data from historical records and monitoring systems. The viability of these monitoring techniques will be considered and the actual measurements will be conducted to continue with the actual maintenance techniques. However, when talking about the introduction of AMTs in a company, one has to recognize, as indicated in the case studies in the previous section, that the type and level of detail of the analysis might determine the applicability to the company. Different routes (from data gathering to maintenance decision making) carry different advantages and types of uncertainty; therefore, the choice varies per situation. The selection of a diagnostic or prognostic route is dependent on the availability of data, the personnel's available skills, and the computing infrastructure (Sikorska et al., 2011). The selection of the type of required input data goes hand-in-hand with the selection of the type of maintenance analysis. However, the availability of data can steer this process. In general, companies need to develop skills in data management. To increase the ease of analysis, data should be stored in a central warehouse. Moreover, sometimes only peaks or average values of data are stored, not the underlying raw data. This could hinder the use of some techniques since essential data characteristics can be lost and small temporal changes cannot be detected. Improving on the long term often requires the availability of reliable data about failures and maintenance activities. Hence, data collection systems have to be improved. Urgency and involvement of the workforce is necessary to ensure that data of good quality is entered in the system and techniques can be based on reliable input data.

Step 3: Selection of Advanced Maintenance Techniques

In step three, the selection of the type of maintenance analysis has to be made. As explained in the background section, techniques can–in increasing level of complexity-be categorized as: experience-based, reliability statistics, stressor-based, degradation-based, and mechanism-based.

The first step for many *Experiencers* is to improve their current maintenance concept with evidence-based data and ensure that prescribed maintenance actions are related to the occurring failure mechanism. Moreover, frequent failures or those with high impacts should be analyzed using a root cause analysis logic to ensure the correct steps are taken to prevent these failures from occurring again. To make sure that effective actions are taken, not only the technical but also the organizational aspects should be regarded, such as the six factors considered in a typical quality analysis as an Ishikawa diagram: equipment, process, people, materials, environment, and management (Ishikawa & Ishikawa, 1982). The second step is to conduct the 'check-act' steps of the plan-do-check-act cycle. *Experiencers* often have many processes described ('plan') to improve the current maintenance, and defined projects are set up to execute ('do') these plans. However, these are often one-time activities. For current FMEA- and RCM-based techniques, this 'check-act' step can be taken by performing for example a maintenance feedback analysis (MFA) (Braaksma et al., 2012). After the determination of the value of the feedback on the current maintenance concept (which is the first step in the MFA procedure), data can be collected from existing systems to improve the concept.

Due to already developed techniques and more available history data, the *Improvers* and *Predictors* have more degrees of freedom available in the development of mature maintenance techniques. For the selection of the techniques, it is important to consider a trade-off between the available data, the available knowledge about the physical behavior of the component, and the working conditions of the asset. The latter can determine whether a data-driven approach can offer reliable predictions. A high amount of variation in the usage and loads on the systems can cause variation in the failure pattern of the equipment, making experience or historical data-driven approaches less accurate. Moreover, the prognostic distance, that is the time horizon that the decision maker obtains from the analysis to undertake actions– the time between indication of a failure in the analysis and the actual failure (Sandborn & Wilkinson, 2007)-of the analysis should be considered. If the prognostic distance of the analysis is not larger than the time required to take action or alter the (production) schedule, the value of the analysis is limited. Uncertainties in the analysis do also play a vital role. Two different types of uncertainty can influence the outcome: randomness which is due to the inherent variability of the system and imprecision due to a lack of knowledge and information about the system (Zio, 2009). As *Figure 1* shows, the maintenance analysis consists of a mix of data, knowledge, and a physical model. Varying operational loading and varying conditions will often cause high uncertainty and randomness of static predictions. Therefore, in these situations, the mix of the three elements will often shift towards physical model. However, companies should only opt for the most advanced techniques where and when necessary, which is decided using a business case.

Step 4: Decision Making

In the final step, the actions taken in the previous steps come together in the decision making process. A first result of the techniques will be the technical results, such as the anomaly detection, diagnosis of the system's state and the prognosis of the remaining capabilities, the remaining life and fault free period. These results are input for the decision-making process, which has to support business or life cycle decisions like the planning of maintenance, the introduction of midlife updates, or the extension of the system's life. Boundaries for decision-making based on the conducted maintenance techniques are created by both internal and external regulations and laws by, for example, limiting the possibilities of data gathering or setting the norm for the accuracy of the prediction.

The difficulty in the decision making process for many companies is in the unknown value of the alternatives that can be considered in the decision making process. The heart of the decision support for maintenance is at the determination of the value of *wait-to-maintain* (Haddad et al., 2014), i.e. the maximum time to be taken before maintenance must be carried out to prevent severe failures. Ultimately, for decisions that consider conducting maintenance actions, the decision support provides a cost-benefit-risk model, which helps to determine if the revenues generated by the asset will compensate the risks of failure; in other words, consuming the RUL, by continuing operation, or throwing it away, by replacing parts (Haddad et al., 2014). For these decisions it is important to acknowledge multi-component dependencies in systems and include these in the predictions (Zio, 2009). When these interdependencies are known and redundancies and options to change operation are considered, maintenance planning and scheduling on a system level can be improved. By including these factors, organizational and business aspects can be included to achieve informed decision making. Ferri et al. (2013) show the benefits of a system-wide perspective for prognostics and health management (PHM) approaches for aircraft maintenance planning. By combining the information of the system architecture with RUL estimates of a PHM system for individual components, a larger time margin for the next required inspection was obtained. This offers the opportunity to include *lean* methods to reduce inventory and work in progress in maintenance departments. Moreover, when the failure behavior is known, rejection norms of equipment can be reconsidered since these often accumulate margins for uncertainty caused by unprecise and unreliable predictions. When (real-time) performance and maintenance data are used in the techniques, assets can be compared and benchmarked and the effect on different types of outputs (products) or usage can be assessed, such as situations causing high loading on the system, quality problems due to different types of throughput, or the detection of 'unknown-unknowns'. Tinga and Janssen (2013) show that when the maintenance policy of a navy frigate with several subsystems is tailored to the actual usage profile of the system, large benefits in terms of total maintenance costs can be achieved. Moreover, they show that the average life of critical subsystems influences the optimal maintenance policy. As the continuing step, these strategic fleet-wide decisions about where and how to operate assets can be supported by the AMTs.

Finally, to improve these decisions, the scope of the maintenance analysis should be widened to the 'socio-technical' system (Zio, 2009). The technical and human performance is influenced by outside factors like the organization and management of industrial activities, and exogenous factors like regulations, market pressure, political pressure, and the safety culture (Zio, 2009).

CONCLUSION

AMTs offer valuable options to improve decisions making for asset management. AMTs help in the decision making process to predict the future state of assets. This includes health assessment, the detection of imminent failures and the prediction of the remaining useful life (Lee et al., 2014), or the probability a machine operates without failures or faults up to a certain time given the current machine condition and operational profile. When the characteristics of these imminent failures are known, better-informed maintenance decisions can be made.

However, many companies experience difficulties in implementing and applying these techniques. As the selection and application is not a trivial task, a long and costly implementation process is often experienced.

This chapter has started by an overview of the various AMTs available. Based on case studies at nine companies in different industries in the Netherlands, a classification of the type of companies that apply these techniques has been outlined. This overview shows a clear distinction between three types of companies: *The Experiencers, The Improvers,* and *The Predictors*. Using cross-case analysis in the case studies, four central themes in the application of AMTs have been deducted.

This provides real experiences to academics about the application of AMTs in industry and sheds a light on the problems that companies experience. This can be worthwhile to developments in this field, as communication between industry and academics has been missing (Jardine et al., 2006).

This chapter has provided a four-step process for the application of AMTs, based on the company typology. For the first step, the project initiation, a selection method for the components to apply prognostics to is proposed. As most research within the field of condition-based maintenance excludes the organizational and managerial aspects (Garg & Deshmukh, 2006; Kerkhof et al., 2014; Veldman et al., 2011), both technical as well as business criteria for the application of AMTs are described in the proposed four-step framework. Within the explanation of each step, several applications of informed decision making are discussed that can be achieved when AMTs are well applied.

The presented steps to implement AMTs are difficult for less mature companies. These concerns often have to make large cultural and organizational steps in data management, so create urgency and the development of maintenance techniques. The experience-based maintenance paradigm has often been set in stone since many years. Creating (culture) changes is therefore difficult to achieve. Once the first steps towards improvement have been made, as for the *Improvers* and *Predictors*, local successes are more easily judged on value. Subsequently, progresses can be continued towards total predictive maintenance.

To create structural improvements, focus of organizations should not only lie on developing the technical aspects of the techniques, but also on the creation of a sense of urgency and the willingness of the total organization to adopt these techniques. Therefore, support on both top management and workforce level has to be achieved. As also Kerkhof et al. (2014) have noted, the difficulties associated with successful and systematic organization of a CBM program within an organizational context is often underestimated.

FUTURE RESEARCH DIRECTIONS

Future research on the usage of prognostic techniques for decision making will include the effects of system-wide interactions, of which a glimpse is already shown by (Ferri et al., 2013). Moreover, the knowledge gained from the techniques can be used to improve many company-wide decisions, as for example is shown by (Tinga & Janssen, 2013). However, the many differences in types of equipment, inherent risks, availability of knowledge, data, and physical models create difficulties in providing general guidelines to develop maintenance techniques. Further research will focus on developing the selection criteria that enable the choice of methods. One specific issue in the implementation of these techniques is the time horizon of the decisions. Currently, many short term decisions can be improved by diagnosis and detection techniques, such as shutting down the system before catastrophic failure or planning maintenance within the next maintenance stop. However, few examples of prognostic approaches that improve decision making on the long term are available. These, however, have the largest potential to improve decision making for asset management.

ACKNOWLEDGMENT

The authors gratefully acknowledge the parties providing funding to this research project. This research is part of the Tools4LCM project, funded by the Netherlands Ministry of Defence and the National Aerospace Laboratory NLR. The research is also part of the MaSeLMA project (Integrated Maintenance and Service Logistics Concepts for Maritime Assets), funded by Dinalog (Dutch Institute for Advanced Logistics). The authors wish to express gratitude to the interviewees of the case studies for their contributions to this chapter.

REFERENCES

Adams, D. E. (2007). *Health monitoring of structural materials and components: methods with applications*. John Wiley & Sons. doi:10.1002/9780470511589

Al-Turki, U. (2011). A framework for strategic planning in maintenance. *Journal of Quality in Maintenance Engineering*, *17*(2), 150–162. doi:10.1108/13552511111134583

Benedettini, O., Baines, T., Lightfoot, H., & Greenough, R. (2009). State-of-the-art in integrated vehicle health management. *Proceedings of the Institution of Mechanical Engineers. Part G, Journal of Aerospace Engineering*, *223*(2), 157–170. doi:10.1243/09544100JAERO446

Biagetti, T., & Sciubba, E. (2004). Automatic diagnostics and prognostics of energy conversion processes via knowledge-based systems. *Energy*, *29*(12), 2553–2572. doi:10.1016/j.energy.2004.03.031

Bo, S., Shengkui, Z., Rui, K., & Pecht, M. (2010, January 12-14). Benefits analysis of prognostics in systems. *Paper presented at the Prognostics and Health Management Conference PHM '10.*

Bousdekis, A., Magoutas, B., Apostolou, D., Mentzas, G., & Chan, H. K. (2015). A Proactive Decision Making Framework for Condition Based Maintenance. *Industrial Management & Data Systems*, *115*(7), 1225–1250. doi:10.1108/IMDS-03-2015-0071

Braaksma, A., Meesters, A., Klingenberg, W., & Hicks, C. (2012). A quantitative method for Failure Mode and Effects Analysis. *International Journal of Production Research*, *50*(23), 6904–6917. doi:10.1080/00207543.2011.632386

Coble, J. B., & Hines, J. W. (2008, October 6-9). Prognostic algorithm categorization with PHM challenge application. *Paper presented at theInternational Conference on Prognostics and Health Management*, Denver, CO, USA.

Cross, M. (1988). Engineering Maintenance Organisation Performance An Assessment of the Evidence from over 200 sites. *Management Research News*, *11*(1/2), 20–23. doi:10.1108/eb027938

Dibsdale, C. (2011). The Development of Maintenance and the Role of Integrated Vehicle Health Management. In I. K. Jennions (Ed.), *Integrated Vehicle Health Management: Perspectives on an Emerging Field* (pp. 9–16). Warrendale, PA, USA: SAE International.

Dibsdale, C. (2015). Holistic Prognostics. In R. Rajkumar & L. Redding (Eds.), *Through-life Engineering Services* (pp. 71–82). New York: Springer.

Druzdel, M. J., & Flynn, R. R. (2002). Decision Support Systems. In A. Kent (Ed.), Encyclopedia of Library and Information Science, (2nd ed., Vol. 67). New York: Dekker, M.

Farrar, C. R., & Lieven, N. A. J. (2007). Damage prognosis: the future of structural health monitoring. *Philosophical Transactions of the Royal Society A: Mathematical, Physical and Engineering Sciences*, *365*(1851), 623-632. doi: 10.1098/rsta.2006.1927

Ferri, F. A. S., Rodrigues, L. R., Gomes, J. P. P., de Medeiros, I. P., Galvao, R. K. H., & Nascimento, C. L. (2013, April 15-18). Combining PHM information and system architecture to support aircraft maintenance planning. *Paper presented at the 2013 IEEE International Systems Conference (SysCon).*

Garg, A., & Deshmukh, S. G. (2006). Maintenance management: Literature review and directions. *Journal of Quality in Maintenance Engineering*, *12*(3), 205–238. doi:10.1108/13552510610685075

Greitzer, F. L., Hostick, C. J., Rhoads, R. E., & Keeney, M. (2001). Determining how to do prognostics, and then determining what to do with it. *Paper presented at theIEEE Systems Readiness Technology ConferenceAUTOTESTCON '01.*

Grubic, T., Redding, L., Baines, T., & Julien, D. (2011). The adoption and use of diagnostic and prognostic technology within UK-based manufacturers. *Proceedings of the Institution of Mechanical Engineers. Part B, Journal of Engineering Manufacture*, *225*(8), 1457–1470. doi:10.1177/0954405410395858

Haddad, G., Sandborn, P. A., & Pecht, M. G. (2014). Using maintenance options to maximize the benefits of prognostics for wind farms. *Wind Energy (Chichester, England)*, *17*(5), 775–791. doi:10.1002/we.1610

Ishikawa, K., & Ishikawa, K. (1982). *Guide to quality control* (Vol. 2). Asian Productivity Organization Tokyo.

Jardine, A. K. S., Lin, D., & Banjevic, D. (2006). A review on machinery diagnostics and prognostics implementing condition-based maintenance. *Mechanical Systems and Signal Processing*, *20*(7), 1483–1510. doi:10.1016/j.ymssp.2005.09.012

Jianhui, L., Namburu, M., Pattipati, K., Qiao, L., Kawamoto, M., & Chigusa, S. (2003). Model-based prognostic techniques [maintenance applications]. *Paper presented at theIEEE Systems Readiness Technology ConferenceAUTOTESTCON '03*, Anaheim, CA, USA.

Kerkhof, R. M. d., Akkermans, H. A., & Noorderhaven, N. G. (2014). Knowledge lost in data: Organizational impediment to condition-based maintenance in the process industry. *Paper presented at the21st EurOMA Conference*, Palermo.

Komonen, K. (2002). A cost model of industrial maintenance for profitability analysis and benchmarking. *International Journal of Production Economics*, *79*(1), 15–31. doi:10.1016/S0925-5273(00)00187-0

Kotter, J. P. (1995). Leading change: Why transformation efforts fail. *Harvard Business Review*, *73*(2), 59–67.

Lee, J., Wu, F., Zhao, W., Ghaffari, M., Liao, L., & Siegel, D. (2014). Prognostics and health management design for rotary machinery systems—Reviews, methodology and applications. *Mechanical Systems and Signal Processing*, *42*(1), 314–334. doi:10.1016/j.ymssp.2013.06.004

Mathur, A., Cavanaugh, K. F., Pattipati, K. R., Willett, P. K., & Galie, T. R. (2001). Reasoning and modeling systems in diagnosis and prognosis. *Paper presented at the the International Society for Optical Engineering.*

Murthy, D., Atrens, A., & Eccleston, J. (2002). Strategic maintenance management. *Journal of Quality in Maintenance Engineering*, *8*(4), 287–305. doi:10.1108/13552510210448504

Sandborn, P. A., & Wilkinson, C. (2007). A maintenance planning and business case development model for the application of prognostics and health management (PHM) to electronic systems. *Microelectronics and Reliability*, *47*(12), 1889–1901. doi:10.1016/j.microrel.2007.02.016

Sikorska, J., Hodkiewicz, M., & Ma, L. (2011). Prognostic modelling options for remaining useful life estimation by industry. *Mechanical Systems and Signal Processing*, *25*(5), 1803–1836. doi:10.1016/j.ymssp.2010.11.018

Singpurwalla, N. D. (1995). Survival in dynamic environments. *Statistical Science*, *10*(1), 86–103. doi:10.1214/ss/1177010132

Tiddens, W. W., Braaksma, A. J. J., & Tinga, T. (2015). The Adoption of Prognostic Technologies in Maintenance Decision Making: A Multiple Case Study. *Procedia CIRP*, *38*, 171–176. doi:10.1016/j.procir.2015.08.028

Tinga, T. (2010). Application of physical failure models to enable usage and load based maintenance. *Reliability Engineering & System Safety*, *95*(10), 1061–1075. doi:10.1016/j.ress.2010.04.015

Tinga, T., & Janssen, R. (2013). The interplay between deployment and optimal maintenance intervals for complex multi-component systems. *Proceedings of the Institution of Mechanical Engineers, Part O: Journal of Risk and Reliability*, *227*(3), 227–240. doi:10.1177/1748006X13480743

Tse, P., & Atherton, D. (1999). Prediction of machine deterioration using vibration based fault trends and recurrent neural networks. *Journal of Vibration and Acoustics*, *121*(3), 355–362. doi:10.1115/1.2893988

Van Noortwijk, J. (2009). A survey of the application of gamma processes in maintenance. *Reliability Engineering & System Safety*, *94*(1), 2–21. doi:10.1016/j.ress.2007.03.019

Veldman, J., Klingenberg, W., & Wortmann, H. (2011). Managing condition-based maintenance technology: A multiple case study in the process industry. *Journal of Quality in Maintenance Engineering*, *17*(1), 40–62. doi:10.1108/13552511111116240

Zio, E. (2009). Reliability engineering: Old problems and new challenges. *Reliability Engineering & System Safety*, *94*(2), 125–141. doi:10.1016/j.ress.2008.06.002

Zio, E., Broggi, M., Golea, L., & Pedroni, N. (2012). Failure and reliability predictions by infinite impulse response locally recurrent neural networks. *Chemical engineering transactions, 26*.

KEY TERMS AND DEFINITIONS

Condition Monitoring: The (continuous) measurement of symptoms of incipient failure such as raise in temperature or vibration.

Detection: The binary process of detecting anomalies in the system, indicating a healthy or a faulty system.

Diagnostics: The process of identifying and qualifying anomalies and damage in the system. Its goal is to determine and identify the cause and effect relation and isolate faulty components, failure modes and failure conditions.

Health Monitoring: The process of (continuously) sensing of the critical failure mechanisms and system response by identifying and quantifying the presence and extent of damage in a system.

Load Monitoring: The process of continuously monitoring stressor data such as temperature or humidity.

Maintenance Decision Making: The process of making decisions for maintenance, such as repair or replace, maintenance planning, lifetime extension.

Prognostics: The process of predicting the future state of the system and determining the remaining capabilities. It usually generates a remaining useful life (RUL) and the probability a system works without failures or faults up to a certain time given a certain system condition.

Usage Monitoring: The process of continuous monitoring usage data, such as running hours, km's or cycles.

Chapter 14
Information Supporting of Decision Making for Energy Management in District Heating

Vira Shendryk
Sumy State University, Ukraine

Olga Aleksenko
Sumy State University, Ukraine

Victor Nenia
Sumy State University, Ukraine

Yuliia Parfenenko
Sumy State University, Ukraine

ABSTRACT

The chapter focuses on implementation of information technologies in energy management of district heating system. It captures the current state of energy management in district heating. The main standards of energy management are described. The specific features of district heating system and the existing tools of energy management in district heating are explored. The chapter studies the existing models, which can be applied in the decision support system for heat energy management. It describes the decision support system "HeatCAM" as a tool of energy management in district heating.

INTRODUCTION

Asset management defines as the "coordinated activity of an organization to realize value from assets". The term "asset" means an item, thing or entity that has potential or actual value to an organization (Williams, 2016). Asset management requires finding ways of ensuring that a company's assets would be most effectively used.

Energy typically accounts for 50% or more of operating costs of every organization (Reducing your ICT energy costs, 2016). Due to increasing prices for energy resources operation costs are growing up. Excessively high operating costs adversely affect the financial stability. For countries with financial instability of the district heating (DH) sector this problem is especially urgent and requires immediate steps.

Energy management (EM) plays an important role in energy saving. To stay competitive every organization need to understand the energy behind asset management (Infor EAM Energy Performance

DOI: 10.4018/978-1-5225-0651-5.ch014

Management, 2016). To make smart decisions on how to optimize energy use, it is necessary EM implementation. Hence EM is required for carrying out asset management.

In the decision making process of DH regulation it is necessary to handle large amounts of heat consumption data monitoring, changes in external factors affecting the needs in heat energy, data of forecasting heat energy demand et al.

For operational processing the data, that are used in the process of making decisions on EM of DH, it is necessary to use modern information technologies. The use of decision support systems (DSS) in the management of district heating system can increase the efficiency of decisions on heat energy demand regulation. The decisions on regulation of the heat energy demand need to be taken on the basis of the current heat energy demand and the forecasting data.

The Chapter will cover the studies on improving energy efficiency of district heating with application of information technologies. The models for information support of district heating management will be analyzed. The decision support system «HeatCAM» (DSS «HeatCAM») for making decision support in EM of DH system operation in the demand-driven mode will be presented. This system will provide the real-time monitoring and making decisions on regulation of heat consumption modes of the particular building or group of buildings on the consumer side.

REALIZATION OF ENERGY MANAGEMENT THROUGH THE IMPLEMENTATION OF ENERGY MANAGEMENT SYSTEM

There are a variety of EM definitions. It is the judicious and effective use of energy to maximize profits (minimize costs) and enhance competitive positions. Another definition for EM is the strategy of adjusting and optimizing energy, using systems and procedures so as to reduce energy requirements per unit of output while holding constant or reducing total costs of producing the output from these systems (Energy Management and Remote Energy Management, 2015).

To carry out the successful EM within the organisation it is necessary to use the EM standards providing an organisational framework to integrate energy efficiency into the management practices.

ISO 50001:2011 is the most general standard in the field of EM which provides a systematic approach to dealing with energy saving and specifies the requirements of an EM system of any organization. The purpose of this standard is to enable organizations to establish the systems and processes necessary to improve energy efficiency, use and consumption (ISO 50001:2011(en) Energy management systems – Requirements with guidance for use, 2015). This International Standard carries out according to the scheme which is illustrated in Figure 1.

The energy policy provides a statement of the high-level overview of management's intent that members of the organization should apply to their work activities and a framework for an organization to set energy objectives and targets and associated energy management action plans to improve its energy performance. The energy planning provides creation an energy profile of organization which is a useful tool to have a closer look at the detailed energy consumption status.

The ISO 50001:2011 EM system model is based on a concept of continuous improvement processes which is called the Plan - Do - Check - Act (PDCA) cycle (Sokovic, Pavletic, Kern Pipan, 2010). The EM actions in accordance with stages of PDCA cycle are presented in Table 1 (Huang, 2011).

The decisions on corrective/preventive action to manage the energy are made on the basis of the continuous energy monitoring, measurement and analysis.

Figure 1. The Energy Management System Model for Standard ISO 50001:2011

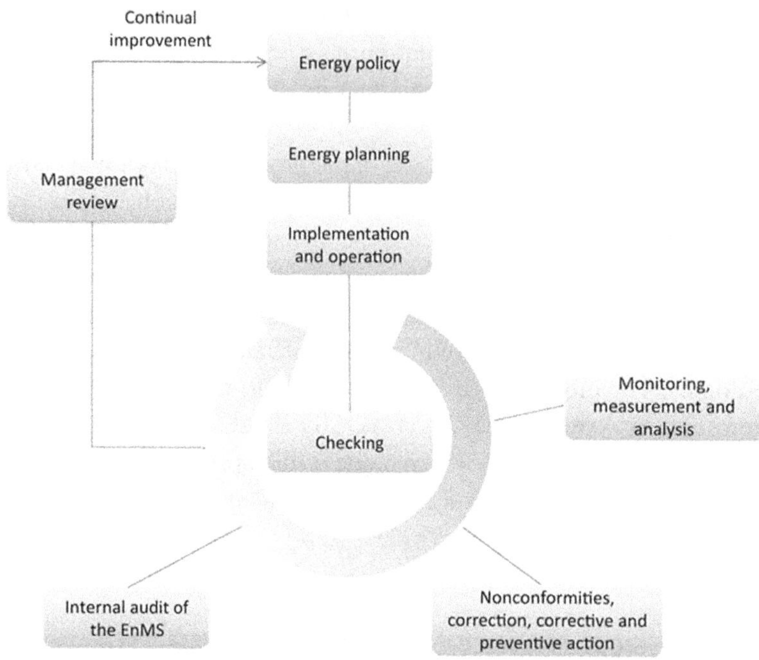

In order to automate the processing information in the implementation of the energy management system model the Energy Management System (EnMS) should be used. The EnMS is the system of computer-aided tools which allows achieving heat energy consumption more effective. Also the organization should establish internal communication in relation to its energy efficiency and the EnMS implementation. According to the ISO 50001, organizations are required to document information that describes the core elements of their EnMS.

To gauge the effectiveness of the EnMS and monitor the actual energy performance, the organization is required to perform regular checking the data of measurement and analysis. The management review assists the organization to achieve continual improvement and to assess the suitability, adequacy and

Table 1. The Energy Management System Model

	Management	Technical
Plan	Policy/goals/targets Resources	Energy review Energy baseline Energy performance indicators
Do	Training Communication Documentation Operational control	Design Energy purchasing
Check	Internal audit Corrective/preventive action	Monitoring Measurement Verifying action plans results
Act	Management review	Energy performance review

effectiveness of the EnMS. When conducting management review, the organization should also take into account the concept of continual improvement, which is achieved by evaluating the energy performance against the policies, objectives and targets (ISO 50001:2011(en) Energy management systems – Requirements with guidance for use, 2013).

The successful implementation of EM in district heating is an important element for every organization which allow to reduce its operational costs. Since the costs for consumed heat energy comprise the bulk of the operational costs, the problem of improving EM of DH system is urgent.

THE CURRENT STATE OF ENERGY MANAGEMENT IN DISTRICT HEATING

The trend of intelligent application of information technologies to manage DH system is rapidly growing. Intelligent DH is the combination of traditional district heating engineering and modern information and communication technologies (Johansson, 2010). This enables to integrate the functions of planning, monitoring and control of the heat supply in a single system.

Now there is a tendency of Smart Cities development and implementation. Smart City is the integrated approach which is based on using intelligent technologies to make cities more efficient, healthier, more sustainable, more livable and cleaner. It is designed to achieve the highest overall efficiency of the energy system. In the context of the Smart Cities design it is necessary to develop models and tools to support of making decisions including the decisions on EM of DH systems to guarantee a proper city development. The development and implementation of EM measures of DH systems must take into account their specific features, which are discussed below.

The Features of District Heating System

District Heating (DH) plays a significant role in the supply of low-carbon heating and hot tap water all over the world. DH technology is most common in the countries with cold climate where the heat demand is large. It is particularly widespread in North, Central and Eastern Europe. In such countries as Iceland, Finland and Sweden the high amount of residential consumers are connected to DH system. In cities like Copenhagen, Helsinki, Warsaw, Vilnius, Riga as much as 90% of residential heat demands are satisfied by DH.

The DH system is a complex technological and socio-economic system that produces heat energy, which is used to support comfort indoor air temperature in heated rooms. The consumers of the heat energy from the district heating systems in cities are usually houses, public sector buildings (organizations of education, science, culture and health), most of commercial non-industrial organization and some industrial enterprises.

The DH system has three main elements: the heat sources, the distribution system and the consumer (Figure 2). The heat carrier (hot water) circulates in pipes of the DH system and transmits the heat energy from the heat source to the consumer in heating systems. DH network as a rule is two-pipes, where there is one supply and one return pipeline.

Regardless of the DH system structure there are heat losses during the heat carrier transportation over the DH network. Heat losses in DH distribution depend on a number of factors such as the length of the system in relation to the heat load, type of insulation and temperature level. Normally, heat losses are in the range of 5-20% annually (Skagestad & Mildenstein). Due to heat losses it is not economically viable

Figure 2. The scheme of the heat transfer from producer to consumer under conditions of district heating

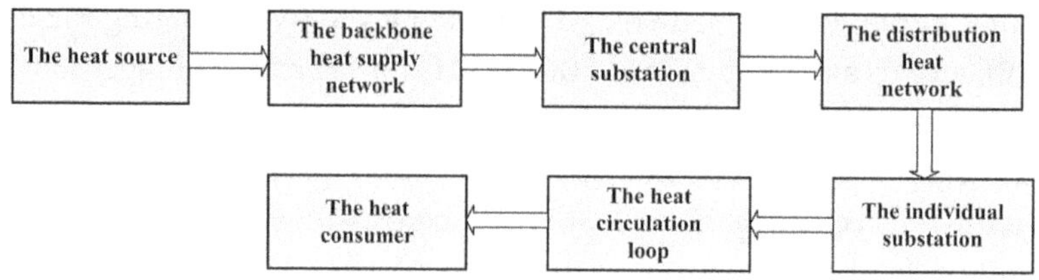

to transfer heat over long distances. Therefore, a DH system of every country is in practice composed of a large number of discrete DH networks.

The benefits of DH by comparison with other heating options are listed as follows (Euroheat & Power Task Force Customer Installations, 2011; the Guidance for district heating companies, 2012; Skagestad & Mildenstein):

- **Equipment and Maintenance:** The usage of DH at the customer end requires the installation equipment with compact size which is simple to run and maintain;
- **High Reliability and Efficiency:** DH systems are built with stand-by heating capacity to ensure that heat is always available at the central plant. The extensive range of primary fuels that can be utilized by DH makes heating supply very reliable. Renewable fuels can also be utilized.
- **Cost:** DH entails moderate investment costs and very low maintenance costs for the end consumer;
- **Environmental Issues:** Using renewable and other low carbon fuels in heat generation, combined heat and power generation from a single fuel source allows to reduce primary energy use greenhouse gas emissions;
- **Comfort for Consumers:** No need to purchase and handle fuels, detect problems in the heating system and eliminate them.

The efficient operation of DH system requires the implementation the EM measures. The management of DH systems is a complex task which requires optimising all the stages of the DH process.

The Organization Aspects of District Heating System Management

The EM of DH system is carried out on the side of DH company and on the consumer side. The DH Company management concentrates on the basic activity concerning operation, modernisation and development of the DH systems. It allows to optimize the heat sources and DH networks, reducing excessive energy losses.

Improving heat generation efficiency can substantially reduce fuel use and heat supply costs. Another way of improving heat supply efficiency is reduction of heat carrier losses from DH pipelines during its transmission.

The EM tasks on the stage of heat transmission are mainly connected with development, modernisation and operation of the DH network. Installation of pre-insulated pipes and optimisation of the heat carrier temperature reduces heat losses and water losses in DH networks.

There are two modes of DH system operation which determine the ability of management of DH operation on the consumer side: the generation-driven mode and the demand-driven mode. In generation-driven DH Systems, which have constant water flow, the heat supply is usually regulated through changes of water temperature supplied from the heat source. The generation-driven mode of DH system operation is not able to provide rational energy use on the consumer side. This causes the impossibility to ensure comfort in heated rooms.

The demand-driven mode, when customers are able to regulate heat use according to individual needs, is the most usable today. It allows the quality and quantity regulation of water supplied from heat source to network, depending on weather conditions and schedule of organization, and automatic control of flow rate. The implementation of the demand-driven mode involves install heat meters in buildings and devices of heating mode regulation (District Heating System Management Guide, 2015).

Information Technology as a Tool for Energy Management in District Heating

Information technology (IT) allows managing and sharing information between people within the EnMS system. Advanced IT can help to integrate all EM functions in one tool for proper EM implementation. IT are widely used in Smart Cities development.

Using IT allows to create an organizational framework and network decision-making processes in the cities in order to establish smart technologies. It adds intelligence to the EM process. This intelligence uses for the analysis of visible and actionable information on energy use and performance and brings new opportunities of data processing. Intelligent EM can use the information provided by new technologies in decision-making. It includes energy information management, fault detection and diagnostics (FDD) and measurement and verification of energy savings (M&V). These three key components compose the Intelligent Energy Management Triangle which is shown in Figure 3 (Granderson, Piette, Ghatikar, & Price, 2009).

Energy Information Management is the presentation of data related to a building's energy consumption that allows to monitor and track energy usage for more comprehensive energy management. FDD allows more efficient operation of DH systems and equipment by identifying inefficiencies or operational problems quickly. This all provides tools to attain the high-performance of DH system functioning. FDD may also display equipment failures or inefficiencies in financial terms in addition to signaling technical issues. M&V allows determining savings from an energy-efficiency retrofit project. It is important in helping to justify or fund energy improvement or equipment purchase decisions. Combining M&V with FDD enhances the ability to detect equipment problems that affect a building's energy usage (Nix, 2011).

Figure 3. The Intelligent Energy Management Triangle

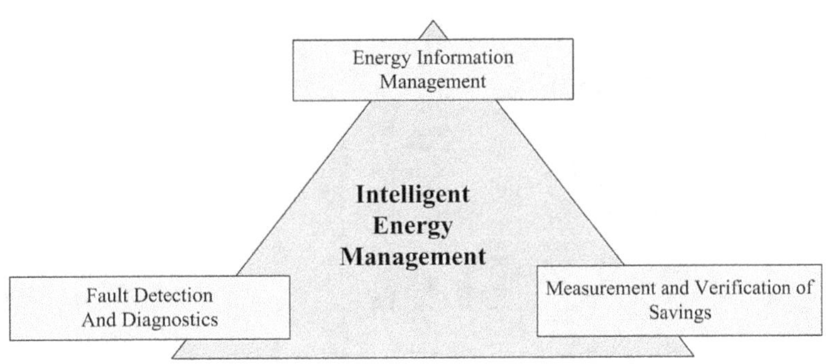

In general case the EM tool based on IT is energy information system (EIS). It is hardware and software complex which consist of monitor software, data acquisition hardware and communication systems used to monitor, store, analyze and display of energy data. Time-series data from meters, sensors, collected during the monitoring, and other data streams are used to energy consumption tracking and performing analysis for detection of abnormal energy consumption.

The EIS provides the follows functions (Granderson, Piette, Ghatikar, & Price, 2009):

- Web accessible hourly whole building data of energy consumption;
- Data collection, transmission, storage and security;
- Graphical/visualization capabilities;
- Weather, energy price signals, and demand response information;
- Energy, advanced and financial analysis;
- Remote control and management.

The EIS serves as the information link between persons who are involved in EM process in organization. The actors of EIS are facility and energy managers, owners and decision makers, operations and technicians (Figure 4).

The EIS operates with monitored data of DH system functioning and data from weather information services. The EIS users are then able to make decision and to take action based on this information. As follows the EIS transforms input data to actionable information. The modern EIS system is web-based. It provides a convenient user interface accessible through a standard web browser and/or mobile devices.

The EIS may be used for information support of EM on the stage of heat supply (for generation-driven regulation) and heat consumption (for demand-driven regulation). Depending on the specific software not every function of the EIS might be available. The comparison review of the top energy management software products is presented in Table 2. In the table are used such reductions, which denote separate functions of software:

Figure 4. The EIS Actors and Stages of Data Transformation

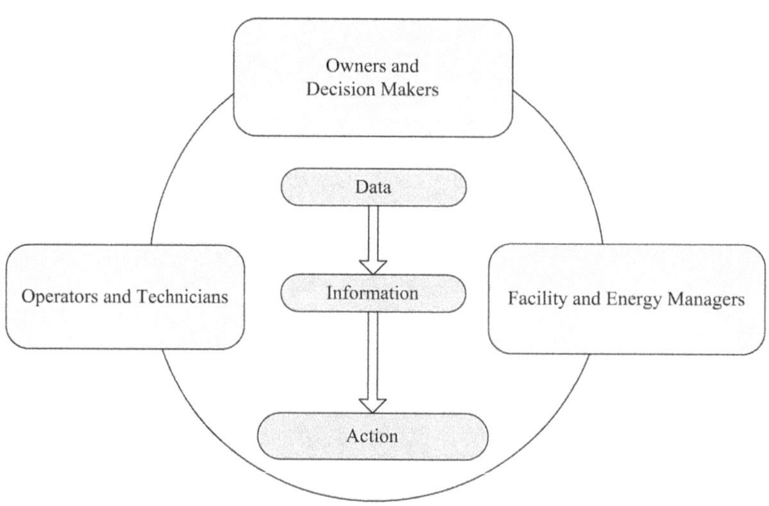

Table 2. The comparison review of software

Software	Heat Supply				Heat Consumption			
	1	2	3	4	1	2	3	4
The MiSO system (Mipro Company, 2015)	+	+	+	+				
The metering solutions of the Wattis energy management company (The Official Website of Wattics Company, 2015).					+	+	+	+
The ePortal software (The Official Website of ePortal: Energy Insight Software, 2015)					+	+	+	+
The Heat Monitoring Software of the ENMAT company (The Heat Monitoring Solution of ENMAT Software, 2015)					+			+
The software of the ENORO company (Enoro Solutions, 2015)					+		+	
The CyberVille Application Platform ((District Heating Network Monitoring, Control and Optimization with CyberVille IoE Application Platform, 2015)	+	+	+	+				
The UltraAssist software of the Landis+Gyr company (The Official Website of Landis+Gyr Company, 2015)					+			

1. Monitoring,
2. Control,
3. Data management and analysis (include forecasting),
4. Reporting.

All tools for EM in DH can be divided into two categories: tools which are provided complex solutions, and tools for realization of concrete tasks (heat metering devices, software for optimization, modeling of DH networks, etc.).

The vast majority of EM tools is commercial and expensive. This makes their usage impossible for low-profit and non-profit organizations. Moreover, most EM tools can work only with specialized heat meters, sensors etc. It causes additional costs using such EM tools.

Thus it is necessary to develop the EIS that would be available for public sector organizations which are non-profit. It should be based on standard data transmission protocols and work with any heat meter with digital output. It must provide online monitoring of the DH system condition and supporting of decisions making on implementation of energy saving measures.

The Analysis and Classification of Information Support Models in Energy Management Process

The EIS is based on models which provides information support to solve the EM tasks. The models for information support of EM in district heating can be classified in four categories (Wernstedt, Davidsson, & Johansson, 2015).

- **By Type, Microscopic or Macroscopic:** Microsimulation generally refers to detailed studies in both time and space of individual entities in the system. Macro-simulation refers to studies where the entities of the system are grouped together and modelled as larger entities.

- **By Approach, Physical or Black Box:** Physical models uses general priori knowledge of the nature of district heating systems, whereas black box models are based on relations determined from measured data.
- **By Method, Dynamic or Steady State:** Dynamic models calculates current state from previous states, whereas steady state models are time independent or assume steady state conditions.
- **By Usage, Design or Operation:** The purpose of design usage is to study different configurations of the DH system. Operational usage generally considers improvements of an existing system with respect to its economy or performance.

There are various models which are intended to different stages of DH process. Simulation models are used to study the heat distribution. Monitoring models are used to check the DH operation parameters on the stages of heat production and consumption. The optimization models are appointed to choose the best parameters of DH system structure and operation. Forecasting models are primarily designed to obtain the future data of the heat demand. The decision support models aid solving different tasks of making decisions on every stage of DH process. These models are described below.

The Simulation Modeling of District Heating Network

The simulation model is intended for research the dynamic properties of DH networks. It involves a full topology modeling and physical modeling of the DH network, taking into account individual pipes, dimensions, material properties etc. Such full models are computationally intensive when considering large DH systems. The DH network can be considered as a collection of nodes connected by directed edges (pipes). The topology model of DH network is complete description of the DH network structure by the graph theory. A graph contains information about pipes, nodes and the relation between them.

The DH network graph in a matrix data structure contains the incidence matrix and the loop matrix (Valdimarsson, 2014).

The dependence of the flow rate and the pressure is nonlinear. The main difference between methods of solving this problem lies in the selection of independent variables. The finite element method is widely used for fluid flow and heat transfer analysis (Gabrielaitiene, Kacianauskas, & Sunden, 2003).

The simulation model for hydraulic and thermal calculation may be presented as a system of equations which defines dependence between parameters of flow in DH network.

As output there are calculated:

- **For Each Branch:** Hydraulic resistance, mass flow, pressure and heat losses
- **For Each Point of Junction:** Pressure and temperature.

This data is used for further analysis and making decisions on regulation of the DH system operation. The scope of the thermal and hydraulic calculation is adjustment for the creation of optimal hydraulic regimes in DH systems, heat carrier distribution between consumers in strict accordance with the thermal load.

The Monitoring of District Heating System Operation

The monitoring in district heating performs checking the quality of DH system operation to detect the deflections from normal heat load mode, the corrosion of DH pipes and to determine the failures of installed equipment in DH plants and substations, etc.

There are several approaches for DH system monitoring. Earlier monitoring was performed by operators in set time intervals. But due to the rapid development of information technologies and perfecting of metering devices it is possible to control the DH system operation in real time. Now most common approaches for monitoring are web-based or agent-based. These techniques are allows to provide remote control of DH system and access to monitoring data in online regime.

In web-based monitoring it is appropriate to use GPS/GPRS wireless technologies for data acquisition. At present GPS is the only satellite system with sufficient availability and accuracy for remote monitoring which allows data transferring for long distances (Peulic, Dragicevic, & Jovanovic, 2013; Parfenenko, Shendryk, Nenja, & Vashchenko, 2014).

The agent-based monitoring technique realizes the decentralized approach to the DH management problem. The multi-agent system manages the operational decisions, e.g., reacts on shortages and performs resource redistribution. This approach to monitoring and control in DH system allows achieving coordination of production and distribution of heat energy (Wernstedt, 2003).

Thus the efficient monitoring performs checking the DH system operation through acquisition of the required data for their further processing in real time. The monitoring data are the basis for making decisions on energy management since they reflect the current state of DH system.

THE HEAT DEMAND FORECASTING

To analyse a large amount of the monitoring data the forecasting techniques are used. Primarily the forecasting techniques in district heating were served to determine the required volume of heat energy supplied from the heating source to the consumers. Now the consumer-oriented forecasting models of heat consumption are increasingly being used (Grzenda, 2012). It allows the regulation of heat energy demand on the consumer side, which reduces the consumption of the heat energy.

The heat demand forecasting is typically divided into short-term, mid-term and long-term according to which time horizons are considered. In recent years a large amount of the short-term techniques for solving the problems of forecasting in heating system has appeared (Buhari, Adamu, 2012). The major methods of heat demand forecasting are the regression analysis method and the method based on using artificial neural networks.

The regression method is used to establish relationships between the amount of the heat energy demand, temperature, flow rate and such parameters as the type of the building, the heated area and the time of day, the day of the week (working day or weekend), and period of the heating season (Chramcov, Baláte, & Princ, 2010; Komass, Sniders, & Laizans, 2012).

Now the forecasting models based on the artificial neural networks are widely used in district heating. The forecasting allows DH customers to plan their own heat consumption and possible local production in smart DH systems. The benefits of using forecasting models for DH producers are better production planning and ability of organization of customer profiles for different customer types (Kontu, 2013).

The Optimization in District Heating

Optimization tasks are solved to find solutions which are optimal or near-optimal with respect to some goals. In relation to district heating these goals may be to minimize cost on heat production, to minimize greenhouse gas emission or to maximize profit of heat generating company. Depending on how the optimization goals can be achieved there are the structural and the operational optimization of DH system.

In general, structural optimization is focused on determination of optimal DH network configuration. The basic factors for determination of the structure of DH system are economy and functionality of the tree pipe networks. In structural optimization the classical problem is how to select the optimal diameter of pipes in which a fluid is transported. In order to find the optimal solution the evolutionary algorithm may be applied (Weber, Marechal, & Favrat, 2006).

The definition of optimal diameter of the pipe is based on the fact that the costs of pipeline installation and heat losses increase strongly with diameter, while the pumping power drops rapidly. The dynamic simulation models of DH networks today are used for this optimization task. To solve this task, the models for energy system analysis such as EnergyPlan and RETScreen International models are also applicable (Hlebnikov, Dementjeva, & Siirde, 2009).

The optimal path of heat carrier with minimal transport costs may be defined by two methods of linear programming, the transport method and the simplex method. The mathematical model for defining the optimal tree path consists of the objective function of capitalized costs that is minimized (Jamsek, Dobersek, Goricanec, & Krope, 2010).

The operational optimization as a rule is intended to the determination of such DH network parameters as the supply temperature and the pump pressure. The temperature optimization models may be based on the SCADA data from the DH network. This constitutes the best baseline for the DH system optimization and encompasses operational changes and any fluctuations in DH system operation during weekends and vacation periods (Temperature Optimization saves millions for district heating, 2015).

In addition, an optimization of the district heating may be the focus on the optimization the existing pumps, and the possibility of implementing an alternative source of energy such as a biofuel system.

THE MAKING DECISIONS ON DISTRICT HEATING MANAGEMENT

The organisation of decision-making process in the management of DH systems aims at realization of corrective and preventive actions which are intended to reduce energy consumption and CO_2 emission. Moreover, the DH consumers must receive the heating services of high quality. Hence the decision making criteria on DH management are:

Social: To satisfy heat energy demand;
Economic: To reduce costs on heat energy production and distribution to consumer;
Environmental: To obtain carbon reduction.

The models of making decision may be single- or multi-criteria. In order to consider the criteria in the complex the models of multi-criteria decision making were developed (Bush & Bale, 2014).

The decision making process can be represented by the following stages:

- Definition of goal;
- Definition of actions;
- Establishment the evaluation criteria;
- Identification and analysis of solutions;
- Selection of the best solution.

The realization of these stages with high effectiveness requires using intelligent information technologies to create DSS. DSS are a class of information systems intended to help decision makers use the necessary data, documents, knowledge and/or models to complete decision process tasks.

The vast majority of modern DSS has user web interface which enables the managers of the district heating services to choose their preferred energy profile online. Thus, a decision on regulation of DH operation may be suggested at once by request of energy manager.

Decisions on regulation of DH operation are made under conditions of uncertainty. Application of fuzzy logic for making decision support provides a convenient framework that can deal with ambiguity or partial process knowledge (Musee, Lorenzen, Aldrich, 2006).

The DSS may be oriented on making decision on the heat generation side and on the consumer side. In making decisions the specific features of heat consumer should be taken into account. For example, most organizations are characterized by the heat energy demand fluctuations not only due to changing weather conditions but also due to changing of their working schedule.

THE DECISION SUPPORT SYSTEM «HEATCAM» FOR HEAT ENERGY MANAGEMENT ON THE CONSUMER SIDE

The DSS «HeatCAM» is a tool of the decision support for the EM of DH system on the consumer side. It is implemented through a three-level client-server architecture of the web-based applications.

The general architecture of the DSS «HeatCAM» is presented in Figure 5 as a UML diagram of components.

It includes the tools of data storage and data manipulation, a web-server, the client's browser, and a set of software modules that implement data processing. The software modules of the DSS «HeatCAM» are combined in subsystems, functionalities of which are listed in the Table 3.

Thus the functionality of the DSS «HeatCAM» is implemented as a set of software applications that are united into the monitoring, modeling, forecasting, mode determination and data visualization subsystems.

The DSS «HeatCAM» is used for EM of the DH system of the Sumy State University in the last three heating seasons. It provides the necessary data to make decisions on regulation of heating modes in dependence of changing weather conditions and working schedule of this university.

The using of the DSS «HeatCAM» can be done through a web-interface from any computer workstation that has access to the internet. The processing of the user's requests that are transmitted with the help of http protocol, is implemented by the web-server Apache. Data, the processing of which is done by the subsystems of the DSS «HeatCAM», are stored in the database and are managed by the DBMS MySQL.

The DSS «HeatCAM» provides monitoring of the the current state of heating mode, short-term forecasting of heat energy demand and also an information support of the decision making on heat energy management of the social and public sector buildings.

Figure 5. The general architecture of the DSS «HeatCAM»

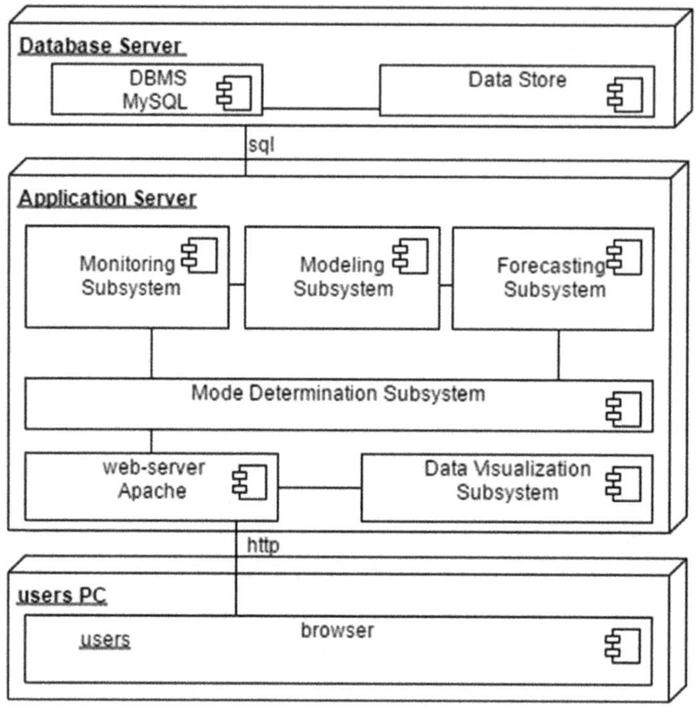

Table 3. The brief description of the DSS «HeatCAM» subsystems

Subsystem	A Brief Description	Input Data of the Subsystem	Output Data of the Subsystem
of monitoring	Data collection from heat meters, sensors and the weather forecast site, their validation and data store in the database. Calculations of the daily monitoring indicators that are based on the data, collected hourly.	Monitoring data collected from sensors, meters and the site of weather forecast.	Monitoring data of heating and the data of the forecast of the meteorological conditions collected daily and hourly
of forecasting	Definition of the forecasted data of heat demand	Data of the forecast of the meteorological conditions and data of the organization schedule for tomorrow, historical data of the daily heat demand	Forecasted value of the volume of the heat energy
of modeling	Carrying out of the numerical calculation of the flow in the network to determine the heat loss and the correction of the monitoring and forecasting data including heat losses	Data of monitoring and forecasting of the heat demand	Refined data of monitoring and forecasting of the heat demand
of mode determination	Determining of the current mode of heating and providing recommendations for management of the heating modes	Data about temperature of the environment, the amount of the heat energy consumed for the current day, the forecasted amount of heat energy	The current heating mode and data of heating modes management
of visualization	Implements the data display in the web-interface of the DSS	Data from the database that are requested by the user	Tables and graphs on the web-page

The group of people, that manages the heating of the organization of social and public sector online has an authorized access to the DSS «HeatCAM». Each user has an access to the needed components of the system.

The key functions of the DSS «HeatCAM» users are presented in Figure 6 as a UML use case diagram.

The administrator has full rights for the web-resource management. He carries out the distribution of the access rights to the web-system between the users, has the right to make and edit all monitoring data and to correct all the output data.

The operator is responsible for data collection from the installed sensors, meters of heat consumption and of the information resources about the weather conditions, and also for entering the collected data to a single database. In the DSS «HeatCAM» the role of the operator is performed by the software modules of the monitoring subsystem, so the monitoring is automated in real time. Data collection of the heat energy monitoring is carried out by a specialized terminal (Parfenenko, Okopnyi, & Nenja, 2012). The data collection of monitoring of the meteorological conditions of the environmental is done by using specialized software – parser that provides the data extract from the weather forecasting website (Shendryk & Vashchenko, 2011).

The role of the person, who makes decisions, usually fulfills the specialist, responsible for energy audits and energy efficiency. This person analyzes an actual level of heat demand and other indicators of functioning of the heating system and does the comparative analysis of the actual and the forecasted heat demand.

The expert on energy saving forms the base of knowledge and rules through which the determination of the current heating mode is carried out and the recommendations for setting the needed mode are given.

Figure 6. The use case diagram of DSS «HeatCAM»

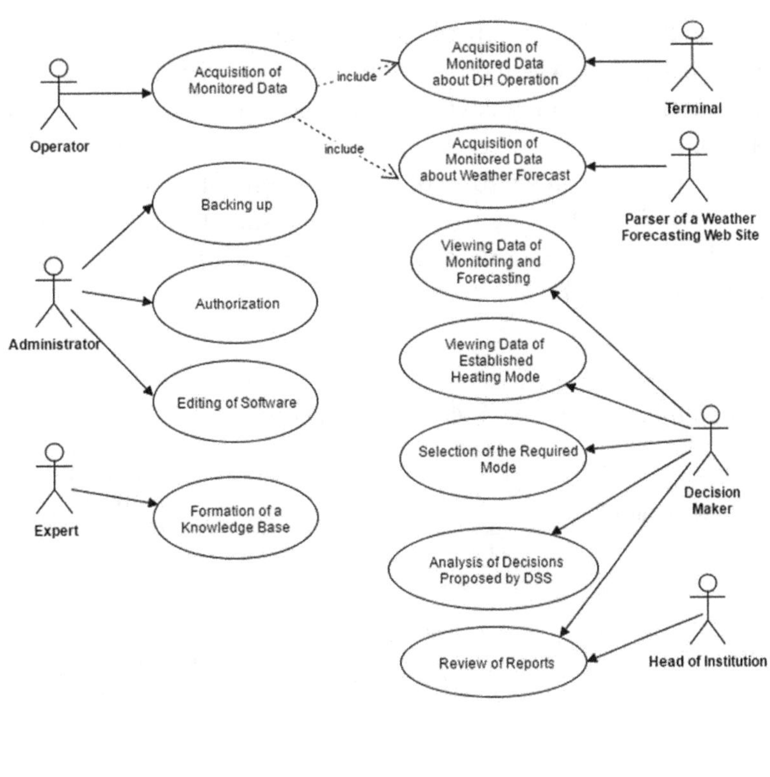

The person, who makes decisions, tracks the current heating mode and decides on the correction of the heating mode on the basis of the recommendations made by the DSS.

The head of the organization has the ability to view the reports on the DH system functioning and on the state of payment for the consumed heat.

The process of the DSS «HeatCAM» functionality can be represented as follows:

$$\mathrm{Proc} \rightarrow O \times H_c \times H_m \times S_c \times C_{ch} \times C_s \times D_s \times Z$$

where H_c – is a set of hardware maintenance, H_m – is a set of meters, from which the indicators are taken, S_c – is a set of software, C_{ch} – is a set of communication channels, C_s – is a set of communications signals, D_s – is a set of data on which the decisions are made, where O – is a set of people who make decisions ; Z – is a set of tasks that must be executed by the DSS «HeatCAM», which is represented in the form:

$$Z = Z_m \cup Z_s \cup Z_p \cup Z_f$$

where Z_m – is a set of monitoring tasks, Z_s – is a set of tasks of thermal network modeling, Z_p – is a set of heat consumption forecasting tasks, Z_f – is a set of tasks of decision formation.

These tasks are performed by the DSS «HeatCAM» subsystems that perform tasks of monitoring, forecasting and making decision support.

The Realization of Monitoring Tasks in the DSS «HEATCAM»

The information for the analysis of DH system of the organization in real time is given by the monitoring subsystem of the DSS «HeatCAM», which performs the following functions: data collection, presented in a specialized form from the accounting devices, their converting into a format suitable for the transmission over the communication channel; archiving of the information in the database; visualization of the results of the analysis of data in the form of charts, graphs, standardized reports for the decision support; providing the possibility of taking the monitoring results from the remote terminals via web-interface, regardless of the software platform. The general scheme of the monitoring process in the DSS «HeatCAM» is presented in Figure 7.

Figure 7. The scheme of monitoring process in the DSS «HeatCAM»

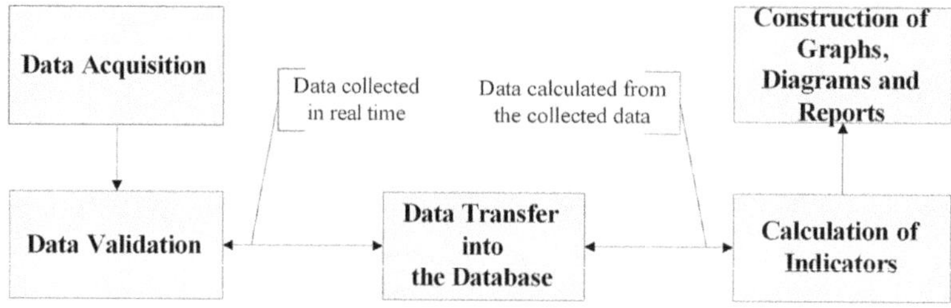

The formal mathematical description of the monitoring of the DH system of the social and public sector in the terms of the theory of sets is presented below. This model has the degree of detalization required for the making decision support.

The set of the monitoring parameters $M_p = M_{po} \cup M_{pi}$ consists of a subset M_{po} that contains parameters that are taken online and the subset M_{pi} containing daily monitoring parameters which are calculated based on the collected data online. The set $M_{po} = M_{ho} \cup M_{wo} \cup M_{wno}$ of the monitoring parameters collected online, consists of the subsets of parameters of the heating M_{ho}, the weather forecasting parameters M_{wo} and the set of temperature data of the environment M_{wno}.

The set of the heating parameters M_{ho} consist of the number of heat meter, the quantity of the consumed heat energy, Gcal; s the temperature of the coolant in the direct pipeline, ° C; the coolant temperature in the return pipeline, ° C; the temperature of the coolant in the pipeline after mixing, the pressure of the coolant in the direct pipeline kgf / cm^2; the pressure of the coolant in the return pipeline kgf / cm^2; the thermal capacity of the coolant in the direct pipeline Gcal / h; the thermal capacity in the return pipeline, Gcal / h; the volumetric flow rate of the coolant in the direct pipeline, m3 / h; the volumetric flow rate of the coolant in the return pipeline, m3 / h.

The set of monitoring of the weather forecast, carried out online, M_{wo} consists of the collected data of the meteorological conditions such as the air temperature, ° C; the wind speed, m / s; the atmospheric pressure, mm. Hg. c. and the air humidity, %.

The set of the monitoring parameters M_{pi} consists of a set of elements similar to the above described elements of the set M_{po} and contains daily data, which are calculated based on hourly data.

The data, collected during the monitoring, after their validation are stored in the database of the DSS «HeatCAM». The monitoring data are displayed on the web-page after the user's request in the form of tables and graphs and they are also used during the report formation for the consumed heat. These data are also used to forecast the heat demand of the buildings of social and public sector.

The Realization of Forecasting Tasks in the DSS «HEATCAM»

The short-term forecasting of the heat energy demand is intended for evaluation of the amount of consumed heat energy by certain time intervals (hour, day, etc.) in future. It allows to set dependence between the heat demand and the factors that affect its fluctuations.

There are two types of factors that affect the heat energy demand. The first type consists of meteorological factors, which determine such conditions as the ambient temperature, the air humidity etc. The second type of factors takes into account the social component which affects the fluctuation of heat energy demand. For public sector buildings the social component is determined by the working schedule of the organization. On weekends, there is no necessity to consume the same amount of heat energy as in the working days. Consequently, the heat energy demand for public sector buildings is changing with changes of these factors. So forecasting should be conducted considering of these features.

The forecasting subsystem of the DSS «HeatCAM» performs the determination of the daily heat energy demand based on archived monitoring data.

The data sets for forecasting include the following data for the previous heating seasons:

- The data of the daily building heat demand in accordance with the date of their monitoring and the type of the day of the week (the working day or the day off);
- The data of meteorological conditions (the ambient temperature, wind speed, the atmospheric pressure, and the air humidity) according to the date of their monitoring.

The technology of artificial neural networks to build the forecasting model was chosen. The process of the neural network model's building can be divided into three stages: the stage of data preparation, the stage of the structural synthesis and the stage of the parametric synthesis. These stages are performed until the forecasting error will not be reduced to the minimum value. The data preparation stage is required for the selection of monitoring data relevant to the normative heat supply modes and its validation. At the stage of the structural synthesis of the neural networks it is necessary to choose more appropriate type neural network architecture.

The following network architectures were used for the structural synthesis of neural network models:

- The nonliniar network (NIO) architecture:

$$\hat{y}(t) = F(x(t-1), x(t-2), ..., x(t-n));$$

- The nonlinear autoregressive network (NAR) architecture:

$$\hat{y}(t) = F(y(t-1), y(t-2), ..., y(t-n));$$

- The nonlinear autoregressive network with exogenous inputs (NARX) architecture:

$$\hat{y}(t) = F(y(t-1), y(t-2), ..., y(t-n), x(t-1), x(t-2), ..., x(t-n)).$$

The input vector of neural network models $x(t)$ consists of components:

- x_1 : The ambient temperature, C;
- x_2 : The wind speed, m/s;
- x_3 : The atmospheric pressure, mm. Hg. c.;
- x_4 : The air humidity, %;
- x_5 : The type of the day of the week (working day or weekend).

The output variable of the proposed neural network models is the daily heat energy demand $\hat{y}(t)$, Hcal. Thus the proposed models represent the dependence between the data about meteorological conditions, data of the organization schedule and the daily building heat demand.

At the stage of the parametric synthesis the training of the neural networks is performed. The procedure of training is the reflection of the monitoring data on the weights of the neural network to obtain the optimal forecasting heat demand data as the output signal of neural network.

The training of the neural network is implemented using Levenberg-Marquardt algorithm for changing the weights coefficients, until the forecasting error will be minimized (Hao Yu, & Wilamowski, 2011; Roweis, S., 2015).

The evaluation of forecasting accuracy of the proposed models has been carried out. The mean absolute percent error MAPE, the mean absolute percent error MAE and the root-mean-square error RMSE were calculated.

To evaluate the forecasting accuracy the data of the heat energy demand of one campus of the university which were collected during the heating season 2014-2015 have been used. The results of forecasting errors calculation are presented in Table 4.

The daily heat energy demand of the public sector building for which the forecasting errors have been calculated is from 20 to 50 Hcal per day. It depends on the weather conditions and on the working schedule of this organization. The value of RMSE and MAE errors doesn't exceed 2 Hcal that is less than 10% from the daily heat energy demand. Thus the obtained results proved the high accuracy of proposed forecasting models. Using the NARX model the highest forecasting accuracy in comparison with other studied models has been achieved.

The Realization of Making Decision Support in the DSS «HEATCAM»

The subsystem of mode determination of the DSS «HeatCAM» is designed to establish the current heating mode and to propose the decision on the regulation of the heating in order to establish the needed mode. In this subsystem the decision support models are used, that are built with the usage of the methods of the theory of fuzzy sets and fuzzy logic, which allows formalizing the expert knowledge on the heat energy management.

The model of the setting of the current mode of heating is presented as expression:

Table 4. The results of computational experiments to evaluate the forecasting accuracy of heat consumption

The Error	The Type of Neural Network	December 2014	January 2014	February 2015
RMSE, Gcal	NIO	6,15	1,74	0,89
	NAR	3,88	1,51	1,75
	NARX	2,19	0,88	0,66
MAE, Gcal	NIO	4,69	1,25	0,37
	NAR	1,78	0,88	1,4
	NARX	1,06	0,68	0,47
MAPE, %	NIO	14,95	4,28	1,48
	NAR	5,16	2,89	5,67
	NARX	3,17	2,34	1,87

$$R_c = \begin{cases} f(ZY1), \text{if } Zw \text{ is } Zw1 \\ f(ZY2), \text{if } Zw \text{ is } Zw2 \\ f(ZY3), \text{if } Zw \text{ is } Zw3 \\ f(ZY4), \text{if } Zw \text{ is } Zw4 \end{cases}$$

where R_c - is the heating mode; ZY1, ZY2, ZY3, ZY4 – are linguistic variables that determine the amount of the consumed heat energy at different temperature conditions, Zw – is a linguistic variable that characterizes the state of temperature conditions of the environment; Zw1, Zw2, Zw3, Zw4 – are terms of linguistic variable Zw.

According to this model, initially we determine the state of the temperature conditions of the environment as a function of the average daily air temperature Zw = F (Xw), where the output variable Zw sets the affiliation of the temperature of the environment to a particular class, Xw – air temperature of the environment, °C.

Depending on the resulting variable Zw, the value of the heating mode *Rc* is determined with the help of the linguistic variables ZY1, ZY2, ZY3, ZY4. The realization of the models of fuzzy inference decision to determine the state of temperature conditions and the current heating mode is carried out according to the algorithm of Sugeno (Sugeno Fuzzy Models, 2015).

The result of the work of the model of the setting of the current mode is the determination of the current heating mode among the possible values "Below necessary", "Acceptable", "Optimal", and "Higher than optimal". The determination of the current heating mode allows you to monitor the compliance with the appropriate temperature conditions in the heated rooms and promptly detect deviations from the planned mode.

The model of the formation decision on heating mode regulation is represented as the function Z = F (ZY, BY), where Z – is the control parameter data, ZY – is the variable, characterizing the current heating mode, BY- is the variable, characterizing the required heating mode.

For fuzzy inference decision to control the current heating mode the Mamdani algorithm is selected (Mamdani Fuzzy Models, 2015).

The result of the model of fuzzy inference decision is a recommendation to regulate the current mode to the necessary one among the possible values "Acceptable" and "Optimum". The example of the rule for fuzzy inference decisions on the heating mode regulation is given in the Table 5.

The energy manager decides to regulate the heating mode based on the proposed by the DSS «HeatCAM» recommendations. The use of the DSS «HeatCAM» allows the energy manager to react promptly to the changing of the influencing factors, according to the need of the organization in the heat energy,

Table 5. The example of the formal fuzzy inference rule for the making decision on control of heating mode

Prerequisite	ZY is ZYM1, BY is BYM1
The formulation of the rule	If ZY is ZYM1 and BY is BYM1 then Z = NC;
Interpretation	If currently the mode "Acceptable mode1" is set and tomorrow this "Acceptable mode1" will be needed, then the change of the regulatory option is "Leave unchanged"

which ensures the compliance of the comfortable temperature conditions in the heated rooms and avoids excess heat consumption.

Thus the DSS «HeatCAM» provides decision support for managing centralized heating of the organizations of social and public sector on the basis of monitoring data, forecasting data of the heat demand of the building and also using expert knowledge to regulate the heating mode.

CONCLUSION

This chapter studies an urgent problem of improving energy efficiency of EM process as one of the stages of asset management for any organizations. It may be achieved through the application of information technologies.

The current state of EM in DH and main EM standards were explored. The analysis of the existing models and tools which can be used for information support of EM process was carried out.

The DSS «HeatCAM» for making decision support on regulation of district heating on the consumer side was presented as a tool which is intended for the information support of the energy management stages in accordance with ISO 50001:2011. The main functions of DSS «HeatCAM» are monitoring of DH operation, short-term forecasting of heat energy demand, DH simulation and decision support on regulation of heating modes. The application of this system allows determining the required amount of heat energy for concrete building. So, it is possible to achieve energy conservation through the elimination excess heat energy consumption. Hence the DSS helps decision-makers to achieve the primary goal of EM on the consumer side which is to reduce energy consumption. Also it enables to evaluate the effectiveness of taken decisions. The obtained research results will be used to development of the EM in DH systems at the Smart Cities implementation.

Future research direction is related to taking into account the dynamic aspect of DH system functioning in EM of DH system operation. This enables to make decisions on regulation of DH system operation based on the current state of DH system, which increases the efficiency of decisions.

REFERENCES

Buhari, M., & Adamu, S. (2012). Short-Term Load Forecasting Using Artificial Neural Network. *Proceedings of the International MultiConference of Engineers and Computer Scientist.* Retrieved from http://www.iaeng.org/publication/IMECS2012/IMECS2012_pp83-88.pdf

Bush, R. E., & Bale, C. S. E. (2014). Realising the Social Benefits of District Heating Through Strategic Planning. *Proceeding of the 14th International Symposium on District Heating and Cooling*, Stockholm, Sweden. Retrieved from http://www.svenskfjarrvarme.se/Global/Konferenser/DHC14/Proceedings/_2.2%20Ruth%20Bush%20REALISING%20THE%20SOCIAL%20BENEFITS%20OF%20DISTRICT%20HEATING%20THROUGH%20STRATEGIC%20PLANNING.pdf

Chramcov, B. Balâte, Ja., Princ, M. (2010). Heat Demand Forecasting for Concrete District Heating System. International Journal of Mathematical Models and Methods in Applied Sciences, 4(4), 231-239.

District Heating Network Monitoring. Control and Optimization with CyberVille IoE Application Platform. (n. d.). Retrieved from http://www.iiconsortium.org/case-studies/Cyberlightning_Fortum_Case_Study.pdf

District Heating System Management Guide . (n. d.). Retrieved from http://projects.bre.co.uk/DHCAN/pdf/Management.pdf

Energy Management and Remote Energy Management . (n. d.). Retrieved from http://www.power-protection.co.uk/energy-management-and-remote-energy-management/

Enoro Solutions . (n. d.). Retrieved from http://www.enoro.com/#!solutions/component_74511

Euroheat & Power Task Force Customer Installations . (2011). District Heating in Buildings. Retrieved from http://www.euroheat.org/Files/Filer/documents/Publications/District%20Heating%20in%20buildings_final.pdf

Gabrielaitiene, I., Kacianauskas, R., & Sunden, B. (2003). Thermo-hydraulic Finite Element Modelling of District Heating Network by the Uncoupled Approach. *Journal of Civil Engineering and Management*, *IX*(3), 153–162. doi:10.1080/13923730.2003.10531321

Granderson, J., Piette, M. A., Ghatikar, G., & Price, P. (2009) Building Energy Information Systems: State of the Technology and User Case Studies. Retrieved from http://eis.lbl.gov/pubs/lbnl-2899e.pdf

Grzenda, M. (2012). Consumer-oriented heat consumption prediction. *Control and Cybernetics*, *41*(1), 213–240.

Hao, Yu., Hao, Yu., & Wilamowski, B. M. (2011). Levenberg–Marquardt Training. In Irwin J. David (Ed.), The Industrial Electronic Handbook: Intelligent Systems. USA: Taylor & Francis Group, CRC Press.

Hlebnikov, A., Dementjeva, N., & Siirde, A. (2009). Optimization of Narva District Heating Network and Analysis of Competitiveness of Oil Shale CHP Building in Narva. *Oil Shale*, *26*(3), 269–282. doi:10.3176/oil.2009.3S.09

Huang, E. G. T. (2011) Understanding the Requirements of the Energy Management System Certification. Retrieved from http://www.sgs.com/~/media/Global/Documents/White%20Papers/sgs-energy-management-whitepaper-en-11.ashx

Infor EAM Energy Performance Management (n. d.). Retrieved from http://www.infor.com/product-summary/eam/energy-performance-management/

ISO 50001 : 2011 (en) Energy management systems – Requirements with guidance for use . (n. d.). Retrieved from https://www.iso.org/obp/ui/#iso:std:iso:50001:ed-1:v1:en

Jamsek, M., Dobersek, D., Goricanec, D., Krope, J. (2010). Determination of Optimal District Heating Pipe Network Configuration. *WSEAS Transactions on Fluid Mechanics*, 3(5), 167-174

Johansson, C. (2010). Towards Intelligent District Heating. Retrieved from https://www.diva-portal.org/smash/get/diva2:833722/FULLTEXT01.pdf

Komass, T., Sniders, A., & Laizans, A. (2012) Statistical Analysis of Municipal Heat Supply.*Proceedings of 11th International Scientific Conference* (Vol. 11, pp. 1 – 8).

Kontu, K. (2013). Forecasting district heating consumption based on customer measurements. Report on the 36th Euroheat & Power Congress. Retrieved from http://www.euroheat.org/Files/Filer/Presentations/20130527_EHP_Congress/130527_1600_1730_Pl_6_Kaisa%20Kontu.pdf

Mamdani Fuzzy Models . (n. d.). Retrieved from http://www.bindichen.co.uk/post/AI/mamdani-fuzzy-model.html

Mipro Company: Water and Energy Management Products and Services . (n. d.). Retrieved from http://www.mipro.fi/eng/businesslines/waterandenergymanagement/productsandservices

Musee, N., Lorenzen, L., & Aldrich, C. (2006). Decision support for evaluating energy demand in vinification processes using fuzzy sets theory. *Journal of Energy in Southern Africa*, *17*(4), 4–18.

Nix, O. (2011). Building Information Technology and Management. Retrieved from http://www.institutebe.com/InstituteBE/media/Library/Resources/Smart%20Grid_Smart%20Building/Issue_Brief---Building_Info_Tech_and_Mgmnt.pdf

Parfenenko, Yu., Okopnyi, R., & Nenja, V. (2012). Creation of Control Instruments for Buildings' Heating. *The Journal of the National Technical University NTU KPI. Information Systems and Networks*, CCIS (Vol. 465, pp. 93–97).

Parfenenko, Yu., Shendryk, V., Nenja, V., & Vashchenko, S. (2014). Information System for Monitoring and Forecast of Building Heat Consumption. Communications in Computer and Information Science, 465 1-11

Peulic, A., Dragicevic, S., Jovanovic, Z., & Krneta, R. (2013). Flexible GPS/GPRS Based System for Parameters Monitoring in the District Heating System. *International Journal of Computers, Communications & Control*, *8*(1), 105–110. doi:10.15837/ijccc.2013.1.174

Reducing your ICT energy costs (n. d.). Retrieved from https://www.jisc.ac.uk/guides/reducing-your-ict-energy-costs

Roweis, S. (n. d.). Levenberg-Marquardt Optimization. Retrieved from https://www.cs.nyu.edu/~roweis/notes/lm.pdf

Shendryk, V. V., & Vashchenko, S. M. (2011). The system of Data Collection, Allocation and Analysis. *Visnyk of the National University Lviv Polytechnic*, *715*, 343–353.

Skagestad, B., & Mildenstein, P. District Heating and Cooling Connection Handbook. Retrieved from http://www.districtenergy.org/assets/CDEA/Best-Practice/IEA-District-Heating-and-Cooling-Connection-Handbook.pdf

Sokovic, M., Pavletic, D., & Kern Pipan, K. (2010). Quality Improvement Methodologies – PDCA Cycle, RADAR Matrix, DMAIC and DFSS. *Journal of Achievements in Materials and Manufacturing Engineering*, *43* (1), 476-483.

Sugeno Fuzzy Models . (n. d.). Retrieved from http://www.bindichen.co.uk/post/AI/takagi-sugeno-fuzzy-model.html

Temperature Optimization saves millions for district heating. (n. d.). Retrieved from http://www.global-download.schneider-electric.com/mainRepository/EDMS_CORP7.nsf/69f5d72c7a0cf811c12573d800389503/a9c3fb166094c67b85257a78005c3f20/$FILE/TCC-Hjoerring-eng.pdf

The Guidance for district heating companies. (2012). The environmental benefits of district heating: using the new Ecoheat4cities label Guidance for district heating companies. Retrieved from http://www.bre.co.uk/filelibrary/rpts/ecoheat4cities/Ecoheat4Cities_WP4_Guidance_for_companies.pdf

The Heat Monitoring Solution of ENMAT Software. (n. d.). Retrieved from http://www.en-mat.com/solutions/heat-monitoring

The Official Website of ePortal: Energy Insight Software. (n. d.). Retrieved from http://eportal.eu/en/

The Official Website of Landis+Gyr Company. (n. d.). Retrieved from http://www.landisgyr.no/

The Official Website of Wattics Company. (n. d.). Retrieved from http://www.eltec-petrol.com/home/

Valdimarsson, P. (2014). District Heat Distribution Networks. Retrieved from http://www.os.is/gogn/unu-gtp-sc/UNU-GTP-SC-18-27.pdf

Weber, C., Marechal, F., & Favrat, D. (2006). Design and Optimization of District Energy Systems. *Proceedings of the 10th International Symposium on District Heating and Cooling*, Lausanne, Switzerland. Retrieved from http://www.lsta.lt/files/events/16_weber.pdf

Wernstedt, F. (2003). Multi-Agent Systems for District Heating Management. Retrieved from http://www.bth.se/fou/forskinfo.nsf/0/d613f6536b050fa6c1256e2800391f91/$file/Lic.pdf

Wernstedt, F., Davidsson, P., & Johansson, C. (n. d.). Simulation of District Heating Systems for Evaluation of Real-time Control Strategies. Retrieved from http://cj.noda.se/papers/ESMC-Wernstedt-DavidssonJohansson.pdf

Williams, W. Asset Management Perspective (n. d.). Retrieved from http://bv.com/Home/news/solutions/Expert-Perspectives/realizing-value-with-iso-55001

KEY TERMS AND DEFINITIONS

Decision Support System: An interactive system, that can produce data and information, in some cases, contribute to the understanding relating to this subject area in order to provide useful assistance in solving complex and ill-defined problems.

District Heating: An economic activity related to supplying heat produced centrally in one or several locations to a non-restricted number of customers

Energy Management: Energy management is the proactive, organized and systematic coordination of procurement, conversion, distribution and use of energy to meet the requirements, taking into account environmental and economic objectives.

Energy Management System: A system of computer-aided tools used by operators of electric utility grids to monitor, control, and optimize the performance of the generation and/or transmission system.

Energy Information System: The software, data acquisition hardware, and communication systems used to store, analyze, and display building energy data.

Forecasting: The use of historic data to determine the direction of future trends.

Information System: An integrated set of components for collecting, storing, and processing data and for delivering information, knowledge, and digital products.

Monitoring: An act of observing something to regulate and control processes in observed system.

Chapter 15
Risks and Uncertainties in the Planning Phase of Offshore Wind Projects

Jannes van der Wal
University of Groningen, The Netherlands

Peter Eecen
ECN, The Netherlands

Jasper Veldman
University of Groningen, The Netherlands

ABSTRACT

Megaprojects are large and complex projects that entail multi-actor management, non-standard technology and processes. This chapter aims to explore offshore wind projects (OWPs) as megaprojects, particular in the planning phase. Based on interviews with 26 experts from a variety of backgrounds in the offshore wind industry in The Netherlands, the risks and uncertainties in the planning phase of OWPs and key factors in the decision making process are explored. A framework is presented that depicts the planning phase of an OWP, as well as ten risks and seven uncertainties that are most common in an OWP. The role of the government and the project structure are further highlighted. The findings of this research allow practitioners to gain a better overview of the planning process of an OWP and can help to improve asset management decision making.

INTRODUCTION

In 2014 the total capacity of wind energy grew with 10.2% in the European Union (Eurobserver, 2014). This rapid growth is driven by the ambitious goals that are set by a large group of countries to increase the use of wind energy by 2020. EWEA (2011) predicts that 14% of the total European electricity demand can be covered with wind power in 2030. The largest part of that wind energy will come from offshore wind farms (EWEA, 2009). Although investment costs are considerably higher compared to

DOI: 10.4018/978-1-5225-0651-5.ch015

onshore wind farms, offshore wind farms generate a greater amount of electricity due to higher wind speeds above sea (Bilgili, Yasarn & Simsek, 2010). The rapid development and deployment of this type of engineering construction is risky, which underlines a strong need for building experience, and a dedicated focus on innovation to optimize the delivery of this type of renewable energy through engineering constructions (Koch, 2012).

Offshore wind projects (OWPs) are frequently described as large and complex projects, and are often referred to as 'megaprojects'. Flyvbjerg, Bruzelius and Rothengatter (2003) describe a megaproject as a project that consists of complex engineering constructions, complex time schedules, high building costs and complicated performance measurements. Megaprojects cover all phases of the asset lifecycle until operations and maintenance, and include the delivery of, for instance, bridges, power plants and public transportation projects (Priemus, Flyvbjerg & Wee, 2008). Due to its size and complexity megaprojects often lead to cost and schedule overruns and are sometimes not finished at all (Flyvbjerg et al., 2003). Although megaprojects become more common, it seems that in practice they are still treated as standalone projects so that the knowledge and lessons learned from one project to another does not take place regularly. Successful megaprojects are therefore copied frequently and implemented in other countries with only minor adaptations to local or project-specific situations (Priemus et al., 2008). For these kinds of reasons, PM Network (2014) stresses the importance of understanding and standardizing project management practices. However, anecdotal evidence suggests that there is a lack of knowledge transfer between megaproject stakeholders after project delivery. Consequently, knowledge is available in the industry but is not generally applied. In the literature, research on standardized project management practice or procedures that can be applied to OWPs or megaprojects is lacking. Therefore, exploratory research into these issues seems essential.

The success of asset delivery is generally measured against the well-known performance measures of project management: cost, time and quality (Cooke-Davies, 2002; Thi & Swierczek, 2010; Lopez del Puerto & Shane, 2014). The first offshore wind turbine was installed about 25 years ago. Since then offshore wind has been frequently described in research, focusing mostly on cost and quality aspects of installation, construction logistics, operations and maintenance (O&M) or commissioning. Investigations into the time aspect of these kinds of megaprojects are rare. Proper project planning is crucial for safeguarding all three performance dimensions (Zwikael & Globerson, 2006). Planning provides a project with the bigger picture of deadlines, allowing to distinguish between short-term and long-term goals over time for the sake of project control. An OWP involves a series of reciprocal actions, decision making and phases that include planning with the authorities, engineers, the public and other stakeholders. Each phase in an OWP can influence successful project deployment as the interaction and decision making between the different stakeholders create complexity, uncertainties and risks (Koch, 2012). Puddicombe (2006) highlights that planning is nothing more than an ongoing process of assessing, restoring and preventing uncertainties and risks in the project. Alessandri, Ford, Lander, Leggio and Taylor (2004) state that managers need to address the critical nature of risk and uncertainty in the decision making process of a project. Without identifying and assessing the risks and uncertainties, decisions made for the project are likely to be sub-optimal (Alessandri et al., 2004). Zwikael and Sadeh (2007) add to Alessandri et al. (2004) and state that risks must be managed throughout the entire lifecycle of the project; starting with the planning phase, when risks must be identified and analyzed. According to Giezen (2012) a better overview of the risks and uncertainties will result in making a project and planning more manageable.

Surprisingly little research has been conducted on megaprojects in general or OWPs when it comes to planning, risks and uncertainties. It is expected that offshore wind energy will play a significant role in the future due to the increasing importance of sustainability and renewable energy (Kuijk, 2013). The number and size of initiated OWPs will increase over the years. As the projects get bigger, they are getting more complex which requires increased attention for decision making processes. Hence an increasing number of project managers (current and future) could encounter difficulties in managing the complexities of an OWP. Identifying the main risks and uncertainties in OWPs, therefore, is crucial as this issue receives little attention in the literature.

In light of the increasing important of offshore wind assets, this chapter aims to explore offshore wind projects (OWPs) as megaprojects, in particular the planning phase. Based on interviews with 26 experts from a variety of backgrounds in the offshore wind industry in The Netherlands, the risks and uncertainties in the planning phase of OWPs and key factors in the decision making process are explored. It is expected that the results will be theoretically and practically relevant for the offshore wind sector, as well as for megaprojects in general.

This chapter is organized as follows. The next section provides a theoretical framework on megaprojects, planning, risks and uncertainties. It concludes with a set of research questions. After that the methodology used for this research is given, followed by the presentation of the results. The results are discussed. In the final section conclusions follow, along with this study's limitations, implications, and directions for future research.

THEORETICAL FRAMEWORK AND RESEARCH QUESTIONS

Offshore Wind Projects as Megaprojects

Flyvbjerg (2009; 2011) defines a megaproject as an engineering construction project that contains complex interfaces, complex decision making and planning, a lot of risk, non-standard technology and design, a multi-actor management process with conflicting interests and costs more than hundred million dollars. According to this definition, delivery of OWPs can clearly be considered as a megaproject. As the offshore wind industry still needs to build experience and expertise OWPs show that associated companies and stakeholders frequently act independent from each other, using their own working methods (Edwards, 2011). Still, Edwards (2011) describes that a level of synergy is present between the offshore oil and gas industry and offshore wind. From a research perspective, however, previous literature about OWPs is often written from a quantitative approach (e.g., maintenance) and rarely from a qualitative approach, investigating, for instance, initiation, planning or permitting procedures.

Main Phases of Offshore Wind Projects

Gerdes, Tiedemann and Zeelenberg (2005) identified seven main phases for the realization of OWPs. See Table 1.

The processes and phases described in Table 1 are outlined as sequential steps. It must be noted though that in practice these phases and processes may run parallel to a certain degree. No literature exists on whether the phases depicted here are valid and what the effect of skipping tasks would be. Adler (1995) identified that the more complex a project is, the more difficult it is to prepare and impose a realistic structure with flexibility to the limitations of the pre-planning of a project. Thus, as more actors get involved in a project, the more difficult it becomes to make decisions, regulate and organize all interests.

Table 1. Phases in an offshore wind project

Phase	Description
Pre-planning	Pre-feasibility study, development of strategies and the project structure.
Detailed planning	Project approval procedure, site investigation, functional requirements of main elements, planning of internal controlling system, financing and insurance arrangements, tender process and the master plan.
Contracting / Financial Investment Decision	
Engineering, testing	Detailed engineering and planning, pre-testing and training.
Production & procurement	Production of elements, quality assurance and control, factory acceptance tests, interface and workflow management and transport to logistic center.
Installation & commissioning	Site preparation, pre-assembly, installation of the foundation, installation of wind turbines, installation of electrical infrastructure offshore and onshore, commissioning of supervisory control and data acquisition systems.
Full operation	Service, maintenance and environmental monitoring.
Re-powering	Replacing old with new technology that, either have a greater capacity or have more efficiency resulting in an increase of generated power (Power Partners, 2009).
Dismantling	Better referable as decommissioning; what needs to happen with the wind turbines when they have reached the end of its useful life (Ferrell & DeVuyst, 2013).

(Adjusted and summarized from Gerdes et al., 2005: 138)

This may provide an explanation why it is nearly impossible to stay within schedule. It also indicates that research into the planning phase is required in order to gain knowledge about this aspect and improve it for the future to support and improve the decision making process for an OWP.

Planning

Project management is the process of planning activities, organizing, directing and controlling resources, procedures and protocols to achieve unique goals for a project within certain time, cost and quality constraints (PMI, 2014; Lechler, Edington & Gao, 2012). Time, costs and quality are of equal importance in a project and need to be fulfilled to deliver the right scope of work (Lopez del Puerto & Shane, 2014). In the literature the contribution of these three dimensions has widely been discussed. Planning provides control over the main steps and activities that need to be performed within a project, and can have a significant contribution to the success and progress of a project (Johnson, Boucher & Robinson, 2001; Megavind, 2010). Planning provides an overview of the complete picture in terms of deadlines, and distinguishes short-term and long-term goals. Brugaard Villmo (2012) emphasizes that the assumptions made in the planning phase are essential to future outcomes and decisions, since little can be adjusted regarding these assumptions and estimations when the implementation starts. Gerdes et al. (2005) also provided a more detailed description of the planning phase, using a subdivision between pre- and detailed planning. See Table 2.

Pre-planning is essential to limit the potential for later delays and cost overruns (Johansen and Wilson, 2006). Whittaker (1999) identified a strong correlation between budget overrun and schedule overrun, indicating the relevance of planning. Although planning is not directly associated with costs, it can make a huge difference when not enough resources are allocated (Puddicombe, 2006). Duncan and Gorsha (1983) argue that project planning is needed to overcome under-costing, overspending and

Table 2. Detail description of the first two phases

Phase	Description
Pre-planning	Pre-feasibility study of: 1. Stakeholder involvement, 2. Technology to use, 3. Supply chain management, 4. Logistics, 5. Economic assessment, 6. Environmental and public impact; Development of strategies for financing, media, stakeholder involvement and approval; Project structure.
Detailed planning	Project approval procedure for the grid connection and the project itself; Site investigation of the wind speed and direction, the (ocean)graphical, chemical, geological and biological influence; Functional requirements of main elements determine the infrastructure, logistics and HSE; Planning of internal controlling system regarding the: 1. Key performance indicators, 2. Quality assurance, 3. Factory acceptance tests and 4. Reporting systems; Tender process for the preparation of documents, elaboration on proposals and negotiations with subcontractors; Financing and insurance arrangements for the financing of the entire project.

Given by Gerdes et al. (2005: 138)

late completion. Although a detailed planning at the beginning of a project does not guarantee success, a lack of planning will most probably result in guaranteed project failure (Dvir, Raz & Shenhar, 2003).

Risks and Uncertainties

Puddicombe (2006) and Kezner (2009) state that planning is an ongoing process of assessing the uncertainties, risks and trust of a project. Novel projects with a long duration, like megaprojects, are commonly plagued with fundamentally unforeseeable events and unknown interactions within the project (Loch & Pich, 2002). The variables that are unknown in advance by the planner involve uncertainty and risk (NOAA, 2014). Risk identification is concerned with recognition of factors that may make the project plan obsolete or suboptimal (Loch, DeMeyer & Pich, 2006). Risk identification can be performed by expressing the impact, the probability and the type (financial, time, quality) of risk (Loch et al., 2006). Uncertainty describes any situation that people are not completely sure about (NOAA, 2014). Uncertainty estimations are often based on the experience of the planner. However, some things are fundamentally unknowable and occur randomly (NOAA, 2014). As it is impossible to identify and treat all uncertainties at the beginning of a project, uncertainties are inevitable, regardless of how much information is gathered before the project starts (Hubbard, 2007).

Megaprojects have a high degree of complexity, especially in the beginning. Accumulating risks result in a higher likelihood of failure (Gerdes et al., 2005). The amount of risk influences the level of planning effectiveness (Zwikael et al., 2014). Within OWPs most stakeholders are not able to accept the accumulated risk alone. This results in combinations of firms and public institutions executing the project. The unique nature of an OWP results in entanglement of risk and uncertainty, since a lot activities have to start from scratch with slim resources to learn from previous experiences. This causes a threat for the optimization of the project, which is for a large part done in the planning phase. It is the primary role of the project planner to identify risks and uncertainties in the project parts that have the greatest potential to cause concern (NOAA, 2014). It is important for the planner to acknowledge the existence of the risks and uncertainties and plan and incorporate them into the analysis (NOAA, 2014).

RESEARCH QUESTIONS

Although every megaproject is unique, Giezen (2012) called for more research on megaprojects. The focus of this chapter is on Offshore Wind Projects (OWPs) as megaprojects. Current literature regarding OWPs fails to address the planning aspect, in particular the risks and uncertainties occurring in the planning of OWPs related to planning. The following research question is addressed here:

What are the risks and uncertainties in the planning phase of an offshore wind project?

This research aims at gaining better insight into the planning of an OWP and the identification of risks and uncertainties linked to the planning phase. Depicting the planning process in a framework could provide practitioners with a better overview of how, what and when phases, tasks and activities need to be performed. In this chapter, such a framework is developed first. As a second step, risks and uncertainties are identified and linked to the framework. In order to answer this main research question, supporting research questions are formulated as follows:

1. *What is the structure of the starting (planning) phase of an offshore wind project?*
   a. *What are key issues in the planning phase of an offshore wind project?*
   b. *How does the developed framework of the pre- and detailed planning fit into practice?*
2. *How do risks and uncertainties influence the planning phase of an offshore wind project?*
   a. *What are the major risks and uncertainties assessed in the planning phase offshore wind projects?*
   b. *Which factors critically influence the (role of) risks and uncertainties in the planning phase?*

As we mentioned before, there is little research on megaprojects in general or OWPs when it comes to planning, risks and uncertainties. It is our intention to contribute theoretically and practically to both the offshore wind sector, as well as megaprojects in general.

METHODOLOGY

The data required for this study has been collected by interviewing experts on their views and opinions (Handfield et al., 1998). The Delphi study is applied as an appropriate research design to describe, discover and gain insight in a research area (Handfield et al., 1998, Akkermans et al., 2003). It provides an interactive communication structure between the researcher and experts, in order to develop themes, directions or predictions about a topic (Neill, 2007). The approach of this research is therefore qualitative with an explorative nature.

Delphi Technique

The Delphi study is well suited as a consensus-building method to collect data from a panel of selected subjects (Hsu & Sandford, 2007). The feedback process in this technique allows interviewees to assess the judgements and information provided by other experts and themselves (Hsu et al., 2007). Another important characteristic of using Delphi is the ability to provide anonymity of the interviewees. The Delphi process consists of four iterative rounds in which the data is gathered. After each round the different opinions and findings regarding a specific topic are identified and summarized for the next Delphi round. The new insights are a point of discussion in the next round. The exact process of the Delphi technique is explained in more detail by Hsu and Sandford (2007).

Interviewing

In order to gather as much information as possibly a large variety of experts has been interviewed. This research captured the opinion, knowledge and expectations of 26 experts via semi-structured interviews (Emans, 2002). The 26 experts (see Table 3) based their knowledge on their (working) experience in the wind energy industry (at least three years). Additionally, the experts needed to have a background in OWPs or be involved in the development or operation of an OWP. All the interviewees are anonymized in order to ensure objectivity (Hsu et al., 2007) and mind the necessary ethical considerations (Emans, 2002).

Data Collection

The interviews are conducted in person and are collected and organized according to the guidelines and procedures of the Delphi Technique. All the interviews were held in the period between August and October 2014. Appointments with the interviewees were at their offices and set between 60 and 90 minutes, depending on the interviewees. Several documents were provided to the interviewee to guide the conversation, and a number of predetermined questions were used to start the conversation, as the interviews were attended by one researcher. The interviews were recorded in order not to lose data. This assured higher quality of the interview transcripts (Karlsson, 2009). The transcripts were completed after the interviews and provided to the interviewees in order to review and revise the transcripts if necessary.

Data Analysis Method

After reviewing and revising the transcripts by the interviewees, the transcripts were used for analysis. In order to perform a clear and efficient analysis, the answers of the interviewees were coded by means of a coding tree. Kwalitan was used as an instrument to highlight and organize the coding of the transcripts, and to get a good overview of all valuable data.

RESULTS

Interviews were held in order to investigate and describe the process up to the financial close of OWPs. In particular the risks and uncertainties during the development phase of the OWP are addressed. The data from the interviews have led to a description of the (decision making) process, a structuring of the

Table 3. Overview of the 26 interviewed experts with their characteristics

Industry	Company	Position	Background	Experience
Science	Research institute	Developer of standards	Onshore	32 years
Science	Research institute	Advisor in R&D	Onshore & Offshore	42 years
Wind energy	Wind farm contractor	Project manager	Offshore	15 years
Wind energy	Windfarm owner	Operations manager	Offshore	9 years
Construction	Contractor	Commercial manager	Offshore	9 years
Consultancy	Advisory bureau	Advisor operations & maintenance	Onshore & Offshore	30 years
Science	Research institute	Researcher in design of turbines and wind farms	Onshore & Offshore	28 years
Wind energy	Industry association	Coordinator offshore	Offshore	12 years
Public administration	Regional government	Project manager	Offshore	4 years
Wind energy	Advisory bureau	Consultant & project manager	Onshore & Offshore	8 years
Science	Research institute	Managing director	Onshore	33 years
Consultancy	Engineering consultant	Consultant asset management	Offshore	3 years
Consultancy	Engineering consultant	Consultant transport and mobility	Onshore & Offshore	3 years
Consultancy	Engineering consultant	Investment advisor	Onshore & Offshore	4 years
Consultancy	Engineering consultant	Global sector director	Offshore	5 years
Utility company	Power company	Business development manager	Offshore	4 years
Consultancy	Financial advisor	Co-owner and director	Offshore	7 years
Facility management	Government	Program manager and consultant	Offshore	7 years
Real estate	Property owner	Department head	Offshore	16 years
Public administration	Government	Project leader policies	Offshore	4 years
Utility company	Power company	Planning and risk engineer	Offshore	3 years
Science	Research institute	Chief Operating Officer (COO)	Offshore	5 years
Financial intermediation	Bank	Team leader project finance	Onshore & Offshore	7 years
Education	Research institute	Director and lecturer	Offshore	9 years
Financial intermediation	Asset consultant	Managing director	Offshore	9 years
Utility company	Power company	Project Manager	Onshore & Offshore	5 years

process and the identification and quantification of the risks and uncertainties. This is described in this section. How to read the section: Observations and interpretations of the experts are used to support the results, and literal quotes are provided in italics.

Importance of the Planning Phase

The preliminary phase is very important, since the possibility of making a mistake in the beginning could result in an offshore wind farm that will not perform optimally during its lifecycle. The first phases of the project determine the rest of the project to a large extent; it can be the decider for the project itself and its future course. Gaining a better understanding of the planning phase is therefore essential. The experts indicate that putting more effort and time in the preparation of a project results in better and more favorable risks, costs and investments. The experts indicate that currently there are a lot of unknowns regarding the development stage of an OWP. Another interviewee stresses the necessity that the first phase, the preparation, is the hardest to plan, i.e., "It is important to develop the steps of the process in advance. Because a lot of problems can be related back to insufficient preparations and insufficient decision making."

Framework

Before the interviews took place, we developed an initial framework for the development phase of an OWP. The interviewees regarded the framework a good representation of practice. Figure 1 depicts the framework that is the updated result after the interviews, illustrating the planning and decision making of an OWP from pre-initiation until financial investment decision (FID). The illustration of this framework clearly depicts to project developers what needs to be done. Experts indicated that all things in the process interweave with each other; each activity cannot be pinned and finished without the next activity in mind.

The initial step in the process of an OWP starts with a government decision. According to an interviewee "the government plays an important role in this entire process, mainly because they have set their goals and policy to generate 16% renewable energy by 2023. The process needs to start with their decision otherwise the project developer cannot even start." Megaprojects often start with a governmental ambition and decision to create an interesting investment climate that will attract long-term investors. Next the initiation phase starts with assessing the business potential, the permit options, possible locations, desk study and subsidy opportunities for renewable energy project incentivisation (SDE: stimulering duurzame energieproductie). There are some prerequisites for an OWP:

Figure 1. Developed framework for the planning process of an OWP (This framework depicts the planning process of an OWP that is initiated with a government decision and concludes with the Financial Investment Decision (FID). The boxes in the framework indicate activities or tasks that require a decision or a certain performance, in order to successfully continue the project. After particular tasks have been done, the project leader needs to decide whether the project can continue by means of decision gates (DGs))

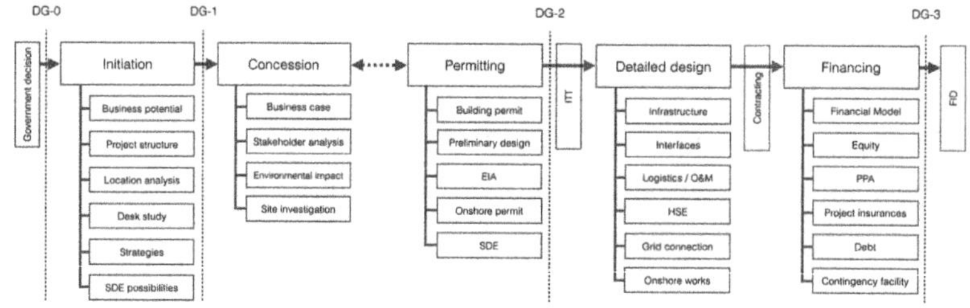

- The location,
- The permit,
- SDE options,
- The grid connection.

These are required to secure and even start an OWP.

If this part of the project is done, the most important foundation of successful continuation of the project is established. However the next phases, commissioning and permitting, are the most prolonged phases as indicated by the interviewees: "the moment the project loses the most valuable time is in the interaction with the government for the different permits and the site investigation". The frequent contact and decision making that is required with the different departments within the government causes the majority of delays. Also the possibility of appeals and objections on the permits adds a lot of uncertainty to the expected planning of the project. The project continues with the invitation to tender (ITT) towards contractors. This process is about issuing the request to commission the elements for the OWP. This includes managing the different interfaces, after which the detailed design phase commences by developing the details of the wind farm based on the previous (coarse) calculations, with more attention to the specifics regarding the preferred contractor. One interviewee explains that "ideally the final detailed design is drawn before FID, however certain aspects, for instance the desired quantity of steel, are unknown at that time. This results in postponing decisions or activities past FID."

Interviewees furthermore indicated that details are essential for the FID, however details frequently become clear past FID. The main reason for this behavior is that the risk of not achieving FID is still present, and each party wants to prevent huge investments without the certainty of continuation. Before a project can continue to its final planning phase, financing, it needs to address the contracting of the (sub)contractors to provide the terms and input for the financing phase. At FID investors want to know who are contracted to check whether the parties involved have a good track record. Certainty measurements are furthermore incorporated at the final stage of financing to secure successful continuation of the project. When this is all arranged the projects moves towards the FID, which is the act of signing all the different contracts and deciding on the different terms and the financing phase. A final point is that the most important moments in a megaproject are the decisions gate(s) (DG). The experts indicated that DGs consist of a process of informed decisions to continue the huge investments for a megaproject. DGs are moments where the project developer seeks permission for commitment to the next step(s) in the project. Thus, project managers decide often on the activities and decisions necessary to get to the next DG and gain approval.

Uncertainties

An uncertain outcome of any event is completely unknown, lacks certainty and cannot be measured or guessed. The main uncertainty that flows through the entire project for an offshore wind farm is the uncertainty of continuation of the project. All the experts emphasize this, as for OWPs "everyone is trying to postpone its commitment, to not be dependent on others until the very last moment." Subsequently, every party has to keep in mind that "showing commitment to the project provides certainty to other parties and firms." The postponement of making decisions by the parties is explainable since the project is uncertain to continue if it does not acquire the permits and the SDE. These elements are part of the government decision and the interviewees state that "if the next government does not want to pursue the same goals as the current government, the perceived uncertainty in their decision making is increasing." Since the duration of the development of an OWP can take between four to seven years, changes of governments are possible. Another great amount of uncertainty comes from all the accompanied parties that are involved in the project. They all have to give their approval for decisions and investments.

Figure 2 illustrates the described uncertainty in decision making for successful continuation of an OWP. Alongside the uncertainty line, an investment line is drawn that indicates the increasing investments of capital expenditure (CAPEX), which includes development expenditure (DEVEX). Since the CAPEX differs a lot between megaprojects, the line only illustrates the increase without tangible values. The purpose of the figure is to illustrate the uncertainty project developers experience throughout the lifecycle of an OWP. The decision uncertainty decreases throughout the project measured at the DGs, since the project continuation is evidently decided at the DGs due to the added value. The figure also stresses the interdependency of the experienced uncertainty relative to the postponement of the CAPEX decisions. The percentages in the figure are stressed by the experts to provide an impression of the experienced remainder of uncertainty:

The experienced uncertainty throughout the planning process is high until FID is achieved. Until that moment the development investments are kept as low as possible since there is no certainty for project continuation.

A key point is that CAPEX postponement will not result in added value, thus evidently the uncertainty will not decrease. A good example explained by the interviewees is the moment "projects perform FEED studies to gain information for a better price indication. However these studies require investments, investments made while the possibility of discontinuation still exists." The more uncertainty is perceived by external parties, the more they will increase their price for the perceived risk. Thus, project developers could invest millions to ultimately discover that the site is not suitable for offshore wind. Interviewees state that "to avoid planning overruns, the project needs to decide on certain items in advance (one year before FID) despite the uncertainty whether the project will continue." If these decisions are postponed too much the project will never get started. It is explainable that every investor would behave this way, and tries to prevent early investments. However in the end someone needs to invest: entrepreneurial risk.

Figure 2. Experienced uncertainty in decision making for successful continuation of the project

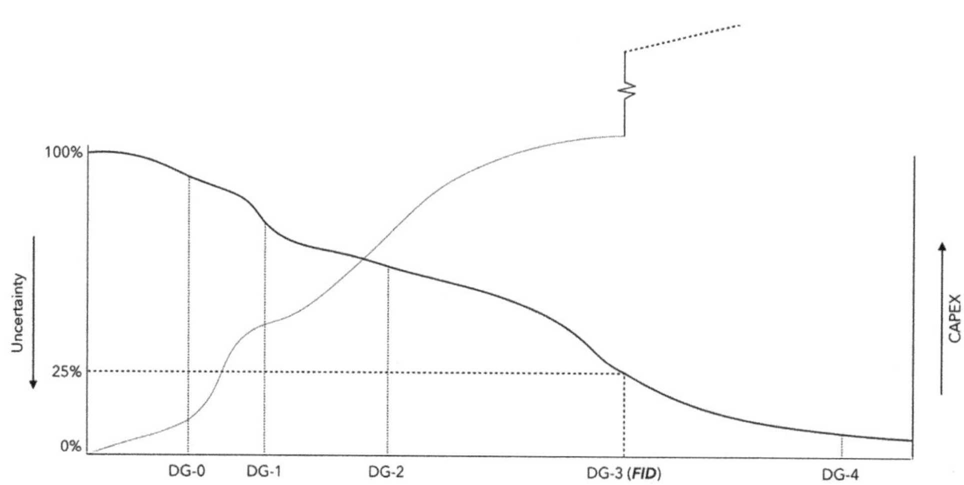

Table 4 summarizes the identified uncertainties of this research with a short explanation. The uncertainties are ranked according to the frequency at which they have been mentioned by the different experts.

These uncertainties affect the planning process in such a way that a lot of the involved stakeholders or parties remain skeptical and hesitant to commit to an OWP. The hesitation is reflected in the tender price; firms incorporate the absence of data or certainty into their price. The skepticism is a result of the postponement of investments from another party, since most investors perceive too much risk or they do not want to commit too early. This results in a vicious circle because the postponement of decisions and investments results in a slow increase of certainty, which evidently results in only a slim number of commitments, ultimately making the project more uncertain. That same uncertainty subsequently is incorporated in the price, resulting in a less beneficial investment decision.

Risks

In the previous sub-section, uncertainty was referred to as the lack of complete certainty. An uncertain outcome of any event is completely unknown and cannot be measured or guessed. Risk, on the other hand, is an uncertain event that, if it occurs, has a positive or negative effect on at least one project objective and can both be measured and guessed. Risks often occur due to underestimations of certain activities or tasks, however when they are analyzed and addressed, they are not risks anymore but events that are going to occur or not. The most underestimated activity in an OWP is interfacing. This concerns the moments in time where different contractors meet each other. If one party is delayed, all the parties having to make decisions after that are also delayed and need to postpone their decisions, investments and activities. According to an interviewee:

Table 4. Identified uncertainties

#	Uncertainty	Mentioned	Explanation
1	Return of investment	21/26	The perceived uncertainty of successful continuation of the project is high at the beginning. Therefore investors suppress their investments to a later stage, due to the increased chance of losing the investment.
2	Ordering Long-Lead Items	19/26	LLI are expensive and need to be ordered before FID to prevent huge planning overruns afterwards. LLIs are also customized for the project, making them useless if the project subsequently does not reach FID.
3	Government shift	18/26	Elections can shift the regime in a country and can result in adverse changes towards prior offshore wind decisions.
4	Not allocating all risks	16/26	Assessing and assigning unknown risks incorrectly, causing higher costs for the involved parties. This results in variation orders with higher costs for the entire project.
5	Site investigation analysis	15/26	If the assigned site is investigated and reveals that the suitability of the site is insufficient despite the huge investments.
6	Entrepreneurial risk	8/26	The postponement of investments and commitments, resulting in a power play strategy of not being dependent on one another. Causes the project to create unnecessary uncertainty due to postponements.
7	Weather conditions	5/26	Legislation for driving piles and heavy weather fluctuations limit the planning and create an intense pressure on the planning that, despite good preparations, is dependent on an uncontrollable element.

The project leader is responsible for the coordination. The more contractors a project has, the more attention it needs to pay to the interfaces and the bigger the risks are for the project with the chance of loss of time and money.

Moreover, not addressing properly who is responsible for each task and not actively sharing information are two other much-identified risks for megaprojects. "Every project is observed as a standalone project, and knowledge is not transferred. Firms are scared to share their mistakes or sensitive information". Unfortunately, no one can learn from their mistakes. It has often been argued that the wind industry lacks the willingness to copy, paste and learn from the oil and gas industry. Observing every project as a standalone project results in a high level of change of project teams. Subsequently, projects are sold to other parties and the developers are unreachable afterwards, resulting in unexplained decisions and strategies for the new management:

Parties need to be involved as early as possible in the decision making. Otherwise the result can be unsatisfying in which the development requires compromises on things that have already been developed, resulting in additional project delay.

The interviewees emphasize that late involvement of parties happens too frequently, and early involvement could reduce avoidable expenses. Otherwise parties could be unsatisfied with certain decisions, which would then require compromises from several sides as they otherwise cannot contribute to the project. This ultimately leads to delays in the project. This relates to another identified risk: insufficient preparations. The preparatory work for all kinds of activities, e.g. onshore activities, is often insufficient because of the underestimations of activities relative to the focus of completion of the project. The underestimation of durations, complexities, decisions and postponement of applications delay the project as these activities are not completed within the set timescale. Project developers tend to only pay attention to the current phase and postpone activities that are currently irrelevant.

As a project wants to control its expenses and is constrained by external factors, it requires tight planning and decision making. However, the effect is that parties expect to begin their tasks within the tight timeframe, resulting in extra costs because of foreseeable delays. This results in another risk: poor contracting. Not addressing all risks can have serious performance consequences. "Without good research the project ends up with additional work and costs. Weak project management that does not quickly enough discover possible gaps or problems will leave the project with many additional costs." Inexperienced project teams or parties negatively influence the decision making since they have no experience or knowledge about (offshore) projects. Table 5 summarizes the identified risks of this research in order of perceived importance.

Main Findings

The explanation why these ten identified risks are most prevalent for the planning phase of an OWP can be related back to the underestimation of the decisions made in a megaproject and the relative inexperience in the offshore wind industry. The corollary of the inexperience is the underestimation of the project in the interfacing, the preparations, the permits and the involvement of important parties. The elements aforementioned are identified as risks for an OWP (see Table 5). The identified uncertainty with the underlying skepticism and hesitations, contributes to the overall risks and uncertainties perceived in an

Table 5. Identified risks

#	Risk	Mentioned	Explanation
1	Poor interfacing	24/26	Certain tasks and deadlines are not correctly assigned to the different parties. Thus, the contractors have to wait for each other, which will lead to delays and cost overruns.
2	Lack information sharing	22/26	Every OWP is observed by the firms as a standalone project in which the firms are reluctant to share viable information about mistakes and lessons learned.
3	Late involvement of important parties	21/26	Choices made earlier in the process cannot be aligned with the expertise of the parties. This results in discomfort since the requirements do not match one another.
4	Long application process permit(s)	20/26	Bureaucracy causes long application processes, eventually resulting in outdated or invalid permits due to the duration between approval and fulfilment.
5	Insufficient preparations	18/26	The project underestimates the importance, complexity or duration of certain activities resulting in not completing these activities within the set timescale, thus delay.
6	Dependent in obtaining SDE	16/26	An OWP relies and requires the SDE. The approval of SDE is moreover dependent on governmental authorities; with a first come, first serve principle.
7	Tight scheduling	15/26	OWPs are often constrained by external factors and their dependencies, resulting in a planning that is too tight with a high possibility of overruns in time and costs.
8	Poor contracting	14/26	The different (sub)contractors are not analyzed on their financial and technical capabilities and as they perform bad that consequently results in delay of the project.
9	Incompetent project team	14/26	People are assigned to a project, however they lack experience, skills or knowledge to execute a megaproject in a good manner.
10	No grid connection	11/26	The commissioning is completed, however the OWP is not connected to the power grid onshore, thus it cannot provide electricity and subsequently generate revenue.

OWP. It is further noteworthy to mention that the permits, the grid connection and the site, which are considered to be basic elements of an OWP, are respectively identified as important risk factors and as an uncertainty factor. It clearly suggests that uncertainties are large during the planning phase of an OWP due to the novelty of the offshore wind industry.

DISCUSSION

This section will first elaborate on two important factors: the project structure and the role of the government. Both are hard to control and the project continuation is truly dependent on them. After that, we will present some additional reflections on the presented results.

Project Structure

The activity 'project structure' refers to the way the project is organized throughout its lifecycle (see Figure 1). The planning and decision process is dependent on the selected project structure. A project is either performed with finance on balance (FOB) or by execution of project financing (PF). Table 6 displays the differences between the two structures.

The main benefit and limitation of the different structures do not direct towards one prevailing structure. However, FOB projects can postpone a lot of activities and decisions past FID since they can better deal with delays in their planning. This is because of the backings of a FOB project, providing them with more certainty. FOB projects tend to be less focused on the economical driver as small cost overruns are not considered fatal. PF projects position most of the financial risks with other parties, which makes PF more self-protective to possible planning and cost-overruns. Thus, performing up to the task is very important to the parties involved in PF projects. Moreover, some experts indicated that PF can be subdivided into pre-construction (PrePF) and post-construction (PostPF) project finance. Figure 3 is developed to illustrate the continuum to what extent a project is financed on balance or is project financed.

Experts indicate that the project developers choose a particular structure, which does not imply that a project is, for instance, completely project financed. Experts indicate that PF projects show less innovation than FOB projects and are more focused on the bankability of the project. As mentioned, PF projects are often special purpose vehicles that are highly dependent on the requirements of the different financiers. These financiers are risk averse, which limits the initiative to introduce new methods, approaches or turbines. FOB projects have more opportunities to incorporate these innovations in their project, since they are less dependable on external parties. "The FID for FOB projects already takes place in DG-2. The project team and board make the decision to make all CAPEX available for ITT and immediately continue with commissioning." This last statement clearly demonstrates the project structure differences and the effect the structure can have on the planning and approach throughout the process.

Changes in the Risk and Uncertainties for a Project Finance Structure

In the following discussion we look at the two financing structures and what their impact (decrease or increase) would be on a particular risk or uncertainty. We thereby take the perspective of a normal OWP in which a financing structure is not yet applied. Table 7 depicts the changes of the identified risks and uncertainties with regard to the PF structure. The PF structure has a noticeable increase (+) of risks and uncertainties. A PF project is more insecure at the beginning due to the lack of assets that guarantee

Table 6. Comparison between the two project structures

Finance on Balance	Project Finance
Long-term focused	Short-term focused
Technical and innovation driven	Financial and bankability driven
Complete own financing	High debt rate
Ability to postpone activities	Fast decision making required
In-house developments	External contracting
Bureaucratic	Flexible
Assets are warranty	Independent project company (SPV)
Lean & standardization	High interchangeability of project teams
Dedicated decisions due to backup	Dependent on requirements financiers
Designer-Owner-Operator (DOO) Model	Many contractual partners

Figure 3. Project structure continuum

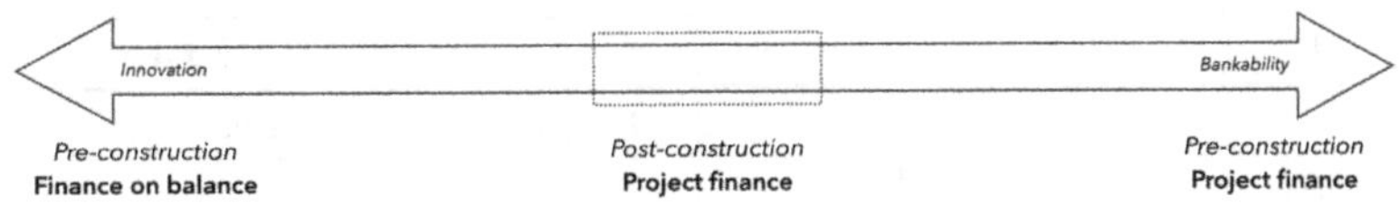

a warranty. PFs furthermore consist of special purpose vehicles with new stakeholders, which often result in an increased risk for 'late involvement of important parties', 'incompetent project teams' and 'insufficient preparations'.

PF projects are financially driven and are characterized by their fast decision making, which stimulates the information sharing with other parties and results in improved allocation of the risks and contracting. The identified effect of poor interfacing is established as indeterminable, considering that the interviewees stress that PF in some cases contributes to tight interface management. Others indicated that the inexperience resulted in the underestimation of the interfacing, resulting in an increase of the risk. The remaining risks and uncertainties neither increase nor decrease, because they are not dependent on the PF structure.

Changes in the Risk and Uncertainties for a Finance on Balance Structure

Table 8 shows the changes of the identified risks and uncertainties with regard to the FOB structure. The FOB structure compared to the PF structure has a greater effect in (-) reductions of risks and uncertainties. The uncertainties of 'return on investment' and 'ordering long lead items' reduce in the FOB because these firms execute their FID at DG-2, which creates more certainty for the rest of the project. The disadvantages of FOBs are that they are often focused on themselves, resulting in 'late involvement of important parties' and 'poor contracting'. The uncertainties of 'return on investment' and 'ordering long lead items' reduce in the FOB because these firms execute their FID. However, some interviewees

Table 7. The identified changes in the risks and uncertainties for a project finance structure

#	Risk	Effect	#	Uncertainty	Effect
1	Poor interfacing	?	1	Return of investment	+
2	Lack information sharing	-	2	Ordering Long-Lead Items	=
3	Late involvement of important parties	+	3	Government shift	=
4	Long application process permit(s)	=	4	Not allocating all risks	-
5	Insufficient preparations	+	5	Site investigation analysis	=
6	Dependent in obtaining SDE	=	6	Entrepreneurial risk	+
7	Tight scheduling	=	7	Weather conditions	=
8	Poor contracting	-			
9	Incompetent project team	+			
10	No grid connection	=			

risk/uncertainty reduced (-)
risk/uncertainty increased (+)
risk/uncertainty dissolved (x)
risk/uncertainty unchanged (=)
risk/uncertainty indeterminable (?)

Table 8. The identified changes in the risks and uncertainties for a finance on balance structure

#	Risk	Effect	#	Uncertainty	Effect
1	Poor interfacing	=	1	Return of investment	-
2	Lack information sharing	+	2	Ordering Long-Lead Items	-
3	Late involvement of important parties	+	3	Government shift	=
4	Long application process permit(s)	=	4	Not allocating all risks	=
5	Insufficient preparations	?	5	Site investigation analysis	=
6	Dependent in obtaining SDE	=	6	Entrepreneurial risk	=
7	Tight scheduling	-	7	Weather conditions	=
8	Poor contracting	+			
9	Incompetent project team	-			
10	No grid connection	=			

risk/uncertainty reduced (-)
risk/uncertainty increased (+)
risk/uncertainty dissolved (x)
risk/uncertainty unchanged (=)
risk/uncertainty indeterminable (?)

indicated that this is not always the case. The remaining risks and uncertainties are neither expected to increase nor decrease, mainly because they are not dependent on the FOB structure.

Governmental Effect

The framework developed in this research starts with a government decision followed by activities that are highly dependent on the interaction with the government: the location, building permit, EIA, SDE, onshore permit and grid connection. Three of the four elements aforementioned are essential for an OWP. The government is observed as the restraining and stimulating factor at the same time, as they decide on the policies and need for renewable energy. One interviewee claims that "the government provides a starting opportunity for the project. The crux is that the support is also required after FID, since an OWP last seven years on average, and there is a high chance the project then has to deal with a new government."

This uncertainty reflects the ignorance of the industry on how to approach this: "the government is not straight in its policy." In the Netherlands for example, a lot is invested in permits that were provided by the government in 2007. However "the government decided to change the rules of the game at full-time (2014), evaporating all the hard work of the parties with the permits." This behavior is sometimes seen as typical Dutch opportunism in which permits are changed so radically that a number of project developers are discarded after their effort and investments. This influences the decision making of future project developers as investors observe a possibility that the effort and investments can be worthless in an instance.

National Differences

Every country has its own government process, hence the development phase differs per country regarding, for instance, when a project developer gets assigned to a parcel or when the ITT can start. Still, "the basics of the process will be the same in every country, only the sequence differs." Nevertheless, we only interviewed experts who have experience with especially Dutch OWPs. The experts indicated that the Dutch setting is comparable to other European countries, independent of the national legislation. Still, a successful (start of an) OWP partially depends on the government and its legislation, and if an international corporation initiates the project it is still bound to the rules of the Dutch game. Some of the risks (#4, #6, #10) and uncertainties (#3) relate to the role of the government. But the remaining identified risks and uncertainties impact the OWP, irrespective of the government's role. Therefore the results have significant practical implications.

Parallel or Linear Framework

Another important observation that needs to be discussed is the linear reproduction of the framework. Interviewees stated that "the process is not linear but parallel. All activities in the framework are repetitive, have interdependencies and are not a tick in the box. Also "every aspect mentioned in the framework is almost treated or discussed at the same time. The focus, however, shifts onwards throughout the project." This explains why this process cannot be linear in practice. Although these statements are endorsed in this research, the process is illustrated as a linear process for discussion and analysis purposes.

Puzzle Game

An aspect that contributes to the experienced uncertainty in the planning phase and decision making of an OWP is the so-called 'puzzle game'. It is described as follows: "The project deals with a high level of uncertainty at the start and throughout its process. It is part of a game, because a too dedicated project plan weakens the position of the negotiator and the continuation of the project. Thus, if everyone knows what the project desires they will demand the highest possible price because the continuation of the project depends on it." It is a big puzzle in which every party has a piece and the entire puzzle only falls into place just before the FID. Such a setting resembles a non-cooperative game, in the sense that information is held privately, and decisions are made relatively simultaneously and non-cooperatively. The dependency clearly illustrates that altering things after FID is extremely difficult as everything is set and cannot be changed. This remark stresses the essence of the framework which can provide the industry with a clear overview of the planning process.

Lessons Learned

The scope of this research is the planning of an OWP until FID has been achieved. However as has already been illustrated the entire project ultimately lasts until decommissioning. "A couple years after installation, firms often look back on the project and assess whether the things went as planned. From this they acquire lessons learned." It is important to note that it takes approximately ten years between the first and the last DG (Brockmann, 2009), which creates the possibility that another project already started, and the same mistakes are made again. A frequent argument of the experts is that there are guidelines written about project planning, decisions and developments. However, these handbooks are for internal use by the respective firms, and this contributes to the second identified risk: lack of information sharing.

Relevance of the Framework

A number of experts discussed that the OWP process is equal to a road construction project. Observations indicate that the framework is to a large extent suitable in practice for megaprojects. Compared to megaprojects from governments, the budget and time overruns of an OWP are highly visible. In OWPs there is just one project manager who either makes a profit or suffers losses and operates in slim margins. This is also frequently the case for megaprojects in general. However the effects and consequences are more hidden since they often consist of multiple project leaders who only contribute their part, and are not responsible for the entire project.

CONCLUSION

This research investigated the risks and uncertainties in the planning phase of an OWP. Current literature regarding OWPs is focused on the costs and quality aspects of a project and lacks research into the planning and decision making aspect. First a framework was developed to gain an overview of the process for an OWP, which could be used to identify the risks and uncertainties with regard to planning and decision making. This research determined that due to the novelty of the industry, a number of risks and uncertainties are the most prevalent. Novelty results in inexperience and repetitive underestimation of a megaproject like OWPs, and results in at least ten identified risks for overruns in a project. Additionally, the novelty creates an underlying skepticism and hesitation towards the initiation of an OWP as a result of the experienced uncertainty. The high dependency on the government contributes to skepticism and hesitation, since the project requires the four basic elements of an OWP (SDE, permits, grid connection and location) from the government.

Since the legislation per country differs, the influence of a government on the planning process of an OWP is significant. This results in a high variety of duration and prevalence of the process, creating interesting and less interesting countries to initiate OWPs. The project structure has been identified and discussed as an element that influences the approach and continuation of an OWP. A project can either obtain a FOB or a PF structure, with the main difference being that the project is financed by equity of the respective firm or otherwise financed mainly (70%) with debt. The latter is highly dependent on the requirement of external parties and financiers that need to be incorporated in the project, whereas the former can determine their course throughout the project. Hence, a lot can be gained in the planning process of an OWP related to the reduction of the risks and uncertainties, and improved decision making and understanding of affiliated elements and procedures.

Limitations

This research only addressed Dutch interviewees. Despite some experts having additionally international market knowledge, this research needs to acknowledge this limitation. Further research is required by performing more Delphi rounds, as the third and fourth iteration is required for the Delphi-technique to achieve international consensus (Custer, Scarcella & Stewart, 1999). Future research should take up the third round and interview a variety of international OWP experts as they can contribute to insights on the risks and uncertainties. With the fourth round we recommend to interview experts who have experience with different sorts of megaprojects (e.g., bridges or tunnels), to eventually improve generalizability of the framework. Also, with objective data of, for instance, cost overruns of different OWPs and other quantitative data, the findings can further be validated. Another limitation of this research is the linear representation of the framework. Further research is required to investigate the planning process in more detail to eventually develop a standard roll-out for an OWP that possible integrates parallel processes and could reduce costs, as indicated by Junginger et al. (2004).

Theoretical Implications

The theoretical implication of this research is that the identified risks and uncertainties, related to the performance of a project, can be identified and translated into critical success factors for an OWP. Hence, further research is needed to quantify the identified risks and uncertainties and their probability, and link this to the likelihood of success in the development of OWPs. As mentioned, a project needs to meet its three internal performance measures: costs, time and quality. This research focused on the timing aspect until FID. Since this research has only indirectly considered cost, further research on the interrelation between the performance dimensions could improve the overall optimization of OWPs. "In an OWP there is a constant optimization between the cost and the costs associated with stretching the planning." Further research is suggested to improve the overall performance of an OWP and gain better insights into the question when project developers can postpone activities and when project costs become too determinant for the overall performance.

At a general level, this exploratory research can be considered as a first step towards theoretical understanding of how the complexities of OWPs may be approached. At the same time, as our interviewees considered the framework as being applicable for our settings as well, it is safe to conclude that our work can also advance megaprojects research in general.

Practical Implications

Throughout the research, interviewees indicated that the tasks and activities were known, while the sequence and duration were not. The framework allows stakeholders in offshore wind to gain a better overview of the planning and decision making for an OWP. Improving the preparation of OWPs and additionally make the stakeholders aware of the most common risks and uncertainties. The framework aims to reduce the experienced uncertainty in OWPs. The framework can eventually contribute to the successful completion of more, longer and complicated megaprojects. During the process, this research has identified that the planning process differs per country due to the government and its legislation. The framework can still be applied to every initiated OWP after adjusting activities and sequences to its national legislation. Interested stakeholders, developers and contractors can use this framework to anticipate on future changes and determine their position in the process of an OWP. At a more general level, the practical implications for, for instance, installation- and electricity companies could be:

1. To get involved as early and quickly as possible in the decision making to make sure everything is thought of, and
2. Create an environment in which information and knowledge is shared to gain insights faster and better.

REFERENCES

Adler, P. (1995). Interdepartmental interdependence and coordination: The case of the design/manufacturing interface. *Organization Science*, *6*(2), 147–167. doi:10.1287/orsc.6.2.147

Akkermans, H. A., Bogerd, P., Yücesan, E., & van Wassenhove, L. N. (2003). The impact of ERP on supply chain management: Exploratory findings from a European Delphi study. *European Journal of Operational Research*, *146*(2), 284–301. doi:10.1016/S0377-2217(02)00550-7

Alessandri, T. M., Ford, D. N., Lander, D. M., Leggio, K. B., & Taylor, M. (2004). Managing risk and uncertainty in complex capital projects. *The Quarterly Review of Economics and Finance*, *44*(5), 751–767. doi:10.1016/j.qref.2004.05.010

Bilgili, M., Yasar, A. & Simsek, E. (2010). Offshore wind power development in Europe and its comparison with onshore counterpart. *Renewable and Sustainable Energy Reviews,* 15(2011), 905-915.

Brockmann, C. (2009). *Mega-projects: getting the job done.* LEAD Conference 2009.

Brugaard Villmo, H. (2012). *Planning and management of megaprojects. Industrial Economics and Technology Management.* Norwegian University of Science and Technology.

Cooke-Davies, T. (2002). The "real" success factors on projects. *International Journal of Project Management*, *20*(3), 185–190. doi:10.1016/S0263-7863(01)00067-9

Custer, R. L., Scarcella, J. A., & Stewart, B. R. (1999). The modified Delphi technique: A rotational modification. *Journal of Vocational and Technical Education*, *15*(2), 1–10.

Duncan, G. L., & Gorsha, R. A. (1983). Project management: A major factor in project success. *IEEE Transactions on Power Apparatus and Systems*, *102*(11), 3701–3705. doi:10.1109/TPAS.1983.317755

Dvir, D., Raz, T., & Shenhar, A. J. (2003). An empirical analysis of the relationship between project planning and project success. *International Journal of Project Management*, *21*(2), 89–95. doi:10.1016/S0263-7863(02)00012-1

Edwards, I. (2011). *Overcoming challenges for the offshore wind industry and learning from the oil and gas industry.* Report from POWER cluster, natural power and the Interreg IVB North Sea Region Programme.

Emans, B. (2002). *Interviewen: theorie, techniek en training* (4th ed.). Groningen: Stenfert Kroese.

Eurobserver. (2014). *Wind energy barometer.*

EWEA. (2009). *Wind at Work: Wind energy and job creating in the EU.*

EWEA. (2011). *Wind in our Sails: the coming of Europe's offshore wind energy industry report.*

Flyvbjerg, B. (2009). Survival of the unfittest: Why the worst infrastructure gets built and what we can do about it. *Oxford Review of Economic Policy*, *25*(3), 344–367. doi:10.1093/oxrep/grp024

Flyvbjerg, B. (2011). Over budget, over time, over and over again: managing major projects. In Morris, P.W.G., Pinto, J.K. & Söderlund, J. (2011) The Oxford Handbook of Project Management (321–344). Oxford: Oxford University Press. doi:10.1093/oxfordhb/9780199563142.003.0014

Flyvbjerg, B., Bruzelius, N., & Rothengatter, W. (2003). *Megaprojects and risk: an anatomy of ambition.* Cambridge: Cambridge University Press. doi:10.1017/CBO9781107050891

Gerdes, G., Tiedemann, A. & Zeelenberg, S. (2005). *Case study: European offshore wind farms a survey for the analysis of the experiences and lessons learnt by developers of offshore wind farms - final report.* Report.

Giezen, M. (2012). Keeping it simple? A case study into the advantage and disadvantages of reducing complexity in megaproject planning. *International Journal of Project Management, 30*(7), 781–790. doi:10.1016/j.ijproman.2012.01.010

Handfield, R. B., & Melnyk, S. A. (1998). The scientific theory-building process: A premier using the case of TQM. *Journal of Operations Management, 16*(4), 312–339. doi:10.1016/S0272-6963(98)00017-5

Hsu, C.-C., & Sandford, B. A. (2007). The Delphi Technique: Making Sense of Consensus. *Practical Assessment, Research & Evaluation, 12*(10), 1–8.

Hubbard, D. (2007). *How to measure anything: finding the value of intangibles in business.* Hoboken, NJ: Wiley.

Johansen, E., & Wilson, B. (2006). Investigating first planning in construction. *Construction Management and Economics, 24*(December), 1305–1314. doi:10.1080/01446190600863160

Johnson, J., Karen, D., Boucher, K. C., & Robinson, J. (2001). *Collaborating on project success. Software Magazine.* February/March.

Junginger, M., Faaij, A., & Turkenburg, W. C. (2004). Cost reduction prospects for offshore wind farms. *Wind Engineering, 28*(1), 97–118. doi:10.1260/0309524041210847

Karlsson, C. (2009). *Researching Operations Management.* New York: Routledge.

Kezner, H. (2009). *Project Management: a system approach to planning, scheduling, and controlling* (10th ed.). New York: John Wiley & Sons.

Koch, C. (2012). Contested overruns and performance of offshore wind power plants. *Construction Management and Economics, 30*(August), 609–622. doi:10.1080/01446193.2012.687830

Lechler, T. G., Edington, B. H., & Gao, T. (2012). Challenging classic project management: Turning project uncertainties into business opportunities. *Project Management Journal, 43*(6), 59–69. doi:10.1002/pmj.21304

Loch, C. H., DeMeyer, A., & Pich, M. T. (2006). *Managing the unknown: a new approach to managing high uncertainty and risk in projects.* New York: John Wiley & Sons. doi:10.1002/9780470172377

Loch, C.H. & Pich, M.T. (2002). Managing project uncertainty: from variation to chaos. *MIT Sloan Management Review,* Winter, 60-67.

Lopez del Puerto, C., & Shane, J. S. (2014). Keys to success in megaproject management in Mexico and the United States: Case study. *Journal of Construction Engineering and Management*, *40*(4), 1–7.

Megavind. (2010), *Denmark - Supplier of Competitive Offshore Wind Solutions report.*

Neill, J. (2007). *Delphi study: research by iterative, consultative inquiry.* Retrieved from available at http://www.wilderdom.com/delphi.html

PM Network (2014). Standardised project management practices = success. *PMI Today,* April(2014), 8-9.

NOAA. (2014). *Coastal ecosystem restoration: risk and uncertainty in environmental restoration programs*, Retrieved from http://coast.noaa.gov/archived/coastal/economics/riskinrest.htm?redirect=301ocm

PMI. (2014). *What is project management.* Retrieved from http://www.pmi.org/About-Us/About-Us-What-is-Project-Management.aspx

Priemus, H., Flyvbjerg, B., & van Wee, B. (2008). *Decision-making on mega-projects: cost–benefit analysis, planning and innovation*. Cheltenham: Edward Elgar. doi:10.4337/9781848440173

Puddicombe, M. S. (2006). The limitations of planning: The importance of learning. *Journal of Construction Engineering and Management*, *132*(9), 949–955. doi:10.1061/(ASCE)0733-9364(2006)132:9(949)

Thi, C. H., & Swierczek, F. W. (2010). Critical success factors in project management: Implication from Vietnam. *Asia Pacific Business Review*, *16*(4), 567–589. doi:10.1080/13602380903322957

van Kuijk, G. (2013), *P13-13 EUROS – Excellence in Uncertainty Reduction of Offshore wind Systems.* Retrieved from http://www.stw.nl/nl/content/p13-13-euros-%E2%80%93-excellence-uncertainty-reduction-offshore-wind- systems

Whittaker, B. (1999). What went wrong? Unsuccessful information technology projects. *Information Management & Computer Security*, *7*(1), 23–29. doi:10.1108/09685229910255160

Zwikael, O., & Globerson, S. (2006). Benchmarking of project planning and success in selected industries. *Benchmarking. International Journal (Toronto, Ont.)*, *13*(6), 688–700.

Zwikael, O., Pathak, R. D., Singh, G., & Ahmed, S. (2014). The moderating effect of risk on the relationship between planning and success. *International Journal of Project Management*, *32*(3), 435–441. doi:10.1016/j.ijproman.2013.07.002

Zwikael, O., & Sadeh, A. (2007). Planning effort as an effective risk management tool. *Journal of Operations Management*, *25*(4), 755–767. doi:10.1016/j.jom.2006.12.001

KEY TERMS AND DEFINITIONS

Delphi Technique: A consensus-building method that consist of four rounds in which data is collected from a selected group of specialists via their judgements. After each round the findings and opinions are summarized to identify certain elements and are then input for discussion for the next round.

Finance on Balance: This is a project funded with equity of the executing party. It often concerns major utility companies, venture capitalists or private equity firms.

Financial Investment Decision: The moment all participating parties come together and agree on the amount of money that is going to be invested in the project. Prior to this moment a.o. lawyers, accountants of the different parties identified and checked whether the involved parties have a good track record.

Megaprojects: A big project that costs more than one hundred million dollars and contains a lot of complexity regarding the interfacing, planning, decision making and non-standard technologies. Megaprojects often are linked to big construction projects for building, e.g., bridges.

Offshore Wind Projects (OWPs): An offshore wind project is the installation of a group of wind turbines far out of sea. Offshore Wind Projects are complex since they are not onshore, which results in high costs for installation, maintenance and decommissioning.

Project Finance: This is a project structure in which the executing party does not have sufficient equity to fund the entire project. Via financial institutions, private investors or SMEs they assemble a loan or a share in the project.

Project Management: The process in a project that consists of planning, organizing, directing and controlling the resources of the project to deal with the complexity of a project. The main constraints of project management are time, costs and quality.

Risks: Risks in a project are uncertainties that a recognized factors that may influence your project plan. By assessing the impact and probability of the risk, risks can be prepared for and safety-measures can be taken.

Uncertainties: Are elements in the project that no one is completely sure about and it is hard to guess if it will occur. Uncertainties are inevitable regardless of the amount of information gathered, since they include elements that are unknowable.

Chapter 16
The Set-Up Process

Manuel Rodríguez Méndez
ESeyPro S.L., Spain

Javier Cárcel-Carrasco
Universitat Politècnica de València, Spain

ABSTRACT

To prepare the machines to make the product that client needs and to give him the product in the necessary moment is the objective of all companies. This means to have all machines prepared to make the product, it means that the productive system has not problems to make the product and this complies the quality standards. To have prepared the machines to make a product at some point it is the objective of set-up process. If this process is correct performed, the machines will make the product without problems, but if the set-up process is not performed in correct way, then the posterior productive step will have many problems. Perhaps, the set-up process is the more important productivity activity. The time necessary to perform the set-up process is the parameter to minimize, with the process cost, and this time says the flexibility level of the productive process of the factory.

INTRODUCTION

The Management has defined production activity in various ways, often depending on the type of business involved. Bueno Jauregui (1991) defined it as "The transformation of some initial goods in other final goods, consumer goods, which are best suited to satisfy certain needs." However, and although all definitions look similar, since it is your goal; what it does them different is the transformation type is performed on the initial goods. However, perhaps the definition that has been used Shingo (1985) is the best to apply to any kind of productive activity: "Production is a network of processes and operations, phenomena that are arranged along two axes intersect."

However, all definitions have a common basis that it explains the production as a set of processes and operations located in space and time, they can transform raw materials in new products to satisfy some specific needs. The definition of Shingo (1985). previously required two concepts that go into its definition are identified: Process and Operations. Shingo (1985) defined the process as the flow of

DOI: 10.4018/978-1-5225-0651-5.ch016

all actions to transform raw materials in new different products, and the operations are specific actions during the processes.

In general, in any productivity process there are a set of parallel activities, previous and subsequent to the production process that they will complement it or they will allow it to make the new product and they will affect to its cost. Activities such as:

1. The receipt and storage of raw materials to be used in the production process. Storage, both after receipt from the supplier and the temporary storage prior to use in the production process.
2. The raw materials transport from the warehouse to the machine, to be used during the process.
3. Inspections of raw materials and intermediate products, to determine the suitability of such materials, and to determine the evolution of the process. Also, inspections of manufactured products, ready to send the final client to ensure that the process has developed the product to be manufactured for him.
4. The storage of processed products, and their subsequent shipment to clients.
5. The stops to resolve faults and perform machine maintenance.
6. Preparation of machines to make in them the product to be manufactured.
7. The management of orders and the production management.
8. Etc.

Business Management, initially oriented only to the process and manufacture management of a product to be delivered to the client demand, had to adapt to the new demands he will demand. These requirements not only they are referred to the management of a product with a low-cost, but it has others connotations:

- A product of high quality.
- The delivery of the product within the agreed period.
- A product made from socially acceptable constraints: clean process safe for worker's productive process without the use of illegal labor.

For this reason, the current Management is not only focused on offering a product for a particular market, but to offer a range of products that will satisfy the desires of potential clients. For this it is necessary that the production means are ready to perform its function at the right time and when the client demands it. This group of activities to prepare the machines or production lines, to make in them certain products, has been denominated set-up operations.

The set-up operations are defined as all activities that are performed on a machine or production lines to adapt, to a new productivity process. And, the set-up time is defined as the necessary time to prepare the machines or the production line.

This process has a triple value. On the one hand, it allows to modify a machine to process in it different goods. And So, we can consider the set-up process as the parameter to evaluate the response capacity of the production system to market needs; hence their time permits evaluate the flexibility of the productivity process. On the other hand, the set-up cost will shape the economic lot, defined as the amount that it's necessary to produce to minimize the sum of the costs of storage and the preparation of the machines.

Traditionally, the set-up activity it is considered an "waste operation" that it does not add value to the product and, therefore its cost must be remove. This is too radical statement about the set-up process, then, too, it should be considered as an activity that it improves the use of machines, allowing to adapt them to market needs, making therefore be delayed its obsolescence. Although an improvement of set-up process may seem an improvement in overall operating costs if we minimize the set-up time of machines, it not always is reduced the unit cost of products. Because, sometimes, it's only an improvement in the category of costs, for example, the stock costs can be reduced (financial costs, storage, handling), but the fixed costs of products an increase (initial outlay), because to accelerate the set-up activity is often required an initial investment in machinery.

The allocation of a value to the time and cost of this activity is very complex. This complexity stems from ignorance and disorder in the execution of operations that make up the set-up process and from the characteristics of these operations. For this, it's difficult to find different cost and times in similar set-up process. This problem was solved, from the production management, setting a time and cost standard for set-up process within the productive system in the factories.

This paper aims to solve some of the problems posed by this activity and change its current image within the Production Management:

1. This paper aims to provide another image of the set-up activity, beyond the simple title of "waste operation", which it has been attributed to this process. Well, now, adaptability of the machine to the variety of products that the market sue, is a key competitive factor, with the saving of investment in new machines.
2. The complexity of some activities and the relationships to develop them, together with the demands of time, cost and quality, which is derived from this adaptation of a machine into a another new, it allows to say that the set-up activity as a process rather than an operation.

Thus, in this paper, it is proposed to define the set-up activity as the group of activities on the production means to transform and to adapt them to new work requirements and the market demands. And the set-up time as the stop time of machines of production lines from the time that the last product leaves the machine and the new product is make in the machine or production line. This time will condition the response of the productive system of a factory to market needs and it is a parameter that allows to assess the response and adaptation of the production unit to market demands

The set-up activity is a process that it requires the implementation of an adequate logistic, not only of the operations involved, but also, of information, of materials and workers who allow their successful carrying out. Like all processes, the set-up activity must be subject to criteria of cost, efficiency and performance, as any production process.

HISTORICAL DEVELOPMENT OF SET-UP PROCESS

The set-up process is an activity that it has always been present in any productive system, from the carpentry workshop and the blacksmith workshop, back in ancient times, they had to prepare his desk and choose the tools to prepare different elements, depending on the element to be manufactured, to modern aircraft industry. However, being an old process, this activity always was associate to a bad activity but a necessary activity that it's better to remove. These changes have been imposed by the evolution experienced by the market. Thus, in the twentieth century, before the seventies, the production used to be marked by high production lots, which forced to maintain large stocks of raw materials, to meet the production and the final products, to anticipate the demand for the product on the market. The time invested in the set-up process was a factor of minimal relevance, as the cost was diluted in the product cost to make it, and its implementation was left to the expertise of the workers, and the only requirement of to be make in a correct way to prevent stops during the manufacturing process by performing a defective set-up process. In addition, not only there was a few set-up process, but also it was unusual for a company to offer several lot of products, which it forced it to perform many set-up process. It is not surprising that Henry Ford, in the early twentieth century, said: "Can you choose any color, but it must be the black color". This phrase offers an approach to philosophy that it was immersed the production system at this time. When the market evolved, decreasing the units in the demand, the factories decreased the quantity of product to manufacture, and increasing the number of new products to maintain a stronger presence in the market. Then, the set-up cost became very important and to conditioner the manufacture. In order to minimize the impact of this cost in the production departments, these began to devote their efforts to minimize this time and cost. They found many problems that revealed the neglect in which the set-up process was submerged:

- There is not a methodology to study this kind of process and its operations.
- The workers impose the conditions to make the set-up process. It is not the machine that impose the work rate. And, too, the workers consider the set-up process as a "waste operation".
- The machines were designed to achieve high production rates, but not to do a quick set-up process.
- An incorrect layout of the tools to change, it forced to do long circuits for the transport of the pieces to put and to remove of the machines.

Actually, all these factors reflected the lack of attention that all production managers, as machine manufacturers, had in the set-up activity during many years and the fact entrust this operation to the skill and ability of maintenance workers; whereas, sometimes, an economic incentive accelerated this operation. Consequently, this led to the set-up process as a marginal operation of all production process and it considered "a waste operation", no more value than that of preparing the machines or production lines to produce a new product, and no other condition that prevent this operation does not cause delays during the production of products, due to improper execution. Early efforts in reducing setup times of machines were resolved by organizational measures, sequencing products requiring common elements for production, thereby changing time is reduced. These first attempts at reducing time were dismissed, because if the lots were too small, and time spent on adjustments had to be made during the process were high, the machine reached the operating speed at the time it was over the production; In addition, these alternatives were increasing stocks of manufactured product, or delay in deliveries.

The evolution experienced by the market, hovering over the concept of customer service, has re-affirmed, yet again, the position of the set-up process in the production management; which, while still being considered "wasteful operation", it has become one of the basic operations of the Flexible Manufacturing, as of the Successful Factory defined by Tompkins (1989) and it must be optimized to minimize its time and cost.

With the arrival of the production models in smaller lots, it started to study the set-up operation and arriving to develop various models for improvement with a single goal: To minimize the time spent performing set-up process. The most important models are:

- SMED Methodology (Single Minute Exchange of Die). This methodology was developed by Shingo (1985).
- OTED Methodology (One Touch Exchange of Die). It was developed by Shingo (1985) as a posterior evolution of SMED Methodology.
- Model Zero Preparing Machines Model. It was proposed by Sekine and Arai (1992).

In the period 1950-1960, in Japan, Shingo (1985) developed, as a consultant, the idea that the set-up time could be reduced. Later, and especially, as a consultant for Toyota Motors, he had the opportunity to implement new ideas to minimize the set-up process time. This accumulation of ideas originated the SMED methodology to reduce the time to prepare the machines to make a new product. Shingo (1985) introduced two important concepts, developing the SMED methodology. He identified, within the set-up process, two types of operations: internal operations and external operations. These concepts have been the basis for all subsequent improvement methodologies about the set-up activities:

- The internal operations. They are all operations that they make up the set-up process and they have to be developed with the machines stopped.
- The external operations. They are those operations of set-up process that the workers can be performed without the machines stopped, for example, prior to stop, or even after shutdown of the machines.

The improving, or eliminating, of the set-up process has been driven by buoyant industrial sectors and highly competitive, for whom customer service is the basis of its business management. These have caused in its suppliers a review of the values in this activity within the global production process, reaching improvement levels considered unreachable at first.

The importance that the set-up process has acquired in recent years, (Harmon and Peterson, 1990; Suzaki, 1991) in the management of production, it is supported by the number of papers published recently and they confirmed the set-up activity as a basic process in the production process works, and a key activity to adapt the factories to the demands of its customers (Reiman & Wein, 1998; Righter & Shanthikumar, 1998) to obtain a competitive product for the customer (Katok et al., 1998a; Hwang and Singh, 1998; Magnanti & Vachani, 1990; Duenyas et al., 1996, 1998). Many authors (Katok 1998b; Macintosh, 2001; Birmingham & Jelenik, 2007; Kusar, 2010; Herr, 2013) consider the set-up process as an operation, but it is a process. A process in which a machine, in an initial state, transforms, by small operations, in a machine that can process a different product.

However, although the market has evolved considerably, the concept still has about the set-up process has not changed, and is still considered a waste operation in many productivity process from the point of view of production management, but, however, from the Management Company should be considered otherwise because:

- The set-up process will allow adapting the production process to customer requirements.
- The set-up time must be regarded as the parameter to evaluate the flexibility level of a production line.
- The availability of a machine or production line that can adapt to all market demands through the set-up operation, it will allow savings in new investment to the company.

Therefore, there is a new image of the set-up process. The image of a process that it allows companies to adapt the production system to the needs of the market. And, like any other process, the set-up activity must be subject to improvements that they allow bring to the market a product at the lowest cost and as soon as possible.

VARIABLES OF A SET-UP PROCESS

The set-up process consists in a set of operations that will adapt the production equipment of a factory to make a new product. Therefore, set-up process, as any production process, is associated to time and cost criteria. It is formed of a set of variables whose control will establish, within reasonable values, the cost and time of set-up process. In addition, acting on these variables, the process can be improved by minimizing the time and cost. The most important variables that can affect the development of set-up process are:

1. **The Design of the Machine:** The design must be focused to set-up operations. For example, the access to interchangeable parts of machines must consider in the design step, because the location of these parts will restrict the access to them and the safety conditions will condition the set-up operation.
2. **The Workers Involved in the Process:** This is the most important variable, while in a production process the production speed is marked by machines, in the set up process, the speed, is marked by workers to participle in the set-up process.
3. **The Products to be Processed:** They are object of change. Its shape, size, etc. are determining factors in the process.
4. **The Organization of the Set-Up Process.**
5. **The Safety Elements in the Machines:** The safety is a very important aspect to consider when the workers perform the set-up process because they are exposed to different situations of risks in the machines: electric risk, blow risk with pieces of the machines, risks derived of the work position in/on or into the machine, risk derived of body position of worker to reach the pieces.

TYPOLOGY OF OPERATIONS IN THE SET-UP PROCESS

The set-up process consists of operations conditioned by the type of product previously processed in the machine and the type of product to be processed further. The operations that make up the process are the following, although its relevance will depend on the type of set-up process and the improvement level:

- The storing of machine pieces and the tools.
- The transport of the tools and pieces of machines.

- Exchanges operations to place or removal pieces in the machines.
- Cleaning the elements (tools and pieces) to put or eliminate of machines during set-up process.
- The fits of pieces in the machines and the machines.

Storage and Transport of Tools to Use in the Different Operations

Both operations are connected because the most transport operations, at least the most important, are between the production unit and the department store. Storing of removed parts of the machine, in right place, is in this operation group.

The parts to be exchanged, and they are not use in the new production process, need to be stored somewhere in the factory. When they are necessary for the production process, they transport from its storage place to the machine where they will use and returned to their storage place after removal of the machine. The factory design does not consider the set-up process, but it should be considered the need for space near machines to locate the necessary tools for the set-up operations. When the plant design considers these operations, it is necessary to know:

- All kind of product that the machine can process.
- The different formats, present and future, in the machines.
- The number interchangeable elements in each format.
- The size of the pieces.
- The frequency of changes

Thus, a piece and tool storage located correctly with respect to machines, will minimize the logistic cost and, therefore, it minimizes the cost and the time of set-up process. The place to locate the pieces of the machine is a very important variable and the area that they need depends on:

- The different formats that the machine can process
- The different sizes of the interchangeable pieces.

When there are several different machines, the identification and the location of interchangeable pieces are important.

The Exchange of Element in the Machines

These operations are very important. They can have conditioned the future production. The exchange operations are conditions by the physical form (shape, size and form) of products, by the piece position in the machine, the type of tool to use or by the fastener system to the machine.

The initial design of the machines and the situation of interchangeable pieces will condition the performance system of workers at this step of set-up process.

The Cleaning of the Machine Pieces and the Tools

The cleaning level of these elements is very important to obtain an optimum set-up process. The cleaning makes easy the exchange of pieces and the correct handling of tools. In filling and package processes, for example, in the pharmaceutical industry, the dirt in the elements lead to the contamination of products. If the elements necessary for the new product are the same they are used in machine, then, during the set-up process, it's necessary to clean the parts in contact with the product.

To use dirty pieces or tools can cause very important problems. They can affect to the quality of new product.

The Fit of Elements and Machines

They are all operations performed on the pieces and placed in a machine to get it the optimal realization of their productive functions. These fittings are to adapt the machine to the physical-chemical of new product to be processed, or to correct the piece positions placed in the machines.

The fittings are an important cause of the high times of the set-up process and that they are not conform to the standards values. The fit operations is between the 15 and 20% of set-up time, processes executed properly. However, uncontrolled processes can be up to 50% of the set-up process time. During the set-up process, it may be normal to perform 2 different types of fittings:

- **Fittings to Place the Elements of Machines:** They take place during the placement of the elements in the machine. These fittings during rectification operations make at the position of the elements in the machine. If the machine has several elements to exchange, these fit operations will make when the worker put each element in the machine.
- **Fittings on the Machine Operation:** Once all the pieces in the machine, it`s necessary to test if the machine works in if the machine works in the correct form.

The workers are the more important cause of the fittings. Therefore, to minimize the influence of the human factor, decreasing the existence of fittings, it`s necessary to make protocol with all set-up operations and their operation time necessary to make every set-up process. The protocol will indicate the technical details of how to put the pieces on the machines and a proper design of elements and link places in the machine to minimize the human factor during de fittings. Another factor that will influence the existence of fittings is inappropriate maintenance of elements.

The characteristics of the product to entry in the machines, usually, is another reason of fittings, especially in the filling machines. It is necessary to adapt the machine to the product specifications (density, viscosity, etc.) during the filling time. These fittings are due to the physicochemical product specifications have not fixed values, but generally, these parameters must be within a range.

TYPOLOGY OF SET-UP PROCESS

The products that cause the set-up process mark the difference between that change and the others in the same machine. Therefore, each set-up process is made up of an operation groups that it characterizes them and, not always, the set-up process, in the same machine, can be compared with other made in the same machine or production equipment because they have different operations that make up each set-up process. Therefore, these processes is classified into three types: Changeover, product change and presentation change.

- **The Changeover:** It is the set-up process originated to exchange of elements in the machine and is due to a variation of the shape o size of the new product.
- **The Product Change:** It is the set-up process carried out in a machine and caused by a change in the physicochemical properties of the product. For example, in filling processes, filling a cosmetic gel after filling a shampoo. This type of set-up process doesn`t involve changes in the machines, except the filling nozzles, but it needs time to fit operations with the elements of filling machine and the machine to adapt the machine to the fluid characteristics: viscosity, density, etc.
- **The Presentation Change:** It is the set-up process caused for a change in the visual aspect of the product. This visual aspect is not a change of dimensions, nor shape, nor size. It's only a change in the appearance of the product. For example, if it only changes the label of the product, or if it changes the colour of the bottle, etc.

Actually, in a production unit is difficult to perform a pure set-up process type, its normal a set-up process prevails over others. This is the reason for it`s difficult to obtain the same set-up time in the same machine or production line in all processes. Those "extras" operations, in a set-up process, are the causes that they impede to compare the set-up times and costs in the same machine. Thus, a changeover always involves a presentation change, but it does not involve a product change, unless vary the material it is made. A product change involves a presentation change, but it doesn`t involve a changeover. A presentation change doesn`t involve a changeover or product change; for example, a change in the way packaging or label for the customer identification.

Any set-up process consists of a operation sets, that it carried out in sequence, allow that the e machine is ready to develop a new product. Each type of set-up process is shaped by its own set of operations. Thus,

- The changeovers involve, basically, exchange parts in the machine, and the closeness of elements in order to adapt the machine to the shape or the size of the new product to work. Normally, during the changeover there are fittings operations to place the elements in the right way.
- The product change involves cleaning operations in all parts of the machines that they will be in contact with the product to fill or pack: hoppers, nozzles, pistons, pipes, etc. This type of change can also suppose, depending of the physicochemical properties of the product, exchange operations in the parts in contact with the product.
- The presentation change involves the exchange of elements that modify the appearance of the final product, such as the position of a label, or the change a bottle, or a change in the condition of packaging. A presentation change rarely involves movements of elements in the machine.

THE METHODOLOGY FOR THE STUDY AND THE IMPROVEMENT OF SET-UP PROCESS

The improvement of set-up process focus on the minimizing the cost and time of the process. The first step is a preliminary study of the process. This study, depending on the circumstances and the kind of set-up process, it is often a complex and tedious work.

- A complex work because not only requires the complete study of various processes to get to know all the operations that it make it up, but also as a process that it is governed by the modus operandi of workers involved in the set-up process.
- A complex work because it requires to modify the elements in the machines, but also it involves changing in the work habits in the workers involved in the process.
- A tedious work because the process generally carries out with the production equipment stopped and usually it takes advantage of this to perform other operations that tend to merge with the set-up operations modifying the study.

The successfully of the set-up process study should lead to know:

- The operations that they shape the process.
- The workers involved in its realization.
- The flow of operations that shape change.
- The others change taking place during operations.

The set-up process study lead to an optimization of the process that it all time takes into account:

1. An improvement of set-up process time.
2. To minimize the set-up process cost.
3. In some cases, when the improvement time is very important, but without an improvement in the process cost, it is necessary to analyse the improvement influence on the cost per unit of product produced.

Any improvement process must involve the participation of workers involved in the set-up process. Especially in those set-up operations make with stopped machines. The write experimental information and footage obtained experimentally are important tools for success in the improvement process. The data statistical studies are not a good tool because every data must be analysed together with the context in which it obtained.

THE TIME OF SET-UP PROCESS

The parameter to measure during the performance of a set-up process is the time necessary to make each set-up operation, Patel et al. (2001). An analysis of these times reveals that each operation has three associated times. For example, the operation of attaching a piece to the machine consists of previous operations such as:

- The time necessary to pick the piece.
- The time necessary to put the piece on the machine.
- The time to pick the right tool for fixing the piece to the machine.

When the piece is on the machine, it is necessary to make other little operations that they take its time:

- The time to separate the tool and the piece.
- The time to put the tool in a suitable place (toolbox), easy to find for later use.

Therefore, in the improvement process is important to break down the operations into smaller operations and see the time associated with each. So that one is the main operation, that it will have associated other operations around it. Thus, the time taken to perform an operation is the sum of the times of all operations that has broken down the main operation:

1. Time spent in conducting previous operations before the main operation, t_a.
2. The time spent in the main operation, t_{op}.
3. The time spent in carrying out following operation to main operation, t_p.

Thus, the time, t_i, necessary to make the operation, i, can be estimated as the sum of these three stages:

$$t_i = t_{ai} + t_{pi} + t_{opi} \tag{1}$$

The analysis of similar operations, in different set-up processes on the same machine, is different depending on the worker who performs and the situation around the set-up process. Further analysis of these data revealed that this time presents the behaviour of a distribution β which allows the study of new set-up processes, with similar operations, it can be assigned to each operating the time whose estimate may performed by calculating the probable time, optimum time and pessimist time of process as Rodríguez Mendez (2003) proposed. Thus,

$$\text{Probable time, } t_{probable} = t_a + t_{op} \tag{2}$$

$$\text{Optimum time, } t_{optimum} = t_{op} \tag{3}$$

$$\text{Pessimist time, } t_{pessimist} = t_a + t_{op} + t_p \tag{4}$$

The distribution β result of developing the equation:

$$\beta = \frac{t_{probable} + 4t_{optimum} + t_{pessimist}}{6} \tag{5}$$

Therefore, substituting equations (2) (3) and (4) in equation (5), the distribution obtained for each operation β, i, of set-up operation:

$$\beta_i = \frac{5t_{ai} + 6t_{opi} + t_{pi}}{6} \tag{6}$$

That establishes:

Set-up operating time, $t_i = \beta_i$ (7)

The equation (6) indicates where we direct to the efforts to minimize the set-up process time. According to it, the significance of the previous and main operations will need a greatest effort and that they will allow a significant reduction of time to improve the set-up process. Hence, improvement efforts focused on organization and element preparation operations will obtain the optimum value of the process time (Claunch, 1996).

IMPROVEMENT OF SET-UP PROCESS

Several models are proposed to improve the set-up activities (Mcintosh et al., 2001; Shingo, 1985). But, Rodríguez Méndez (2003) suggest to improve the set-up process focus on the minimizing the time and costs following the next steps:

1. **The Complete Study of the Set-Up Process:** The purpose of this study is to identify all activities carried out during set-up process. The operations necessary to make the change. To identify the operation sequence. To identify all activities that it will be deleted. Then, all necessary and identify operations to group them according to the type of set-up operations established.
2. **To Organize All Operations:** With the set-up operations elaborates a protocol where with the operation to identify the operation time, to identify the pieces necessaries to put on the machine. It is necessary to identify the storage location of each of piece.

In this step there is necessary to establish a first approximation about what operations are external and internal, as Shingo S. defined.

3. **All Organization and Preparation Operations Must Become External Operations:** This will minimize the cost and the time of set-up process.
4. **A Redistribution of the Activities that Make Up the Internal Operations to Level the Efforts:** This step will optimize the time and cost of process and it allows eliminating unnecessary elements and workers of workplace. It`s very important to analyse if it`s necessary several people working in parallel.
5. **To Identify the Micro-Operations of the Main Operation**: From the protocols, to identify the micro-operations of the main operation, with the time required for its completion.
6. **To Identify the Critical Operations of the Set-Up Process, the Internal and External Operations:** The PERT methodology helps to identify these operations.
7. **To Minimize Execution Time of Some Operation, t_i, Minimizing Any Variable t_{opi}, t_{ai}, or T_{pi}.:** The identification of the operations that make up the set-up process to minimize execution time of some operation, t_i, minimizing any variable t_{opi}, t_{ai}, or T_{pi}.

The elimination of the main operation represents obviously the elimination of previous or posterior operations. However, not always it`s possible to improve the main operations, but it`s possible to improve the previous and posterior operations. The minimizing of these operations it`s possible to obtain an improvement of operation time. For example, if the main operation is to fix a piece on the machine, if is not possible to improve this operation, perhaps it`s more easy to improve the previous operations modifying the fastening system with other system easier or more quick that it doesn`t need a tool to fix the piece on the machine.

Also, it`s possible to minimize the operation time, minimizing t_{opi}, without to increase the number of previous and posterior operations. This is possible if:

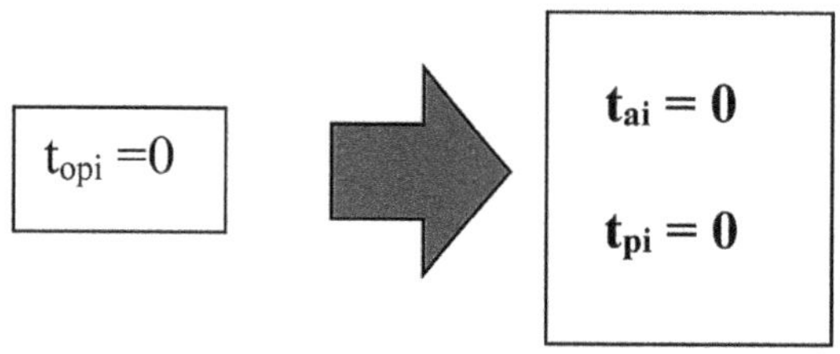

a. To make the operation with several workers in parallel. However, this action must be careful studied to avoid an increase in the set-up operation cost. This operation type causes situations where many workers can delay the set-up operation to interfere their works.
b. A study of the movement of workers during the execution of exchange operations of the pieces, allows to eliminate or minimize their movements and to obtain a minimizing of times, t_{ai} t_{opi}, or T_{pi}.
c. The technical improvements should be focused on:
   i. Modify the design of the pieces to exchange during the process, especially in the part focus on the machine, simplifying the piece to reduce the time for carrying out operations necessary to fix it on the machine.
   ii. Adopt fastening systems similar to propose by Shingo S.

It`s necessary to eliminate, or minimize, the fittings operations. This is possible using reference scales, on the pieces or on the machines where the pieces are placed. The improvement is obtained focusing the efforts to minimize the time t_{ai}.

8. **The Difficulty May be Properly Identifying which is the Main Operation**: The person who studies the process of improvement may apply this definition with all the flexibility you want.

This stage is the most creative of all and allows you to modify or eliminate any operation so that the variable, equation (1), t_{opi} is null, doing $t_i = 0$ because the previous and posterior operations are removed.

In steps 7 and 8, it`s important to analyse the area of machine pieces. If this area is similar in all pieces then it`s possible to use the same tool with them and, in consequence, to minimize the positioning time on the machine. For example, if all bolts need the same key to fix them on the machine it`s possible to reduce the time to look for the tool.

9. **To Reach to the Desired Improvement Level It`S Necessary to Restart a New Phase of Improvement Model, Applying It to the New Set-Up Process Generated:** If this has been a major improvement over the previous process, it must allow learning time to the workers involved in its implementation, in order to obtain more consistent improvement values during the new step.

10. **To Audit the Set-Up Process**: Sometimes, it is necessary to check every set-up process of the factory. The objectives of this audit are several:
    a. To detect deviations in the work way of workers. It is necessary to correct the deviations. With this, the production management is sending to workers the message "the set-up activity is a very important process like the productivity process".
    b. To update the work protocols about the set-up process. To look for new actions to improve the process.

During each stage should be to examine whether there has been any change in the PERT for the analysis of internal and external operations, in order to seek new critical steps where to focus the improvement efforts. Always looking for the two steps outlined in the process: A time reduction step followed by another step of cost reduction through technical improvements and efforts redistribution.

Any improvement will be composed of a flow of information between the workers involved in the set-up process because any modification in the work protocol will require an adaptation period to adaptation to new set-up process.

CONTROL OF THE IMPROVEMENTS ACHIEVED

Performing a set-up process depends on the speed of workers involved in it, so, in time, there is some relaxation during the process and will result in increase of the time and cost process. For this reason, sometimes it`s necessary to examine the set-up process, in order to avoid discrepancies between the actual execution times and standard, which were established. The frequency of the controls is given by:

- The duration of the products on the market.
- The occurrence of deviations in the realization times of set-up process.
- The emergence of new products that require new protocols.

These controls are necessary to check on the following four important aspects:

1. The control of protocols.
2. To control the warehouse of tools and pieces necessary to make a set-up process.
3. The control of internal operations during the set-up process and, then, to contrast the values obtained with the protocols.
4. The transport control of pieces and tools.

CONCLUSION

The set-up process is a very old activity in all industrial processes. But, being an old activity it is the most forgotten activity in the industrial processes. Only, at the end of the twentieth century, the production management saw the necessity to study and improve it. There are several reasons to say that the set-up process is, perhaps, the more important activity within of productive process:

- The set-up time gives the level of process flexibility.
- The set-up process carries out in a right way it will allows a quite productivity process without problems leads from the pieces in the machines.
- A machine where it`s possible to make many products it reduces the capital investment in new machines.
- If this set-up process is performed correctly, it can make a product of high quality, without to modify the production ratio and not delaying the completion date

The set-up process is very important because increase the competitive of the companies and it is the measure of the flexibility of their production processes; for that reasons, the production managers will improve and it will control its operations to avoid the increasing of the time and the cost.

The improvement of set-up process is a complicate activity because it needs knowledge of the machines and it is necessary the participation of workers in that process. In the first step is necessary to identify, and then to separate, the next step is to eliminate the activities not necessaries for the process, then to classify the necessaries operations in external and internal operations. The next step is to perform the external operations with the machines working and the internal operations when the machines are stopped. And finally, to reduce the time and the cost of the set-up process.

REFERENCES

Birmingham, F., & Jelinek, J. (2007). *Quick Changeover Simplifies: The Manager's Guide to Improving Profits with SMED*. New York, USA: Productivity Press.

Bueno Jauregui, R. (1991). *Las Claves de la Gestión de Producción*. Madrid, Spain: Editorial Alción.

Claunch, J. W. (1996). *Set-up Time Reduction*. Chicago, USA: Irwin Professional Publishing.

Duenyas, L., & Van Oyen, M. (1996). Heuristic Scheduling of Parallel Heterogeneous Queues with Set-Ups. *Management Science*, *42*(6), 817–829. doi:10.1287/mnsc.42.6.814

Duenyas, I., Gupta, D., & Olsen, T. L. (1998). Control of a Single-Server Tandem Queueing System with Setups. *Operations Research*, *46*(2), 218–230. doi:10.1287/opre.46.2.218

Harmon, R. L., & Peterson, L. D. (1990). *Reinventar la Fábrica*. Madrid, Spain: Ciencias de la Dirección.

Herr, K. (2013). *Quick Changeover Concepts Applied*. New York, USA: Productivity Press.

Hwang, J., & Singh, M. R. (1998). Optimal Production Policies for Multi-Stage Systems with Setup Costs and Uncertain Capacities. *Management Science*, *44*(9), 1279–1998. doi:10.1287/mnsc.44.9.1279

Katok, E., Lewis, H. S., & Harrison, T. P. (1998a). Lot Sizing in General Assembly Systems with Setup Costs, Setup Times, and Multiple Constrained Resources. *Management Science*, *44*(6), 859–877. doi:10.1287/mnsc.44.6.859

Katok, E., Serrander, T., & Wennstrom, M. (1998b). Setup Reduction through Process Improvement in Aluminum can Manufacturing. *Production and Inventory Management Journal, Fourth Quarter*, 77-82.

Kusar, J, Berlec, T, Zefran, F, & Starbek, M (2.010). Reduction of Machine Setup Time. *Journal of Mechanical Engineering*, *56*(12), 833–845.

Magnanti, T. L., & Vachani, R. A. (1990). Strong Cutting-plane Algorithm for Production Scheduling with Changeover Costs. *Operations Research*, *38*(3), 456–473. doi:10.1287/opre.38.3.456

Mcintosh RI, Culley SJ, Mileham AR, &d Owen GW (2001). *Improving Changeover Performance*. Oxford, Great Britain: Butterworth-Heinemann.

Patel, S., Dale, B. G., & Shaw, P. (2001). Set-Up Time Reduction and Mistake Proofing Methods: An Examination in Precision Component Manufacturing. *The TQM Magazine*, *13*(3), 175–179. doi:10.1108/09544780110385528

Reiman, M. I., & Wein, L. M. (1998). Dynamic Scheduling of a Two-Class Queue with Setups. *Operations Research*, *46*(4), 532–547. doi:10.1287/opre.46.4.532

Righter, R., & Shanthikumar, J. G. (1998). Multiclass Production Systems with Setup Times. *Operations Research*, *46*(3), S146–S153. doi:10.1287/opre.46.3.S146

Rodríguez Méndez, M. (2003). *El Proceso de Cambio de Útiles. La Flexibilidad de una Fábrica*. Madrid, Spain: FC Editorial.

Shingo, S. (1985). *A Revolution in Manufacturing: The Smed System*. Cambridge, USA: Productivity Press.

Sekine, K., & Arai, K. (1992). *Kaizen for Quick Changeover: Going Beyond SMED*. Cambridge, USA: Productivity Press.

Suzaki, K. (1991). Competitividad en la Fabricación en la Década de los 90. (Técnicas para la Mejora Continua). Madrid, Spain: TGP, S.A.

Tompkins, J. A. (1989). *Winning Manufacturing: The How-to Book of Successful Manufacturing*. Norcross, GA: Industrial Engineering and Management Press.

Chapter 17
Assets Management and Risk Control

María de Lourdes Eguren Martí
Universitat de Barcelona, Spain

ABSTRACT

Risk management and internal control is a subject that has increased its relevance due to the recent financial scandals on companies like Enron and Worldcom, and the increment of fraud cases and financial misstatements around the globe. In line with this, several initiatives have been defined or required in order to control the risk exposure in the company processes, including the development of internal control standards like Sarbanes-Oxley (SOX). One of the factors to be considered when taking optimum decisions is risk. Due to this, in this chapter the key concepts over risk assets management will be exposed, including a practical example under SOX framework as well as a system's approach and value management perspective.

INTRODUCTION

Globalization, uncertainty and the high volume of information which is currently available in the financial world, are factors which affect any decision making process under the current environment; considering enterprises are open systems, the turbulence of the environment generates an impact from risk management perspective (Febles & Oreja, 2008) . In this sense, to have into consideration all the variables and elements which can have incidence into the matter of study, is a key aspect for a proper decision making, as there are factors like reputation risk, information availability, process design among others which increase the intrinsic complexity given by the environment (Leisinger, y otros, 2010). In addition to this knowledge, is important to understand this needs to come together with having in place the proper procedures and methodology, otherwise, this will open the door for not taking into account, all the aspects that must have being considered as well as lack of consistency over the applied criteria's; in this sense the need of an integrated approach considering existing variables, open the door to defining new methodologies for an improved management, with a more efficient focus and adapted to reality (Porter & M, 2011)

DOI: 10.4018/978-1-5225-0651-5.ch017

As stated above, current turbulence of the environment and the occurrence of several financial scandals (like Enron explained in detail in a further section), have increased the firm´s interest on risk management procedures and requirements. As a consequence, academic researches on this topic are growing as well, however these researches are still under an early step of evolution, which give opens the door for new and original developments in this area (Pfister, 2009)

More specifically when considering decision making on asset management, one of the key aspects to have into consideration for being able to take optimum decisions, is an adequate management of the risk considering all the types of risks that can be applicable over asset management under a holistic approach (Ittner & Oyon, 2014)

In order to assure risk assessment is done effectively, the methodology and clear understanding of the nature of each of them complement the knowledge over the assets management process. With this being said, we can consider that the following aspects need to be considered from risk perspective, in order to contribute with an optimum decision making by understanding entity objectives, business processes, organizational resources, structure roles and responsibilities (Arens, Elder, & Beasley, 2014):

- **Process:** A deep knowledge over the applicable processes and procedures is required as mentioned above.
- **Information Availability:** As in any decision making process, the quality and completeness of information will be directly correlated with the level of the decisions taken, from an optimal point of view. External information for stakeholders, provided by the company will be affected as well by the quality of the internal information coming from proper risk management (Feng, Li, & McVay, 2009)
- **Understand the Risks:** Risk mapping and a proper identification of the potential risks, is the first basic step for defining an efficient risk strategy. Due to this, to have the right understanding on risk meaning and proper approach over risk coverage are key aspects to be able to capture the main exposures and be able to define the proper controls. The relationship between risk and processes is fundamental, only one risk may be really relevant affecting all processes and areas(Scandizzo, 2005)
- **Internal Control:** Once risks have been defined, the related measures in order to have them "under control" need to be implemented. In this sense, a proper internal control environment adapted to each of the cases (e.g. company size, legal requirements, etc.) must be defined, taking into consideration the available resources, applicable laws, financial impact of the risks, ability to adapt in case of changes over the process or strategy among other factors that will be detailed in this chapter. Management of internal control, is usually referenced to an "internal control framework", recent studies evidence companies select and adapt their internal control frameworks according to their needs, by taking into account specific constraints, contingencies and firm's characteristics. (Jokipii, 2010)
- **Change Management:** Processes are constantly subject to changes. In this sense, the ability to capture these changes and adapt to the new situation and potential new risks, is basic in order to generate the adequate information for decision making. Management control systems, can support the firm on adapting to changes (from a paradigm point of view or by generating new scenarios), or become a growing issue for the firm. (Kloot, 1997)
- **People Management:** Risks and people are two elements that can't be separated. Several risks can come from aspects not directly related to the people involved in the process (as it could be the

case over uncontrolled aspects coming from the environment). However, a significant amount of the risks that finally are not detected and generate a financial impact, come from the fact of having human errors not detected, issues on process design, lack of training or fraud. "to conclude that internal control is effective in any category of objectives - operations, financial reporting, or compliance, requires that a substantial number of other control elements were present, such as the ethics and competence of people and management's philosophy"(Simmons, 1997). In fact, several internal control frameworks, like COSO have a dedicated section to those controls which are common to all the company (Entity Level Controls) covering aspects like people hiring process, performance management, on boarding procedures, training plans, job description definition among other aspects.

Each of the mentioned topics will be detailed in the body of this chapter, together with a practical application under Sarbanes- Oxley (SOX) framework. The reason of selecting this framework instead of other internal control approaches is due to the fact that a significant amount of companies are obligated by Law to comply with SOX requirements, so this could be of help to the readers of this chapter. In addition to this, SOX approach is based on a flexible and process oriented framework, which can be adapted to different companies with independence of their complexity and dimensions.

However, risk management can't be properly performed without having into consideration that resources are limited, a margin needs to be achieved and the probability of implementing non-added value control instead of helping to minimize the risks can generate additional risks. Due to this, value management needs to be considered in junction when defining an asset management internal control environment. Recent studies evidence the relationship between internal control and enterprise value (LIN, ZHENG, & WANG, 2007), moreover firm´s governance is directly linked with company culture, values and ethics, which are supported by those firm elements which assure firm transparency, as it's the case of auditing processed, reporting structures and internal control.(Wieland, 2005)

Having these aspects into consideration, the objective of this chapter is to present with a practical approach the elements and the methodology, in order to implement a proper internal control framework, sustainable and adaptable to each company or case.

More specifically, chapter will covers those elements included in Figure 1.

ASSETS MANAGEMENT AND RISK CONTROL

Issues, Controversies, Problems

Risk management, is inherent to human activities. When it comes to enterprises risk management, the relation of this concept with internal control activities has a direct relation and this link has been increased in the recent years:

The concept of risk has become central to corporate governance and linked to the idea of internal control. In the process, the meaning of both terms has shifted. The shift is symptomatic of important changes in the way corporations govern themselves. (Spira & Page, 2001, p. 641)

Figure 1. Chapter Layout
Source: own elaboration

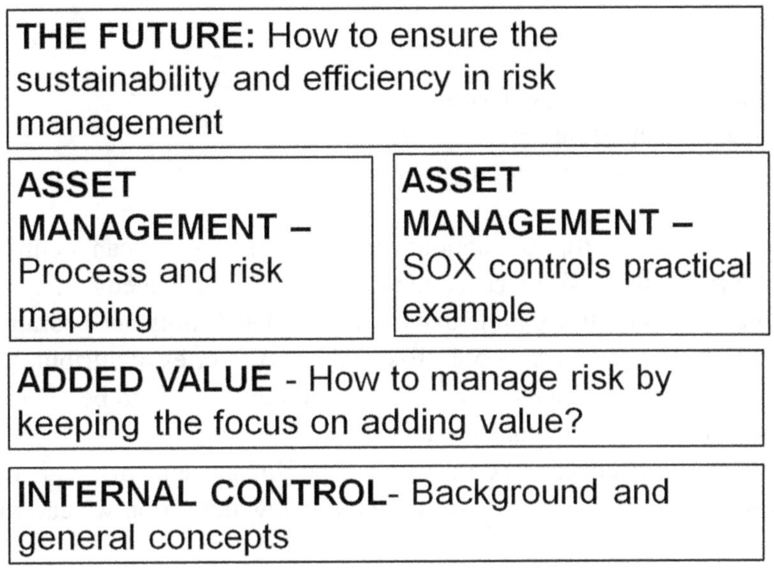

Having this into consideration and focusing the discussion over this aspect, internal control detractors and supporters had increased their differences in their postures especially after the Sarbanes-Oxley implementation in 2002.

Main critics on internal control are related to the increase of costs, empiric evidence on risk contention and advantages for the company (Ribstein, 2002). The increased audit cost, the lack of quantitative evidence showing a relationship between internal control and risk occurrence as well of "real" advantages for the company, are the main aspects defended by this position.

By other hand, defenders support their opinion over the financial impact of scandals like Enron's, the evidence of a need of an improvement on internal control structures, as well mandatory nature of Sarbanes-Oxley Act.

Both positions have valid and fair positions regarding this topic; however the added value which can be generated by a proper internal control structure is an aspect which is generally missed on the existing literature. The same happens when we refer to the human factor that represents the implementation of an internal control model.

In line with this point of view, we will cover all those aspects, given the tools to management to perform a proper risk management implementation on assets management processes.

Internal Control: Background and General Concepts

In order to be able to define the proper measure for managing risks over assets, in this section the main concepts over internal control will be defined and clarified, setting up the basis for the next topics in this chapter.

According to the Committee of Sponsoring Organizations of the Treadway Commission (COSO, 2013), Internal Control can be defined as:

"Internal control is a process, effected by an entity's board of directors, management and other personnel, designed to provide reasonable assurance regarding the achievement of objectives in the following categories:

- Effectiveness and efficiency of operations
- Reliability of financial reporting
- Compliance with applicable laws and regulations"

If we read carefully each of the topics mentioned in this definition, we can identify the main aspects to be considered for defining the basis on risk and internal control management.

Internal control is a process itself, meaning that is not an isolated entity but a complete set of elements which together are defined over a common objective and under a series of attributes. According to this statement, we can consider that internal control is a system that's needs to be understood and defined according to its own characteristics and objectives.

Going to the definition, a system can be defined as a set of linked elements, with a common objective (Von Bertalanffy, 1989). If we consider this split, on the below definition we can make the following deductions:

- Internal control is process, which can be modeled as a system.
- Internal control main element is people. In this sense, every employee of the company must be an active part of the internal control process, in order to align with the "Tone of the Top" principle, a definition that we'll develop in detail afterwards. In addition to this, all other resources (material and immaterial) which are related to the company need to be considered as will be an active part of the process, as for example fixed assets, IT applications, brands, etc.
- The objective of internal control process is to assure a level of assurance over a sub set of objectives defined. If we analyze this in detail, this is related with the concept of risk which has been introduced in the previous section. Risk can be minimized and monitored under a certain level of probability; however, it is not possible to prevent risk in all scenarios as there will always be the possibility of having a risk becoming a reality. Due to this, internal control needs to minimize the exposure on risk, by aligning over the mentioned sub-objectives. In this line:
    - Risk must be controlled, in a efficient way assuring an effective risk coverage according to available resources, business specifics and real exposure.
    - Internal control process must assure that the financial reporting presented by the company, is reliable and adjusted to the real situation of the company and its processes.
    - Internal control activities must be defined in a way that they not only comply with risk coverage and internal control principles, but also with the company statements and external legislation and regulations.

In this line, internal control is not:

- A written procedure
- A law or a quality standard.
- A set of controls
- A finance topic / issue.

- A theory with no link to operations and business
- Sarbanes-Oxley, we will detail this aspect. As an introduction, Sarbanes-Oxley is a framework in order to manage internal control.

So, if we summarize the above, we can say that internal control is a set of activities defined to manage the risk that comes from the processes within the company, across all functions and involving all the personnel with independence of their job.

Having this is mind; let's go back to the risk concept and its meaning under a internal control framework. According to the Concise Oxford English Dictionary (The Concise Oxford Dictionary, 2012), risk can be defined as a "hazard, a chance of bad consequences, loss or exposure to mischance".

As can be observed, this definition only reflects the "negative or undesired" side of the event, however risk can also be considered as an opportunity and result of this "unpredicted events". In this, sense the effect of the uncertainty should be considered in junction to the added value which can be generated thru an adequate risk management. This is a key aspect to have into consideration for taking proper management decisions as well as from an efficiency and effectiveness perspective.

In order to make a proper risk management, the different typologies and factors which influence risk need to be considered according to the scope and effects of the exposure, understanding the business and risks applicable to every level of the organization (Heise, Strecker, & Frank, 2014). Some of the main risks that can be considered are the following:

- Financial
- Operational
- Reputational
- Other

Financial risk includes all the cases related to the probability of having an unpredicted, unexpected or not desired event with effects over the financial statements of the company. In this category we can include specific examples as credit risk with the associated probability of default as well as other examples like over estimation of sales, losses not accounted for the period, etc. In general terms in this typology we will include all the risks with a direct effect on finance. As we will see, Sarbanes-Oxley will be focused on these cases.

Operational cases are related to the potential occurrence of any event which affects the operations of a process. In this case, we can include a wide typology, for example IT disruptions, process breakdowns, natural disasters, among others. It is important to remark, that some operational risks can generate a financial risk due to the effect over the processes behind. This is the reason why process design and systems approach is a key factor on risk management and risk approach.

With reputational risk, we can consider all those cases were the external and internal perception of the company can generate an unexpected impact over the company. There are several well-known examples over the effect of market information over a certain subject with direct effect over the sales or shares of a company, as an illustrative example we can comment about a very well-known bottled water company from Spain, that after facing some rumours over water contamination significantly reduced their sales and never recovered although it was shown rumours were not demonstrated. Other examples can be linked to corporate social responsibility with negative effects on shareholders' perception, with direct financial effects on the company value and strategic positioning (Porter & Kramer, Strategy and Society: The link between competitive advantage and corporate social responsibility, 2006).

Other risks include a wide variety of diverse situations that can affect a specific event. This variety over risks is part of the risk nature itself, unpredictable at some cases, difficult to measure and with an impact that in a lot of cases can't be measured.

Once we have defined the risk concept, we need to understand the measures on how risk can be controlled, in general terms, "controls" are the elements defined to cover this aspect.

Controls can be defined as a task or process designed to mitigate the probability of occurrence of a specific risk.

From a process perspective, applicable steps for control performance should include the following steps:

- **Control Preparer:** Person who executes and performs the control.
- **Control Reviewer / Approver:** Approver or reviewer according to process.
- **Control Owner:** Person responsible to assure control is working properly.

Based on this, controls should comply with a series of attributes in order to be aligned with the objectives:

- **Objective**: Clearly defined in order to minimise / mitigate the risks. SMART criteria can be used in order to be sure, proper objectives are defined considering the following criteria's:
    a. Specific
    b. Measurable
    c. Achievable
    d. Realistic
    e. Time-Bound

- **Owner**: Person who is responsible for control performance according to the defined parameters.
- **Scope**: Area and subject of application of the control.
- **Description**: How the control is performed. Should be clear and easy to understand.
- **Evidence**: Items that demonstrate that the control has been performed. It should include who has performed it, who has reviewed, who has approved and corresponding dates. Reports are a key element in order to evidence control performance.
- **Frequency**: How often the control is performed (e.g. monthly).

In this sense, internal control needs to have evidence on control and processes performance. If evidence is not available, from internal control perspective is the same as not having the control in place. Control performance cannot be evidences, if documentation is not available. There are different types of evidence that should be considered when designing a control:

- **Event Based:** Usually includes transaction level evidence like invoices, purchase orders, credit assessments, etc.
- **System Parameters Check:** To assure configuration is aligned with requirements, so process and embedded controls work properly. Can be done by checking system set-up or process walk-through by trying to generate an issue.

- **Procedures:** Written procedures, policies and authorization tables.
- **Reports – Key for the Company:** When tested, a significant part of the process is covered. Due to this, completeness of these reports is a key piece in effectiveness testing process.

Another key concept, which needs to be defined, is the roles definition over processes and controls. The following cases needs to be taking into consideration:

- **Roles:** Functional description of the tasks performed by a person.
- **Functional Roles**: Based on process and job description with independence of scope
- **System Roles**: Roles granted in system or IT applications, based on requirements. This must be aligned with functional roles. For example, in the case of a company using SAP as ERP, the assignation of the roles from the system according to functions must be the applicable criteria.

Roles clarity, allow to have clear processes, minimize the probability of having controls or processes with no owner, coordinate changes and action plan on a centralised way and with an expert in the area.

Written procedures are another of the aspects, to be considered when defining a control. Policies, standard operating procedures, workflows and other type of documentations, support on describing, regulating and controlling the company activities which they are referred to.

In this sense, is responsibility of every employee to know each of the written procedures is applicable to their activities. Management leadership is in this line a fundamental aspect in order to assure the proper training, distribution, of these procedures to their teams as well as for assuring they are periodically updated to reflect the business and process changes.

There are a few concepts that must be remarked:

- **Responsible**: Each written procedure must have a responsible, who should be expert in the area and in charge of updating the procedures according to the changes received by business or other sources.
- **Relation between Procedures:** All the procedures must be related, as they are not isolated elements. As long as these references are included in this procedures, more opportunities will be created to improve processes and create added value.
- **Internal Control References:** Specific references to the internal control check points, should be include improving the efficiency during control assessment. The fact of having internal controls linked to operational procedures increase the awareness, as well as the supervision and execution tasks for the related controls. If the basis is clear, operations, efficiency, and effectiveness are improved.
- **Legal Aspect:** Written procedures should be aligned with applicable laws and legislation.
- **Roles:** Functions and roles should be defined and linked into written procedures, in a way to assure the three action "who-what-how" are related and formalized with transparency and clarity.

So, summarizing we may say that risk, controls, evidence, roles and written procedures are the core elements to be considered when defining and describing internal control process as shown at Figure 2.

After defining controls, roles and evidence, control type understanding needs to be achieved as well in order to assure a proper control and risk management. In general terms, there are two types of controls, preventive and detective:

Figure 2. Key aspects on internal control
Source: own elaboration

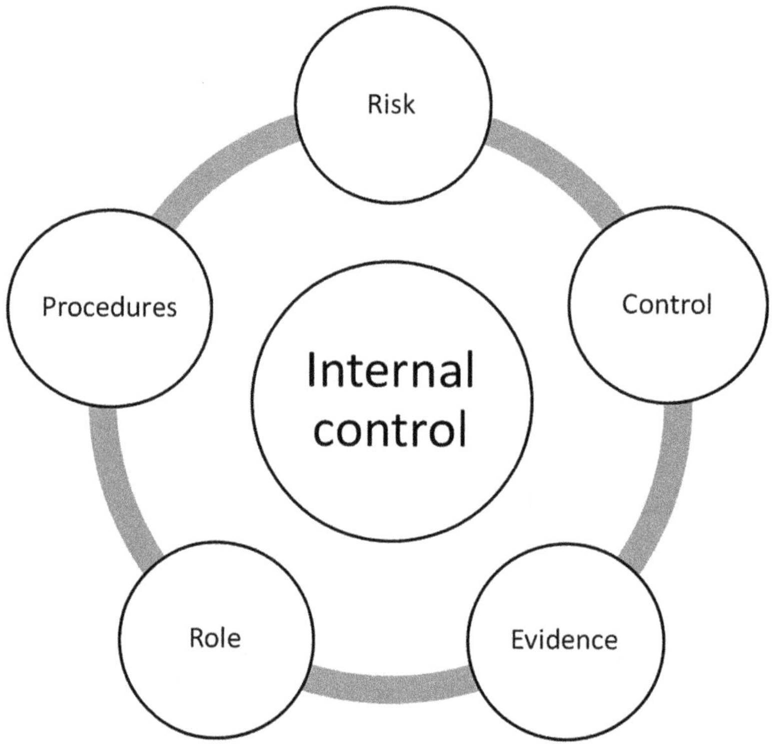

- **Preventive**: Focus on preventing errors and exceptions. Designed for identifying the risk in advance to its occurrence. Preventive approach is efficient and once implemented have less impact from a workload perspective. IT controls tend to be preventive in nature.
- **Detective**: Designed to identify issues after it has occurred. This controls are also called "monitoring controls" as according to their design support business in process monitoring, giving also additional information like performance indicators, trend, etc. This type of control stops issues before they start being recurrent and minimizing the impact, so mitigating the risk. Common examples include exception reports, reconciliations, and periodic audits among others.

In the Figure 3, we can find one example of the application of the criteria's below, showing how preventive controls are built as well as detective or monitoring control.

Preventive controls must be implemented instead of detective controls as possible. Preventive controls, avoid risk occurrence, are usually embedded in the process itself and are the aspects to be considered from resources perspective, as once implemented do not require additional changes unless process changes.

By other hands, detective control requires ad-hoc work to define, implement and execute the control on the defined frequency. This type of controls needs to have a careful design as well to assure added-value is generated as possible and "bureaucratic" controls are avoided.

This is one of the topics that generate more controversies over internal control. If the implemented actions are not aligned with business and embedded in the processes, they will be considered as an imposed requirement with no added value for the company.

Figure 3. Preventive vs. detective control example
Source: own elaboration

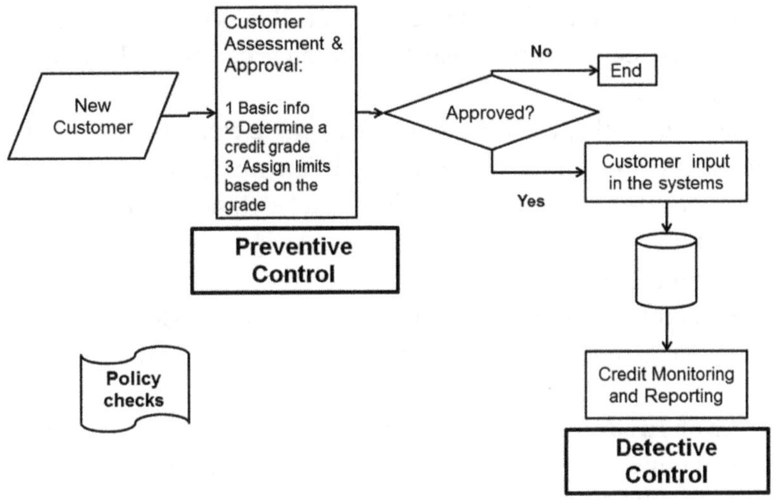

How to Manage Risk by Keeping the Focus on Adding Value?

According to the Oxford Advanced Learner's Dictionary, value can be defined as:

1. The regard that something is held to deserve; the importance, worth, or usefulness of something
2. Principles or standards of behavior; one's judgment of what is important in life
3. The numerical amount denoted by an algebraic term; a magnitude, quantity, or number
4. Music. The relative duration of the sound signified by a note.
5. Linguistics. The meaning of a word or other linguistic unit.
6. The relative degree of lightness or darkness of a particular colour

As we can see from the definition below, value is related to perception, relationship, understanding and calibration.

If we extend this concept to the internal control activities, it can be observed that value is related to the stakeholder's perception over internal control risk, linked to the values and mission of the company and how these values are cascaded to each process, as well as the financial materiality behind each process and risk; considering not only the exposure of the specific controls but also to the potential effect over company value.

Within the existing methodologies, one of the most popular and recognized at strategic and process level is the analysis of the value chain from Porter, which provides a useful model for analyzing the advantages held by the company, allowing fitting the sources of differentiation from the side of demand and supply (Martínez Marín & Eguren Martí, 2013). Under this vision, the company can be conceived as a series of activities, each contributing to the creation of value. In other words, the value chain of a company incorporates a number of activities, processes, resources and objectives that relate to each other provide "value" to products manufactured or services provided, represented by the amount that buyers are willing to pay for "goods and / or services" that the company provides. (Porter M., 1985)

As can be seen in the Figure 4, all the activities and processes of the company must be included when analyzing how the company is generating value. Moreover and likewise to the "Internal Control" characteristics, the value chain methodology presents the attributes to be considered as a system, so that the existing multidisciplinary tools over this field of knowledge can be applied to their analysis (Eguren Martí & Castán Farrero, 2011).

According to (Fernandez, Avella, & Fernandez, 2006) value chain is defined as the set of activities performed by the company with the aim of transforming raw materials into finished products that are then distributed to final consumers, and these activities can be classified into two main groups primary activities and support activities.

Primary activities are directly involved in the production process of the company, covering the physical transformation and creation of the product or services and the after-sales activities (Guerras & Navas, 2015). More specifically the activities included in this area are the following:

- **Inbound Logistics or Input Factors:** Including all the activities related with the receipt, storage, inventory control and internal distribution of raw and auxiliary materials, as well as any element received for developing the services to be provided.
- **Operations / Production**: Covering the transformation of the raw material into products or services.
- **Outbound Logistics**: Warehousing and distribution of finished product or services to the customer.
- **Marketing and Sales**: Activities aimed at selling the product.
- **After-Sales Service**: Activities performed after the sale, in order to assure the service level to the customer and proper feedback on the company performance and perception. Quality and claims management, are among the elements to be considered in this category.

The supporting activities are those that complement the primary activities and are not applicable to a specific area but to all the company, including the following:

Figure 4. Porter's Value Chain Model
Source: adapted from Porter, 1980

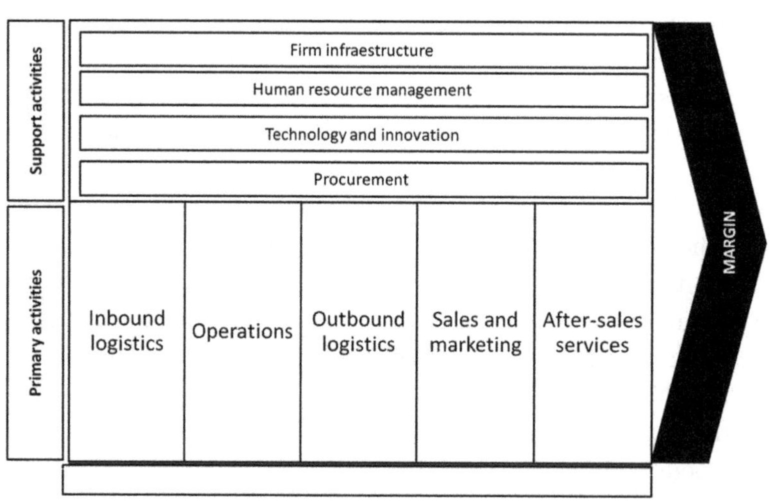

- **Procurement**: Acquisition of productive and non-productive material including services and advertising)
- **Technological Development (IT)**: Activities involved in the development and improvement of new products, research & development (R&D), and innovation.
- **Human Resource Management**: Activities that include all the administration of the human resources of the company (recruitment, training, etc).
- **Firm Infrastructure**: Company business activities oriented management, planning and control of all primary and support activities that sustain it.

Considering this activities, the value chain model considers the capacity of the company of creating competitive advantages thru each of its activities, using two complementary approaches: cost reduction and product differentiation.

A competitive advantage by cost reduction is achieved when the costs of the product or service offered are lower than the offered by the competence for a similar product or service and with an equivalent quality (Guerras & Navas, 2015). According to this strategy, if two competitors offer the same product at a certain price, the one with the lowest cost will have a greater range and therefore increased profitability.

By the other hand, a competitive advantage by product differentiation can be obtained when a product or service has a set of attributes that makes it unique for the customer. (Guerras & Navas, 2015).

In the past, both strategies were considered separately, however recent researches evidence the fact that additional competitive advantage can be created by joining both approaches (Hamel & Prahalad, 1989)

So, going back to the initial subject of this section, risks management will need to be managed in a way that allows to the company to create competitive advantages in comparison to the competence.

In this sense, a system's approach will allow to assure the needed flexibility as well as the completeness required in order to capture the value generators in line with the requirements for risk factor. In this line, the following aspects should be taken into consideration:

- Completeness of the scoping by mapping risks vs. the applicable activities according to the value chain model. This aspect will be detailed in the next section.
- Control level analysis in order to identify value generator elements for each process or cycle, based on the key variables shown on Table 1.
- Mission, vision and company values and its application on processes and people that will be explained with the "Tone of the Top" premise.

Table 1. Internal control and value management strategy

Cost Reduction	Product Differentiation
Lower audit costs Reduced timing for controls execution and performance	Automated controls Differentiated process design
Innovation Information with added value for the business Efficiency, assuring cost reduction and differentiation.	

Source: own elaboration

In summary, risk management should be approached from a system focus in order to be able to identify all the opportunities that may come from how are controls designed, performed, executed and monitored in order to create a competitive advantage considering risk coverage perspectives.

Asset Management Process and Risk Mapping

As previously commented, risk mapping vs, process and activities is one of the key steps to be performed before starting a risk analysis project.

We can summarize the key steps to perform this process as follow:

- **Process Knowledge:** A full understanding of the applicable must be required to the person responsible of risk management. A basic step for this is to perform personal interviews with the process owners as well as checking the available documentation like written procedures.
- **Process Documentation:** Risk manager, must document the processes in order to be sure to have a proper tool to perform the process analysis and risk identification.
- **Process Analysis:** Based on process documentation.
- **Identification of Risks:** Having in mind the type of risk, project objective, resources and reasonability.
- **Controls Design:** Considering all the control attributers mentioned on previous sections. Some examples will be shared on the Sarbanes-Oxley practical example section.

Once this is done, controls should be implemented and monitored to complete the cycle over internal control management.

Coming back to assets management and according to the steps mentioned above, the first step to perform risk mapping is to understand the processes around asset management. Each company will have their own particular aspects that will need to be considered and documented in particular, and which are a relevant aspect on this topic. However, there are several core processes that can be considered when analyzing assets management:

- Acquisition
- Accounting / Reporting
- Disposal
- Safeguarding
- Plus specific processes on: Leases and Intangibles management.

Acquisition management includes:

- Acquisition approval by management in order to authorize the investment from a financial, operational and technical perspective.
- Acquisition analysis from accounting point of view (asset or expense). Based on this, asset type (and related useful life) must be defined, as well as the required cost elements to manage depreciation and investment status.
- Creation of all the elements in the company IT system.

- Definition of the schedule for managing the asset, including the date for the asset to be placed in service (so changed from construction in progress to the adequate asset count account and initiate depreciation).

Accounting/Reporting management includes:

- Monthly review of current projects and investments in order to:
  - Review of construction in progress assets in order to change (when applicable) to the corresponding asset type and initiate depreciation.
  - Identify assets subject to disposal
  - Other actions related to assets monitoring (errors, wrong asset grouping, etc)
- Depreciation run on existing assets
- Year-end accounting activities for assets

Disposal management includes:

- Identification of assets subject to disposal. Two cases can be considered: scrapping or disposal of the assets or sale.
- Disposal approval by management
- Accounting of the activity
- Physical activities (scrapping or delivery to customer)

Safeguarding management includes:

- Access to the company facilities
- Assurances
- Process participant's definition
- Operational monitoring over assets.

Having this high level process description, the risk mapping exercise must understand each of the processes behind and detect the key risks that may arise according to process flow. In this risk mapping exercise, we will focus over the financial risks in order to set-up the basis for the Sarbanes-Oxley case example that will be explained in the following sections.

In line with this, the following risk mappings have been provided for each of the processes:

- Acquisition, please refer to Figure 5
- Accounting / Reporting, please refer to Figure 6
- Disposal, please refer to Figure 7
- Safeguarding, please refer to Figure 8

Figure 5. Risk mapping of assets acquisition
Source: own elaboration

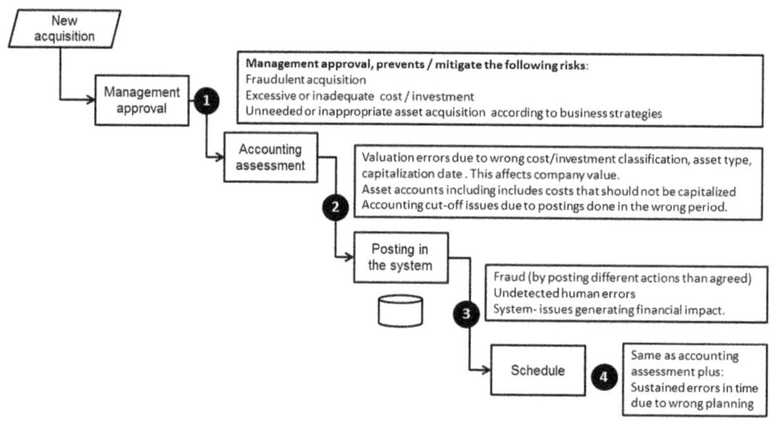

Figure 6. Risk mapping of assets reporting and accounting
Source: own elaboration

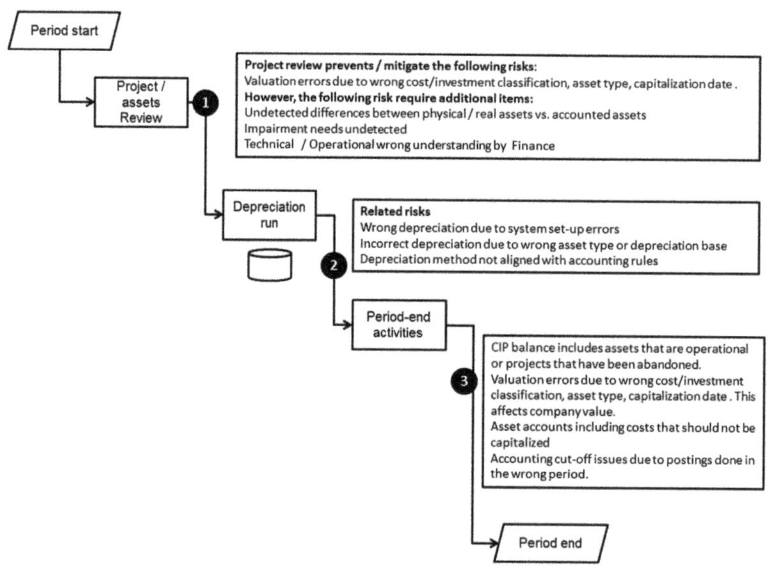

Definition and Implementation of Internal Controls for Asset Management: A Practical Example under SOX Act.

SOX Background and Requirements

As mentioned by the United Exchange Commission (U.S. Securities and Exchange Comission, 2013):

On July 30, 2002, President Bush signed into law the Sarbanes-Oxley Act of 2002, which he characterized as "the most far reaching reforms of American business practices since the time of Franklin Delano Roosevelt." The Act mandated a number of reforms to enhance corporate responsibility, enhance financial

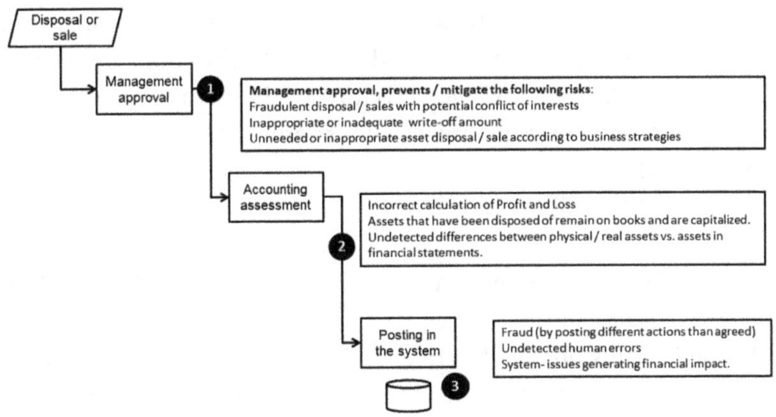

Figure 7. Risk mapping of assets disposal
Source: own elaboration

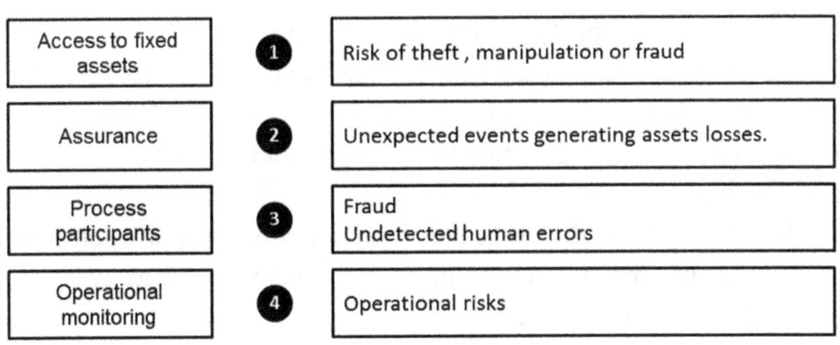

Figure 8. Risk mapping of assets safeguarding
Source: own elaboration

disclosures and combat corporate and accounting fraud, and created the "Public Company Accounting Oversight Board," also known as the PCAOB, to oversee the activities of the auditing profession.

The reason behind the signature of this law was related to the financial scandals occurred in several companies like Enron and WorldCom which had a big impact as can be seen below:

- **WorldCom:**
  - Financial impact: $100 billion.
  - Incorrect capitalization of operating expenses.
  - CEO sentenced to 25 years in jail. CFO sentenced to 5 years in jail.
  - Controller, Accounting Director and Accounting Managers all plead guilty.
  - 300.000 people lost their jobs
- **Enron:**
  - Financial impact: 73 billion.
  - Off Balance Sheet transactions utilizing Special Purpose Vehicles to hide billions of losses.
  - CEO sentenced to 24 years in jail. CFO sentenced to 6 years in jail.

- 15.000 people lost their jobs.

Public companies are required to comply with Sarbanes-Oxley Act, which dramatically changed the approach over financial risk management, based on purely accounting and limited reviews, over a complete framework which affects the whole activities of the company and their processes design, impacting the relationship between different areas as for example IT control governance and IT control quality (Li, Lim, & Wang, 2007)

SOX Act. is composed by 11 sections. Below you will a summary of the most relevant ones:

- Section 404: of the Sarbanes-Oxley acts requires an annual internal control report in addition to the financial report.
  - Includes detailed results on the effectiveness of controls and processes
  - External audit firm should attest and report on the assessment on the effectiveness of the internal control procedures and structure
- Section 201: Services outside the scope of Practice of Auditors
  - Limit the activities that can be offered from the external auditor responsible for giving a opinion over SOX.
- Section 302: Corporate responsibility for Financial Reports
  - CEO and CFO must approve and authorize the financial reports to certify internal controls are in place to record, process, summarize and report financial data.
  - Up to $5 million in fines and 20 years in jail for CEO and CFO.
- Section 409: Disclosures
  - Timely disclosure of material changes to financial condition or operations.
- Section 806: Protection for employees who provide evidence of fraud
  - Whistleblower policies and protection

All the sections below are strongly related and depended of a key concept named "Tone of the Top". This term is used to refer to the top management's commitment towards assuring ethics, compliance, integrity, ethical behavior and cascading this message into all the levels of the company, who should retrofit this principle.

Although the execution of this framework is mandatory for a very specific set of companies, since it was defined it has also changed the mindset of internal control management, making its general principles applicable over "non SOX" companies who use or adapt this framework in order to manage their internal risks.

Having this into consideration, the specific steps required to manage internal control thru SOX will be explained in the next sections, by studying the specific case of assets management. This section, does not intend to be a case of study presentation, but a practical example of the most common SOX controls according to the author´s professional experience.

Scoping

The first step on SOX is to make first assessment over the business areas and business unit which will be a priority for SOX purposes. Although SOX will be applicable for the whole company, internal and external audit will concentrate on those areas were risk exposure is higher due to materiality and other business reasons.

Once this is defined, a work plan must be defined and agreed with Financial Managers (usually the CFO) in order to coordinate all the internal control activities for the year assuring they are aligned with the deadlines required by SOX. In the specific case of Assets Management, the following aspects must be considered:

- Materiality
- Existence of intangible and/or leases
- Nature of the area. Is strategic for the business?
- Do you expect business changes in the year that will require new assets acquisition or disposals? For example, a plant expansion, a new product line, etc.

An internal communication must be done in order to have all the involved team aware of the next actions.

Risk Assessment and Control Definition

The first step to be performed in this area is a risk mapping over the applicable process, as explained on the previous section.

Once this is done, the controls should be designed based on SOX standards in order to cover the identified risks. Some of the most common controls in this area cover the following:

- **Acquisition Approval:** Assure proper management level of approval according to company policies, including capitalized interest calculation, asset type class and other relevant aspects.
- **Asset Recording:** Assure new assets are accounted under the corresponding asset type, related useful life, and depreciation method. Assure the are monitoring controls in place (e.g. annual reviews), to validate the accuracy of the existing assets.
- **Asset Identification:** Assure fixed assets can be physically identified in relation to the recorded information.
- **Fixed Asset Count:** Validate that an existing policy exist over this subject.Review fixed assets counts are done based on the agreed policies and adjustments made over identified differences
- **Construction in Progress (CIP):** Monthly review of CIP accounts, to assure all applicable projects are included and only projects are posted in this account and are transferred when asset is in place to the corresponding asset account.
- **Expenditures - Monitoring:** Variation analysis of approved budget and actual PPE related capital expenditures.
- **Fixed Assets Review:** Monthly controls to validate performed transactions (acquisitions, transfers, disposals and adjustments) are accurate and performed according to agreed procedures.
- **Depreciation Method:** Assure depreciation method and useful life linked to the asset classes are reviewed to assure a proper depreciation calculation. If performed thru a system, this control will require making a set-up check.
- **Depreciation Monitoring:** Perform analytical review on actual depreciation amount vs. previous months or budget in order to identify anomalies or any potential issues.

- **Accounts Reconciliation:** Perform monthly fixed and intangible accounts review vs. general ledger accounts.
- **Disposal Approval:** Assure proper management level of approval according to company policies, including profit / loss analysis, reasons for disposal and any other relevant aspects for the business.
- **Impairment Analysis**: Validate that an existing policy exist over this subject. Assure there are procedures in place in order to identity trigger events that can lead to an impairment action. Perform as a minimum an annual impairment check and assure proper postings are done after this analysis.
- **Segregation of Duties:** Check that a proper segregation of duties has been defined over the personnel involved with assets management, including a functional and system accesses approach.

In addition to the mentioned controls, it's recommended to have written procedures over the mentioned areas: acquisition, recording, CIP, disposal, physical count, impairment, accounts reconciliation and monitoring activities.

Implementation

Once controls have been designed they should be implemented.

In order to succeed with this activity, is important to clearly define the control attributes explained at the beginning of this chapter:

- People: Control preparer, reviewer, approver and owner
- Objective
- Scope
- Description
- Evidence
- Frequency
- Written procedures

A project leader should be nominated to manage control implementation and monitoring in order to assure proper implementation and internal control understanding. This function can generate into a full-time position depending on company size and business complexity.

Monitoring

SOX methodology clearly defines the steps to be performed in order to monitor and assure that internal control environment can be reliable. This is the main steps:

- **Scoping**: Priority areas, testing plan for the year, key events expected for the year
- **Self-Assessment and Management Testing:** To be performed by the company including:
    - **Walkthrough:** Control description update according to current existing procedures, interview to control performers, control design test .Usually as part of this process, the person who is performing the walkthrough traces the process with one evidence of each step in order

to assure control is designed according to what has been agreed and documented. This allows detecting design deficiencies before performing the full-sample operational test.
    - Testing of the applicable controls according to sample size selection method, control type, and specific applicable attributes. This testing must be done over a full sample and in line with agreed deadlines. There are two events to have in mind when planning the testing:
        - **Interim Testing Period:** Usually covers from January to September. Must be coordinated with the external auditors, to assure a efficient work
        - **Year End Results:** Which covers the testing required covering the period between October and December, annual controls performed at year end as well as the controls which deficiencies which have been remediated and need to be re-tested by year end.
- **Action Plan and Remediation:** For those controls considered as deficient, a detailed action plan must be defined, so control will change thru the following status:
    - **Failed**: control with deficiencies detected on design or operational level
    - **Remediated**: Action plan implemented, but not tested
    - **Closed**: Remediation actions tested, by testing a full sample set of evidences
- **External Auditor Assessment and Testing:** As commented before external auditors will make an opinion on internal control effectiveness. In practice this means, auditors will make their own walkthrough and testing over selected controls, also aligned with the two main periods mentioned before: interim and year-end.

SOLUTIONS AND RECOMMENDATIONS

As have been explained, risk management, requires to define and implement a internal control process. Depending on the way this process is implemented, it can impact the company from different ways including an increased cost over controls implementation, execution and audit.

Due to this, the question that arises is: *How to ensure the sustainability and efficiency in risk management?*

As a result, after researching over several dictionaries, we may say that sustainability is related with two main concepts:

- The fact that a process can be maintained during time by itself with no significant external intervention
- Development that meets the needs of the present without compromising the ability of future generations to meet their own needs. (Presidio Graduate School, 2006), in other words the fact of keeping a process in place with an ecological approach, not harming the environment or depleting the existing resources.

So as can be observed, sustainability is related to resources, time, impact, and continuity and planning.

From a management / enterprise perspective, there are three areas with their corresponding attributes which needs to be taken into consideration (Presidio Graduate School, 2006)

- *Social Criteria:*
    - *Socially desirable*

  - *Culturally acceptable*
  - *Psychologically nurturing*
- *Financial Criteria:*
  - *Economically sustainable*
  - *Technologically feasible*
  - *Operationally viable*
- *Environmental Criteria:*
  - *Environmentally Robust*
  - *Generationally Sensitive*
  - *Capable of continuous learning*

In this sense, internal control sustainability is strongly related with the ability on generating added value to the company at the same time it assures proper risk management.

The acceptance of the internal control by employees, is related to the "Tone at the Top" concept mentioned above and the added value that these processes can generate to the business if are adequately implemented. Open communication, internal control training, will support on the process of having a "social" acceptance over the internal control structure and future changes.

From Financial criteria, the sustainability of internal control is related with the ability to use existing resources on defining efficient controls in line with the existing information systems, minimizing the exposure on financial risks and updating according to the requirements.

As internal control is a system, the capacity of learning and retrofitting is a key element which needs to be monitored and take the opportunity of improving current processes or detecting new competitive advantages.

In addition, to the above a proper resource resources management and planning need to be in place to assure cost contention, value creation as well as proper projection in time.

From a resources perspective, there are two key areas to have in mind:

- **Team Capacities / Human Resources:** Probably the most relevant aspect to consider, as internal control is a system driven by people. Know-how maintenance, clear role definition, proper training plans in internal control, motivation on the team for internal control performance are the key aspects to consider.
- **Other Resources:** All the needed additional resources required in order to maintain the internal control structure, including information systems.

From a planning perspective, internal control needs to be considered from strategic planning level to the operational level assuring to have all the involved personnel as responsible in the process.

FUTURE RESEARCH DIRECTIONS

Risk management, and more specifically internal control management is an area under continuous changes and suffering frequently updates. From an academic perspective is a relatively new area to explore and open to new researches. Due to this, most part of the existing documentation and books have been written based on empirical experience rather that purely scientific approach. This opens the possibility for interested researchers, covering a wide typology of risk related topic as:

- Quantitative analysis of risk management
- Qualitative aspects on risk management (like reputation)
- Literature review on internal control management
- Information systems relationship with internal control management.
- Case of studies on a selected sample of companies or sectors

More specifically, under assets management criteria, future investigations in the lines stated above are also applicable to the specific asset management flows.

CONCLUSION

In this chapter, several aspects over risk and internal control management have been discussed, starting from a general approach to set-up the required basis to the specific case of assets management.

From this analysis, we can summaries the main aspects as follows:

- Risk cannot be fully avoided, but its probability of occurrence can be minimized or mitigated by setting up the proper controls
- Risk is generally perceived as a negative event; however, it can also be considered as an opportunity depending on the events.
- Internal control is a system, constituted by people with the objective of managing the internal risk coming from the company activities.
- Internal control can be used as a value added generator if a proper design and approach is assured.
- SOX is an internal control framework, applicable to a set of companies whose principles can be used for managing internal control in companies which are not required to do so.
- The "Tone at the Top" principle must be in place for proper internal control management
- The control's defined under SOX framework on asset's management cover the risks identified during the risk mapping process.
- The sustainability of the internal control system depends on the ability to use resources, manage people, perform proper planning, and maintain added value.
- Internal control it is an area with great opportunities for the development of new researches.

REFERENCES

Arens, A., Elder, R., & Beasley, M. (2014). *Auditing and assurance services-An integrated approach; includes coverage of international standards and global auditing issues, in addition to coverage of.* Boston: Aufl.

COSO. (2013). United States: Internal Control - Integrated Framework.

De Quevedo, E., De La Fuente, J., & Delgado, J. (2005). Reputación Corporativa y Creación de Valor. Marco Teórico De Una Relación Circular. *Investigaciones Europeas de Dirección y Economía de la Empresa*, *11*(2), 81–97.

Eguren Martí, M. L., & Castán Farrero, J. M. (2011). La cadena de valor bajo un enfoque sistémico y sus implicaciones en el análisis de la creación de valor en la empresa. *Proceedings of the XXI Congreso Nacional de la Asociación Científica de Economía y Dirección de la Empresa (ACEDE)*. Barcelona: ACEDE.

Febles, J., & Oreja, J. (2008). Factores externos e internos determinantes de la orientación de la cultura estratégica de las empresas. *Investigaciones Europeas de Dirección y Economía de la Empresa*, *14*(1), 13–32. doi:10.1016/S1135-2523(12)60009-4

Feng, M., Li, C., & McVay, S. (2009). Internal control and management guidance. *Journal of Accounting and Economics*, *48*(2-3), 190–209. doi:10.1016/j.jacceco.2009.09.004

Fernandez, E., Avella, L., & Fernandez, M. (2006). Estrategia de Producción (2nd ed.). Madrid, España: Editorial McGraw Hill.

Guerras, L., & Navas, J. (2015). *La dirección estratégica de la empresa. Teoría y aplicaciones*. Thomson Reuters.

Hamel, G., & Prahalad, C. (1989). Strategic intent. *Harvard Business Review*, 1989, 63–76. PMID:10303477

Heise, D., Strecker, S., & Frank, U. (2014). ControlML: A domain-specific modeling language in support of assessing internal controls and the internal control system. *The International Journal of Accounting*.

Ittner, C., & Oyon, D. (2014). *The Internal Organization of Enterprise Risk Management*.

Jokipii, A. (2010). Determinants and consequences of internal control in firms: a contingency theory based analysis. *Journal of Management & Governance*.

Kloot, L. (1997). Organizational learning and management control systems: Responding to environmental change. *Management Accounting Research*, *8*(1), 47–73. doi:10.1006/mare.1996.0033

Leisinger, K. M., Blasco, J., Trigo, J., Petitbò, A., Álvarez, J. M., & Argandoña, A. P. (2010). Hacia una nueva ética económica global. Innovación vs Status Quo. Barcelona, España: Editorial Huygens.

Li, C., Lim, J., & Wang, Q. (2007). Internal and external influences on IT control governance. *International Journal of Accounting Information*.

Lin, Z., Zheng, J., & Wang, S. (2007). Study on Relation of Internal Control and Enterprise Value: An Empirical Analysis from Shanghai and Shenzhen Stock Markets. *Journal of Finance and Economics*.

Martínez Marín, J. E., & Eguren Martí, M. L. (2013). Maritime Transport: A Theoretical Analysis Under a System's Approach. *Journal of Maritime Research*.

Pfister, J. (2009). *Managing Organizational Culture for Effective Internal Control: From Practice to Theory*. Berlin, Germany: Physica-Verlag. doi:10.1007/978-3-7908-2340-0

Porter, M. (1985). *Competitive Advantage: Creating and Sustaining Superior Performance*. Nueva York, USA: Free Press.

Porter, M., & Kramer, M. (2006). Strategy and Society: The link between competitive advantage and corporate social responsibility. *Harvard Business Review*. PMID:17183795

Porter, M., & M., K. (2011). The Big Idea. Creating Shared Value. How to reinvent capitalism and unleash a wave of innovation and growth. *Harvard Business Review*.

Presidio Graduate School. (2006). The dictionary of sustainable management. Retrieved from http://www.sustainabilitydictionary.com/

Ribstein, L. E. (2002). Market vs. Regulatory Responses to Corporate Fraud: A Critique of the Sarbanes-Oxley Act of 2002. *The Journal of Corporation Law*, *28*(1), 1–74.

Scandizzo, S. (2005). Risk mapping and key risk indicators in operational risk management. *Economic Notes*, *34*(2), 231–256. doi:10.1111/j.0391-5026.2005.00150.x

Simmons, M. (1997). *COSO based auditing*. Internal Auditor.

Spira, L. F., & Page, M. (2001). Risk management The reinvention of internal control and the changing role of internal audit. *Accounting, Auditing & Accountability Journal*, *16*(4), 640–661. doi:10.1108/09513570310492335

The Concise Oxford Dictionary. (2012). Oxford: Oxford University Press.

U.S. Securities and Exchange Comission. (2013). The Laws That Govern the Securities Industry. Retrieved from http://www.sec.gov/about/laws.shtml#sox2002

Von Bertalanffy, L. (1989). Teoría General de los Sistemas (7th ed.). New York: George Braziller.

Wieland, J. (2005). Corporate governance, values management, and standards: A European perspective. *Business & Society*, *44*(1), 74–93. doi:10.1177/0007650305274852

KEY TERMS AND DEFINITIONS

Control: A task or process designed to mitigate the probability of occurrence of a specific risk.

Exposure: Set of activities, which includes financial exposure that has a probability to be affected by a defined risk.

Fraud: Deliberated use of current knowledge on processes and activities, in order to obtain profit by misuses or misapplication of the company resources or assets.

Internal Control: System composed of a group of activities defined to manage the risk that comes from the processes within the company, across all functions.

Risk: The potential exposure to an unexpected event or unintended situation or outcome.

Role: Set of functions assigned to an individual in a specific process or activity.

SOD: Segregation of duties. Principle that has the objective of preventing the risk of errors or fraud. One person cannot perform two functions that if performed together can compromise the "independence" premise.

SOX: Sarbanes-Oxley Act.

Value Management: Process defined with the objective of managing the value applicable under the selected scope, having into consideration that value includes those actions defined in order to control cost as well as those actions defined to create value (differentiation, innovation, perception aspects among others).

Chapter 18
Reliability Based Maintenance of Industrial Assets

A. Syamsundar
Visakhapatnam Steel Plant, India

ABSTRACT

Most industrial assets today are complex repairable systems. Maintenance decisions on these systems / assets are still made on an empirical basis based on the experience and understanding of the maintenance engineers / managers. This leads to sub-optimal decisions and will lead to inefficient and ineffective maintenance. Reliability is a key attribute of such assets. Reliability centred maintenance as a maintenance philosophy and reliability engineering as an engineering discipline have been well developed over the past several decades. However, these are yet to find a place in maintenance decisions of industrial assets on a regular basis. This chapter deals with how reliability centred maintenance and reliability of repairable systems can be combined together to develop reliability based maintenance of industrial assets in an objective way to improve the efficiency and effectiveness of industrial asset maintenance. The methodology is illustrated with a simple case study.

INTRODUCTION

Most industrial assets today are complex repairable systems. Maintenance decisions on these systems / assets are still made on an empirical basis based on the experience and understanding of the maintenance engineers / managers. This leads to sub-optimal decisions and will lead to inefficient and ineffective maintenance.

Reliability is a key attribute of such assets. It not only helps to improve the availability of the assets but also helps to reduce operational losses, safety and environmental protection issues due to unwanted failures. If maintenance of industrial assets can be performed keeping their reliability in view it will lead to better and safer utilization of industrial assets and will also help to optimize the maintenance efforts thus reducing maintenance costs.

This chapter introduces the concept of reliability based maintenance of industrial assets. It emphasizes the importance of reliability as an attribute for maintaining industrial assets and presents the methodology

DOI: 10.4018/978-1-5225-0651-5.ch018

to be adopted to attain reliability based maintenance of such assets. An industrial asset, its maintenance and reliability are introduced. The reliability centered maintenance process and quantitative assessment of reliability of repairable systems are described. These are then applied to a part of an industrial asset to illustrate the process of reliability based maintenance of industrial assets.

INDUSTRIAL ASSET, MAINTENANCE, AND RELIABILITY

An industrial asset, its reliability and maintenance are presented in this section.

Industrial Asset

Industrial assets considered here are defined in a narrow sense as equipment deployed by the industry to convert inputs to outputs which can be then marketed by the industry as goods or services.

Industries in the fields of mining – minerals, diamonds etc., transportation – space, airlines, bus, railways, shipping, automobiles, trucking etc., manufacturing – automobiles, tools, equipment, instruments, electronics, etc., construction – roads, buildings, stadiums, dams, bridges etc., production – chemicals, metals, pharmaceuticals, petroleum etc. can be considered under the above categories. Equipment used in these industries will fall under the definition of industrial assets considered here.

Industrial assets can be specialised, tailor made for specific operations or general which can be used in multiple industries. Some examples of industrial assets in different types of industries are provided in Table 1. These contain equipment used for specific and general purposes.

Such equipment comprises systems such as mechanical, electrical, structural, civil, electronic, computers, instrumentation, telecommunications etc. A single system can contain sub-systems, assemblies, sub-assemblies and components. A typical example of industrial equipment along with its sub-systems, assemblies and components is given in Table 2.

This equipment is usually employed for a lifetime and is a complex system which is repairable.

Table 1. Examples of industrial assets

Sl. No.	Industry	Some Equipment
1	Mining	Excavators, loaders, dumpers, draglines, shovels, rigs
2	Manufacturing	Lathes, milling, drilling, boring, shaping machines
3	Power Plant	Generators, boilers, turbines, coal handling, transportation and disposal equipment
4	Power Transmission	Cabling, transmission line towers, transformers
5	Iron and Steel Industry	Raw materials handling, coke ovens, sintering machines, pelletising machines, blast furnaces, steel making facilities, casting facilities, rolling facilities
6	Transport	Aircraft, ships, trains, buses, automobiles, bikes, trucks
7	Defence	Armour, tanks, battleships, aircraft
8	Communication Networks	Cables, nodes, hardware, software

Table 2. Industrial equipment electric overhead travelling crane

Sl. No.	Systems	Sub-Systems	Assemblies/Components
1	Mechanical	Long Travel Assembly	Gearbox, Brake, Drive Wheel Assembly, Non-Drive Wheel Assembly, Floating Shaft Assembly
		Cross Travel Assembly	Gearbox, Brake, Drive Wheel Assembly, Non-Drive Wheel Assembly, Floating Shaft Assembly
		Main Hoist Assembly	Drum Assembly, Brake Assembly, Hook Assembly, Ropes
		Auxiliary Hoist Assembly	Drum Assembly, Brake Assembly, Hook Assembly, Ropes
2	Electrical	LT Electrics, CT Electrics, MH Electrics, Current Collector Assembly	Motors, Cables, Electrical Panels, Switches
3	Structural	LT Girder Assembly, CT Trolley Assembly, MH Platform Assembly	Long Travel Girder, Cross Travel Frame, End Carriage, Balancer Body

Reliability

Assets once manufactured and put to use or kept idle are subject to deterioration over a period of time. When this deterioration crosses a threshold the asset will fail to perform its intended function. Thus reliability of an asset is an attribute of the asset whereby it continues to perform its function satisfactorily for a given period of time under stated conditions of operation.

Failures of assets are influenced by a wide variety of factors, its inherent design, manufacturing, erection and commissioning, setup, operation, maintenance and its environment. Some of these factors are controllable, others are not. This creates a random environment under which the assets function and influences its deterioration which can take place at different rates under different circumstances. Thus the failures of assets happen at random moments in time.

Thus reliability is quantified as the probability of the survival of an asset for a specified period of time under a specified set of operating conditions and environment. This is expressed mathematically as;

$$R(t) = P(t > T) \quad (1)$$

where $R(t)$ is the reliability of the asset at a specified period of time t

T is the random variable of time to failure

$P(t > T)$ is the probability that the asset survives up to time t without failure

Maintenance

Setting up an industry entails large capital expenditure and it has to work on a continuous basis to sustain the industry in its endeavour to supply the desired goods or services to its customers.

As already stated, industrial assets tend to fail. The assets have to be brought back to their functional state at the earliest so that the working of the industry is not interrupted. The work carried out on the industrial asset to bring it back from failure to its functional state or to postpone its failure is termed as maintenance. Maintenance carried out on an industrial asset can be termed as industrial maintenance.

The maintenance normally carried out on an industrial asset can be broadly divided into two categories – corrective and preventive. Maintenance carried out on failure is termed as corrective maintenance. Maintenance carried out prior to failure in order to postpone it, is called preventive maintenance. Preventive Maintenance can be routine, scheduled – either age or time based and condition based. In routine preventive maintenance jobs like inspection of the equipment, cleaning, lubrication, bolt tightening etc. are taken up. In age based and time based preventive maintenance the maintenance is carried out either at a specified age of the asset or at scheduled time intervals. Condition-based preventive maintenance is carried out when specified condition monitoring parameters reach an unacceptable level/threshold.

Another type of maintenance activity which can make a difference to the reliability of the asset is proactive maintenance. This can be of two types – predictive and design out i.e., modifications on the asset. As already stated the reliability of an industrial asset can be quantified using probability theory. In predictive maintenance, the time to maintenance is predicted based on the information supplied by the system and using stochastic models to analyse this information, estimate or predict the probability of failure and the possible time of maintenance. When no maintenance action is possible to ward off the failure and the effects and consequences are high it is preferable to redesign the existing asset to accommodate maintenance actions as listed above. Types of maintenance are listed in Figure 1.

Scheduled preventive maintenance can be minor – repair or replacement of one or two items, medium – when a number of items are replaced and major where large assemblies, sub-systems are repaired or replaced.

Apart from this capital repairs are carried out where capital items are replaced and renovations/upgrades of equipment can take place. Opportunistic maintenance can be carried out in a window created by a breakdown of some other equipment or a break in operations. Sometimes maintenance activities are deferred leading to deferred maintenance.

Various philosophies of maintenance have evolved over a period of time like terotechnology, total productive maintenance, total quality maintenance, reliability centered maintenance etc. which incor-

Figure 1. Types of maintenance

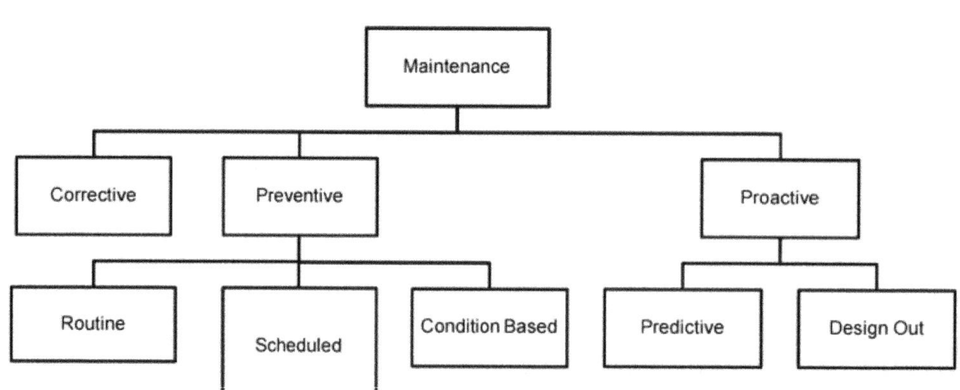

porate the above types of maintenance. Each of these philosophies is derived with particular objectives in mind and encompasses the types of maintenance listed above.

Terotechnology has been defined as a combination of management, financial, engineering and other practices applied to physical assets in pursuit of life cycle costs (LCC). Total productive maintenance has been designed to go with just in time philosophy. It calculates overall equipment effectiveness, eliminates six big losses, has a scheduled maintenance program, carries out autonomous maintenance and carries out small group activities like total quality management. Total quality maintenance measures overall process effectiveness, combines operation and maintenance activities, focuses on condition monitoring activities and is based on Deming cycle. Reliability centered maintenance is a two-part process. Failure modes, effects and consequence analysis is carried out in the first part followed by a decision diagram in the second part where maintenance actions are decided based on the consequences of failure.

Of these, reliability centered maintenance is based on setting up the maintenance scheduling in a cost effective manner with reliability of the asset at its core and is described in the next section.

RELIABILITY CENTERED MAINTENANCE

The philosophy of maintenance termed as reliability centered maintenance emerged out of the commercial airline industry in the 1960s and 70s. The first application of this concept was on Boeing 747 through a document known as MSG – 1 prepared by maintenance steering group in 1968. It was first brought to public notice through the report of Nowlan and Heap (1978) sponsored by the United States Department of Defense. It spread from commercial aircraft industry to nuclear power and defense sectors, and then to chemical, oil, and shipping industries. Some books describing this philosophy and its application are Anderson and Neri (1990), Moubray (1991), August (1999), ABS (2004), Smith and Hinchcliffe (2004), and Bloom (2005). Three key standards that define the criteria for an RCM process and provide a guide to RCM are IEC 60300-3-11, SAE JA1011 and SAE JA1012 respectively.

The strict classical approach to RCM defined in Moubray (1991) is an elaborate and time consuming methodology. Several variants to this methodology have been developed by various practitioners while applying the principles to different industries. To cut down the time required to develop the classical approach, RCM derivative approaches like streamlined RCM and PM (preventive maintenance) optimisation have been developed which cut down the detailed analysis required in the first step of the RCM process. The classical approach, though elaborate and time consuming, is a better approach as it addresses every one of the risks associated with all the failure modes and provides optimum reliability of the assets.

Reliability centered maintenance essentially derives the maintenance requirements of assets based on its functional failures and then decides on the appropriate maintenance actions to be taken based on the effects and consequences of these failures. This ensures that the maintenance actions mitigate the probability of failure and thus the reliability becomes the focus of the maintenance strategy deployed on the asset in a cost effective manner.

This is carried out in a two-part process. In the first part a failure, modes, effects and consequences analysis (FMECA) of the assets are carried out. Based on this exercise decision logic is employed to decide the appropriate maintenance actions on the asset.

The various steps in the first part of the RCM analysis are as follows:

1. The boundaries of the asset to which the RCM analysis is to be applied are defined clearly. An example of an asset taken up for RCM analysis is given in Figure 2.
2. The system, its sub-systems and components of each sub-system which form the least replaceable or repairable units are listed out as given in Table 3.
3. All the functions of the asset with their performance standards/ranges as per the design and what constitutes functional failure of the asset are listed as given in Table 4.
4. The various failure modes associated with each functional failure are listed as given in Table 5.
5. The effect of each failure and the consequences of the failure based on its effects on the asset or its sub-systems and assemblies are listed. The consequences are classified into four categories hidden, safety, operations and others including maintenance. This is listed as given in Table 6.

In the second part of the RCM analysis, decision logic is applied to generate the maintenance tasks to address each and every failure mode based on the consequences of the failure. The tasks must be applicable i.e., they address a particular failure mode and must be effective i.e., they help to reduce the probability of failure in a cost effective manner. The steps in this process are:

1. The failure modes which have serious consequences like safety and environmental are selected. These are to be addressed by predictive maintenance tasks based on condition monitoring which can be online or offline depending on the risk involved.
2. In case condition monitoring is not possible for these failure modes, the particular component, assembly or sub-system are to redesigned to accommodate condition- based maintenance as these failure modes cannot be allowed to occur.
3. The failure modes which lead to operational consequences are selected. These can be addressed by scheduled preventive maintenance tasks based on age/ time.

Figure 2. Asset / system boundaries

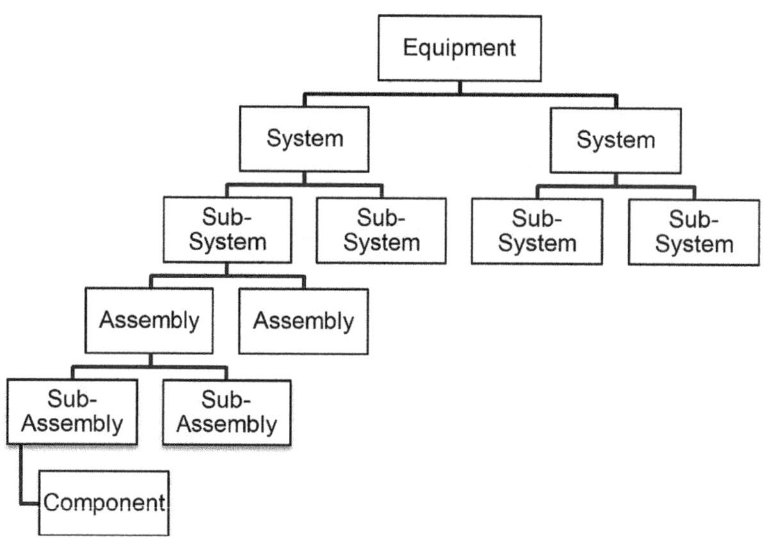

Table 3. System, its sub-systems and components

Asset / Equipment			
System			
Sub-system	Assembly	Sub-Assembly	Components

Table 4. Functions and associated performance standards of the asset

Equipment			
System			
Sub-System			
Assemblies/ Sub-Assemblies/Components	Functions	Performance Standards	Functional Failure

4. The failure modes which lead to only maintenance or other consequences are selected. These can be addressed by scheduled preventive maintenance based on age/time or be allowed to fail and corrective maintenance performed afterwards.
5. Once the maintenance tasks are decided, next the frequency of the tasks i.e., the interval at which these are to be performed is decided. This is done based on past experience with same or similar systems or through analysis of failure data.
6. The maintenance tasks are grouped based on their frequency and a maintenance repair cycle for the asset is generated.

Table 5. Functional failures and failure modes of the asset

Equipment		
System		
Sub-System		
Assemblies/Sub-Assemblies/Components	Functional Failures	Failure Modes

Table 6. Functional failure modes, failure effects and failure consequences of the asset

Equipment					
System					
Sub-System					
Failure Modes	Failure Effects	Failure Consequences			
		H	S	O	Ot/M

Table 7 lists the selected maintenance types, tasks and their interval.
Table 8 shows the maintenance repair cycle of the asset.

RELIABILITY OF REPAIRABLE INDUSTRIAL ASSETS

Repair/replacement can be carried out on an industrial asset at any level – equipment level or its system, sub-system, assembly, sub-assembly or component levels. While carrying out a quantitative assessment of the reliability of the industrial asset all of its levels are treated as an industrial system in this section for the sake of simplicity. A repairable industrial system consists of two or more components which can be repaired to restore its functioning on failure.

Modeling of the reliability of a repairable system begins with the modeling of its failure process. The failure process consists of a series of failures; the system being brought back into operation after each failure. As the time to failure is uncertain, the sequence of failures can be treated as a sequence of random point events occurring at specified points in time leading to their modelling by point processes which are stochastic processes.

The development in time of a point process is governed by its intensity process. The conditional intensity process $\lambda(t \mid H_{t^-})$ provides a general framework for the modelling of the failure processes with the intensity $\lambda(t)$ conditioned on its history H_{t^-} . The history of the process characterized by H_{t^-} is the collection of events that have taken place before time t which in the basic case are the times of the series of failures occurring on or before time t .

Table 7. Functional failure modes, maintenance actions of the asset

Equipment			
System			
Sub-System			
Function			
Functional Failure			
Failure Modes	Maintenance Action		
	Type	Task	Interval

Table 8. Maintenance repair cycle of the asset

Asset / Equipment		
System		
Sub-System		
Assembly/Sub-Assembly/Component	Frequency	Maintenance Task

All the basic functions characterizing the system reliability can be obtained once the conditional intensity function is known. These are:

Conditional failure density function

$$f\left(t \mid H_{t^-}\right) = \lambda\left(t \mid H_{t^-}\right)\exp\left(-\Lambda\left(t \mid H_{t^-}\right)\right) \tag{2}$$

Conditional failure distribution function

$$F\left(t \mid H_{t^-}\right) = 1 - \exp\left(-\Lambda\left(t \mid H_{t^-}\right)\right) \tag{3}$$

Conditional reliability function

$$R(t \mid H_{t^-}) = \exp(-\Lambda(t \mid H_{t^-})) \tag{4}$$

where $\Lambda\left(t \mid H_{t^-}\right) = \int_0^t \lambda\left(s \mid H_{s^-}\right) ds$.

Corrective Maintenance

Corrective maintenance is the maintenance carried out on an asset on failure. The most commonly used stochastic processes to model the failure process of a repairable system are the renewal process, the non homogeneous poisson process and the homogeneous poisson process.

The renewal process is generated with the assumption that on failure the repairable system is replaced with a new one every time or a perfect repair is performed leading to a system which is same as new (SAN). For a renewal process the conditional intensity function is a function of the time since the last failure as given by:

$$\lambda\left(t \mid H_{t^-}\right) = \lambda\left(t - T_{N_{t^-}}\right) \tag{5}$$

where $T_{N_{t^-}}$ is the time of the last failure before t.

To account for trend which is normally observed in the failures of any industrial repairable system a non homogeneous poisson process (NHPP) has been postulated. This process is generated under the assumption of a system which is repaired as little as possible in putting it back to functional use thus leaving it in the same condition as it was before failure i.e., same as old (SAO) and looks at the trend of failures of a repairable system in a monotonous increasing or decreasing way. For an NHPP the conditional intensity function is a function of the chronological age of the system as given by:

$$\lambda(t \mid H_{t^-}) = \lambda(t) \tag{6}$$

The homogeneous poisson process (HPP) is a special case of both the renewal process with the times between failures distributed exponentially (SAN) as well as an NHPP (SAO) process with the failure intensity taken as a constant, i.e., a shape factor equal to one. For a HPP the conditional intensity function is a constant as given by:

$$\lambda(t \mid H_{t^-}) = \lambda \quad (7)$$

These processes are mathematically simple and highly tractable and have been widely applied to repairable systems as on date therefore, most often regarded. Ascher and Feingold (1984) and Rigdon and Basu (2000) deal admirably with the application of Poisson and Renewal processes to the problems of repairable systems.

The renewal process and the NHPP model the extreme cases of failure processes, maximal (SAN) and minimal repair (SAO). In general repairable systems exhibit behaviour in between these two extremes and imperfect/general repair processes are postulated to generalise the modelling of failure processes. The concept of a repair indicator has been developed to help generate general failure processes.

A repair indicator indicates the system state by means of its age or failure intensity and conveys the effectiveness of the repair actions taken on the system. Such a repair indicator incorporated into the stochastic process of failures of the system provides a tractable general model of the failures of a system.

Some of the more common models proposed in this area are given in Malik (1979), Brown, Mahoney and Sivazlian (1983), Kijima and Sumita (1986), Kijima, Morimura and Suzuki (1988), Kijima (1989), Chan and Shaw (1993) and Doyen and Gaudoin (2004). These models are not considered here.

Preventive Maintenance

By carrying out preventive maintenance timely it often happens that the incipient failures get postponed. Preventive maintenance is of two types, scheduled – age / time based or condition based.

Scheduled Preventive Maintenance

Scheduled time based PM is carried out at fixed times while age based PM is carried out at a fixed age after the previous failure irrespective of the condition of the system. Stochastically age/time based-PM censors the failure but is independent of the failure mechanism. Age/time based PM can be incorporated into all the models given above.

For a PM and CM both having a renewal density, the conditional intensity of the PM-CM process is as given by (Doyen & Gaudoin, 2011):

$$\lambda\left(t \mid H_{t^-}\right) = \lambda\left(t - C_{K_{t^-}}\right) \quad (8)$$

This is not a renewal density as the failure intensity is not a function of $\left(t - T_{N_{t^-}}\right)$.

For a PM having a renewal density and CM having trend, the conditional intensity of the PM-CM process is given by (Doyen & Gaudoin, 2011);

$$\lambda\left(t \mid H_{t^-}\right) = \lambda\left(t - D_{M_{t^-}}\right) \tag{9}$$

Condition Based Preventive Maintenance

Condition based preventive maintenance is carried out when the condition being monitored deteriorates to an unacceptable condition and reaches a threshold. This is termed as a soft failure wherein the repairable system is still functional but not as per design and may fail anytime in contrast to a hard failure where the system breaks down. Thus stochastically CB-PM carried out on reaching a soft failure state, censors the hard failure leading to a competing risks scenario which implies dependence between the CB-PM and the failure mechanism.

Lindqvist, Langseth and Stove (2006) proposed a repair alert model wherein the crew performs preventive maintenance timely based on the condition of the system thus avoiding costly failure. They propose a cumulative repair alert function combined with imperfect repair to model the CB-PM process. This model has been used by Dijoux, Doyen and Gaudoin (2005) with an ARA_1 model. These models are not considered here.

Proactive Maintenance

Based on the analysis of the failure process of the system proactive maintenance in the form of predictive maintenance can be carried out on the system. Apart from this wherever a suitable maintenance task cannot be found based on the RCM analysis and the consequences are severe a change in design may be required for the corresponding assembly/sub-assembly.

Predictive Maintenance

The models which best fit the failure data along with their parameter estimates can be used to generate the quantities of interest for maintaining the asset on the basis of its reliability like the reliability at various quantiles, mean time between failures, expected number of failures and the expected time to next failure etc. These quantities of interest can be used to generate the maintenance schedules for the asset.

The expected number of failures is given by:

$$E(N_t) = \int_0^t \lambda(s)ds \tag{10}$$

Design Out Maintenance

Wherever a suitable maintenance task cannot be found based on the RCM analysis and the consequences are severe a change in design may be required for the corresponding sub-system/assembly/sub-assembly/component either to remove the cause of failure or permit to take a maintenance action.

Multiple Repairable Systems

All the models presented above deal with a single repairable system. If the number of failures of a single repairable system is less and a number of similar systems are available, these can be considered and analysed together. This can be carried out using the collection of k univariate processes of the conditional point process intensity of each system as given by:

$$\lambda_j(t \mid H_{t^-}) \tag{11}$$

where $j = 1, 2, 3, ..., k$ represents the number of systems.

Estimation of Parameters of the Model

The most common and widely used method of inferring the parameters of the stochastic process from the data is the method of maximum likelihood. Inference using this method for repairable systems is presented in Lindqvist (2006) and in the appendixes E and F of Lawless (1982).

The likelihood function of $\lambda(t \mid H_{t^-})$ for a single system is given by:

$$L(t) = \prod_{i=1}^{n} \lambda(t \mid H_{t^-}) \exp\{-\int_0^t \lambda(s \mid H_{s^-}) ds\} \tag{12}$$

where n = number of failures in (0,t]

The likelihood function for a number of systems considered together based on equation (12) and under the assumptions that the systems are independent of each other and an identical probability distribution for the failures of each system is given by:

$$\prod_{j=1}^{k} L(t) = \prod_{j=1}^{k} \prod_{i=1}^{n} \lambda_j(t \mid H_{t^-}) \exp\{-\sum_{j=1}^{k} \int_0^t \lambda_j(t \mid H_{s^-}) ds\} \tag{13}$$

where i=1,2,3...n and j=1,2,3....k and k = number of systems

Testing for Model Fit

Goodness of fit tests based on the asymptotic distribution of the estimators or the log likelihood ratio are required to test the models for assessment of the model fit.

A simple check for arriving at the model with the best fit is to choose the model whose function has the maximum log likelihood value among the alternative models.

A better check for correct fit is obtained by using a penalized likelihood criterion such as the Akaike likelihood criterion (AIC). To counter the fact that the log likelihood ratio test, especially when applied to large data sets, favours the more complex models Akaike (1973) proposed to penalize the log likelihood of a model with the number of parameters in the model to obtain a better criterion for comparison. The criterion is given by:

$$AIC(k) = -2\ln L + 2k \tag{14}$$

where k = the number of parameters of the model

The above methods can be used appropriately for checking the model fit of the renewal, HPP and NHPP models.

CASE STUDY OF RELIABILITY- BASED MAINTENANCE OF INDUSTRIAL ASSETS

A simple case study is used to illustrate the implementation of the methodology given above. The long travel gearbox of an electric overhead travelling crane is considered for study. The electric overhead travelling crane is used in most industries for shifting of material within a bay. There are four such identical gear boxes for the crane which together drive the crane along the bay. As the four gearboxes are identical and perform the same function under similar circumstances they are considered together for this study. The gearboxes are in operation for a period of 2374 days (six and a half years) at the time of the study.

The gearbox is taken for routine preventive maintenance every fortnight and inspection, cleaning, lubrication and bolt tightening are carried out. The defects logged during inspection are fed to a central maintenance data repository. Decisions are taken based on the nature of the defects such as to give clearance for the operation of the gearbox or to carry out corrective maintenance wherever necessary at the same or next scheduled inspection time or an appropriate opportune time in between. The current maintenance repair cycle of the gearbox is presented in Table 9.

Reliability centered maintenance and analysis of reliability of repairable systems are applied to the gearbox in an integrated manner to arrive at a reliability based maintenance schedule.

FMECA part of the RCM analysis is carried out for the gearbox. The first step in this process is to define the system boundary. This is carried out in Figure 3.

The gearbox assembly, its sub-assemblies and components are listed out as given in Table 10.

All the functions and the functional failures of the gearbox assembly are listed in Table 11.

The various failure modes associated with each functional failure are then listed as given in Table 12.

The effect of each failure and the consequences of the failure based on its effects on the gearbox are listed in Table 13 with Y indicating yes, there is an effect. The consequences are classified into four categories hidden, safety, and others including maintenance.

Before proceeding to the decision logic part of the RCM exercise the frequency of the maintenance intervals are to be decided. In the initial stages of the equipment operation the maintenance intervals are decided based on past experience or experience with similar items elsewhere. Once failure data on the system are available these can be analysed, and measures of interest to the maintenance engineer estimated. Based on these results the maintenance intervals can be modified.

As all maintenance carried out is on failure as indicated by inspection, the various models of CM are applied to the failure data. There are four gearboxes per crane. The failure data of each gearbox is considered separately. As the failure of each gearbox is independent of the others, the gearboxes and

Table 9. Current maintenance repair cycle of the gearbox assembly

Asset / Equipment		Electric Overhead Travelling Crane
System		Mechanical
Sub-System		Long Travel Mechanism
Assembly	Frequency	Maintenance Task
Gear Box	Fortnight	1. Gearbox is cleaned externally of dust and muck 2. Gearbox is thoroughly checked and all defects observed are logged 3. All loose bolts are tightened 4. Oil level in the gearbox is checked and topped up if found low

Figure 3. Long travel mechanism gearbox and its sub-assemblies

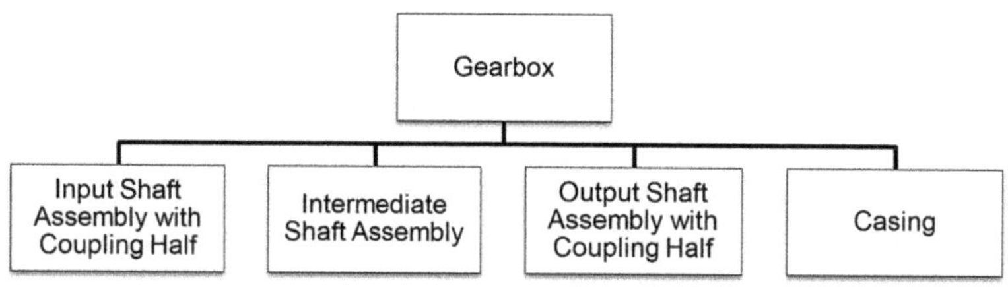

Table 10. EOT Crane, mechanical system, long travel Mechanism, its assembly gearbox and its sub-assemblies/components

Asset / Equipment		Electric Overhead Travelling Crane	
System		Mechanical	
Sub-system	Assembly	Sub-Assembly	Components
Long Travel Mechanism	Gear Box	Input Shaft Assembly	Input Shaft
			Gear
			Bearings
			Spacers
			Coupling Half
		Intermediate Shaft Assembly	Intermediate Shaft
			Intermediate Gear
			Gear Pinion
			Bearings
			Spacers
		Output Shaft Assembly	Output Shaft
			Gear Pinion
			Bearings
			Spacers
			Coupling Half
		Casing	Casing Top
			Casing Bottom
			Bolts & Nuts
			Gaskets

Table 11. Functions and functional failures of the gearbox assembly

Equipment	Electric Overhead Travelling Crane		
System	Mechanical		
Sub-System	Long Travel Mechanism		
Assembly	Functions	Performance Standards	Functional Failure
Gearbox	Transmit Torque from Motor to Drive Wheel Assembly	Torque to be transmitted – 553.9 kgm	Does not transmit torque
	Integrity of gearbox to be maintained		Integrity not maintained
	Adequate lubrication to be provided to bearing and gears	Splash lubrication with sump capacity 6.5 litres	Inadequate lubrication

their operating and maintenance conditions are identical and no two gearboxes have been found to fail at the same time, the failure processes are estimated using equation 13.

The estimated parameters, the expected number of failures in the next year and the time for attaining 0.9 and 0.5 reliabilities after the last failure are enumerated in Table 14.

While homogeneous poisson process (HPP) model provides the best fit for input shaft assembly, a non homogeneous poisson process with a power law process baseline (NHPP- PLP) and non homogeneous poisson process with a log linear process baseline (NHPP- LLP) provide the best fits for intermediate

Table 12. Functional failures and failure modes of the gearbox

Equipment	Electric Overhead Travelling Crane	
System	Mechanical	
Sub-System	Long Travel Mechanism	
Assembly	Functional Failures	Failure Modes
Gearbox	Does not transmit torque	Input Shaft Assembly Failed
		Intermediate Shaft Assembly Failed
		Output Shaft Assembly Failed
		Input Coupling Half failed/bolts sheared
		Output Coupling Half failed/bolts sheared
		Gearbox casing damaged
	Integrity not maintained	Joint bolts/studs failed/loose
		Inspection cover open/damaged
		Bearing covers not tightened/loose
		Foundations bolts loose
		Casing cracked
		Leakage through joints
	Inadequate lubrication	Low oil level due to leakage
		Contaminated oil
		Oxidised oil

Table 13. Functional failure modes, failure effects and failure consequences of the Gearbox

Equipment	Electric Overhead Travelling Crane				
System	Mechanical				
Sub-System	Long Travel Mechanism				
Failure Modes	Failure Effects	Failure Consequences			
		H	S	O	Ot/M
Input Shaft Assembly Failed	Torque is not transmitted			Y	Y
Intermediate Shaft Assembly Failed				Y	Y
Output Shaft Assembly Failed				Y	Y
Input Coupling Half failed/bolts sheared				Y	Y
Output Coupling Half failed/bolts sheared				Y	Y
Gearbox casing damaged				Y	Y
Joint bolts/studs failed/loose	Integrity of the gearbox is not maintained		Y	Y	Y
Inspection cover open/damaged				Y	Y
Bearing covers not tightened/loose				Y	Y
Foundations bolts loose				Y	Y
Casing cracked			Y	Y	Y
Leakage through joints				Y	Y
Low oil level due to leakage	Inadequate lubrication			Y	Y
Contaminated oil				Y	Y
Oxidised oil				Y	Y

shaft assembly and output shaft assembly & structural integrity respectively. The last three assemblies exhibit a deteriorating trend.

In the case of input shaft assembly, the expected number of failures in the next year is 1.2 per assembly i.e., 4.8 for all four assemblies, 3.3 for each intermediate shaft assembly i.e., 13.2 for all four assemblies, 2.2 for each output shaft assembly i.e., 8.8 for all four output shaft assemblies, and 0.1 for structural integrity i.e., 0.4 for structural integrity of all four gearboxes. The deteriorating trend is responsible for the high number of expected number of failures for intermediate and output shaft assemblies during the next year. Usually Input shaft assembly of a gearbox is more vulnerable than intermediate and output shaft assemblies and should exhibit more failures. This is not the case here and reasons for this are to be thoroughly investigated and necessary actions are to be taken to arrest this trend.

Considering the above results, we can see that at the present time a reliability of 0.9 is attained at 32.3 days from last failure for input shaft assembly, 15.4 days for intermediate shaft assembly, 28.1 days for output shaft assembly and 263.4 days for structural integrity. As the last three assemblies indicate a deteriorating trend, this may further come down unless the trend is arrested. Hence routine preventive maintenance work, (PM-R) scheduled inspection of 15 days for the intermediate shaft assembly, one month for input and output shaft assemblies, and six months for the structural integrity are proposed in the RCM schedule to start with.

In the decision logic part of the RCM process, the maintenance types and tasks are selected using the decision logic and their intervals decided based on the above analysis.

Table 14. Estimates of the model parameters for the failure data of gearbox internals along with some maintenance indicators

Sl No	Assembly/ Sub-assembly	Model	Scale	Shape	Expected No. of failures in next one year per unit	No. of days at which the last failure has occurred	Time for 0.9 Rel. in Days after last failure	Time for 0.5 Rel. in Days after last failure
1	Gearbox							
1.1	Input Shaft Assembly	HPP	0.00326	----	1.2	2246	32.3	212.3
1.2	Intermediate Shaft Assembly	NHPP – PLP	1.2403e-10	3.1599	3.3	2246	15.4	97.4
1.3	Output Shaft Assembly	NHPP – LLP	-9.19395	0.001596	2.2	2246	28.1	165.5
1.4	Structural Integrity	NHPP – LLP	-5.73243	-0.00097	0.1	2022	263.4	----

If one gearbox is down, clearance can be given for the crane to run after disconnecting it, so run to failure i.e., corrective maintenance is taken as the predominant mode of maintenance and no scheduled time/age based preventive maintenance is considered. The scheduled inspections followed by corrective maintenance at the same or next scheduled inspection time or an appropriate opportune time in between address the failure modes and hence are applicable. If by carrying out corrective maintenance in between inspection schedule intervals there is not much loss of production then this will form the most cost effective way of changing the failed components by extracting their full life.

The maintenance actions for each functional failure mode are listed in Table 15.

However, in case the production loss is higher while carrying out corrective maintenance as and when required then a reliability based scheduled preventive maintenance can be contemplated for changing the gearbox internals. Considering the above results, we can see that at the present time a reliability of 0.5 is attained at 212.3 days from last failure for input shaft assembly, 97.4 days for intermediate shaft assembly, and 165.5 days for output shaft assembly. For the structural integrity reliability of 0.5 occurs after a very long period. This period can be considered as a starting point for change of internals after inspection of their condition. There is also a possibility of incorporating a predictive maintenance schedule for the gearbox based on the prediction of the expected times to next failure.

The maintenance repair cycle of the gearbox is generated by grouping all the maintenance tasks by their frequencies / time intervals. This is shown in Table 16.

It can be seen from Tables 9 and 16 that the maintenance repair cycle of the gearbox shows an improvement after the reliability based analysis. The maintenance jobs are more clearly defined and their scheduling is more scientific.

DISCUSSION AND FUTURE WORK

The results of the case study were reviewed along with a group of operators and maintainers and their engineers/managers. The issues that have emerged out of the review are:

Table 15. Functional failure modes, maintenance actions of the gearbox assembly

Equipment	Electric Overhead Travelling Crane		
System	Mechanical		
Sub-System	Long Travel Mechanism		
Failure Modes	Maintenance Action		
	Type	Task	Interval
Input Shaft Assembly Failed	PM-R	Condition to be checked	4 weeks
Intermediate Shaft Assembly Failed	PM-R	Condition to be checked	2 weeks
Output Shaft Assembly Failed	PM-R	Condition to be checked	4 weeks
Input Coupling Half bolts sheared	PM-R	Condition to be checked	4 weeks
Output Coupling Half bolts sheared	PM-R	Condition to be checked	4 weeks
Input Coupling Half failed	PM-R	Condition to be checked	3 months
Output Coupling Half failed	PM-R	Condition to be checked	3 months
Gearbox casing damaged	PM-R	Condition to be checked	6 months
Joint bolts/studs failed/loose	PM-R	Bolts to be checked and tightened	2 weeks
Inspection cover open/ damaged	CBM	Inspection cover to be checked and replaced	4 weeks
Bearing covers not tightened/ loose	PM-R	To be checked and tightened	2 weeks
Foundations bolts loose	PM-R	To be checked and tightened	2 weeks
Casing cracked	PM-R	Condition to be checked	6 months
Leakage through joints	PM-R	Input/Output Shaft Assembly alignment to be checked and corrected and leakage arrested	4 weeks
Low oil level due to leakage	PM-R	Leakage through shaft seal, inspection cover and joints to be checked and arrested	2 weeks
Oxidised Oil	CBM	Oil to be checked for oxidation and changed if necessary	6 months
Contaminated Oil	CBM	Oil to be checked for contamination and changed if necessary	12 months

1. It was seen that the maintenance schedule was arrived at in a more scientific way after performing a failure; modes, effects and consequences analysis, quantitatively assessing the reliability of the least replaceable assemblies and the integrity of the gearbox and framing the maintenance program in a more systematic manner using decision logic.
2. This could lead to the incorporation of a scheduled preventive maintenance program for the assemblies based on their reliabilities.
3. It was pointed out that the schedule generated for the gearboxes may not hold good under varying environmental factors / operations. Hence it was decided to perform the quantitative calculations of reliability once every six months to start with and make appropriate changes in the schedules if necessary.
4. It was also resolved to automate the calculations and RCM analysis through developing appropriate software to make this possible.

Table 16. Maintenance repair cycle of the gearbox assembly

Asset / Equipment		Electric Overhead Travelling Crane
System		Mechanical
Sub-System		Long Travel Mechanism
Assembly	Frequency	Maintenance Task
Gear Box	Fortnight	1. Intermediate shaft assembly condition to be checked 2. Gearbox casing Joints studs/bolts to be checked and tightened 3. Bearing covers to be checked and tightened 4. Foundations bolts to be checked and tightened 5. Oil leakage through shaft seal, inspection cover and joints to be checked and arrested
	Month	1. Input Shaft Assembly condition to be checked 2. Output Shaft Assembly condition to be checked 3. Input/Output Coupling Bolts condition to be checked and replaced if necessary 4. Gearbox Inspection cover to be checked and replaced 5. Input/Output Shaft Assembly alignment to be checked and corrected and leakage arrested
	Quarter	1. Input/Output coupling half teeth condition to be checked
	Half-Year	1. Gearbox Casing condition to be checked 2. Oxidation of oil to be checked and oil changed if necessary
	Year	1. Contamination of oil to be checked and oil changed if necessary

5. The revised maintenance schedule was integrated into the computerized maintenance management system (CMMS) which was in place in the organisation. Spares procurement/repair program was put in place in the CMMS based on the quantitative assessment of reliability.
6. The reliability of intermediate shaft assembly was found to be low and that of output shaft assembly lower than that of the input shaft assembly. Also the reliabilities of both these assemblies were found to be deteriorating. It was agreed to carry out a detailed root cause failure analysis of these assemblies to identify the reasons for their more frequent failures and to carry out the necessary modifications to improve their reliability.

RCM is basically a static exercise in framing a PM program. This needs to be made dynamic i.e., in real time by revising the program based on changing circumstances in the field and integrating the program with other information systems such as operational processes or maintenance management systems. Such approaches are just beginning to take shape.

The strategy of reliability based management should fit into the latest international standards on engineering asset management such as ISO 55000. Aladon a leading purveyor of RCM methodology has already begun to address this issue by developing RCM III program to fit the requirements of ISO 55000.

As the future evolution of maintenance is going to be web-based and through e-technologies, reliability based maintenance needs to be integrated into such systems to form a part of future maintenance strategies.

CONCLUSION

Reliability centered maintenance as a philosophy schedules maintenance with an aim to enhance the reliability of the asset under consideration at optimum cost. This methodology is mostly qualitative and decides the maintenance intervals based on past experience and age-exploration methods. A quantitative way of deciding the maintenance intervals is obtained using the principles of reliability engineering, especially the methods of reliability of repairable systems. This is based on data and relies on the failure and other data generated by the asset. This complements the reliability centered maintenance philosophy and in tandem helps in maintaining an asset at optimum reliability.

These techniques are still not in vogue in the industry in general and need to be applied to carry out asset maintenance on a sustainable basis and help to reduce maintenance costs and failure consequences.

REFERENCES

ABS. (2004). *Guidance Notes on Reliability-Centered Maintenance*. Houston, TX, USA: American Bureau of Shipping.

Akaike, H. (1973) Information theory and an extension of the maximum likelihood principle *Proceedings of theInternational Symposium on Information Theory*, Akademia Kiado, Budapest.

Anderson, R. T., & Neri, L. (1990). *Reliability-Centered Maintenance: Management and Engineering Methods*. New York, NY: Elsevier Science Publishing Co. doi:10.1007/978-94-009-0757-7

Ascher, H., & Feingold, H. (1984). *Repairable systems reliability. Modelling, Inference, Misconceptions and Their Causes*. New York: Marcel Dekker Inc.

August, J. (1999). *Applied Reliability-Centered Maintenance*. Tulsa, Oklahoma: PennWell.

Bloom, N. B. (2005). *Reliability Centered Maintenance Implementation made simple*. New York, NY: McGraw Hill.

Brown, J. F., Mahoney, J. F., & Sivazlian, B. D. (1983). Hysteresis repair in discounted replacement problems. *IIE Transactions*, *15*(2), 156–165. doi:10.1080/05695558308974627

Chan, J.-K., & Shaw, L. (1993). Modelling repairable systems with failure rates that depend on age and maintenance. *IEEE Transactions on Reliability*, *42*(4), 566–571. doi:10.1109/24.273583

Dijoux, Y., Doyen, L., & Gaudoin, O. (2005) Simultaneous assessment of the efficiency of preventive and corrective maintenance for repairable systems in a competing risk framework. *Presented at 5th European safety and Reliability Conference,* Tri City (Gdynia-Sopot-Gdansk), Poland

Doyen, L., & Gaudoin, O. (2004). Classes of imperfect repair models based on reduction of failure intensity or virtual age. *Reliability Engineering & System Safety*, *84*(1), 45–56. doi:10.1016/S0951-8320(03)00173-X

Doyen, L., & Gaudoin, O. (2011). Modelling and assessment of aging and efficiency of corrective and planned preventive maintenance. *IEEE Transactions on Reliability*, *60*(4), 759–769. doi:10.1109/TR.2011.2171115

International Electrotechnical Commission. (2009). *IEC 60300-3-11 Dependability Management – Part 3:11 Application guide – Reliability centered maintenance.*

Kijima, M. (1989). Some results for repairable systems with general repair. *Journal of Applied Probability, 26*(1), 89–102. doi:10.2307/3214319

Kijima, M., Morimura, H., & Suzuki, Y. (1988). Periodical replacement problem without assuming minimal repair. *European Journal of Operational Research, 37*(2), 194–203. doi:10.1016/0377-2217(88)90329-3

Kijima, M., & Sumita, U. (1986). A useful generalisation of renewal theory: Counting processes governed by non-negative Markovian increments. *Journal of Applied Probability, 23*(1), 71–88. doi:10.2307/3214117

Lawless, J. F. (1982). *Statistical Models and Methods for Lifetime Data*. New York: John Wiley and Sons.

Lindqvist, B. H. (2006). On the statistical modeling and analysis of repairable systems. *Statistical Science, 21*(4), 532–551. doi:10.1214/088342306000000448

Lindqvist, B. H. Langseth, H. & Stove, B. (2006) Modelling of dependence between critical failure and preventive maintenance: the repair alert model, *Journal of Statistical Planning and Inference (Special issue on competing risks)*, 136, 1701–1717

747. Maintenance Steering Group. (1968). *Handbook: Maintenance Evaluation and Program Development 1 (MSG-1)*. Washington, DC: Air Transport Association.

Malik, M. A. K. (1979) Reliable preventive maintenance scheduling. *AIIE Transactions,* 11, 221-228

Moubray, J. (1997). *Reliability-Centered Maintenance*. New York, NY: Industrial Press.

Nowlan, F. S., & Howard, F. H. (1978). *Reliability Centered Maintenance*. Springfield, VA. U.S. Department of Commerce, National Technical Information Services, Report Number AD/A066-579.

Rigdon, S. E., & Basu, A. P. (2000). *Statistical Methods for the Reliability of Repairable Systems*. New York: John Wiley & Sons Inc.

Smith, A. M., & Hinchcliffe, G. R. (2004). *RCM – Gateway to World Class Maintenance*. Burlington, MA, USA: Elsevier Butterworth-Heinemann.

Society of Automotive Engineers. (2009). *SAE JA1011 Evaluation Criteria for Reliability-Centered Maintenance (RCM) Processes.*

Society of Automotive Engineers. (2011). *A Guide to the Reliability-Centered Maintenance (RCM).*

KEY TERMS AND DEFINITIONS

Industrial Assets: Industrial assets considered here are defined in a narrow sense as equipment deployed by the industry to convert inputs to outputs which can be then marketed by the industry as intermediate or finished products.

Industrial Maintenance: Maintenance carried out on industrial assets.

Reliability: An attribute of the asset whereby it continues to perform its function satisfactorily for a given period of time under stated conditions of operation.

Reliability Centered Maintenance: Maintenance carried out on an asset keeping in view the reliability requirements of the system at optimum cost.

Reliability of Repairable Systems: Reliability of repairable systems is characterised by the mean time a system operates between its failures.

Repairable System: Any system which can be restored to its functional state on failure by repairing or replacing one or more but not all of its components.

APPENDIX

Notation

N_t Number of corrective maintenance actions

T_i ith corrective maintenance time, $T_0 = 0$

M_t Number of preventive maintenance actions

D_j ith preventive maintenance time, $D_0 = 0$

K_t Number of maintenance actions i.e., $K_t = N_t + M_t$

C_k kth maintenance time, $C_0 = 0$

$\lambda\left(t \mid H_{t^-}\right)$ Conditional intensity process given the previous failure history H_{t^-}

Chapter 19

Burning Assets? Designing Wood Fuel Energy Strategies in Mozambique:

2MBio a Novel Participatory Model to Promote Creativity and Knowledge as Strategic Assets

Ricardo Martins
CEG-IST, Portugal

ABSTRACT

Getting wood for cooking, heating, processing into charcoal and/or selling is a fundamental part of most household survival strategies in Developing Countries. Entangling in complex and dynamic ways local and global ecosystems, poverty, technology and business Wood Fuel Energy Systems (WES) are fundamental and require simple to use design tools to support the strategic and optimised used of available socio-ecological resources/assets. However, there are very few tools able to support relevant actors (e.g. charcoal makers, experts, policy makers) in that task. To bridge that gap the 2MBio, a participatory conceptual design tool to support the strategic design of WES, is introduced and its practical results in Mozambique presented. The 2MBio explicit in a simple and intuitive layout the set of necessary and sufficient resources/assets required to produce comprehensive and meaningful WES designs/strategies, which represent in themselves a strategic asset, while further stimulates knowledge and creativity as a tacit asset.

INTRODUCTION

Bioenergy (the energy from biomass) represents an estimated 10% of all energy consumed in the world, 50% of all energy consumed in Developing Countries (DCs) (IEA 2006) and around 70% in Mozambique (Vas *et al.* 2011). Besides being central to the life of millions of households, bioenergy systems based on

DOI: 10.4018/978-1-5225-0651-5.ch019

wood, i.e., wood fuel energy systems (WES), also poses complex and crucial social, technical, economical and ecological challenges for sustainability, livelihoods, resource management and policy making (§2.1). Despite this complexity, most policies and approaches developed for WES are based on top-down deliberate strategies, and limited economic views on resources. Particularly preeminent is the "energy transition paradigm" which labelling the current WES practices and technology as "traditional", "dirt" and damaging for the environment, promotes a "technological leap" towards technologies deemed to be "modern", clean and "green". In basic terms, these rationalist decision-making approaches assume that complexity can be simplified and that best decisions can be made based on the optimised use of assets seen as constant and objectively understood by all actors equally (§2.2). Thus, most political and technical strategies in DC had been focusing on "HOW" to make this transition to take place, rather than work with "WHAT" make people do, think and act while dwelling and making strategic decision everyday on WES. There are more contextualised and constructivist alternatives to this positivist paradigm, like the Community Based Natural Resources Management (CBNRM), the Participatory Modelling or other generic design methodologies. However, these are not specific to WES and being too contextualised fail to provide a common base for comparison and generic strategy development.

In face of these gaps and the lack of useful strategic design tools this chapter reframed WES policies and practices as a strategic design problem on complex socio-ecological systems (§3), and consequently applied strategy thinking, systems thinking, design thinking and more contextualised views on assets to justify and specify a novel tool for the strategy design of WES, the 2MBio (§3).

The 2MBio is a novel participatory conceptual design tool, the, which favours participation, learning, knowledge sharing and innovation. Presented in a simple intuitive low-tech layout of 13 boxes in a paper (Figure 5), the 2MBio makes explicit "WHAT" are the necessary and sufficient basic (ontological) conceptual design elements (the 13 boxes) one has to think about when tanking about the conceptual design of biomass energy systems. Focusing on "what to consider" within a structured perspective (the layout of 13 boxes), the 2MBio is not normative or prescriptive, neither it is totally subjective. Rather, like a basic set of LEGO™ blocks shared by all, the 2MBio simultaneously provides:

1. A formalised common space for dialogue;
2. Allows participants to express their knowledge and experience on the "WHY" and "HOW" to address problems and solutions; and
3. Facilitates participatory conceptual design of comprehensive and integrated strategies, policies, projects and solutions on WES.

In practical terms the 2MBio is used in participatory design workshops, where the participants are invited to write or draw on those design elements what to do, how and why, making explicit and available to discussion the tacit and implicit knowledge.

The chapter ends with the presentation and analysis of several results of a study conducted in Mozambique, where graphical evidence is provided on how the use of the 2MBio expanded the perspectives on assets, design and strategy around WES, produced a number of innovative, comprehensive and meaningful conceptual designs of sustainable WES for Mozambique (§3.2), and further shows how the 2MBio once filled by the users constitutes, in itself, a strategic asset for future designs and strategies (§2.3).

STRATEGY DESIGN FOR WOOD FUEL ENERGY SYSTEMS: LINKING ASSETS, RESOURCES, ISSUES, THEORY AND INITIATIVES

Wood Fuel in Mozambique: Relevance, Issues, and Contexts

Mozambique, a Southeast Africa Developing Country (DC), has consistently been in the 23 least developed countries since 1988. A sparsely populated country (25.7 people per km^2), Mozambique has around 25million inhabitants (INE, 2007), from which 44-79% are poor or live in severe poverty (OPHI, 2011; PARPA, 2001; UNDP, 2014) and 70% live in rural areas (despite the over 3% annual urbanisation rates) (INE 2007), where over 95% of all farms are with smallholders practicing a subsistence agriculture (Hanlon & Cunguara, 2010). Considering that the Mozambican road density is around 3.8km per 100km2 (WB-WDB), it easy to perceive that Mozambique is composed mostly of poor rural communities living dispersed in a vast territory.

In terms of energy, a net exporter of electricity and gas Mozambique is well endowed with energy resources and assets, which include (Vaz *et al.*, 2011): Cahora Bassa with an installed capacity of 2075MW; an estimated over 8 billion tonnes of coking coal; proven reserves of 127.4 billion m^3 of natural gas; and a potential to produce around 3.1 million barrels of oil equivalent per day of sustainable biofuels, i.e., without encroaching on food crops or endangering biodiversity. However, Mozambique remains: one of the 20 lowest energy consumption countries, 0.415-0.425toe per capita (WRI 2010); among the 10 lowest national electrification rates, 17%-20% in urban areas and less than 2% of rural areas (Cumbe *et al.*, 2005); and heavily dependent on wood fuel energy (WF), mostly as firewood in rural areas and charcoal in the cities.

Indeed, available data from 1988 to 2011, Figure 1, shows that WF represents consistently over 77% of total energy consumption in Mozambique. This WF is almost entirely dedicated (an average of 86.3%) to supply households (which in turn represent 77.8% of the national energy demand) to fuel the energy demanding and time consuming cooking of Mozambican cuisine (e.g. Cumbe *et al.* 2005). Moreover, for the period considered in Figure 1, the WF consumption also increased steadily at an average rate of 1.6% per year, a trend that is expected to continue in the near future.

As in other DCs with relative abundance of free access to forests (e.g. Tanzania, Kenya, Rwanda) the overwhelming consumption of WF in Mozambique has been associated with a number of interlinked and relevant socio-ecological issues, including: deforestation and/or forest degradation; health hazards; gender bias; socio-economic influence on technological choices; and historical backgrounds.

WF production is definitely an important factor in deforestation and forest degradation in Mozambique, since: the resource is "free"; there are no plantations for energy or suitable forest management; local earth-mound charcoal kilns have only 15% efficiency in weight; consumption is done in technically inefficient three stones stove or open iron stones; and urban demand is high. However, deforestation is a complex and contextual process involving a number of dynamic, interwoven and sometimes reinforcing causes including, in Mozambique (e.g. Sitoe *et al.* 2012): the use of land for agriculture; human caused fires; natural calamities; logging; and the crescent economic activity around important railways or major roads. Therefore, while WF could be associated with deforestation and forest degradation, the degree of importance and dynamics of such association is disputed and contextually dependent.

The use of WF is also problematic for human health. Producing, collecting and using WF could result in burning and serious occupational health risks (WHO 2002). However, the major concern is the indoor air pollution (IAP) which results from the emission of small particles and several chemicals dur-

Figure 1. Total energy consumption in Mozambique from 1990 to 2011 according to energy source and economical sector
[Source: Martins 2014]

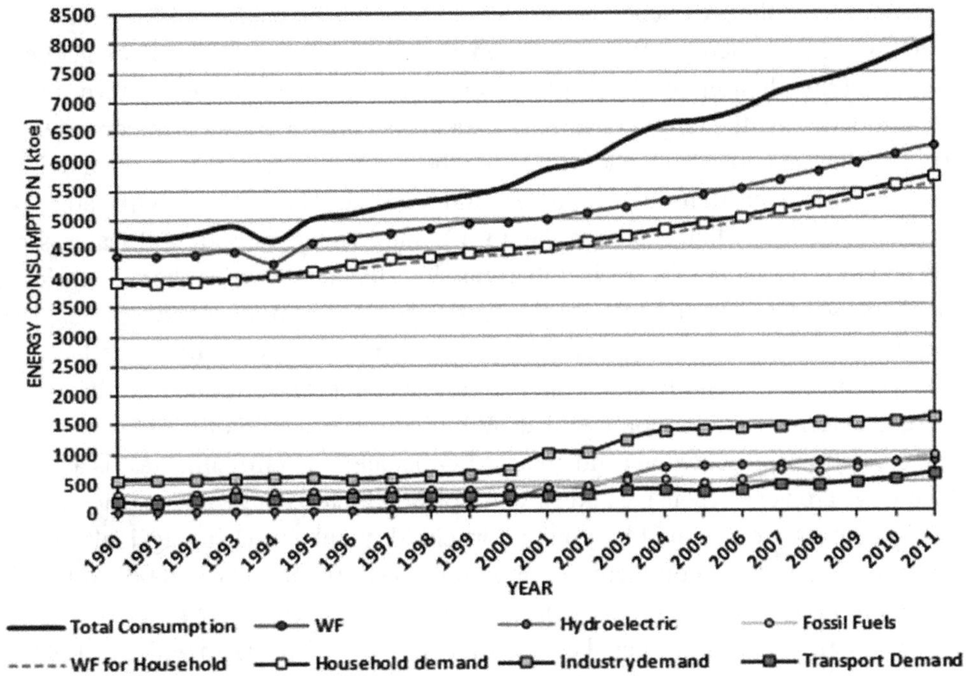

ing incomplete combustion of WF in insufficiently ventilated areas. While this is a complex issue and other factors might be relevant (e.g. fuel type, technology used, household layout, user behaviour) (e.g. Jetter *et al.* 2012), WF has been consistently associated with IAP, which in turn is responsible for 16% of outdoor particle air pollution in the world (Smith *et al.* 2012) and nearly 4 million premature deaths in 2010 (Lim *et al.* 2012). Conversely, when WF, usually the sole affordable fuel for poor households, becomes scarce health might be affected as well, since households to save fuel might prefer fast to cook, but less nutritive food (Brouwer *et al.* 1997); stop boiling water (Plummer 1999), or rely on even more pollutant fuels (Jagger & Shively, 2014). In Mozambique, in rural areas, and in many households of urban areas, cooking is done in the open air, which drastically reduces IAP. Research also found that in Maputo, cars, not WF are the responsible for the high levels of gases in the city (Cumbane, 2007). Nevertheless, research suggests that firewood users in Maputo were found to have significantly more cough symptoms than other groups (Ellegard, 1996) and that that in-house smoke caused by WF increased the risk of otitis in small children (da Costa *et al.* 2004).

The WF gender dimension arises from the fact that, despite the central role women have on WF management at the household level, and the health risk they endure daily (e.g. cooking in confine kitchens, carry heavy loads of wood over several kms) their participation on energy strategies, decision or design of technology, at any level is rather low, if not inexistent (e.g. Cecelski, 2004). However, research also points that men and women establish different relation with energy (e.g. Tragett, 2012). Also, it is not entirely clear that this "decision gap" is the result of women being women, or rather the consequence of the household being poor (Wamukonya 2004). In Mozambique, traditionally the role of women in household decision (including energy) is quite confined and secondary to men, however the Author also found that while charcoal is a physically demanding activity, several women are involved in the charcoal business, individually, with other women, in charcoal associations and even hiring men.

Another perspective on the prevalence of WF in the Mozambican energy mix is related with the lack of sound technological and energy options in the market. Providing natural gas or grid electricity to remote and dispersed rural households has high capital costs and low investment return since these poor households tend to have small energy consumptions and low and/or irregular income (e.g. Zerriffi & Wilson, 2010). High initial costs also hinder rural poor the access to alternative off-grid RETs. However, when RETs and/or grid connections are available, rural households are served with rather poor quality installations and service (daily blackouts are the norm), and manage only to boil water, since cooking is rather costly and too demanding for the installations. In urban areas, while there is a wider range and better quality of energy services and technological options at more competitive prices, reality is not much different. Due to the unstable or unreliable source of income poor households lack the financial capacity to pay a contractual monthly fee or the equipment cost for utilities and available RETs even if these are, on the long term, more economical than WF. Electricity is used only to boil water or very fast meals for the same reasons presented in rural areas. LPG (Liquefied Petroleum Gas) is available, but the canister and the stove are very expensive and LPG is considered too dangerous, particularly if there are children around or stoves are operated by young household members. Using agricultural and forestry residues is currently not a viable commercial option due to (Batidzirai *et al.* 2006): difficult logistics, high dependence on timber processing; low energy gain, and competency with other uses (e.g. animal feed, fertiliser and energy). As for dedicated crops, after an initial enthusiasm with ethanol from cassava and biofuel from Jatropha the outcome was quite disastrous.

However, alone, these market gaps cannot justify the current energy consumption pattern in Mozambique and should, in fact, be understood within a wider socio-ecological and historical context and political dynamics, which include:

- **The Abundance and Easily Accessible and Non-Monitored Wood Resources:** Mozambique has over 50% of forest cover, from which 67.2% has a high timber volume (over of 32m3/ha) and is located outside the protection and conservation zones (Marzoli, 2007). However, according to Mozambican law, only those trees that cannot be used for anything else (class 4) or are damaged are suitable to produce charcoal. This does not prevent the fact that Maputo is being mostly provided with dense charcoal with high calorific value charcoal produced from the extensive forests of class 1 Mopane (*Colophospermum Mopane*) in Gaza Province. While probably over 50% of Mopane is hollow (damaged), being locally available cost-free, with vague/ ambiguous ownership and complete lack of monitoring it is impossible to screen-out the healthy Mopane trees from charcoal production.
- **The Lack of Local Legitimated Forest Management**: With very few exceptions, Civil War and some governmental measures, weakened traditional leaders' authority and legitimacy to implement and enforce local forest management regulation and disrupted the social fabric in rural areas. Consequently, this legitimacy void was filled by poor farmers and/or ex-soldiers looking for fast income.
- **The Lack of Income Generating Activities Outside Farming in Rural Areas**: With the end of the civil war in 1982 in order to provide an income generation activity to thousands of ex-soldiers with limited professional skills, very few bonds to their zone of origin the government and some

NGOs actively promoted the production of charcoal. The plan incorporated low state investment (wood was free and technology was local) and crescent energy demand from cities, but was possibly the most immediate and easy option. Even today, if there are not big farming investments (and normally there are not) the low initial investment business of choice is, consistently, charcoal production.

- **Climate Change and Low Food Production**: The biggest charcoal production area for Maputo is Mabalane, which is also characterised by rather difficult conditions, including (Levy & Kaufmann, 2014): acute lack of rain and water; high frequency of floods, draught, pests, widespread human and animal deceases; very poor soils in most areas; limited access to market and credit facilities; lack of information and infrastructures; economical and physical isolation; and neglected state investment in smallholder agriculture. While some of these hardships have been associated with global climate change (WB, 2010), wars and social disruption (Levy & Kaufmann 2014) farmers (98% of the population) in Mabalane typically produce 50% of their annual food consumption (FAO, 2005), face repeated crop failure, have limited means to change production conditions (Levy & Kaufmann, 2014).
- **Crescent Monetization of the Rural Areas**: Partially motivated by the efforts of the Government to fulfil the Millennium Development Goals, partly due to the increase of commercial activity all over Mozambique (see Hanlon & Cunguara, 2010), turned the need for cash income much more necessary than in previous decades. For instance, the need to improve education levels brought to rural houses hold new costs (e.g. with pencils or notebooks) which require cash to be acquired.

Consequently, a combination of easy access to available cost-free forestry resources, lack of good quality energy services and affordable technology, need to cope with food insecurity and generate cash income seems to facilitate the prevalence of the culturally embedded WF as the logical energy option for most Mozambicans. Notably, a remarkable network of over 3 million people (Cuvilas el al. 2010), mostly made of uneducated people, created, operate and manage (mostly though informal channels and networks) an estimated 800million € per year business (Martins, 2014), that is able to covert natural resources in distant forests into charcoal available in suitable amounts, prices and quality in convenient locations throughout the cities. To cope with difficult conditions farmers in the remote areas turn to their natural assets to produce charcoal with the associated impacts on forest resources and livelihoods described above. At this point, the main question posed in this chapter, "are the farmers in Mozambique burning assets?" had gained a complex and contextualised socio-ecological frame, which will be explored from a strategy design perspective in the next sections.

Current Strategies and Modelling Approaches on WES: A Critical View

The previous analysis (§2.1) shown that decisions on WES are the expression of complex interactions embedded in specific local (e.g. poverty) and global (e.g. climate change) socio-ecological contexts and dynamics. Thus, within the focus of this chapter, it is relevant to know which strategies and modelling approaches exist to support the strategic decision in WES.

Most WES strategies are framed by the Energy Transition Paradigm (ETP): a set of normative and prescriptive assumptions that perceive energy problems in DCs to be caused by a vicious mix of poverty and the (so considered) dirt, hazardous, inefficient and traditional WF and technologies (Martins 2014: 63-80). Thus, ETP endorses the switch to the (so considered) clean, efficient and modern energy carriers and renewable energy technologies (RETs) like solar panels. Strongly marked by deterministic, normative and prescriptive technocratic views, the ETP conceives WES as a set of resources and technologies, aligned in a chain of operations, and ruled by mechanic, linear and deterministic cause-effect dynamics, Figure 2.

The outcome of ETP in several DCs has been the top-down National Forest Strategies which impose the temporary, permanent, nationwide or localised prohibition to cut trees (e.g. Rwanda, Tanzania, Zimbabwe, and Kenya). Complementary or alternatively National Energy Strategies are defined to promote, through special fiscal benefits, funding or subsiding, RETs and technologies like the improved cooking stoves (ICS) which have been implemented almost in all sub-Saharan Countries. Alongside, a number of projects and/or programmes by NGOs, governmental bodies or International Agencies promote the engagement of charcoal makers in other equally profitable activities. In less degree the private sector has also been included or stimulated to participate in the dissemination of new technologies, being the ICS in Kenya a commonly cited example. The final goal is to give modern energy services and change energy habits in rural and urban populations while improving their living standards.

In Mozambique the ETP is clearly stated in the National Energy Strategy and in a number of official documents, and a national organization FUNAE (National Energy Fund) is managing the ICS programme. Despite the eminent relevance (§2.1) until a few years ago, WES and WF were mostly absent in governmental priorities and were rather seen as a problem to avoid. This reality has changed and currently there is a Biomass Strategy in place (BEST, 2012) and specifically a WF strategy has been defined (MIREME, 2014). This tendency mirrors the regional trends in East-Africa to consider WES as an opportunity once the sector is fully formalised, closely monitored and new technologies are implemented. Accordingly, the WF strategy promotes the introduction of technically more efficient charcoal kilns, better monitoring along the entire supply chain, improved forest management practices, afforestation and the implementation of a certification process.

The "modelling version" of ETP are the predictive mathematical models (PMM) which assume that design WES constitutes a solvable problem in deterministic terms, consider actors as rational decision makers in homogenous groups (the "community", the "women") and address problems/solutions from a disciplinarily expert perspective (Martins, 2014). Most PMM are global optimization models which integrate forest and/or agriculture biomass, processing technology and infrastructure (e.g. roads) to (Martins, 2014): optimise WF supply chain or WES configurations; and simulate WF flows. There are very few PMM for Mozambique (Nhamtumbo *et al.* 2001) and none addresses charcoal. This absence of PMM and the recent strategic change in Mozambique could be a direct consequence of the limited success of ETP approaches (as already exposed in §2.1). Indeed, unlike ETP and PMM assume, WES complexity, uncertainty, multiplicity of preferences (besides the economic) and individuality (among communities) cannot be described or calculated by predictive equilibrium models and optimisation based on closed systems. Moreover, efficiency, a common measure on PMMs and ETP, becomes ambiguous

Figure 2. A linear perspective of WES linking forest resources with people consumption
[Source: The author based on Goldemberg & Johansson, 1995].

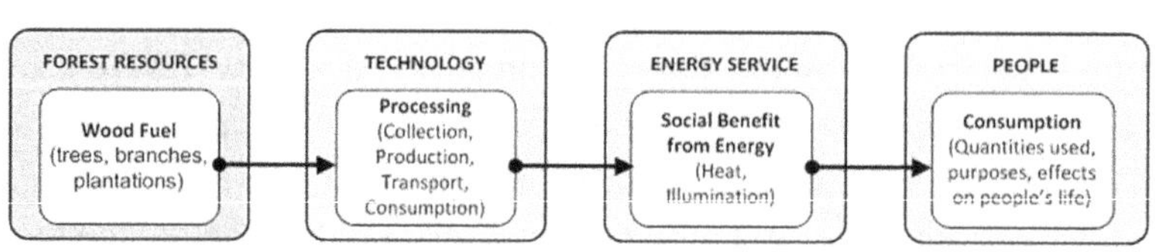

when moved away from technical metrics. In other words, ETP and PMM reflect more a perspective on "how reality should be" according to the policy makers and modellers, than actually on "what reality is" according with who lives it.

Departing from the normative and prescriptive ETP, more constructivist and contextualised paradigms propose the use synthetic and creative strategic thinking to identify and create emergent strategies from complex contextual circumstances (e.g. Pisapia *et al.* 2016). For example, the Community Based Natural Resource Management (CBNRM) assumes that people are not moved exclusively by economic or opportunistic behaviour, but rather have valuable knowledge, sound practices and sophisticated systems for forest management that ecologically and socially benefits and suits their context (e.g. Weir *et al.* 2013). However, while supporting the development by the community of their own forest management strategies and conceived in Mozambique to address the charcoal issue, CBNRM was never used to address WES and in Mozambique it become a rather bureaucratic non-operational process. On the modelling side, constructivist and contextualised paradigms have been represented by the participatory modelling tools (PMT). PMT is an umbrella term to define a wide range of techniques, methods and approaches to elicit perspectives from participants and produce relational models to explore, simulate and decide on alternatives. Among the most well know are: Group Model Building (e.g. Andersen *et al.* 1997); Mediate modelling (e.g. van den Belt 2004); Companion Modelling (e.g. Souchère *et al.* 2010); Participatory Systems Mapping (Sedlacko *et al.* 2014). Remarkably, in order to support the constructivist paradigm some PMM tried to include fuzzy logic or agent based modelling to simulate real humans and cope with unpredictable events (e.g. Vera *et al.* 2010; Wolbert-Haverkamp *et al.* 2014). Additionally, other generic design tools to produce creative designs (e.g. scenario building, brainstorming, Soft systems Methodology), could be used in the design of strategies for WES. However, while some of the PMT addressed forest resource management (Purnomo *et al.* 2003; Standa-Gunda *et al.* 2003) there are no PMT, modified PMM or generic design tools specifically conceived to address WES. Moreover, these models are mostly computer based and cannot be used by WES actors with low literacy levels and lack of computers and/or electricity, i.e., basically all WES actors in Mozambique. Moreover, all these approaches can only produce one-time experience, i.e., they lack a common framework that could be (re) used consistently and repeatedly, allowing to comparisons across experiences, contexts or even trials.

The brief review conducted here, besides describing the gap of models to support design and strategy on WES also demonstrated that:

1. The tools available specifically for WES design are too rational and positivist and thus cannot properly deal creatively with the complexity of WES;
2. Constructivist based tools address complexity and promote creativity, but lack the structure or the layout to facilitate the use by all WES actors across disciplines or design exercises.

Therefore, to bridge these gaps and further support different WES actors in the task of design creative strategies for WES novel tools is required. A tool that could support both charcoal makers and decision makers in designing the burning of trees as part of a complex WES in rural Mo- zambique as a contextualised act of strategy.

The Arguments for a New Tool to Support WES Strategic Design

To address the gaps identified in §2.2 the present work implemented a design thinking approach and assumed the definition of a tool to support strategic design in WES as a design task. Consequently, in line with current trends on design and strategy research (e.g. Funtowicz & Ravetz, 1993; Taylor, 2015), the departing point to create such tool was integration at different levels: vertically between design and strategy; and horizontally, between different trends within design and strategy.

The vertical integration between design and strategy is quite intuitive. Besides all conceptual controversy, strategy is about making the matching between the organization or people's values, mission and scope with available assets/resources through specific activities/operations at several levels at given contexts (Gane, 2007); design creatively translate ideas or visions (individual or collective) into tangible/intangible artefacts considered suitable to a given purpose, need or desire (Flusser, 2010). However, strategy seem to have a narrower aim of provide some form of "competitive advantage", "leverage" or "stand-out result", while design targets creativity in a ore broader aesthetic, engineering and business perspective. Consequently, there was an appropriation of design by strategy to produce such leverage. Indeed, design become a powerful strategic tool (Kotler & Rath, 1984) and is increasingly used by organizations as a core capability (i.e. a resource) to enhance performance and to unlock product and business innovation (e.g. Battistella *et al.* 2012). Design thinking provides strategy with innovative products, processes and services, as well as, promotes creativity, leaning dynamics and knowledge sharing, changing thus the way organizations conceive, develop and manage resources and assets. Conversely, strategic design could be defined as a creative way (a specific thinking style) to address strategic problems.

The horizontal integration reflects current tendency to integrate (or bypass) the common divide between rationalist/positivist and the constructivist stances, or between analytical convergent conventional thought and the synthetic creative thought process. Relating on the different stands on strategy and knowledge, Heracleous (1998: 486) concludes that "strategic thinking and strategic planning are interrelated in a dialectical process, where both are necessary for effective strategic management, and each mode on its own is necessary, but not sufficient", Figure 3A. Indeed, presently the term strategic thinking integrates the two distinct thought processes: planning and thinking (Steptoe-Warren *et al.* 2011). Likewise, current trends on design thinking (e.g. Buxton 2007) also promote design process that integrate divergent and convergent approaches, Figure 3B, i.e., that promotes the dynamic reasoning and interplay between exploring opportunities and setting the problem ("*What is the right thing to build?*") and making decisions and solving the problem ("*How do we build this?*"). In general terms this integration takes the pragmatic stance that it is impossible to control complexity, as it is impossible to be continuously innovating and adapting to contexts. Notably, this tendency *within* design and strategy disciplines also implies integrative tendencies on how resources (Figure 3C) and knowledge/learning are conceptualised and operated on (Figure 3A-B).

Remarkably, the resource conceptual integration brings two important results. Fist, every kind of resource can be defined through, or in relation with, each other, e.g., capitals are "the confluence of capabilities to be and act" (Bebbington, 1999). Secondly, "resources" (or assets) are contextual and multidimensional. Different people, in distinct contexts adopt different perspectives (e.g. economical, ecological) and attribute different properties, behaviours and values to the same "resource" for different reasons and different purposes.

Finally, both horizontal integration within design and strategy and vertical integration across design and strategy disciplines are not independent or context-free, but strongly shaped by (and shaping) the

Figure 3. (A) The integrative processes occurring in strategy; (B) Design integration process; and (C) Resource conceptual integration
[Source: The Author after (A) Heracleous, 1998; (B) Buxton, 2007: 142; (C) Jurow, 2006; Scoones, 1998; Sen, 1997]

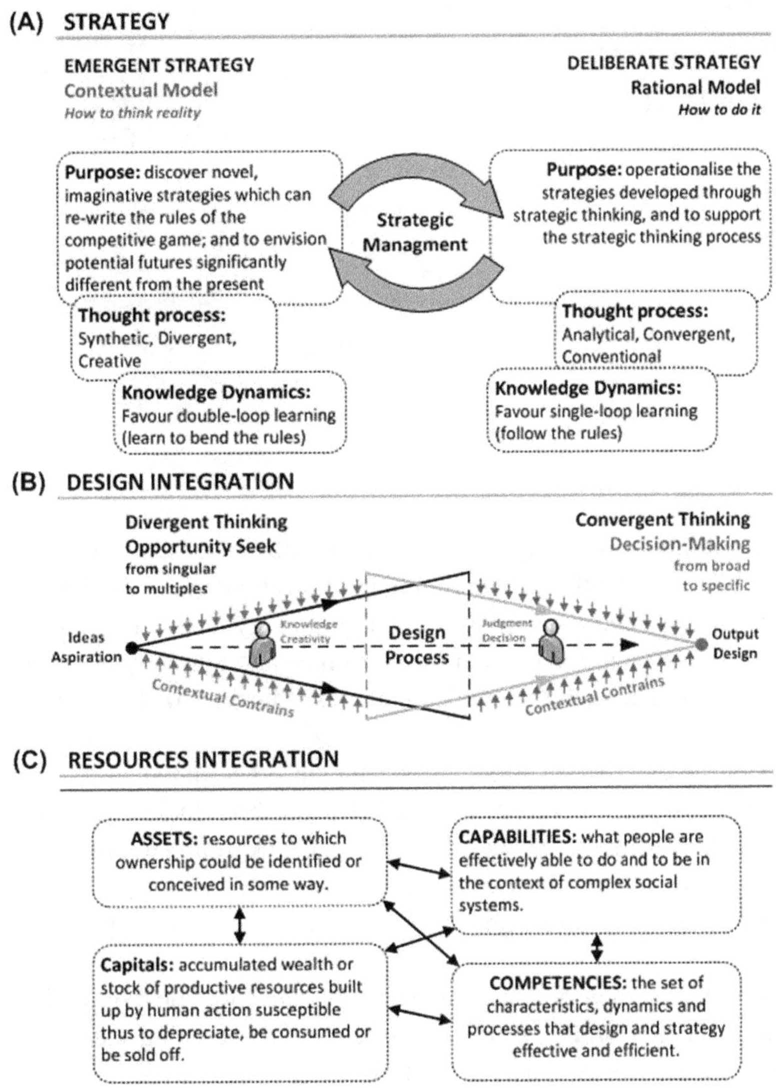

dialogue between designers/strategist and the socio-ecological context they are imbedded into. Thus, strategic thinking emerges as a specific kind of design thinking, or as "a way of solving strategic problems that combine a rational and convergent approach with creative and divergent thought processes" (Bonn, 2005: 337). Moreover, both horizontal and vertical integration should also integrate specific socio-ecological contexts where they are conducted, Figure 4. In this regard Strategy Design defines the design process that creates strategies that fulfil given design purposes, context constraints and designer skills. In other words, the outcome of strategy design is a strategy, conceived as a "design artefact", i.e., a resource of any kind.

Figure 4. Strategic Design thinking as a multidisciplinary integration of design thinking, multi-loop learning dynamics, and contextual flows and stocks within specific socio-ecological contexts
[Source: The author]

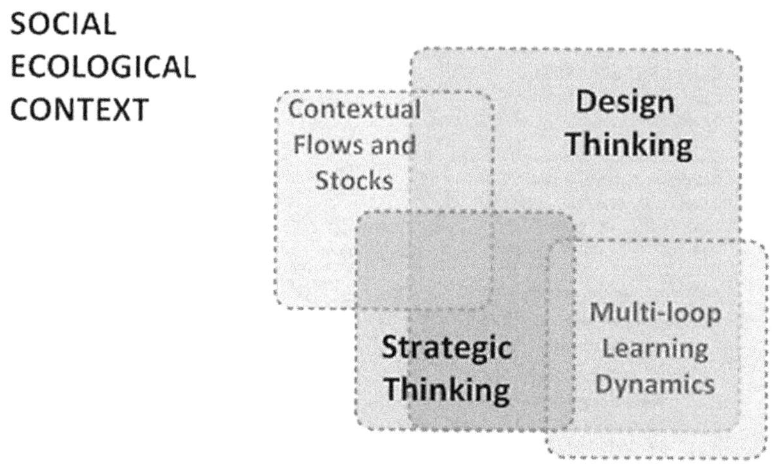

In this sense the resulting thinking style that supports Strategy Design, Strategic Design Thinking, becomes a "hub of hubs", i.e., should be able to integrate different views on design and strategy (each a hub of perspectives already) and consequently specify strategy design tools for different areas, e.g., WES. In practical terms, this implies a number of interlinked and mutually re-enforcing characteristics to be considered in the design of such tool, which includes (Martins 2014: 104-116):

- Encouraging both the creative, divergent, synthetic and the analytical, convergent mindset. The tool should facilitate different levels of abstraction to see both the big picture and the operational implications, and allow different people from different background to produce and explore innovative strategies and to plan for their realisation.
- Support different perspectives, identification and characterization of resources, as well as, their role in the strategy.
- Be centred on people and their practices, imagination and strategies. In particular, the design tool should address the energy strategies of the users (i.e., those using the tool) by providing non-normative and non-prescriptive space of creativity.
- Provides spaces for knowledge systems to interact in co-operative learning and to be used in improved and more contextual resource management practices. Stimulate learning, reflection, sensemaking and knowledge exchange on the process of doing strategy, promoting dialogue and learning-by-doing, social learning, focusing on participatory knowledge construction and innovation. The idea is to integrate organisational and personal perspectives, decision-making styles and knowledge on the decision-making process, promoting, thus, a continuous dialogue between actors and reality, which is the same to say, a learning process involving diverse knowledge and mental models and the act of design/make decisions.
- Promote the participation and participatory design of strategies as a means to address complexity, reduce cognitive load, democratise decision-making and increase innovation and analytical potential in the design of the strategy.

- Promote visualization as an intuitive, pro-active and facilitating means to promote exploratory, interpretative and analytical actions to elicit represent and analyse the perspectives and concepts that participants have of, and for, strategy.
- Be adapted to the specific socio-ecological reality for which the strategy is being designed and implemented.

Besides establishing the design parameters to build the novel tool presented in this chapter, the 2MBio (§3), these characteristics fundament a new theoretical and operational vision on what strategic tools should be and support in optimum asset management. Rather than defining an optimum use of resources to attain a predefined goal following a rational, deterministic logic evaluated through some quantified metrics, the perspective presented here defends that tools should be designed to support people in "local optimisation" in terms of what they actually do, think and create. The intent is not to tell how and why to optimise the assets, but rather create the spaces of participatory creativity and analysis for people to explore and create informed optimised use and strategic decisions in dialogue between them, the actual strategy and the socio-ecological system they are embedded in. In the next sections this new perspective on strategic design tools and resource management and optimization will become clear for the case of wool fuel energy systems in Mozambique.

A NEW APPROACH: THE 2MBio EXPERIENCE

In order to address the design parameters derived in §2.3 the present work removed focus on the how to use resources as means to an end (the premise of rational design/strategy) or how to use resources to adopt to contextual behaviours, (the premise of constructivist design/strategy) to what are the basic set basic blocks that managers, designers or strategists have to play with, to design strategies. Moreover, this work focused on the conceptual stage of design process cycle, since this is the best option to involve creatively the WES actors with lower levels of technical skills and literacy. Indeed, results from engineering, quality management and product design show that although representing only a small fraction of the project cost (~5%), the decisions taken in conceptual design have a huge impact in all design costs (85-95%), as well as on the improvement in the design performance (e.g. Yang & Shi, 2000). The result of this combined approach is the 2MB, the WES conceptual design metamodel, which characteristics (§3.1) and practical use (§3.2) will be discussed in this section.

Introducing a New Strategy Design Tool: The 2MB

The 2MBio is a metamodel that delivers WES conceptual designs and strategies. Generically, a metamodel is an explicit model of the constructs and rules needed to build specific models within a domain of interest (InfoGrid, 2011) and can be used to graphically illustrate concepts, conditions and constructs and their relations to each other (Henderson & Venkatraman, 1999). Therefore, the 2MBio can also be viewed as:

- A design ontology (an abstract representation of the design logic of WES in DCs) being design logic of the abstract comprehension of the way stakeholders consider energy systems design: "what they consider and how they consider it".

- A graphical representation of the design logic composed by a set of distinct concepts/constructs and relations used to structure and specify WES for DCs.

The 2MBio is the result of coding and explicit wide array of mental models that different WES actors hold on WES conceptual design using reflexive ontological analysis as the modelling methodology to extract, classify and relate the WES conceptual building blocks, i.e., the WES design elements (DEs). This ontological analysis was conducted for a period of six months in the south of Mozambique, using data from over 200 titles in the literature, interviews and participatory design exercises with 150 participants, including charcoal makers; farmers; WES entrepreneurs; technology providers; NGOs and experts; government authorities; users/consumers. From the 150 interviewees over 80 have been selected to participate in Participatory Design Workshops (PDW) where the 2MBio was further tested and refined. This methodology allowed to involve all knowledgeable actors in the Mozambican WES as co-modellers in all modelling design stages. Consequently, the identified 13 DEs, Figure 5, represent the minimum and necessary set of conceptual building blocks that designers *think about when they think about conceptual design of WES* in Mozambique.

The 2MBio is visual, simple to use, modular and non-computer based, non-normative and non-prescriptive metamodel. In physical terms the 2MBio is presented as a simple set of well-defined and easy to understand 13 Des (visually presented as boxes in a piece of paper, see Figure 5) that once filled jointly by participants allow user to express the specification of a conceptual design model suitable to be further analysed, worked upon and communicated. Remarkably, the 13 DEs in the 2MBio are also the resources (assets, capabilities, responsibilities add/or assets) designers/strategists have to consider to produce comprehensive WES systems designs and strategies. In order to do so, users of the 2MBio are

Figure 5. The final layout of the 2MBio. In the real 2MBio used on practical work, there was plenty of space for writing or drawing.
[Source: Martins 2014]

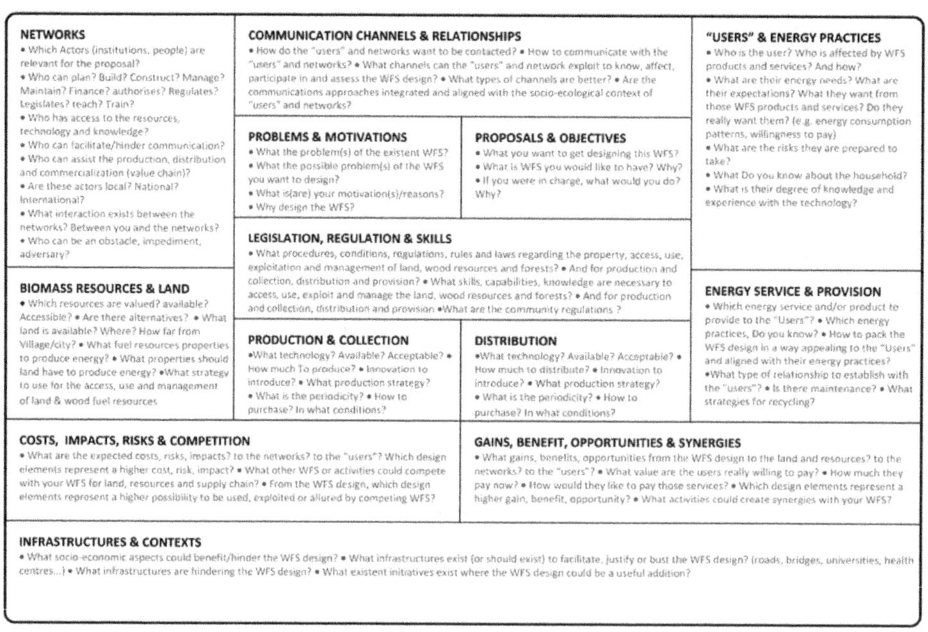

stimulated to use their creativity, knowledge and experience as a further tacit asset to be incorporated in the final output: the comprehensive specification of a WES strategy and/or design. Furthermore, the output of the design process with the 2MBio is always a piece of paper with a structured specification of the WES strategic design and thus, constitutes in itself, a resource for further refinement and strategic use, again based on the knowledge and creativity of the users. Providing a set of well-defined necessary and sufficient set of design elements that describe the logic of WES conceptual design, the 2MBio operates as a visual, modular and interactive learning device, in many ways similar to a box of LEGO™. Alone or in group the designer can play around with these blocks (the boxes drawn in paper) and create their own WES conceptual models limited only by his/her creativity, knowledge, experience, bias and the pieces supplied.

Finally, note that the idea of building and developing a standard framework for characterizing a field of human activity is not new, as Osterwalder (2004) already presented the "business model ontology" that describes, in a structured way, elements and sub-elements of the business model (see also Battistella *et al.* 2012). However, while Osterwalder *et al.* (2005) basically encapsulated capitalism to design business models, the focus in this work was the conceptual strategic design of WES from a wider socio-ecological perspective. This is a quite novel and innovative approached to the field of WES conceptual design of strategies, projects and policies and so far, to the best knowledge of the Author, there is nothing similar. How this approach worked in practical terms is the purpose of the next section.

Illustrative Examples and Results

Using the definition of creativity as the link between innovation and value in design, this work could compare results from similar design conditions with and without the use of 2MBio based on the seven PDW mentioned in §3.1.

Defined as "the act of either applying some creative idea, or creatively applying a familiar idea, in such a way as to create value fulfilling a set of stated and unstated design requirements" (Milne & Leifer, 1999) innovation is also an act of conceptual combination since "making new links between concepts is a powerful driver of creativity" (Morris 2009). Thus creative design tools should support the creation of new associations/combinations and the finding of relationships and similarities between concepts and items previously unconnected (Hedblom 2013). In the context of this work these creative combinations and associations have been parameterized as:

- **Comprehensiveness:** An increase in the number of design elements considered and specified implies an increase of creativity in WES conceptual design, since new combinations have been identified.
- **Innovative Ideas in Design Elements:** An increase on the number and variety of specification in the design elements (design aspects) implies an increase of creativity in WES conceptual design since new aspects of the combination have been identified.
- **Interaction Between Design Elements:** An increase in the number of conceptual links between design elements implies an increase of creativity in WES conceptual design since new association and more integrated design is produced.

The first parameter, comprehensiveness was measured comparing the results of the individual interview (in urban areas) and small group interview (rural areas) used to produce the 2MBio (§3.1), and the results of the actual use of the 2MBio in a PDW with the same interviewees in PDW, Figure 6.

For each PDW participant it is possible to see how coincident in the aspects mentioned are in relation to each design element by him/her, and the aspects mentioned by his/her group for the same design element, i.e. it is possible to check for the presence of new ideas, taken as an expression of added creativity. In simple terms this comparative analysis intends to quantify the degree of conceptual overlap between the individual WES design and the team design using the 2MBio as a common comparative ground. The outcome of the comparison can give an indication of how much the 2MBio is comprehensive and thus how useful the 2MBio is to assist designers to expand their design thinking and creativity. This comparison is only possible because the 2MBio can actually provide that common ground of comparison and because the individual interview with experts explored a design challenge similar to the one offered to each team member in the various PDW (§3.1).

The results showed (Figures 7-8) that between the two extremes of conceptual overlapping between 0% (individual 2MBio did not considered this design element) and 100% (the individual interview and the team interview are the same), there are always one or several members of the Teams that ignore, overlook or consider only implicitly or indirectly some design element of the design aspect. These results indicate that the 2MBio design tool facilitates a clear and easy view over the WES conceptual design big picture, meaning, a more comprehensive and parsimonious perspective revealing new design elements, or highlighting design elements not explicitly or directly considered previously, which could also facilitate the participatory creativity of team members.

To analyse the 2MBio effectiveness towards innovative ideas in design elements a similar approach was conducted, but this time for each design element, looking at the design aspects considered by the Team and by the individual expert. The design aspects are what each Team wrote or drew in each design element of the 2MB, i.e., the description of that design element made by the Team. The purpose is to identify if the contents defined by the group for each design element (the design aspects) in the 2MBio are just the conjugation of previous ideas or perspectives held previously by each group member or if the 2MBio actually facilitated the creation of new ideas and perspectives. i.e., design aspects, for that design element.

With the definitions provided in Figure 9, and once again translating into 2MBio the several WES conceptual design description produced in the individual interviews it is possible to compare the design

Figure 6. Comparison between the individual 2MBio with the team 2MBio to assess comprehensiveness [Source: Martins 2014]

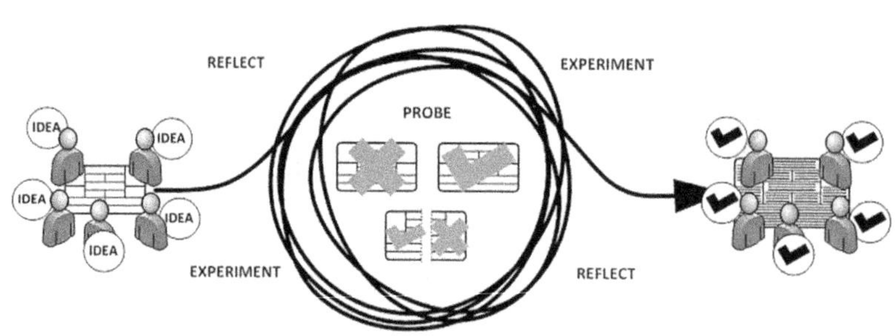

Figure 7. Percentage of overlapping between individual WES conceptual designs and team WES defined in PDW1 and PDW2
[Source: Martins 2014].

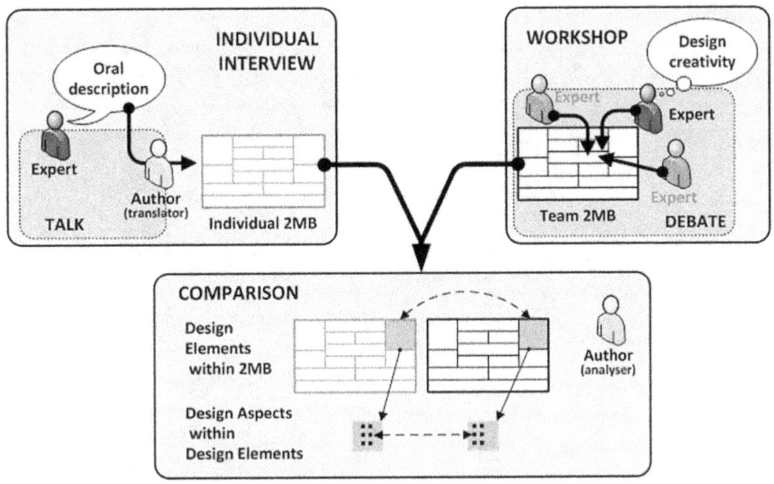

aspects of all the design elements both in the individual WES design and the team design using the 2MB, Figure 10.

Again the results show the potential of the 2MBio design tool as a facilitator of innovation in participatory design, since most design elements present an innovation way over 0%. Therefore, compared with the designs generated individually, each team produced designs with completely new aspects for most design elements. In other words, more than simple conjunction of ideas to produce a whole as the sum of the parts, 2MBio actually produced a creative integration, that is, the generation of synergies, where the whole is more than the sum of the parts.

Besides these analyses the 2MBio was evaluated through an anonymous questionnaire provided at the end of each PDW. The results, Table 1, confirm the visual comparison provided in Figures 7-8, 10 and several assertions done in §4.2, that is the 2MBio does facilitate the creative and participatory WES conceptual design, as well as, dialogue and learning dynamics.

Figure 8. Percentage of overlapping between individual WES conceptual designs and team WES defined in PDW3 to PDW7
[Source: the first Author]

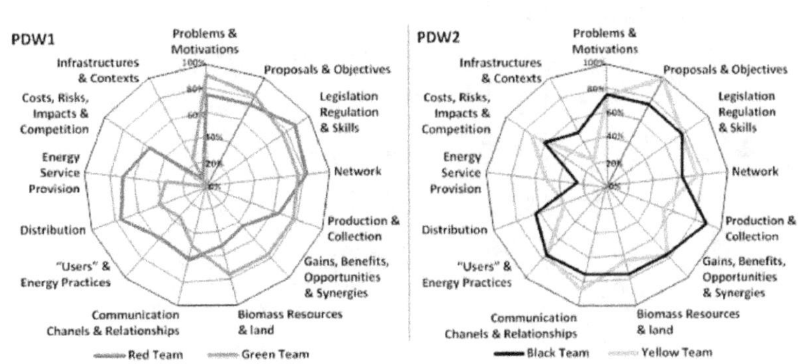

Figure 9. Description and examples for each of the three design aspects overlapping possibilities in a generic design element "X" in the 2MBio
[Source: Martins 2014]

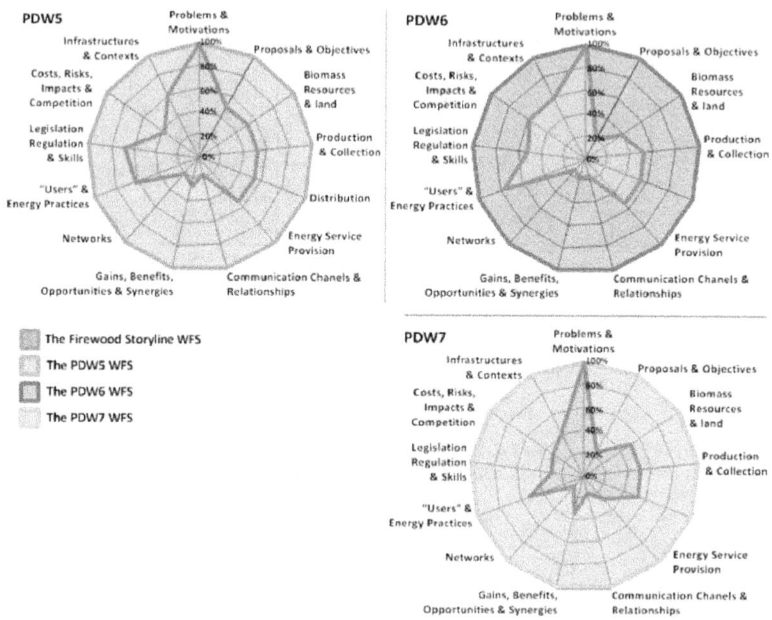

Figure 10. Degree of innovation in each design element and number of team members that identified extra design aspects for the four teams in the PDW1-2 (for definitions see Figures 4 and 5)
[Source: Martins 2014]

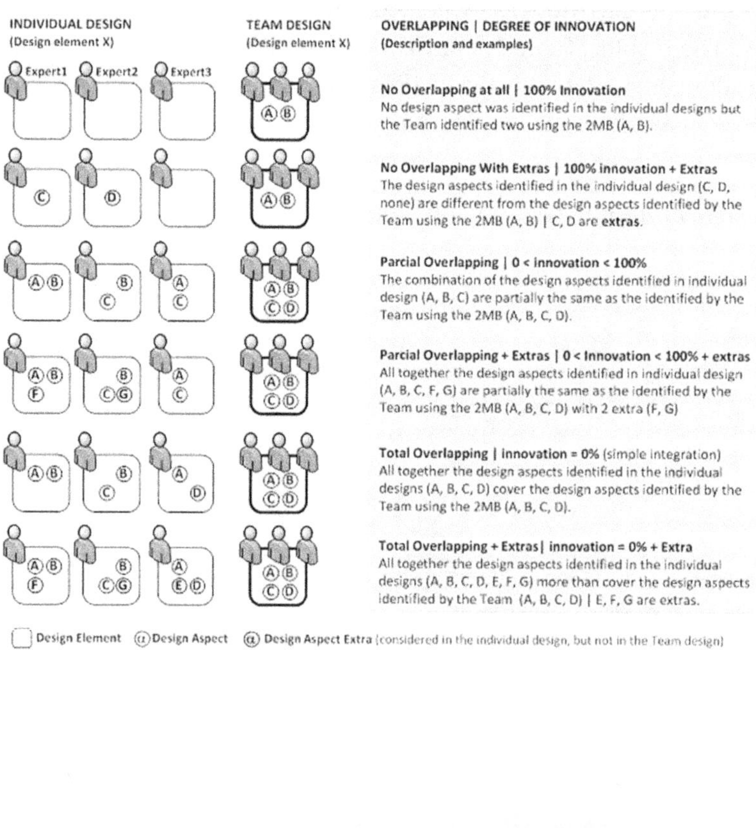

Table 1. Average and mode (most frequent values) results for the evaluation of the 2MBio utility, usefulness and relevance in relation with the major learning, dialogue and objectives of 2MBio produced in the Urban PDW involving 19 participants

2MBio is Useful for...	By Facilitating...	Results [%]	
		Avg	Mode
Design/Dialogue	WES participatory design	85.9	80
Dialogue	Dialogue on WES design	90.6	100
Communication	Communication of design ideas on WES	88.2	80
Learning/Dialogue	Understanding of other's WES conceptual design	82.4	100
Learning	Organisation of thinking on WES conceptual design	81.3	100
Creativity/Learning	Thinking creatively on WES conceptual design	77.6	80
Learning/Creativity	A more informed strategic planning of WES	82.4	80
Learning/Dialogue	Analysis/vision of WES *big picture* & interactions	85.9	100
Dialogue/Learning	Participatory WES design with rural populations	80.0	100

(original values on a 1-5 scale, being 1= nothing; 5= very much; AVG- Average)
[Source: Martins 2014]

Finally, the interaction between design elements as an indicator of the 2MBio ability to promote creativity in WES conceptual design, was measured comparing the number of interactions (arrows in Figures 11 and 12) between design elements. This analysis was done only for the rural PDW where it was possible to identify across the several small group interviews a relatively stable WES conceptual design, named here, WES story line (the interviewees were invited to expose their views on WES conceptual design through story telling).

Figure 11. The Tinonganine Charcoal Storyline as described by the 2MBio following input from rural group interviews in Tinonganine also showing the multiple interactions (transparent double arrowed lines) between design elements (white boxes- Des not considered; Light gray Boxes- Des only indirectly on implicitly considered; Dark gray- Des explicitly mentioned; [TEXT] author notes)
[Tinonganine Actors & Martins 2014]

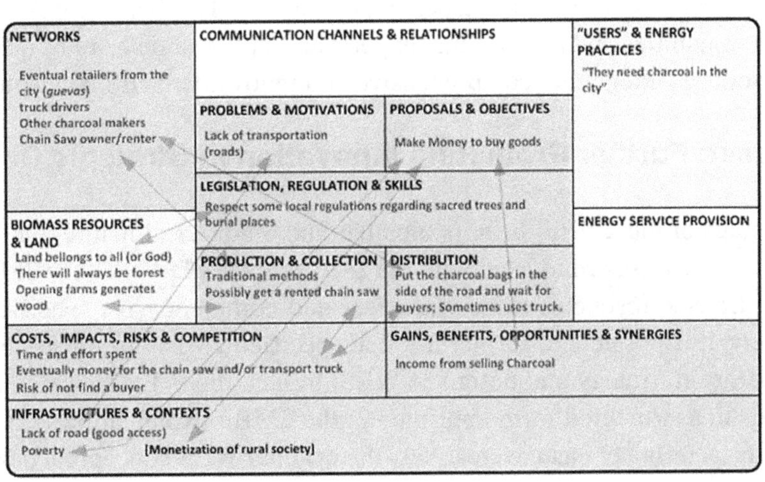

Figure 12. The WES from the PDW4 in Tinonganine with the representation of interactions (transparent red double arrowed lines) between design elements (similar figures could be produced for the WES design from the PDW3)
[Source: Martins 2014]

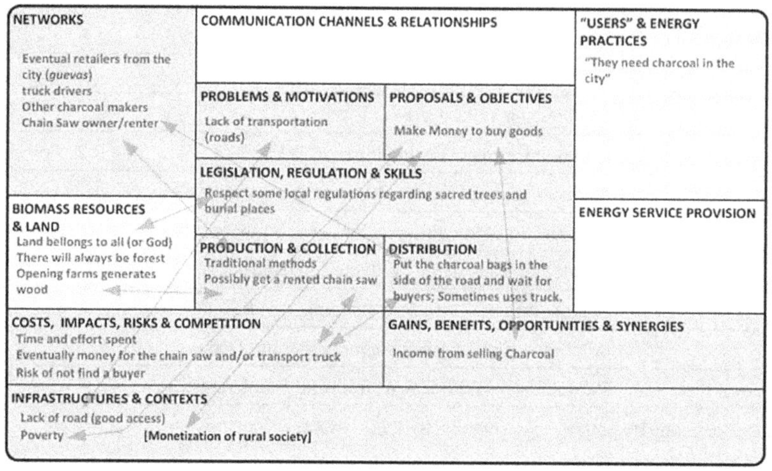

In resume, the 2MBio facilitated in both PDWs an increase in the quality of the WES design by considering a wider number of relevant design elements, generating more design aspects for each design element and interactions between the design elements. These improvements were motivating and motivated by an active dialogue/debate between the participants stimulated mostly by the suggestive questions and the willingness to address them. During the 3:30h to 4:00h debate/dialogue it was possible see how each new design element inspired creativity, inquiring and learning. Triggered to go "beyond the road", the participants considered forest management to address their living conditions, learning more about the WES with the support of the 2MBio. This was clearly stated in a number of occasions by the participants verbally and through surprise expressions, another indicator of creative leap/insight.

Finally, complementing all these results are the body language and expressions of the 2MBio users in PDW. It was rather frequent to see, both with rural communities and experts alike, the "ahh" moment, the "creative leap" moment of insight where one realises with surprise something new and meaningful/valuable that was never before considered, i.e., personal discovery. The fact that WES conceptual designs produced by rural communities and professional experts are on the same level of quality is a definitive proof that 2MBio produces integrated, comprehensive and innovative WES conceptual designs.

2MBio as an Asset: Further Promoting Innovation in Strategic Design

A remarkable advantage of the 2MBio is in its modular and ontological nature. Once in possession of the output of the design exercises conducted with the 2MBio in the PDW, i.e., the 2MBio with each box (design element or strategic resource) filled with clear and comprehensive specification of the WES design and/or strategy, it is possible to further use that specified 2MBio to illustrate, inspire, combine and integrate new ideas on strategy and design of WES. In fact, the WES conceptual design produced by the teams remains in a structured form imprinted in the 2MBio which allows different designers to further analyse, criticise, refine or change creatively the original WES conceptual design, in what could be understood as a learn/design process. The possible creative combinations with the conceptual design produced with the 2MBio include, for instance (Figure 13): the combination of different design elements from different WES designs (on 2MB); the integration of different design aspects within the same design element (on 2MB) from different WES designs (on 2MB); or a combination of the two processes.

Regardless of the process used, refinement is above all a continuous dialogue between probing and reflection, where some designs are rejected, other accepted for further refinement, others only partially used until a satisfying design is reached, Figure 14.

In practical terms this could also imply the investigation of the 2MBio platform as a physical object (i.e. an a piece of a paper with boxes in it) that can be simultaneously a simple and visual support to communicate WES conceptual designs, as well as, a design artefact itself able to be manipulated in continuous circles of dialogue and refinement among the participants till reaching a reasonable degree of agreement or, alternatively, the identification of the design elements, design aspects or reasons behind persistent discordances. This is a research that would be somewhere in the border between management sciences and semiotics. Furthermore, there is also the need to further investigate how to combine accepted/refined WES conceptual designs with other participatory models/tools (downstream in the decision process) for implementation of envisioned/designed solutions.

Figure 13. Possible processes to create innovative WES conceptual design using the 2MBio
[Source: Martins 2014]

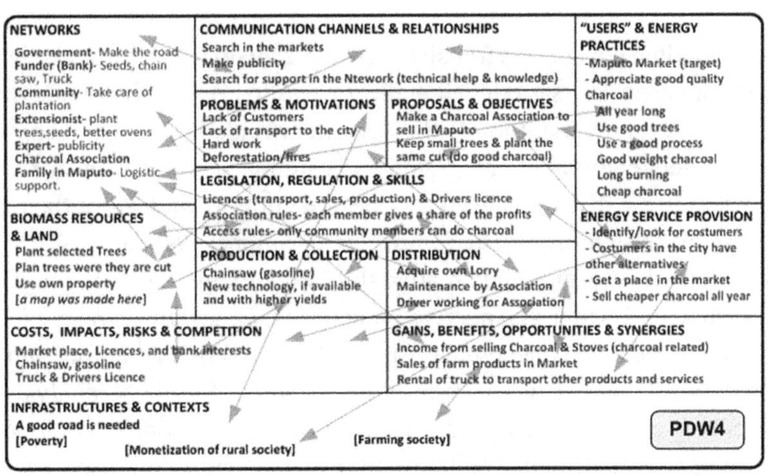

Figure 14. The refining design process
[Martins, 2014 after Osterwalder & Pigneur, 2010]

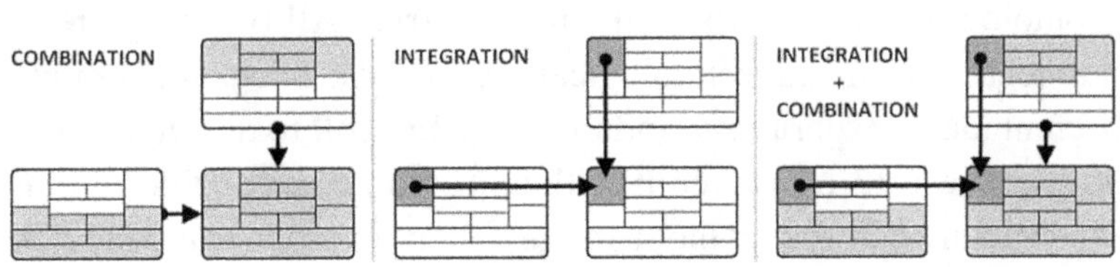

CONCLUSION

Considering the crucial importance on WF in Mozambicans' livelihoods and the complexity of WES, defining tools that can effectively support strategic design of WES is both an urgent need and a considerable challenge. Available tools for WES design and/or strategic decision making are normally based on strategic planning and thus unable to address complex contexts, while alternative tools are not specific to WES and contextualised, and thus unable to provide a common structure to facilitate comparison and learning across realities. In simple terms current tools are either focused on "how to do it", or on "how to understand it". While these are relevant cognitive processes, they fall short in promoting creative "competitive advantages" in real world situations. In order to bridge those gaps this work integrated design thinking, strategic thinking and a contextualised view on resources/assets to focus on identifying what resources and what kind of resources the users need to consider in their strategic design to develop a common space for participation, knowledge share, learning and creativity for all actors involved in existing WES (from charcoal makers to policy makers). The outcome of this exercise was the development and testing of a novel participatory tool for strategic deign of WES, the 2MBio.

The results confirm that the 2MBio may positively overcome the challenge of producing a creative design tool able to deal with the WES complexity in three different ways:

1. The 2MBio can convert all the verbal information provided in design/strategy interviews (individual or group) into the body of the 13 design elements, which means that the 13 design elements are sufficient (§3.2).
2. Comparing the individual designs (as translated by the 2MBio from the interviews) with the 9 WES conceptual design specifications produced in the 7 participatory design workshops (involving over 90 participants), results show that the 2MBio always provided (with no exception) more comprehensive, interconnected, elaborated, integrated and original specifications for the WES conceptual design. In the where the 2MBio was used as a design support tool More than the combination of individual conceptual designs (expressed in previous interviews) the 9 WES conceptual design specifications generated synergies and creative thinking among the participants, resulting in more integrated, interactive and wider perspective on conceptual design of WES (§3.2).
3. 2MBio can extend the creative conceptual design of WES using its modular and ontological nature to promote learning and creative design processes through refinement, analysis and creative combination of WES conceptual design produced in PDW, i.e., the once specified (the boxes filled) the 2MBio becomes itself a design artefact to be refined, a strategic asset for further analysis and decision making (§3.3).

Finally, some notes are relevant regarding the applicability of the 2MBio so far. While developed and tested in Mozambique for WES, the ontological nature of the 2MBio makes it potentially useful for other contexts and bioenergy systems in general. Regardless of the geographic area or kind of bioenergy system in place, the ontological basic block will still hold true. All bioenergy requires land, a supply chain, and is subject to legal, social and technological constraints. The context might differ, the participants might have different ideas, experiences and interests, but still to design a comprehensive strategy, they will have to consider these 13 Des (even if with adjusted names and definitions), and if more are required, the modular structure can accept them and test them in other realities to check their ontological robustness. Indeed, more than anything, the great merit and strength of the 2MBio is the possibility to provide a common ground for participatory exploration, learning and knowledge share, facilitating thus more comprehensive problem structuring, more informed decisions and a learning device to foster innovation and creativity. This way, the 2MBio fills a gap in the conceptual problem/solution structuring in WES in DCs (and elsewhere), promoting a design where creativity can be explored in respect with the socio-ecological and technological nature of WES. In this perspective, the initial question, "are farmers burning assets?", should be analysed in terms of: what the different actors understand by "assets"; what contextual aspects (social, economic, technological, ecological and others) are influencing, being influenced and providing meaning for that burning; and what other resources/assets (the Design elements in the 2MB) could be mobilised to integrate that "burning" into a coherent, comprehensive, innovative and participatory designed meaningful strategy. The 2MBio addresses all this issues promoting the use of the users' knowledge and creativity (the tacit asset), i.e., integrates knowledge, creativity and learning in the strategic design of WES.

Finally, the 2MBio can be used along a large spectrum of users. While this chapter refers the experience with charcoal makers, experts and some governmental officials, recently the 2MBio was considered to support the design of a NAMA (National Appropriate Mitigation Action) by MIREME (The Ministry of Mineral Resources and Energy in Portuguese). A NAMA specifies a formal project that must be aligned with the National Emission Reduction Strategy, or any Environmental and/or Energy Strategy, submitted by a given government in order to obtain funding or technical assistance from the international community. Therefore, the 2MBio is finding its way through higher levels of decision making while being useful for charcoal makers and practitioners on more local basis.

REFERENCES

Andersen, D., Richardson, G., & Vennix, J. (1997). Group Model-Building: Adding More Science To The Craft. *System Dynamics Review*, *13*(2), 187–201. doi:10.1002/(SICI)1099-1727(199722)13:2<187::AID-SDR124>3.0.CO;2-O

Batidzirai, B., Faaij, A., & Smeets, E. (2006). Biomass And Bioenergy Supply From Mozambique. *Energy for Sustainable Development*, *10*(1), 54–81. doi:10.1016/S0973-0826(08)60507-4

Battistella, C., Biotto, G., & De Toni, A. (2012). From Design Driven Innovation To Meaning Strategy. *Management Decision*, *50*(4), 718–743. doi:10.1108/00251741211220390

Bebbington, A. (1999). Capitals And Capabilities: A Framework For Analyzing Peasant Viability, Rural Livelihoods And Poverty. *World Development*, *27*(12), 2021–2044. doi:10.1016/S0305-750X(99)00104-7

BEST. (2012): Mozambique Biomass Energy Strategy, ECO-Consult e MARGE para o FUNAE, Maputo.

Bonn, I. (2005). Improving Strategic Thinking: A Multilevel Approach. *Leadership and Organization Development Journal*, *26*(5), 336–354. doi:10.1108/01437730510607844

Brouwer, I., Hoorweg, J., & van Liere, M. (1997). When Households Run Out Of Fuel: Responses Of Rural Households To Decreasing Fuelwood Availability, Ntcheu District, Malawi. *World Development*, *25*(2), 255–266. doi:10.1016/S0305-750X(96)00100-3

Buxton, W. (2007): Sketching User Experience: Getting The Design Right And The Right Design. San Francisco, CA, USA: Morgan Kaufmann & Elsevier.

Cecelski, E. (2004). Re-thinking gender And energy: Old And new directions. ENERGIA/EASE Discussion Paper, Energy, Environment and Development. Retrieved from www.energia.org

Cumbane, J. (2007, December 5). Urban Environmental Management Needs And Priorities In Mozambique. *Proceedings of the SANORD Pre-Conference Workshop, Research and Knowledge for Development in Southern Africa*, Universidade Eduardo Mondlane, Department of Environmental and Geographical Sciences, Maputo.

Cumbe, F., Sharma, D., & Lucas, C. (2005). The Status of Clean Cooking Fuelsin Mozambique. Sydney: University of Technology. Retrieved from www.saee.ethz.ch

Cuvilas, C., Jirjis, R., & Lucas, C. (2010). Energy Situation in Mozambique: A Review. *Renewable & Sustainable Energy Reviews*, *14*(7), 2139–2146. doi:10.1016/j.rser.2010.02.002

da Costa, J., Navarro, A., Neves, J., & Martin, M. (2004). Household Wood And Charcoal Smoke Increases Risk Of Otitis Media In Childhood In Maputo. *International Journal of Epidemiology*, *33*(3), 573–578. doi:10.1093/ije/dyh071 PMID:15105407

Ellegard, A. (1996). Cooking Fuel Smoke and Respiratory Symptoms Among Women in Low-income Areas in Maputo. *Environmental Health Perspectives*, *104*(9), 980–985. doi:10.1289/ehp.96104980 PMID:8899378

FAO. (2005). Special report: Crop and Food Supply Assessment Mission to Mozambique, Food and Agricultire Organization, Rome. Retrieved from www.fao.org

Flusser, V. (2010). *A Filosofia Do Design, A Forma Das Coisas*. Lisboa: Relógio D'água. [In Portuguese]

Funtowicz, S., & Ravetz, J. (1993). Science For The Post-Normal Age. *Futures*, *25*(7), 739–755. doi:10.1016/0016-3287(93)90022-L

Gane, M. (2007). Strategy. In M. Gane (Ed.), *Forest Strategy, Strategic Management and Sustainable Development for the Forest Sector* (pp. 266–392). Dordrecht: Springer.

Goldemberg, J., & Johansson, T. (Eds.). (1995). Overview. In J. Goldemberg, & T. Johansson (Eds.), Energy As An Instrument for Socio-Economic Development (pp. 1-9). New York: United Nations Development Programme (UNDP).

Hanlon, J., & Cunguara, B. (2010). *Poverty is not being reduced in Mozambique*. LSE Crisis States Research Centre.

Hedblom, M. (2013). *A Cognition-Based Definition of Creativity and A Proposition for Approaching Creativity Artificially* [Master Thesis]. Institute of Computer and Information Science, Linköping University.

Henderson, J., & Venkatraman, N. (1999). Strategic Alignment: Leveraging Information Technology for Transforming Organisations. *IBM Systems Journal*, *38*(2/3), 472–484. doi:10.1147/SJ.1999.5387096

Heracleous, L. (1998). Strategic Thinking Or Strategic Planning? *Long Range Planning*, *31*(3), 481–487. doi:10.1016/S0024-6301(98)80015-0

IEA. (2006). Energy For Cooking. In *Developing Countries, in IEA: World Energy Outlook 2006* (pp. 419–446). Paris: International Energy Agency, Organisation for Economic Cooperation and Development.

INE. (2007). *Third General Population and Housing Census- 2007*. Maputo: Instituto Nacional de Estatística.

Infogrid. (2011). InfoGrid Web Graph Database. Retrieved from http://infogrid.org

Jagger, P., & Shively, G. (2014). Land Use Change, Fuel Use And Respiratory Health In Uganda. *Energy Policy*, *67*, 713–726. doi:10.1016/j.enpol.2013.11.068 PMID:24535892

Jetter, J., Zhao, Y., Smith, K., Khan, B., Yelverton, T., DeCarlo, P., & Hays, M. D. (2012). Pollutant Emissions And Energy Efficiency Under Controlled Conditions For Household Biomass Cookstoves And Implications For Metrics Useful In Setting International Test Standards. *Environmental Science & Technology*, *46*(19), 10827–10834. doi:10.1021/es301693f PMID:22924525

Jurow, S. (1996). Core competencies: Strategic thinking about the work we choose to do. *Journal of Academic Librarianship*, *22*(4), 300–302. doi:10.1016/S0099-1333(96)90122-6

Kotler, P., & Rath, A. (1984). Design: A Powerful But Neglected Strategic Tool. *The Journal of Business Strategy*, *5*(2), 16–21. doi:10.1108/eb039054

Levy, C., & Kaufmann, B. (2014). Charcoal for Food Livelihood Diversification In Two Peasant Communities In Mozambique, in Socio-Economic Insecurity. In K. Fakier, & E. Ehmke (Eds.), Emerging Economies: Building New Spaces (pp. 68-82). Routledge.

Lim, S., Vos, T., Flaxman, A., Danaei, G., & Shibuya, K. et al.. (2012). A Comparative Risk Assessment Of Burden Of Disease And Injury Attributable To 67 Risk Factor Clusters In 21 Regions, 1990-2010: A Systematic Analysis For The Global Burden Of Disease Study 2010. *Lancet*, *380(9859)*, 2224–2260. PMID:23245609

Martins, R. (2014). *The Wood Fuel Energy Systems Conceptual Design Metamodel, A Novel Approach To Participatory Design Contextualised To The Case Of Mozambique* [PhD Thesis]. Imperial College London.

Marzoli, A. (2007). *Relatório do inventário florestal nacional*. Maputo: Direcção Nacional de Terras e Florestas. Ministério da Agricultura.

Milne, A., & Leifer, L. (1999, August 24-26). The Ecology Of Innovation In Engineering Design. *Proceedings of the International Conference On Engineering Design ICED '99*, Munich.

MIREME (2014): Estratégia De Conservação E Uso Sustentável Da Energia Da Biomassa Para O Período 2014-2025, (ECUSEB 2014-2015) Ministério da Energia (actual MIREME), Maputo [in Portuguese].

Morris, R. (2009). *The Fundaments of Product Design*. AVA Publishing SA, Lousanne.

Nhantumbo, I., Dent, J., & Kowero, G. (2001). Goal Programming: Application In The Management Of The Miombo Woodland In Mozambique. *European Journal of Operational Research*, *133*(2), 310–322. doi:10.1016/S0377-2217(00)00300-3

OPHI. (2011). *Country Briefing: Multidimensional Poverty Index At A Glance*. Oxford: Oxford Poverty and Human Development Initiative.

Osterwalder, A. (2004). The Business Model Ontology- A Proposition In A Design Science Approach [Doctoral Thesis]. Lausanne University, Lausanne.

PARPA. (2001). *Action Plan For The Reduction Of Absolute Poverty 2001-2005, (PARPA), Strategy Document For The Reduction Of Poverty And Promotion Of Economic Growth*. Maputo: Mozambican Government. [in Portuguese]

Pisapia, J., Jelenc, L., & Mick, A. (2016). The Foundations Of Strategic Thinking: Effectual, Strategic, And Causal Reasoning. In *Neostrategic management* (pp. 45-55).

Plummer, M. (1999). To Fetch A Pail Of Water. *Natural History*, *108*(1), 56–65.

Purnomo, H., Yasmi, Y., Prabhu, R., Hakim, S., & Jafar, A., Suprihatin. (2003). Collaborative Modelling to Support Forest Management: Qualitative Systems Analysis at Lumut Mountain, Indonesia, Small-scale Forest Economics. *Management and Policy*, *2*(2), 259–275.

Scoones, I. (1998). *Sustainable Rural Livelihoods. A Framework For Analysis, IDS WORKING PAPER 72, Institute of Development studies*. Brighton: Sussex University.

Sedlacko, M., Martinuzzi, A., Røpke, I., Videira, N., & Antunes, P. (2014). Participatory Systems Mapping For Sustainable Consumption: Discussion Of A Method Promoting Systemic Insights. *Ecological Economics*, *106*, 33–43. doi:10.1016/j.ecolecon.2014.07.002

Sen, A. (1997). Editorial: Human Capital And Human Capability. *World Development*, *25*(12), 1959–1961. doi:10.1016/S0305-750X(97)10014-6

Sitoe, A., Salomão, A., & Wertz-Kanounnikoff, S. (2012). The context of REDD+ in Mozambique- Drivers, agents and institutions. CIFOR Occasional Paper 79, Centre For International Forestry Research, Bogor.

Smith, K., Balakrishnan, K., Butler, C., Chafe, Z., Fairlie, I., . . . (2012). Energy And Health. In T. Johansson, N. Nakicenovic, A. Patwardhan, & L. Gomez-Echeverri (Eds.), Global Energy Assessment: Toward a Sustainable Future (pp. 255-324). Cambridge, Cambridge University Press/Laxenburg, Austria: Int. Inst. Appl. Syst. Anal. doi:10.1017/CBO9780511793677.010

Souchère, V., Millair, L., Echeverria, J., Bousquet, F., Le Page, C., & Etienne, M. (2010). Co-Constructing With Stakeholders A Role-Playing Game To Initiate Collective Management Of Erosive Runoff Risks At The Watershed Scale. *Environmental Modelling & Software*, *25*(11), 1359–1370. doi:10.1016/j.envsoft.2009.03.002

Standa-Gunda, W., Mutimukuru, T., Nyirenda, R., Prabhu, R., Haggith, M., & Vanclay, J. (2003). Participatory Modelling To Enhance Social Learning, Collective Action And Mobilization Among Users Of The Mafungautsi Forest, Zimbabwe. *Small-scale Forestry*, *2*(2), 313–326.

Steptoe-Warren, G., Howat, D., & Hume, I. (2011). Strategic Thinking And Decision Making: Literature Review. *Journal of Strategy and Management*, *4*(3), 238–250. doi:10.1108/17554251111152261

Taylor, P. (2015). Constructivism. In R. Gunstone (Ed.), Encyclopedia of Science Education (pp. 218-224). SpringerLink.

Tragett, C. (2012). *From romanticized Women to politicized Gender: How has the literature addressed Gender and Natural Resource Use?* [Master Thesis]. Imperial College London, London.

UNDP. (2014). *Human Development Report 2014, Sustaining Human Progress: Reducing Vulnerabilities And Building Resilience*. New York: United Nations Development Program.

van den Belt, M. (2004). *Mediated Modelling: A System Dynamics Approach To Environmental Consensus Building*. Washington: Island Press.

Vaz, K., Storey, P., Cipriano, A., Cuamba, B., D'Addario, P., Falcão, P., & Lehman, N. (2011). Green Investments. In *Mozambique*. Maputo: Banco de Moçambique, Agence Francaise de Development.

Vera, D., Carabias, J., Jurado, F., & Ruiz-Reyes, N. (2010). A Honey Bee Foraging Approach For Optimal Location Of A Biomass Power Plant. *Applied Energy*, *87*(7), 2119–2127. doi:10.1016/j.apenergy.2010.01.015

Wamukonya, N. (2007). Solar Home System Electrification As A Viable Technology Option For Africa's Development. *Energy Policy*, *35*(1), 6–14. doi:10.1016/j.enpol.2005.08.019

Weir, J., Crew, D., & Crew, J. (2013). Wetland Forest Culture: Indigenous Activity For Management Change In The Southern Riverina, New South Wales. *Australasian Journal of Environmental Management*, *20*(3), 193–207. doi:10.1080/14486563.2013.819303

WHO. (2002). *Addressing the Impact Of Household Energy And Indoor Air Pollution on the Health Of the Poor: Implications for Policy Action And Intervention Measures, World Health Organisation*. Geneva: WHO.

Wolbert-Haverkamp, M., Feil, J., & Mußhoff, O. (2014). The Value Chain Of Heat Production From Woody Biomass Under Market Competition And Different Intervention Systems: An Agent-Based Real Options Model (Discussion Paper 1407). Department of Agricultural Economics and Rural Development, Georg-August-University, Goettingen.

World Bank. (2010). *Economics of Adaptation to Climate Change Mozambique*. Washington: The World Bank.

World Bank. (n. d.). World Bank World Data Base. Retrieved from http://data.worldbank.org

WRI. (2010). Country Profiles- Mozambique, World Resources Institute. Retrieved from http://earthtrends.wri.org

Yang, Y., & Shi, L. (2000). Integrating Environmental Impact Minimization Into Conceptual Chemical Process Design- A Process Systems Engineering Review. *Computers & Chemical Engineering*, *24*(2-7), 1409–1419. doi:10.1016/S0098-1354(00)00384-7

Zerriffi, H., & Wilson, E. (2010). Leapfrogging Over Development? Promoting Rural Renewables For Climate Change Mitigation. *Energy Policy*, *38*(4), 1689–1700. doi:10.1016/j.enpol.2009.11.026

Chapter 20
No Fault Found Problems in Asset Management

Samir Khan
Coventry University, UK

ABSTRACT

Within aerospace and defence sectors, maintaining asset availability during operational service has become more important than quality of service throughout the system life cycle. This requires organisations to establish cost effective strategies to manage uncertainties within their value led services e.g. maintenance activities. In large organizations, it is not always apparent whose decision affects the outcome the most. Often, accountability moves away from the designated organization personnel in unforeseen ways, and depending on the decisions of individual decision makers, the structure of the organization, or unregulated operating procedures may change. This can have far more effect on the overall reliability leading to inadequate troubleshooting, repeated down-time and reduced availability. This chapter focuses on discussing the No Fault Found problems in aviation and highlights the drivers that influence its decision making process. It further articulates the contents of tacit knowledge and discusses the knowledge gaps with current management policies.

INTRODUCTION

A set of businesses whose activities can be classified as the operation, maintenance, service, manufacturing, etc, can be regarded as an *industrial network* (Low, 1997). They can be analysed in terms of the patterns of their actors, activity links and resource links. A supply chain is a good example of such a network where numerous interactions exist – which can group together suppliers, system environment, services and customers, all into a cohesive entity. These *actors* (belonging to the entity) within the supply chain, coordinate their activities and collaborate together in order to improve the efficiency of the whole system in general. However, as technological complexity increases, the likelihood of system interruptions at any point in the network increases as well (Singh, 1997). This indicates that modern technology is strongly reliant on the correct functioning of these technical assets within the overall network. This reliance has made it susceptible to their failure. Any interruption – due to component degradation or

DOI: 10.4018/978-1-5225-0651-5.ch020

incorrect use, is of concern not just to the operator (the user of the system) but also to everyone involved i.e. the manufacturers, suppliers and maintainers of the equipment. This can have adverse effects on safety, availability, and reputation; directly reducing the cost-effectiveness of all elements of the supply chain (Khan et al, 2015).

Interruptions can occur in different forms. Gradual degradation, given the time taken to source and fit a new electronic component, is rather trivial. However, a component breaking during operation is more alarming, as it impedes the ability of the system to perform its function until the component is replaced. Another form of interruption (in the same category) is that of reported faults where the root-cause of the problem cannot be diagnosed. In these situations, a suspected component is swapped, only for it to be found that the fault has not gone away. Furthermore, when the removed component is makes its way through the supply chain to the supplier to be tested (for functionality), it is found to be functioning as expected. This phenomenon has been given the name "No Fault Found (NFF)" and is the subject of this chapter.

Successful maintenance activities require one hundred per cent effort to ensure customer's expectations with service reliability and availability (Zeithaml et al, 1990). However, the occurrence of faults where the root cause cannot be determined, usually described as NFF can be a disruptive mechanism on realizing customer demands. As new technology and innovative designs diffuse within system service, together with an aggressive operating environment, a variety of unpredicted failure modes can manifest themselves. These faults are often unanticipated by design engineers and the traditional diagnostic systems may not be able to diagnose them. As the aerospace and railway industries continuously strive to improve their performance, by delivering more reliable vehicles, with a higher availability, operational pressures reduce the time available for diagnostic investigations. This results in on-speculative replacements and hence higher levels of NFF (James et al, 2002).

The implications are therefore, operationally and economically detrimental – with erroneous diagnoses procedures, maintenance activities can penalise organisations in terms of lost man-hour costs, machine downtime and unavailability of system assets. It can further damages reputation and business relations with anyone within the component supply chain. Some studies have even suggested that equipment having an in service life of around 20 years, the operating and service/maintenance activities accounts for about 60 – 80 per cent of the total whole life cycle cost of the system (Pickthall, 2014). Over the years, organisations have established strategies to manage uncertainties, like NFF, within their maintenance activities. Most often, information processing and decision aiding systems are employed but, it is not always apparent whose decision affects the outcome the most and hence accountability moves away from the designated people in unforeseen ways[1].

This chapter focuses on the aerospace industry's decision making ability with regards to maintenance activities, in particular with NFF events. It examines accounts reported by maintenance professionals in order to articulate the underlying dynamics within organisations. The following sections with help the reader with:

- An understanding of the NFF problem
- The influential factors that obscures the issue
- The impact of NFF decision making

THE CASE OF NO FAULT FOUND

The world is surrounded by many products and devices that are getting increasingly complex with sophisticated functionalities. Despite better reliability, more efficient design and effective products, we find an important drawback in the difficulty to distinguish the root cause of many faults that are now growing due to the complex integration and interdependencies among items (Caniato, 2015).

"No fault Found" is a term that catalogues suspected faulty products that have no faults found on further investigation. In order to address the scale of the problem we need to become familiar with a sequence of events which result in what we will describe in this paper as a 'diagnostic failure' (often reported as NFF or a similar descriptive term). This sequence begins during operational service when a Built-In-Test (BIT) fails or the operator reports the possibility of an error. Independent functionality tests will then be ordered on the suspect Line Replaceable Unit (LRU) at 1st line maintenance to look to verify the fault/failure, if it cannot be repeated a failure to diagnose the problem will be recorded. If the functionality tests fail, then further off-line tests within the maintenance shop/depth will be used to diagnose the system fault to a group of Shop Replaceable Units (SRUs) that are suspected of being the source of the LRU failure. Depending upon the accuracy of the diagnosis at this level, ideally only one SRU will be called out; less precise diagnostics may call out two, three or more SRUs. The called out SRU(s) are then sent back to the depot for functional testing using Automatic Test Equipment (ATE). If the SRU passes at this stage a diagnostic failure will be recorded. There are then two possible scenarios, either the SRU may be healthy and falsely replaced, or it is probably faulty and the diagnostic testing is inadequate (Ungar, 2007).

Such scenarios will evidently have a negative impact on system requirements, which at the top level may include safety, dependability, availability as well as negative implications for Whole Life Cycle (WLC) costs. All major industries which operate complex engineered systems (in particular those which are electrical or electronic) suffer from such failures within the diagnostic process as described. These industries, which a small selection, include military/civil aerospace, automotive, energy and consumer electronics. It is clear that a great many of the diagnostic failure events which are of interest occur in avionics, electrical and electro-mechanical, but initial research shows that mechanical systems also give rise to similar difficulties but are far less known. Figure 1 captures and presents a non-exhaustive list of instances found in literature that classify events/factors as NFF (or with a similar terminology variant). It indicates that various NFF sources - which are not known – can influence some drivers which can lead to a diagnostic failure. Once an issue is recorded, it will eventually present itself as a fault that can be attributed to various reasons. In Electronics systems, loose connections probably cover most of the faults that have been published as NFF (Jones & Hayes, 2001; Line & Krishnan, 2007). Some of the more common and well known include poor solder joints, faulty electrical components, damaged PCBs and problems with internal wiring. In addition, electronic failures are not considered as static (or random) events, but a process of mechanical and material changes. These changes will not always lead to a loss of operation for the system, even though their components are out of specification. This is because electronics have an inherent self-compensating aspect which makes the task of diagnostics difficult. In addition to this self-compensation, degradation of failure modes will manifest differently depending upon the operating environment and may offset one another depending on circuit configuration (Line & Krishnan 2007).

Figure 1. Various NFF sources, drivers, and faults

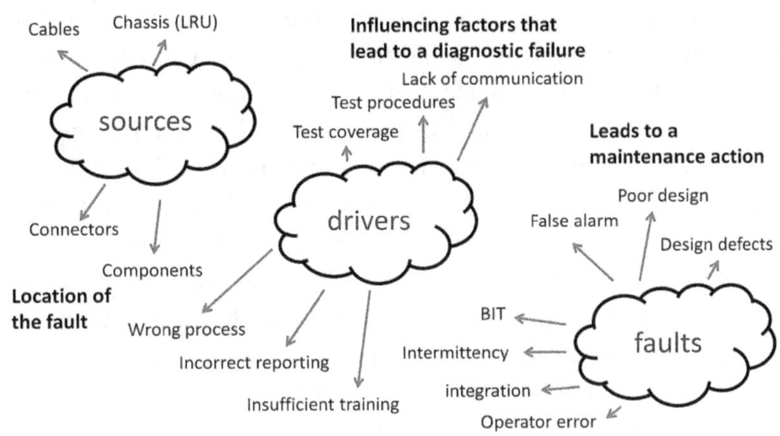

What rapidly becomes evident when reviewing the associated literature, coupled with discussions with industrial organisations, is that there are no standards in place to ensure correct identification, reporting and mitigation of these problems. Adopting standards help industries (and research) overcome technical barriers by promoting organisational success through better workflow paradigms and maintenance strategies. Since diagnosis failure is a multidisciplinary issue, establishing a formal methodology, process, criteria and practice may help reduce the consequences of such failures. However, one interesting point in the discussion is where exactly do these diagnostic failures occur within the maintenance process? Most of the definitions for these phenomena do not provide any consideration for this and therefore leads to the assumption that the level of test is unimportant. For example the following definitions can be extracted from the information provided in published research:

A reported fault for which the root cause cannot be found (Khan et al, 2015).

No Fault Found is a reported failure that cannot be confirmed, recognised or localised during diagnosis and therefore cannot be repaired (Roke, 2009).

The inability to replicate field failure during repair shop test/diagnosis (Kirkland, 2011).

A failure that may have occurred but cannot be verified, or replicated at will, or attributed to a specific root cause, failure site, or failure mode (Qi et al, 2008).

Removals of equipment from service for reasons that cannot be verified by the maintenance process (shop or elsewhere) (ARINC Report 672, 2008).

It should be noted that these are only a few abbreviated examples of a huge number of definitions (or descriptions) which continue to grow. The lack of a single common descriptive and standard term indicates that, even among researchers, practitioners and other experts, the phenomena is not well understood. This can be an output from a failed diagnostic process, which may comprise of a sequence of interlinking events – perhaps at different maintenance levels. Or perhaps it is a human factors issue.

In any case, NFF will play a disruptive role in the system life cycle. From a financial point of view, this causes a burden to almost everybody associated with the through life support service i.e. from the operators and customers, to the manufacturers and their suppliers. To make matters worse, the direct investment of resources and time to investigate NFF events on the business is not easily quantifiable (Erkoyuncu et al, 2016). These can be costs such as those incurred within the supply chain, maintenance performance, as well as indirect effects such as customer perception and nugatory maintenance efforts. Even under warranty, components impose a further "testing bill" to the operator, where the common practice is to charge back additional cost of any NFF item to its service organization. Although, these additional costs can help to motivation the service organization to only return components that they are confident about having issues. However, some investigations into the high part failure rates have revealed that service engineers intentionally cause damage to parts that were being exchanged. The motivation for such practice – so that they are not liable for the cost of any NFF (Khan et al, 2015).

Human psychology plays an instrumental role in such processes, and hence, the human will always continue to occupy a central role in its decision making. The structure of an organization might affect the human decision maker's ability to work effectively under time pressure in a stressful environment, such as the one experienced by air traffic controllers. However, an organization at times exhibits dynamic phenomena that were not anticipated during the creation of the organization structure. There is much anecdotal evidence from engineering managers that their organizations act in unprecedented ways; away from decision makers who were assigned specific responsibilities to manage lower echelons. Changes in organization structure, such as access to decision support systems, can change the sensitivity of the performance measures to the actions of different decision makers. Furthermore, the choice of strategies on the part of these decision makers affects which one has most impact on performance. This shift in responsibility can be viewed from both positive and negative perspectives.

In the former, it is desirable to mitigate the chances of a failure and maintain system performance. Responsibilities must therefore be able to move within the organisation hierarchy through a large scale system. This is to adapt to any structural changes in the system. However, from a negative perspective, shifting responsibilities can lead to unforeseen (or undesirable) situations. For example, moving away from accountability executives (who should hold the positions of responsibility in favour of subordinates), can cause the performance of an organization to deteriorate if, what were seen to be efficient and effective means of processing information, are modified. Even if the structure is designed so that the overall task is performed without overloading employees, it can result in a wide range of performance depending on the strategies chosen by the decision makers (Booth, 2015). Therefore, it is important to ensure that strategies that are mutually acceptable are most regulated and standardised within the industry. In the midst of these problems are the maintenance managers who will manage their resources according to their organisation's ethos. Therefore, accountability becomes an indispensable part in any fundamental solutions for controlling NFF.

In practice, there are predominately three identifiable levels of where NFF events can occur (Khan et al, 2014a):

- The equipment level where the equipment does not function correctly and no problems are found with the unit during its maintenance testing.
- The second is the board level where an equipment does not function correctly and during its maintenance testing, a problem is identified and hence it is sent for repair. However, when subsequently tested, the board functions correctly.

- The third is the component level where a removed board is subsequently tested and a faulty component is identified. However, after removing this component and testing it individually, is it is found to be functioning correctly. After the component has been replaced the board operates normally.

Following on from any of the above three events, a maintenance activity is usually carried out by the Original Equipment Manufacturer (OEM). It is still necessary not to neglect the fact that at some level an event had occurred; may this be a false alarm or component degradation, which has resulted in a maintenance action as described above (ending with a NFF decision at different maintenance levels). Typically, there are some NFF root causes to look out for:

- **Discrepancies and Faults in Test Procedures:** Such errors in process and procedures might exist but they are not known and lie unidentified but nevertheless generate NFF.
- **Incorrect Fault Reporting:** Communication and having a common understanding and interpretation of the fault is sometimes just not achievable. People's perceptions and interpretation often mean the fault will not be properly diagnosed and found. There are also clearly human factors that are involved here.
- **Wrong Processes Applied:** This may be deliberate through misinterpretation and presumed symptoms but also could be just lack of proper training.
- **Incomplete Documentation:** This cause might be processes that were not comprehensive when first devised.

In any case, reasons that are often cited are similar to those recognised at the individual level, which affects an individual's behaviour, such as:

- **Lack of Communication**: At the personal level, perhaps between maintenance personnel when changing shift, poor communication may cause the new shit to misdiagnose the problem. Similarly lack of communication between experts in the organisation means that vital information is not passed on to those who might then solve a regularly reported fault but on different equipment.
- **Not Following the Correct Process:** Technicians can be known to take short cuts because they 'know best' or they may make repeated assumptions which are incorrect. Operational pressures often cap the time available to line maintenance personnel for troubleshoot and are probably the major drivers for such behaviour.
- **Workforce Behaviour:** Within the workforce behaviour there is a reliance on norms that is prevalent - 'we have always done it this way and it always works'. People are reluctant to admit their behaviour, procedures and culture might be part of the problem. But whilst it will always get the equipment back on line, it may have involved changing three units where only one is truly at fault but now they have generated two NFFs in the system with all the attendant costs in the maintenance and supply chain.

By now, it should be evident to the reader that managing NFF incorporates several areas such as human factors, maintenance lines, skills and experience, operational pressures, technological capabilities, contractual agreements and the costs associate with the repair or investigation. Good system design would warrant that all these areas must be recognised and established at the inception of any project; prior to its commencement, in order to allow most of the maintenance requirements (which were integrated into the system) are going to be geared towards fulfilling them. It is therefore essential that caution is taken to define a specific policy to meet out the maintenance requirements, which can help define the necessary arrangements for future support.

MANAGEMENT STRUCTURE FOR DECISION MAKING

I suppose the question that must be raised at this point is: why is NFF a difficult problem to tackle? Figure 2 illustrates that the primarily the root causes - which begin a sequence of events through various levels of maintenance – will blend with other factors such as organisational/behavioural/cultural and technical abilities to result in the final outcome of NFF. These factors should be defined as separate to the root causes and should be considered as the 'drivers towards diagnostic failure'.

There is a drive within UK aerospace to start a cultural change. There is a definite need for managerial level policies which need to be established. Many have expressed their belief that the NFF issues hinder the consequences of a problem. If the result of a test is described as NFF what does this really mean to the strategic level/management and how is it perceived in their mind-set? Firstly, we should consider the two main outcomes of a resulting NFF:

1. There has been no evidence of faults during testing so the unit can be certified as serviceable as there probably never was a fault.
2. There has been no evidence of faults during testing but the test coverage may be inadequate so it is best to replace the unit just to be on the safe side.

Figure 2. NFF drivers leading to a diagnostic failure

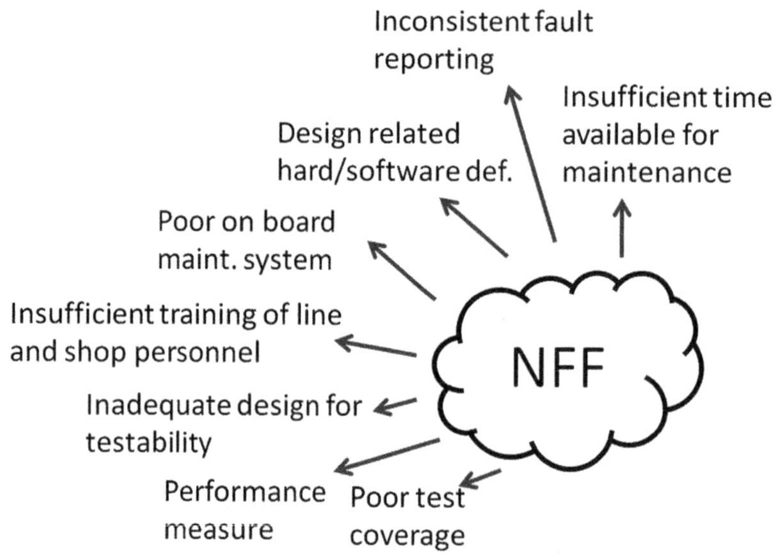

This is the current practice, one way or another there will be a repeat NFF event – either because a faulty unit re-enters service or the wrong unit has been replaced. It needs to be acknowledged that the NFF phenomenon is not a single event – it is a sequence of events which begin with a warning or alarm (detected fault) on board the main equipment. Resulting in a series of actions at various maintenance levels until finally a decision is made to add the NFF label and perform one of the two actions above. These two actions however are missing the key point in solving the problem – something has occurred to result in that final decision - this is the root cause – the initial fault, whether that is intermittency, a false alarm, over sensitive BIT thresholds or systems integration problems, has not been realised. There, the management of NFF events will become a stable cognitive process, with variations in action patterns; such as maintaining ongoing focus on reported faults, simplified diagnostic process, current organisational culture and expert knowledge.

In the organizational context, it might be important to analyse the various facets of a system function and how its elements work together to reveal how NFF manifests itself during maintenance activities. For example, at the top-most level, a legislative body would force the law and requirements that will determine the local activities for inspection and repair, which personnel will have to follow. At this point, organizations are responsible for the quality control[2] and quality assurance[3] of the maintenance system and any troubleshooting process. Failure to carry out these two activities can cause maintenance errors and inefficiencies – resulting in financial issues.

Typically, an organisation's maintenance plan would typically support actions at three levels (Schmidt & Wilhelm, 2000):

1. **Strategic Level:** Priorities and critical targets are established in accordance with business goals. The strategic level is represented by senior management.
2. **Tactical Level:** Resource requirements to achieve the maintenance plan are determined which include requirements, planning and scheduling. The tactical level is represented by mid-level management.
3. **Operational Level:** Maintenance tasks performed in the scheduled time. The operational level is represented by the maintenance staff.

Generally, the staff at the operational level, which encompasses the 'work on the ground', have a good understanding of the NFF phenomena; testing and repair work takes place here and the operational level personnel are the ones who will identify a NFF. This is due to the nature of the problem as it appears primarily during system operation and hence the on-field personnel are the first ones to experience its consequences. It is at this stage that NFF has the potential to economically affect the system operation due to incorrect fault diagnoses, wastage of resources and unproductive time utilization that adds to maintenance costs, downtime and unavailability of the system. It can further damage the reputation and relationships within the supply chain, which is where the tactical level will get involved as they experience the shortage of spares, and the time their maintainers spend looking for faults that cannot being isolated. Due to time pressures, the tactical level will need to make decisions as to whether to allow their staff to keep searching for the symptom-to-cause relationship of the reported fault in order to remove the NFF label; alternatively, they must accept the NFF and send the equipment back through the certification loop, or order further investigations by sending the equipment to a deeper level of maintenance. At the strategic level, NFF events do not inflict an immediate financial impact, due to a lack of benchmarks, and hence they struggle to understand the long term consequences that NFF events inflict on engineering practices. However, decisions made at the organisation's strategic level directly influence the tactical and operational performances. It is suggested therefore that if the cost of NFF at the strategic level were clear, it would enable NFF resolution to become an integrated part of the continuous improvement strategy of the organisation. This has been the subject of many discussions, and helps clarifies why the NFF phenomenon has not been able to attract much attention for resolution, despite being a known issue for many decades (Khan et al, 2014a).

An effective way to control a process is by making use of policies. These guidelines help in achieving objectives and indicate a course of actions to cope with situations as they take place. Organisations seldom define NFF related policies, and hence they rarely exist in any written form. Although, some best practices have intermittently appeared in the past decade:

- **ARINC 672:** One of the major projects focused on the lifecycle of avionics equipment was the ARINC 672 initiative. Starting in 2005, this project was carried out by several people involved with NFF phenomenon in the US aviation industry. The aim was to develop reduction strategies and potential solutions of NFF. The final output was published in June 2008, a report which defines a process to identify NFF issues, analyse possible root causes and propose a resolution (ARINC 672, 2008 and Meyer, 2012).
- **Reliability Enhancement Methodology and Modelling (REMM):** REMM was a research initiative which involved several UK aerospace manufacturers and academic partners. Its objective and focus was improving the general reliability throughout the product life cycle, based on previous information to make future decisions. Nevertheless, it provided new knowledge on the field of NFF phenomenon and suggested best techniques (Jones et al, 2004).

However, these approached have not received widespread adoption across industries (Khan et al, 2015). Perhaps their implementation requires sufficient knowledge regarding the business and the ability to accommodate existing requirements within available means, which may explain why most organisations employ ad hoc approaches to tackle the NFF problems in their business.

Therefore, there is a clear knowledge gap to understand what constraints the NFF decision making process. Perhaps establishing a policy to can help bridge this gap and regulate the problem.

NFF Decision Making

So what influences individuals when making decisions within the NFF process?

Let's explore this question in the context of accountability, the culture of the organisation; the experience of individuals in making NFF related decisions and the operational environment:

- **Accountability:** High accountability leads to people making more accurate decisions, whereas on the other hand according to the second proposition "intense accountability leads to feelings of high responsibility which are relieved by "ranching[4]" decisions up in the organization" (Roberts et al, 1994). This implies that perceived high accountability is a characteristic that enhances the decision making upwards within an organisation. Therefore, NFF investigations should not just be a peripheral activity within an organisation, and this must be reflected within its senior management in order to establish mutual perceptions with regards to the consequence of NFF on an organisation. Managers participate in (or facilitate) the decision-making process for the allocation of resources, the development and implementation of strategic plans, the establishment of intervention and control strategies. Due to their role, managers implement strategies and practices that aim to improve standards and other related tasks. Aviation managers, in particular, may be required to establish and identify approaches for absolute control, reduce incidences of aviation accidents, increase levels of aviation operations, and mitigate all variables that are likely to result in undesired and damaging outcomes that compromise safety. It should be clear by now that the influence of NFF, on maintenance plans and system availability, is far more evident to maintenance managers. This is supported by several cogent arguments which indicate the complexities within commercial contracts, organisational bureaucracies and the lack of adequate metrics for costing the impact of NFF units. Therefore, the policy must recognise the role of senior management as a vital function in the need to improve supporting actions and budgeting for NFF reduction.

- **The Culture of the Organisation:** An organisation's cultural evidently has similarity with human factors but tend to describe the corrective aspect rather than the individual. They have been recognised to be exceedingly bureaucratic and difficult in their response to changes; hence not recognizing NFF as a problem (Pickthall, 2014). However, most agree that one of the most significant contributory factors of NFF events are attributed to the behaviour, skill sets and communication between an organization's technicians, engineers and management personnel.

"The problem here lies more at the human level as there are so many human failings related to the variety of ways that faults are reported, the ways maintenance manuals that are written and presented, and the ways troubleshooting tests are designed. Adding the mix of training, expertise and experience that each engineer has in troubleshooting will affect how a company approaches NFF events. However, there are often insufficient resources to repair items on time, as well as not enough information, training and tools." (Khan, 2014a).

Furthermore, reasons that are often cited are similar to those acknowledged at the individual levels – that affect individual behaviours:

1. Lack of communication
2. Not following the correct process
3. Workforce behaviour
4. Lack of training
5. Operational pressure

The experience of individuals in making NFF related decisions: Troubleshooting processes are affected not only by training and tools, but are also heavily dependent upon experience (Campbell and Reyes-Picknell, 2015). This is important as increased levels of system complexity are a major cause of NFF events and the experience of maintenance personnel is critical to provide system familiarity. When the system is complex, unless the maintainer is knowledgeable or experienced, they will simply send the whole unit for repair rather than carry out further troubleshooting to identify the component at fault. In operational conditions, there can often be a lack of expert knowledge for fault diagnosis because of the lack of experts being available, perhaps due to different shifts, sickness, or a holiday, which then compounds the problem. Khan et al (2014b) summarised the current challenges of on-site experience:

1. To store this experience-based knowledge, and deliver it at the time and place that the same problem symptoms occur, so that it can be re-used to help solve the problem on the first attempt.
2. To deliver that knowledge in a form that is useful to experts and less-experienced technicians alike.
3. To share this knowledge so that everyone benefits from the experience of others.
4. To integrate the knowledge access with the existing troubleshooting tools so that it becomes part of the usual troubleshooting workflow.

Operational environment: Environment interactions can be broad based in the context of:

1. **Physical Environment:** Includes the physical environment as the workplace, such as the maintenance hangar or workshop
2. **Working Conditions:** Includes working patterns, management structures, training and company organizational structure.

Environment related implications, cannot be ignored as they potentially have the most significant impact on the behavior of maintenance staff and influence their ability to undertake effective fault diagnosis. Aviation maintenance is generally undertaken in a fast-moving environment where engineers are regularly challenged by time pressures, limited supervision and difficult working conditions, which can result in human error. Lack of time and associated pressure is a major issue within aviation maintenance due to the penalties such as financial and reputational, if the aircraft is not available for its role of carrying fare-paying passengers, which is the primary source of income for operators. Decisions about NFF events can be influenced by time pressure and high uncertainty (Khan et al, 2014b). The more complex a problem, the higher in the organisational hierarchy decisions tends to be made about them. However, the fact is that to solve and make decisions about a problem, it is essential to understand it – this is why many engineers agree that their management hierarchy does not recognize the gravity of the NFF phenomena, as no metrics exist which can estimate and present its impact to them. Some authors argue that there are three characteristics that influences decision making within organisations; these include high accountability, low familiarity (or routine with the situation) and high political stake. On the other side of the coin, high time pressure was a key characteristic for situations where the decisions making process takes place at the bottom of the management hierarchy (Kruke & Olsen, 2012; Bigley & Roberts, 2001).

All these arguments are related to how decisions actually move in industrial organisations under operational pressure – it is a notable gap in current NFF knowledge regarding characteristics that limits its decision making process. Making use of the NFF policies (defined earlier) can help bridge this gap and regulate the five processes which are being used to mitigate NFF problems.

DISCUSSIONS

From the preceding section, some factors standout about what influences decision making. These include high competition, operational pressures, customer satisfaction, number of options and deliberate oversights during maintenance procedures. Let's extend our discussions on these factors:

- **High Competition and Under Time Pressure:** In a civil aviation environment, operators are under tremendous pressure to get the downed aircraft back in the air. Therefore, availability of the system is paramount to the business – perhaps more than quality of service. Regardless of how contracts are agreed, investigating NFF issues results in nugatory efforts, delays and damages to reputation. Decision making during this phase is usually managed on the tactical level (or higher), and there may be no transfer of decision downwards on the organisational hierarchy at this stage. In these instances, the decision making is driven and focused by considerations of the competitive situation at hand, as well as the strategic importance. Another key driver is that the organization is cautious about customer satisfaction and hence will focus on addressing NFF reports. Customers are becoming more assertive about the cost and resolution of these problems.

  - Current competition amongst suppliers means that if operator (or system provider) has existing equipment from various other OEM's, they could easily switch to other options and stop doing business with their current supplier in the future – unless the current supplier addressing the NFF issues.

- **Key Stakeholder Criticism:** Even though NFF is not explicitly recognised within their contracts, maintenance managers make conscious efforts to keep key stakeholders informed about the rogue units on a regular basis to avoid any criticism. Such practices can help (as bi-weekly bulletins) foster knowledge sharing and to open up a communications channel between their organisation and its customers. The purpose is to share information in order to secure that all key stakeholders were acting based on the same information.
  - The ambition to keep a wide forum of stakeholders informed also indicates a desire to avoid self-reproach in case situations should derail at a later stage – by being able to argue that many people were aware of what was going on and could have taken action. Therefore, avoidance of reproach can be a driving force behind the information sharing as it is based on personal experience of the situation and not in the organisational culture.
- **Lack of Alternatives:** Maintenance personnel are often limited to the written procedures set out in service manuals. In these situations, operators are usually advised to return questionable units back to the OEM for further testing. A challenging part in aerospace maintenance can be attributed to its equipment design. With a drive towards a more electric aircraft, there is a drive towards more functional integration of built in tests. Unfortunately, much of the technical skills required to manage these assets can be limited in isolating all faults – at least to a level where only the offending equipment requires removal. In such situations, there are no alternatives as the management applications that cope with fault investigations are limited and do not provide the required visibility. Some engineers have voiced this concern and expressed their need for the additional fault detection alternative within their standard testing, including for intermittent faults, transient faults, false alarms (James et al, 2003). However, there seems to be no universal agreement on how the capability should be classified and processed, or even the level of granularity that must be included.
- **Deliberate Oversight:** So why maintenance teams do not formally escalate an increase in NFF situations to their strategic level? Engineers have emphasized that although their teams endeavoured to manage the situation at their operational level and in cooperation with customer as far as possible, at times, some personnel become defensive. In these situations no one wants to accept responsibility. Therefore, there is a lack of understanding the group benefit rather then individual blame. It should be pointed out that the consequences of such practices can accumulate by a mix of other problems:
  - Personal relations within the team

- Lack of overall system knowledge
- Insufficient personal competence together with low motivation

When there are no alternative options but to return the unit (there are no obvious technical deficiencies that showed in the unit acceptance test), the operational level made their decisions without escalation (or even informing higher management about it). Similarly, in light of the technical, organisational and personnel related issues, maintenance personnel maintained the decision making within their teams. The situation is made worse when there is not enough time to verify and validate even a small fraction of a system faults during formal testing.

These discussions indicate a general organisational outlook on NFF decision making as shown in the Figure below:

Although organisations promote and encourage delegation of decision making and support, they expect rigid structures and procedures for decision making to be implemented. It was also noted that these structures and procedures are not fully utilised, i.e. there is a considerable difference between what is expected to be done and how it is actually done.

Company values and culture provide a general policy with guidelines for risk-, issue- and non-conformity management in commercial projects, as well as for managing customer claims and warranty issues – all decisions with a financial impact are required to follow the general authorisation routine. In the authorisation routines, specific monetary limits are set for the undersigning at different organisational levels. The picture that emerges here is that the decision making is being controlled, subject to rigid structures and seem to transfer important decisions towards the top level management.

From an NFF point of view, there are no obvious indications on instances where this type of understanding has succeeded. What this study does confirm is that accountability is an important characteristic. However, the perceptions of individuals involved in the situation (who subject to high accountability) play an important role. Table 2 summarises the NFF decision making drivers.

It should be clear that an organisation's culture will heavily influence all factors identified above.

Many engineers (including the interviewees) have expressed that with complexity and high functionality come benefits but also frustration when things do not work the way they are expected to. Training (or at least clear guidance) is thus often more a necessity than a 'nice to have' option. The younger generation seems reluctant to read a manual, the older generations struggle to find the manual on-line. Overall this analysis illustrates that NFF events are not just technical issues but strong cultural, gender and experience related influences. Aspects of human behavioural understanding are therefore required to really solve the NFF issue. Therefore, the most prominent driver for enhancing/constraining the NFF decision making process is training.

Table 2. NFF decision making drivers

What Enhances NFF Decision Making?	What Constraints NFF Decision Making?
• Having alternatives • Key stakeholder involvement • Promoting group value instead of individual efforts • Strategic level policy • Informal networks • Adequate training • Organisation ethos	• High political stake (varying business interests) • Non-routine situations (inexperience) • Business pressures • Lack of management attention • Resource limitations • Lack of training • Organisation ethos

The next section will continue the discussion from Section 2 to map the NFF process map within organisations.

MAPPING PROCESSES

The concept of Petri Nets has been used to study the dynamics of decision making organizations. This is carried out by using *tokens* in the places of a net and then analysing these as a result of the structure of the net and the protocols in its description (Boettcher, 1982). These models are effective in illustrating the flow of information within organizations; where the *tokens* represent information carriers that wait to be processed at *places*. These *places* are conditions which must be met before the information held in them can be processed. Consecutively, *transitions* are events that process and transform this information. This transformation could include analysis, synthesis, transfer from one point to another, or computation (Reisig, 2012).

Levis (1984) described how to express decision making organizations by a mathematical framework. This work is based on the assumption that within these organizations there are a group of decision makers who execute clear tasks; each one limited by his bounded logic[5]. When carrying out a task, the decision makers may be faced with several options – these can be represented as different algorithms to process information, decision aids, training, etc. The approach used to select from these options can have varying effects on the performance of the organization, therefore, the fundamental constituent for defining an organization is to model the interacting decision. Understanding system behaviour in the context of NFF requires the ability to construct models that allow for the diagnosis and identification faults and root causes or at a higher-level, organizational responsibility that can give rise to NFF issues.

The author would like to explore how the NFF decision making process can be captured using the concept of Petri nets. Consider Figures 4-6 that consists of three decision making entities. The variables used in the images are listed in Table 3. In the figures, the NFF decision making process has been divided into four phases: problem assessment, collection and fusion, command and response selection. These phases take place within each organisation i.e. operator, OEM and supplier.

The Operator (NFF1) must make a decision if they want to spend their resources (e.g. resources, maintenance costs, etc) or not in the troubleshooting process to search for the root cause of the problem. This indicates the initial dilemma when there is an "arising" – this has been denoted by the transition/switch TSL. Depending on how the contracts have been setup, the operator also has the option to either send the suspected unit back to their system providers (e.g. the OEM, integrators – NFF2) or discard it.

Discarding a unit can be expensive and hence is not the preferable choice. If the unit is not sent back, the operator will use their own experience and troubleshooting strategies to investigate the problem and make a decision. If unit is sent back to the system provider, they need to make sure they also send over extra situational information – such as the environmental conditions, basic tests carried out, etc.

When the NFF2 (Figure 5) receives the unit, they need to make the decision to either:

1. Do extra tests to find the problem
2. Return the unit back to the operator if they cannot find anything wrong with the unit
3. Send the unit to the supplier for further tests
4. Discard the current unit and send the operator a new one

Figure 3. Organisational levels and drivers for NFF

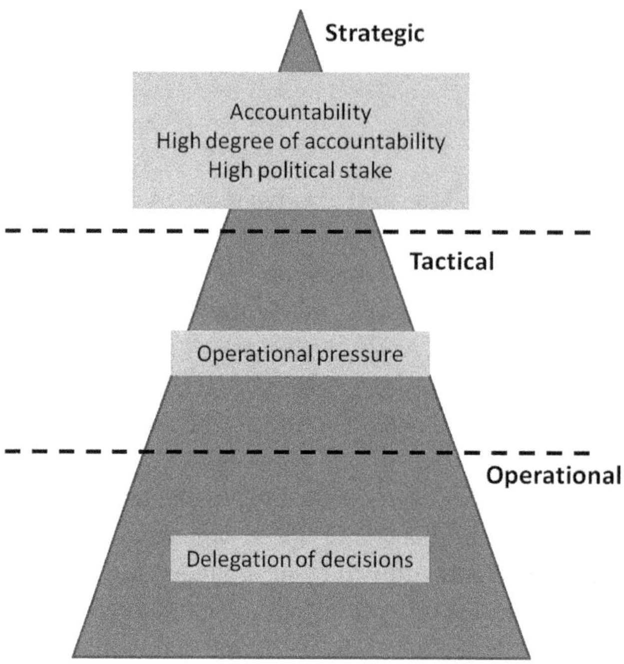

Table 3. variables used in Figure1-3

Variables	Definition
z	is an assessment of the situation that we have an NFF component
P(u)	is the number of diagnostics you run to clear the problem
z1	Operating environmental information
z2	Simulative environmental information
z3	Unit environmental information
z'	revised assessment
v1,v2,v3	management and business pressure
Y1	send unit to system provider
Y2	do not send unit, solve problem yourself
Y3	do nothing/ignore
W1	Send unit to supplier
W2	Send unit back to operator
W3	Do extra tests on unit
W4	discard unit and send new one to operator
U1	send unit back to system provider
U2	Do extra tests
U3	discard unit and send new one to system provider

Figure 4. The NFF process map at the operator

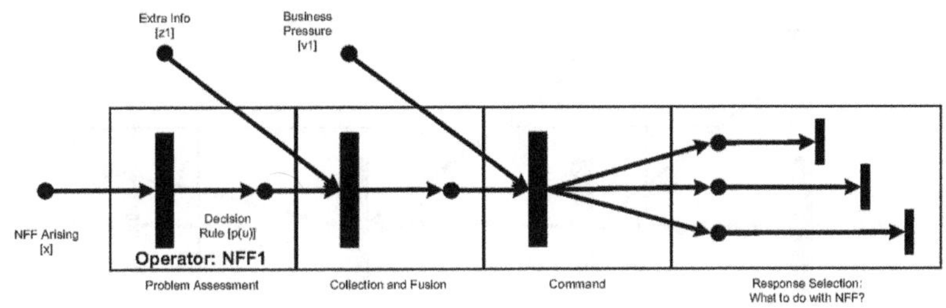

The first two options are standard practices. The latter two options are expensive solutions to problem – which carries no guarantee that it will address the root cause of the NFF arising at the operator[6].

Depending on the organisation's reputation and how their business relations are with each other, it is likely that the choices might be evident to some members of the organisation. However, there is still a high degree of ambiguity surrounding the phenomenon.

Note: The further up you go with NFF1, NFF2… NFFx, the more difficult is becomes to relate the symptom to the root cause. This is will also incur greater costs. This is not only due to the business interests of each organisation, but also the lack of information surrounding the problem context – unless these organisations have some communication channels open.

The studies show how organisations manage integrated NFF related information flow around and how this can play an important role in their accountability for industrial maintenance delivery. This accountability concerns the ability to promptly take actions without reducing the quality of decisions. The ability in supply chains is impacted by partnership uncertainty which is due to partner involvement, relationships and behaviour. Partnering impacts on industrial maintenance targets when the services are delivered as part of networks and the sharing of information is central to long term thinking inside networks. Within major organisations that provide aerospace services, the potential for partnership uncertainty is confronted primarily through an emphasis on contractual agreements for technology platforms supported by industrial maintenance and on strategizing service phase dialogues between customers and partners through the use of bespoke and third party information systems.

Figure 5. The NFF process map at the system provider

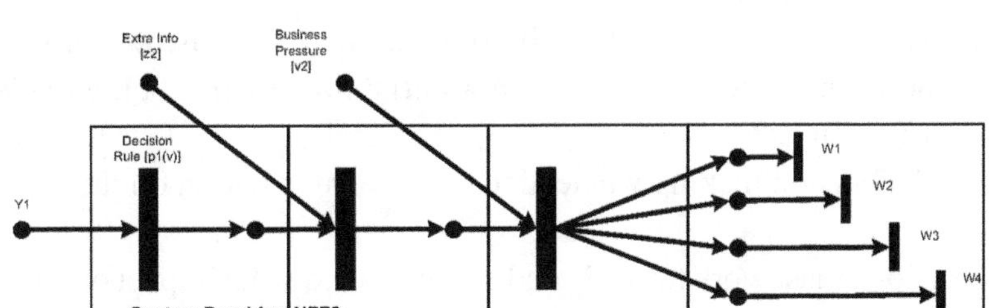

Figure 6. The NFF process map at the supplier

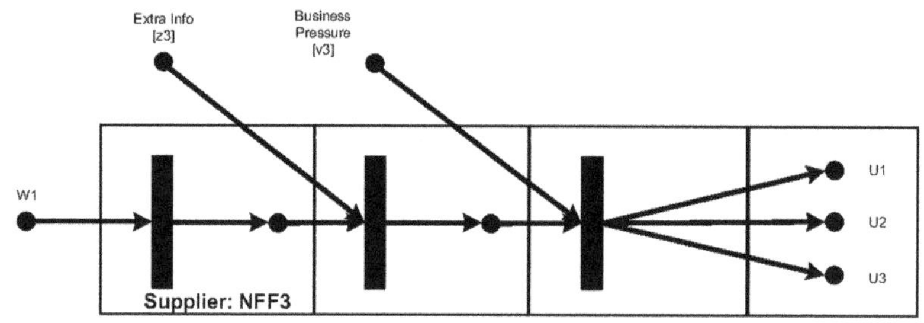

CONCLUSION

The phenomenon which is the subject of this chapter has been, and is in industry, commonly classified as a 'diagnostic failure'. This however indicates a closed-loop system with identifiable symptom → failure → decision relations; identifying and road mapping this relationship will be a key challenge in improving diagnostic success. This will require a complete fundamental understanding of the phenomena including clear distinctions between root cause faults, root cause sources and the influencing factors (drivers) covering the entire maintenance process. What has become clear is that the lack of management decision making in this area results in different maintenance regimes. This obfuscates the scale of the problem – a common policy would be able to provide coherence when approaching the problem. Regardless of how irregular maintenance practices may appear to be, firm policies and guidelines can be instituted to get things under control. These requirements should be analysed, simplified and reduced to bare essentials. Only then can unnecessary complicated procedures be avoided.

This Chapter detailed the NFF problem, how organisations make decisions to control it, and how impact information flows within organisations. There is a definite need for strategic proactive thinking, as well as collective industry wide cooperation, in order to deal with the NFF issue. Having specific policies may be the answer that can help dictate management training programs on controlling some of the attitudes and practices towards NFF events. However, all mitigation attempts need to be carried out within existing organisational cultures. If they are aiming to improve customer relations (or win more contracts), they will aim at addressing even the most minute fault issues as a demonstration of their attention to detail and hence overspend on the total cost within the troubleshooting process.

The key messages from the chapter can be summerised as:

- NFF is not a new problem, but all industries appear to still be roughly in the same place in their ability and willingness to tackle the issue
- The financial impact of NFF still needs to be pinned down. If it is results in financial issues in through-life support of an asset, then why are not initial investments being made up front in order to achieve long term savings?
- To understand NFF decision making we need to capture information on the influential factors that sound the issue.
- Management decisions must work towards understanding how the impact of NFF changes throughout the life-cycle of the product

- There is a tendency for 'blame culture' surrounding NFF, which hinders real progress towards a solution. The downstream effect throughout the supply chain needs investigating
- There is a need to improve data capture and knowledge transfers within organisations. This should also include operating environment and usage details.

What is clear is that there is a need for solutions that can easily be standardised and adopted by industry as a whole. These concepts need joined efforts from manufacturers, maintenance organisations and regulators in order to mutually control the root cause precursors. It also aims to increase awareness of the subject area which will reflect as improvements in maintenance manuals, management attitudes, and financial justifications. In conclusion, this chapter discusses a current hot topic in the aviation industry – the regulation and control of the amount of time, resources and budget to be spent when faced with NFF issues. It discusses many organisational issues and gaps that will require extensive research efforts in due course. But it is important to have presented the current perspectives about where we stand today and indicate possible routes for the future. In this context, I hope I have succeeded.

REFERENCES

672 Guidelines for the reduction of No Fault Found (NFF). (2008). *Proceedings of theAvionics Maintenance Conference.*

Bigley, G. A., & Roberts, K. H. (2001). The incident command system: High-reliability organizing for complex and volatile task environments. *Academy of Management Journal, 44*(6), 1281–1299. doi:10.2307/3069401

Boettcher, K.L., & Levis, A.H. (n. d.). Modeling the interacting decisionmaker with bounded rationality.

Booth, S. A. (2015). *Crisis management strategy: Competition and change in modern enterprises.* Routledge.

Campbell, J. D., & Reyes-Picknell, J. V. (2015). *Uptime: Strategies for excellence in maintenance management.* CRC Press.

Caniato, F., & Größler, A. (2015). The moderating effect of product complexity on new product development and supply chain management integration. *Production Planning and Control, 26*(16), 1306–1317. doi:10.1080/09537287.2015.1027318

Erkoyuncu, J. A., Khan, S., Hussain, S., & Roya, R. (2015). A framework to estimate the cost of No-Fault Found events. *International Journal of Production Economics,* 173, 207–222.

James, I., Lumbard, D., Willis, I., & Goble, J. (2003). Investigating no fault found in the aerospace industry. Proceedings of the Reliability and Maintainability Symposium (pp. 441-446). doi:10.1109/RAMS.2003.1182029

James, I. J., Marshall, J., & Walls, L. (2002). Improving design for reliability with in-service data analysis. *Proceedings of theReliability and Maintainability Symposium '02* (pp. 417-422). IEEE. doi:10.1109/RAMS.2002.981677

Jones, J., & Hayes, J. (2001). Investigations of the occurrence of: No-faults-found in electronic systems. *IEEE Transactions on Reliability*, *50*(3), 289–292. doi:10.1109/24.974126

Jones, J., Marshall, J., Aulak, G., & Newman, B. (2003). Development of an expert system for reliability task planning as part of the REMM methodology. *Proceedings of theReliability and Maintainability Symposium, 2003. Annual* (pp. 423-428). IEEE. doi:10.1109/RAMS.2003.1182026

Khan, S., Phillips, P., Hockley, C., & Jennions, I. (2014). No Fault Found events in maintenance engineering Part 2: Root causes, technical developments and future research. *Reliability Engineering & System Safety*, *123*, 196–208. doi:10.1016/j.ress.2013.10.013

Khan, S., Phillips, P., Hockley, C., & Jennions, I. (2015). *No Fault Found: The Search for the Root Cause*. Published by SAE International. doi:10.4271/R-441

Khan, S., Phillips, P., Jennions, I., & Hockley, C. (2014). No Fault Found events in maintenance engineering Part 1: Current trends, implications and organizational practices. *Reliability Engineering & System Safety*, *123*, 183–195. doi:10.1016/j.ress.2013.11.003

Kirkland, L. V. (2011, September). Why did we add LabVIEW applications to our ATLAS TPSs?Proceedings of*AUTOTESTCON* (pp. 266–271).

Kruke, B. I., & Olsen, O. E. (2012). Knowledge creation and reliable decision-making in complex emergencies. *Disasters*, *36*(2), 212–232. doi:10.1111/j.1467-7717.2011.01255.x PMID:21992233

Levis, A. H. (1984). Information Processing and Decision-Making Organizations: A Mathematical Description. *Large Scale Systems*, *7*, 151–163.

Line, J. K., & Krishnan, G. (2008). Managing and predicting intermittent failures within long life electronics. *Proceedings of theAerospace Conference'08* (pp. 1-6). IEEE. doi:10.1109/AERO.2008.4526629

Low, B. K. H. (1997). Managing business relationships and positions in industrial networks. *Industrial Marketing Management*, *26*(2), 189–202. doi:10.1016/S0019-8501(96)00091-0

Meyer, B. (2012). Unravelling the Mysteries of NFF. *Air Transport and Operations:Proceedings of the Third International Air Transport and Operations Symposium 2012* (p. 354). IOS Press.

Pickthall, N. (2014). The Contribution of Maintenance Human Factors to no Fault Founds on Aircraft Systems Engineering. *Procedia CIRP*, *22*, 59–64. doi:10.1016/j.procir.2014.07.013

Qi, H., Geneson, S., & Pecht, M. (2008). No-fault-found and intermittent failures in electronic products. *Microelectronics and Reliability*, *48*(5), 663–674. doi:10.1016/j.microrel.2008.02.003

Reisig, W. (2012). *Petri nets: an introduction* (Vol. 4). Springer Science & Business Media.

Roberts, K. H., Stout, S. K., & Halpern, J. J. (1994). Decision dynamics in two high reliability military organizations. *Management Science*, *40*(5), 614–624. doi:10.1287/mnsc.40.5.614

Roke, S. (2009). *Harrier No Fault Found reduction* [MSc Dissertation in Engineering Business Management]. Faculty of Engineering and Computing Coventry University.

Schmidt, G., & Wilhelm, W. E. (2000). Strategic, tactical and operational decisions in multi-national logistics networks: A review and discussion of modelling issues. *International Journal of Production Research*, *38*(7), 1501–1523. doi:10.1080/002075400188690

Singh, K. (1997). The impact of technological complexity and interfirm cooperation on business survival. *Academy of Management Journal*, *40*(2), 339–367. doi:10.2307/256886

(1982). Systems, Man and Cybernetics. *IEEE Transactions on*, *12*(3), 334–344.

Ungar, L. (2007). Design for diagnosability guidelines. *IEEE Instrumentation & Measurement Magazine*, 2007, 24–32.

Zeithaml, V. A., Parasuraman, A., & Berry, L. L. (1990). *Delivering quality service: Balancing customer perceptions and expectations*. Simon and Schuster.

ENDNOTES

1 For example, it depends on the decisions of individual decision makers, on the culture of the organization and its operating procedures.

2 Carrying out inspections and auditing actions from regulatory bodies, e.g. the Civil Aviation Authority.

3 The function includes checking engineering change orders, auditing and investigating maintenance activities and components for errors, and examining records.

4 To cause something to rise (or fall) as a step in what is perceived as an irreversible process

5 These limits are on the amount of information that a human can understand and process.

6 It could be that there is no technical problem with the suspected unit, and the NFF event is related to a human factors problem at the operator's end.

Compilation of References

(1982). Systems, Man and Cybernetics. *IEEE Transactions on, 12*(3), 334–344.

672 Guidelines for the reduction of No Fault Found (NFF). (2008). *Proceedings of theAvionics Maintenance Conference.*

747. Maintenance Steering Group. (1968). *Handbook: Maintenance Evaluation and Program Development 1 (MSG-1).* Washington, DC: Air Transport Association.

Abkowitz, M. (2008, October 1). 10 lessons learned the hard way. *Vanderbilt Magazine.* Retrieved from http://www.vanderbilt.edu/magazines/vanderbilt-magazine/2008/10/lessons-learned-the-hard-way/

ABS. (2004). *Guidance Notes on Reliability-Centered Maintenance.* Houston, TX, USA: American Bureau of Shipping.

Adams, D. E. (2007). *Health monitoring of structural materials and components: methods with applications.* John Wiley & Sons. doi:10.1002/9780470511589

Adler, P. (1995). Interdepartmental interdependence and coordination: The case of the design/manufacturing interface. *Organization Science, 6*(2), 147–167. doi:10.1287/orsc.6.2.147

Ajmal, M. M., & Koskinen, K. U. (2008). Knowledge transfer in project-based organizations: An organizational culture perspective. *Project Management Journal, 39*(1), 7–15. doi:10.1002/pmj.20031

Akaike, H. (1973) Information theory and an extension of the maximum likelihood principle *Proceedings of theInternational Symposium on Information Theory,* Akademia Kiado, Budapest.

Akkermans, H. A., Bogerd, P., Yücesan, E., & van Wassenhove, L. N. (2003). The impact of ERP on supply chain management: Exploratory findings from a European Delphi study. *European Journal of Operational Research, 146*(2), 284–301. doi:10.1016/S0377-2217(02)00550-7

Al-Atat, H., Siegel, D., & Lee, J. (2011). A Systematic Methodology for Gearbox Health Assessment and Fault Classification. *International Journal of Prognostics and Health Management.*

Alessandri, T. M., Ford, D. N., Lander, D. M., Leggio, K. B., & Taylor, M. (2004). Managing risk and uncertainty in complex capital projects. *The Quarterly Review of Economics and Finance, 44*(5), 751–767. doi:10.1016/j.qref.2004.05.010

Alexander, D. (2005). Towards the development of a standard in emergency planning. *Disaster Prevention and Management: An International Journal, 14*(2), 158–175. doi:10.1108/09653560510595164

Allan, D. M. (2005). Asset management challenges in ageing power systems. *Proceedings of 2005 International Symposium on Electrical Insulating Materials* (ISEIM '05) (pp. 265–268). ISEIM. doi:10.1109/ISEIM.2005.193395

Almeida, A. T., & Bohoris, G. A. (1995). Decision theory in maintenance decision making. *Journal of Quality in Maintenance Engineering, 1*(1), 39–45. doi:10.1108/13552519510083138

Al-Najjar, B., & Alsyouf, I. (2003). Selecting the most efficient maintenance approach using fuzzy multiple criteria decision making. *International Journal of Production Economics*, *84*(1), 85–100. doi:10.1016/S0925-5273(02)00380-8

Al-Najjar, G. (1999). Economic criteria to select a cost-effective maintenance policy. *Journal of Quality in Maintenance Engineering*, *5*(3), 236–247. doi:10.1108/13552519910282692

Alsyouf, I. (2006). Measuring maintenance performance using a balanced scorecard approach. *Journal of Quality in Maintenance Engineering*, *12*(2), 133–149. doi:10.1108/13552510610667165

Al-Turki, U. (2011). A framework for strategic planning in maintenance. *Journal of Quality in Maintenance Engineering*, *17*(2), 150–162. doi:10.1108/13552511111134583

Amaratunga, D., Haigh, R., Sarshar, M., & Baldry, D. (2002). Application of the Balanced Score-Card concepts to develop a conceptual framework to measure facilities management performance within NHS facilities. *International Journal of Health Care Quality Assurance*, *15*(4), 141–151. doi:10.1108/09526860210431805

Andersen, D., Richardson, G., & Vennix, J. (1997). Group Model-Building: Adding More Science To The Craft. *System Dynamics Review*, *13*(2), 187–201. doi:10.1002/(SICI)1099-1727(199722)13:2<187::AID-SDR124>3.0.CO;2-O

Anderson, R. T., & Neri, L. (1990). *Reliability-Centered Maintenance: Management and Engineering Methods*. New York, NY: Elsevier Science Publishing Co. doi:10.1007/978-94-009-0757-7

Aratani, L. (2015, May 14). Metro resumes old strategy to prevent more smoke incidents in tunnels. *Washington Post*. Retrieved from https://www.washingtonpost.com/local/trafficandcommuting/metro-resumes-old-strategy-to-prevent-more-smoke-incidents-in-tunnels/2015/05/14/ee73c3b4-fa7b-11e4-9030-b4732caefe81_story.html?hpid=z2

Arbonies, A., & Calzada, I. (2004). El poder del conocimiento tácito: Por encima del aprendizaje organizacional. *Intangible Capital*, *6*(0), 1–17.

Arens, A., Elder, R., & Beasley, M. (2014). *Auditing and assurance services-An integrated approach; includes coverage of international standards and global auditing issues, in addition to coverage of*. Boston: Aufl.

Arora, R. (2002). Implementing KM: A balanced score card approach. *Journal of Knowledge Management*, *6*(3), 240–249. doi:10.1108/13673270210434340

Ascher, H., & Feingold, H. (1984). *Repairable systems reliability. Modelling, Inference, Misconceptions and Their Causes*. New York: Marcel Dekker Inc.

Asiedu, Y., & Gu, P. (1998). Product life cycle cost analysis: State of the art review. *International Journal of Production Research*, *36*(4), 883–904. doi:10.1080/002075498193444

August, J. (1999). *Applied Reliability-Centered Maintenance*. Tulsa, Oklahoma: PennWell.

Auzanneau, F. (2010). Diagnostic Filaire: Détection, localisation et caractérisation de défauts dans des réseaux filaires complexes. In Recherche et Innovation des Techniques de l'Ingénieur (pp. 1–11).

Azadivar, F., & Shu, V. (1999). Maintenance policy selection for JIT production systems. *International Journal of Production Research*, *37*(16), 3725–3738. doi:10.1080/002075499190013

Azhar, S. (2011). Building Information Modelling (BIM): Trends, Benefits, Risks, and Challenges for the AEC Industry. *Leadership and Management in Engineering*, *11*(3), 241–252. doi:10.1061/(ASCE)LM.1943-5630.0000127

Bäcklund, J., Blom, A.F., & Beevers, C.J. (1982). Fatigue-Thresholds-Fundamentals and Engineering (Vol 1 & 2). UK: ApplicationsEngineering Materials Advisory Services Ltd.

Baez, J. (2007). *Investigación cualitativa.* Madrid: Esic Editorial.

Bahoque, E., Gómez, O., & Pietrosemoli, L. (2007). *Gestión del Conocimiento en la Industria de la Construcción: Estudio de un caso. Revista Venezolana de Gerencia, 12(39), 393 – 409.*

Baines, T., Lightfoot, H. W., Evans, S., Neely, A., Greenough, R., Peppard, J., & Tiwari, A. et al. (2007). State-of-the-art in product-service systems. *Journal of Engineering Manufacture*, *221*(10), 1543–1552. doi:10.1243/09544054JEM858

Baker, B. A., Manan, A., & Husband, T. M. (1997). Simulating maintenance work in an engineering firm: A case study. *Microelectronics and Reliability*, *16*(5), 571–581. doi:10.1016/0026-2714(77)90288-8

Bana e Costa, C. A. De Corte, J. M., & Vansnick, J. C. (2006). On the Mathematical Foundation of MACBETH. In J. Figueira, S. Greco & M. Ehrgott (Eds.), Multicriteria Decision analysis: State of the art surveys (pp. 409-444). New York, NY: Springer.

Bana e Costa, C. A., & Chagas, M. P. (2002). A career choice problem: An example of how to use MACBETH to build a quantitative value model based on qualitative value judgments (Working Paper LSEOR 02.53). The London School of Economics and Political Science.

Bana e Costa, C. A., Da Costa-Lobo, M. L., Ramos, I. A., & Vansnick, J. C. (2002). Multicriteria Approach for Strategic Town Planning: The Case of Barcelos. *International Journal of Management Sciences*, *30*(3), 227–242.

Bana e Costa, C. A., De Corte, J. M., & Vansnick, J. C. (2011). MACBETH (Measuring Attractiveness by a Categorical Based Evaluation Technique). In J. J. Cochran, L. A. Cox Jr, P. Keskinocak, J. P. Kharoufeh, & J. C. Smith (Eds.), *Encyclopedia of Operations Research and Management Science*. New York, NY: John Wiley & Sons. doi:10.1002/9780470400531.eorms0970

Bana e Costa, C. A., Nunes da Silva, F., & Vansnick, J. C. (2001). Conflict dissolution in the public sector: A case-study. *European Journal of Operational Research*, *130*(2), 388–401. doi:10.1016/S0377-2217(00)00037-0

Bana e Costa, C. A., Oliveira, C. S., & Vieira, V. (2008). Priorization of bridges and tunnels in earthquake risk mitigation using multicriteria decision analysis: Application to Lisbon. *Omega*, *36*(3), 442–450. doi:10.1016/j.omega.2006.05.008

Bana e Costa, C. A., & Vansnick, J. C. (1994). MACBETH - An interactive path towards the construction of cardinal value functions. *International Transactions in Operational Research*, *1*(4), 489–500. doi:10.1016/0969-6016(94)90010-8

Bana e Costa, C. A., & Vansnick, J. C. (1999). The MACBETH approach: basic ideas, software and an application. In N. Meskens & M. Roubens (Eds.), *Advances in Decision Analysis* (pp. 131–157). Dordrecht: Kluwer Academic Publishers. doi:10.1007/978-94-017-0647-6_9

Barberá, L., Crespo, A., Viveros, P. & Arata, A. (2013). The Graphical Analysis for Maintenance Management Method: A Quantitative Graphical Analysis to Support Maintenance Management Decision Making. *Journal of Quality and Reliability Engineering International*, 29(1), 77–87. DOI:10.1002/qre.1296

Barberá, L., Crespo, A., Viveros, P., & Stegmaier, R. (2012, January-March). Advanced model for maintenance management in a continuous improvement cycle: Integration into the business strategy. *International Journal of System Assurance Engineering and Management*, *3*(1), 47–63. doi:10.1007/s13198-012-0092-y

Barragán, A. (2009). Aproximación a una taxonomía de modelos de gestión del conocimiento. *Intangible Capital*, 5(1), 65-101. doi:10.3926/ic.2009.v5n1.p65-101

Barsom, J. M. (1971). Fatigue Crack Propagation in Steels of Various Yield Strengths. Transactions of the ASME. *Journal Engineering for Industry,* Series B, 93(4). doi:10.1115/1.3428061

Barsom, J. M. (1974). *Fatigue Behavior of Pressure-Vessel Steels. WRC Bulletin, 194*. New York: Welding Research Council.

Barsom, J. M., & Rolfe, S. T. (1977). *Fracture and Fatigue Control in Structures. Applications of Fracture Mechanics* (p. 07632). Englewood Cliffs, New Jersey: Prentice-Hall, Inc.

Bashiri, M., Badri, H., & Hejazi, T. H. (2011). Selecting optimum maintenance strategy by fuzzy interactive linear assignment method. *Applied Mathematical Modelling*, *35*(1), 152–164. doi:10.1016/j.apm.2010.05.014

Batidzirai, B., Faaij, A., & Smeets, E. (2006). Biomass And Bioenergy Supply From Mozambique. *Energy for Sustainable Development*, *10*(1), 54–81. doi:10.1016/S0973-0826(08)60507-4

Battistella, C., Biotto, G., & De Toni, A. (2012). From Design Driven Innovation To Meaning Strategy. *Management Decision*, *50*(4), 718–743. doi:10.1108/00251741211220390

Bebbington, A. (1999). Capitals And Capabilities: A Framework For Analyzing Peasant Viability, Rural Livelihoods And Poverty. *World Development*, *27*(12), 2021–2044. doi:10.1016/S0305-750X(99)00104-7

Bechhoefer, E., & He, D. (2012). A Process for Data Driven Prognostics. *Proceedings of MFPT 2012 Conference*.

Ben Hassen, W., Auzanneau, F., Incarbone, L., Pérès, F. & Tchangani, A. P. (2013a). OMTDR using BER estimation for ambiguities cancellation in ramified networks diagnosis. In *Intelligent Sensors, Sensor Networks and Information Processing* (pp. 414-419).

Ben Hassen, W., Auzanneau, F., Pérès, F., & Tchangani, A. P. (2014). Ambiguity Cancellation for Wire Fault Location based on Cable Life Profile. *Proceedings of IFAC World Congress*, Cape Town, South Africa. doi:10.3182/20140824-6-ZA-1003.02411

Ben Hassen, W., Auzanneau, F., Incarbone, L., Pérès, F., & Tchangani, A. P. (2013b). On-line diagnosis using Orthogonal Multi-Tone Time Domain Reflectometry in a lossy cable.*Proceedings of IEEE International Conference on Systems, Signals and Devices*. doi:10.1109/SSD.2013.6564144

Ben Hassen, W., Auzanneau, F., Incarbone, L., Pérès, F., & Tchangani, A. P. (2015). Distributed Sensor Fusion for Wire Fault Location Using Sensor Clustering Strategy. *International Journal of Distributed Sensor Networks*, *2015*, 1–17. doi:10.1155/2015/538643

Ben Hassen, W., Auzanneau, F., Pérès, F., & Tchangani, A. P. (2012). A distributed diagnosis strategy using bayesian network for complex wiring networks.*Proceedings of IFAC Workshop on Advanced Maintenance Engineering, Services and Technology*. doi:10.3182/20121122-2-ES-4026.00031

Ben Hassen, W., Auzanneau, F., Pérès, F., & Tchangani, A. P. (2013c). Diagnosis sensor fusion for wire fault location in CAN bus systems. *Proceedings of IEEE Sensors*, *2013*, 1–4. PMID:25126153

Benedettini, O., Baines, T., Lightfoot, H., & Greenough, R. (2009). State-of-the-art in integrated vehicle health management. *Proceedings of the Institution of Mechanical Engineers. Part G, Journal of Aerospace Engineering*, *223*(2), 157–170. doi:10.1243/09544100JAERO446

Bengio, Y., Lamblin, P., Popovici, D., & Larochelle, H. (2006). Advances in Neural Information Processing Systems: Vol. 19. *Greedy layer-wise training of deep networks*.

Berdinyazov, A., Camci, F., Sevkli, M., & Baskan, S. (2009) Economic analysis of maintenance policies for a system Diagnostics for Electric Machines. *Proceedings of theIEEE International Symposium onPower Electronics and Drives (SDEMPED '09)*.

Bertolini, M., & Bevilacqua, M. (2006). A combined goal programming-AHP approach to maintenance selection problem. *Reliability Engineering & System Safety*, *91*(7), 839–848. doi:10.1016/j.ress.2005.08.006

BEST. (2012): Mozambique Biomass Energy Strategy, ECO-Consult e MARGE para o FUNAE, Maputo.

Bevilacqua, M., & Braglia, M. (2000). The analytic hierarchy process applied to maintenance strategy selection. *Reliability Engineering & System Safety*, *70*(1), 71–83. doi:10.1016/S0951-8320(00)00047-8

Biagetti, T., & Sciubba, E. (2004). Automatic diagnostics and prognostics of energy conversion processes via knowledge-based systems. *Energy*, *29*(12), 2553–2572. doi:10.1016/j.energy.2004.03.031

Bigley, G. A., & Roberts, K. H. (2001). The incident command system: High-reliability organizing for complex and volatile task environments. *Academy of Management Journal*, *44*(6), 1281–1299. doi:10.2307/3069401

Bilgili, M., Yasar, A. & Simsek, E. (2010). Offshore wind power development in Europe and its comparison with onshore counterpart. *Renewable and Sustainable Energy Reviews,* 15(2011), 905-915.

Binney, D. (2001). El espectro de la gestión del conocimiento: El entendimiento del panorama de la GC. *Diario de la Gestión del conocimiento*, 5(1), 33-42.

Birmingham, F., & Jelinek, J. (2007). *Quick Changeover Simplifies: The Manager's Guide to Improving Profits with SMED*. New York, USA: Productivity Press.

Birolini, A. (2007). *Reliability Engineering* (5th ed.). Springer Berlin Heidelberg New York.

Blanco, P. (2013, April 27). Carreteras sin cuidar o reparar. *El Pais.* Retrieved from http://politica.elpais.com/politica/2013/04/27/actualidad/1367084371_564522.html

Bloom, N. (2006). *Reliability Centered Maintenance. Implementation made simple*. New York: McGraw-Hill Inc.

Bloom, N. B. (2005). *Reliability Centered Maintenance Implementation made simple*. New York, NY: McGraw Hill.

Bo, S., Shengkui, Z., Rui, K., & Pecht, M. (2010, January 12-14). Benefits analysis of prognostics in systems. *Paper presented at the Prognostics and Health Management Conference PHM '10.*

Boettcher, K.L., & Levis, A.H. (n. d.). Modeling the interacting decisionmaker with bounded rationality.

Boin, A., & McConnell, A. (2007). Preparing for critical infrastructure breakdowns: The limits of crisis management and the need for resilience. *Journal of Contingencies and Crisis Management*, *15*(1), 50–59. doi:10.1111/j.1468-5973.2007.00504.x

Bonn, I. (2005). Improving Strategic Thinking: A Multilevel Approach. *Leadership and Organization Development Journal*, *26*(5), 336–354. doi:10.1108/01437730510607844

Booth, S. A. (2015). *Crisis management strategy: Competition and change in modern enterprises*. Routledge.

Bousdekis, A., Magoutas, B., Apostolou, D., Mentzas, G., & Chan, H. K. (2015). A Proactive Decision Making Framework for Condition Based Maintenance. *Industrial Management & Data Systems*, *115*(7), 1225–1250. doi:10.1108/IMDS-03-2015-0071

Braaksma, A. J. J. (2012). *Asset information for FMEA-based maintenance*. Groningen, The Netherlands: Faculty of Economics and Business. University of Groningen.

Braaksma, A., Meesters, A., Klingenberg, W., & Hicks, C. (2012). A quantitative method for Failure Mode and Effects Analysis. *International Journal of Production Research*, *50*(23), 6904–6917. doi:10.1080/00207543.2011.632386

Braun, S. (2011). The synchronous (time domain) average revisited. *Mechanical Systems and Signal Processing*, *25*(4), 1087–1102. doi:10.1016/j.ymssp.2010.07.016

Brockmann, C. (2009). *Mega-projects: getting the job done.* LEAD Conference 2009.

Broek, D. (1982). Elementary Engineering Fracture Mechanics. Hingham, MA, USA: Kluwer Academic Publishers. doi:10.1007/978-94-011-9055-8

Brouwer, I., Hoorweg, J., & van Liere, M. (1997). When Households Run Out Of Fuel: Responses Of Rural Households To Decreasing Fuelwood Availability, Ntcheu District, Malawi. *World Development*, *25*(2), 255–266. doi:10.1016/S0305-750X(96)00100-3

Brown, J. F., Mahoney, J. F., & Sivazlian, B. D. (1983). Hysteresis repair in discounted replacement problems. *IIE Transactions*, *15*(2), 156–165. doi:10.1080/05695558308974627

Brown, R. E., & Humphrey, B. G. (2005). Asset management for transmission and distribution. *IEEE Power and Energy Magazine*, *3*(3), 39–45. doi:10.1109/MPAE.2005.1436499

Brugaard Villmo, H. (2012). *Planning and management of megaprojects. Industrial Economics and Technology Management.* Norwegian University of Science and Technology.

BSI PAS 55. (2008). Publicly Available Specification, London.

Bueno Jauregui, R. (1991). *Las Claves de la Gestión de Producción.* Madrid, Spain: Editorial Alción.

Bueno, E. (2002a). Enfoques principales en Dirección del Conocimiento (Knowledge Management) y tendencias. En R. Hernández (Ed.), Gestión del Conocimiento: Desarrollos teóricos y aplicaciones. Cáceres: Ediciones La Coria, Fundación Xavier de Salas.

Bueno, E. (1999). *Gestión del conocimiento y capital intelectual' Experiencias en España.* Comunidad de Madrid- Euroforum Escorial Madrid.

Bueno, E. (2000). *La dirección del conocimiento en el proceso estratégico de la empresa: información, complejidad e imaginación en la espiral del conocimiento* (pp. 55–66). Euroforum Escorial.

Bueno, E. (2002b). *La sociedad del conocimiento: un nuevo espacio de aprendizaje de las personas y organizaciones, en La Sociedad del Conocimiento.* Presidencia de la Generalitat Valenciana, Valencia: Monografía de la Revista Valenciana de Estudios Autonómicos.

Buhari, M., & Adamu, S. (2012). Short-Term Load Forecasting Using Artificial Neural Network. *Proceedings of the International MultiConference of Engineers and Computer Scientist.* Retrieved from http://www.iaeng.org/publication/IMECS2012/IMECS2012_pp83-88.pdf

Bureau of Transport Economics (BTE). (2001). Economic Costs of Natural Disasters in Australia.

Busch, N. E., & Givens, A. D. (2012). Public-Private Partnerships in Homeland Security Opportunities and Challenges.

Bush, R. E., & Bale, C. S. E. (2014). Realising the Social Benefits of District Heating Through Strategic Planning. *Proceeding of the 14th International Symposium on District Heating and Cooling*, Stockholm, Sweden. Retrieved from http://www.svenskfjarrvarme.se/Global/Konferenser/DHC14/Proceedings/_2.2%20Ruth%20Bush%20REALISING%20THE%20SOCIAL%20BENEFITS%20OF%20DISTRICT%20HEATING%20THROUGH%20STRATEGIC%20PLANNING.pdf

Buxton, W. (2007): Sketching User Experience: Getting The Design Right And The Right Design. San Francisco, CA, USA: Morgan Kaufmann & Elsevier.

Cahill, B., Menzel, K., & Flynn, D. (2012). *BIM as a centre piece for optimised building operation. eWork and eBusiness in Architecture, Engineering and Construction.* Taylor & Francis Group.

Calder, S. (2010, December 19). Snow at Heathrow and Gatwick ruins plans of thousands. *BBC News*. Retrieved from http://www.bbc.com/news/uk-12030233

Campbell, J. D., & Jardine, A. K. S. (2001). *Maintenance Excellence*. New York, NY: Marcel Dekker.

Campbell, J. D., & Reyes-Picknell, J. V. (2015). *Uptime: Strategies for excellence in maintenance management*. CRC Press.

Caniato, F., & Größler, A. (2015). The moderating effect of product complexity on new product development and supply chain management integration. *Production Planning and Control*, *26*(16), 1306–1317. doi:10.1080/09537287.2015.1027318

Cárcel Carrasco, F. J. (2014a). *La gestión del conocimiento en la ingeniería del mantenimiento industrial: Investigación sobre la incidencia en sus actividades estratégicas. Omnia Science*. doi:10.3926/oms.197

Cárcel Carrasco, F. J. (2014b). *Planteamiento de un modelo de mantenimiento industrial basado en técnicas de gestión del conocimiento. Omnia Science*. doi:10.3926/oms.198

Carnero, M. C. (2005). Selection of diagnostic techniques and instrumentation in a predictive maintenance program. A case study. *Decision Support Systems*, *38*(4), 539–555. doi:10.1016/j.dss.2003.09.003

Carnero, M. C. (2006). An evaluation system of the setting up of predictive maintenance programmes. *Reliability Engineering & System Safety*, *91*(8), 945–963. doi:10.1016/j.ress.2005.09.003

Carnero, M. C. (2007). Model for the Selection of Predictive Maintenance Techniques. *INFOR*, *45*(2), 83–94.

Carvalho, R., & Lourenço, J. C. (2002). A multicriteria model for assigning new orders to service suppliers. *European Journal of Operational Research*, *139*(2), 390–399. doi:10.1016/S0377-2217(01)00367-8

Castanier, B., Grall, A., & Bérenguer, C. (2005). A condition-based maintenance policy with non-periodic inspections for a two-unit series system. *Reliability Engineering & System Safety*, *87*(1), 109–120. doi:10.1016/j.ress.2004.04.013

Cavalcante, C. A., & Lopes. (2015). V. R. S. Multi-criteria model to support the definition of opportunistic maintenance policy: A study in a cogeneration system. *Energy*, *80*, 32–40.

Cecelski, E. (2004). Re-thinking gender And energy: Old And new directions. ENERGIA/EASE Discussion Paper, Energy, Environment and Development. Retrieved from www.energia.org

Cegarra, J. & Moya, B. (2003). Orientadores del capital relacional. *Cuadernos Administración*, 16(26), 79-97.

CEN. (2001). *Maintenance Terminology. European Standard, EN 13306:2001*. Brussels: European Committee for Standardization.

Chan, F. T. S., & Prakash, A. (2012). Maintenance policy selection in manufacturing firms using the fuzzy MCDM approach. *International Journal of Production Research*, *50*(23), 7044–7056. doi:10.1080/00207543.2011.653451

Chang, S. E., McDaniels, T. L., Mikawoz, J., & Peterson, K. (2007). Infrastructure failure interdependencies in extreme events: Power outage consequences in the 1998 Ice Storm. *Natural Hazards*, *41*(2), 337–358. doi:10.1007/s11069-006-9039-4

Chan, J.-K., & Shaw, L. (1993). Modelling repairable systems with failure rates that depend on age and maintenance. *IEEE Transactions on Reliability*, *42*(4), 566–571. doi:10.1109/24.273583

Charmaz, K. (2006). *Constructing grounded theory. A practical guide through qualitative analysis*. London: SAGE.

Charniak, E. (1991). Bayesian Networks without Tears. *AI Magazine*, *12*(4), 50–63.

Chen, J., Chen, T. H. Y., Vertinsky, I., Yumagulova, L., & Park, C. (2013). Public–private partnerships for the development of disaster resilient communities. *Journal of Contingencies and Crisis Management*, *21*(3), 130–143. doi:10.1111/1468-5973.12021

Chen, M. Y., & Chen, A. P. (2006). Knowledge management performance evaluation: A decade review from 1995 to 2004. *Journal of Information Science*, *32*(1), 17–38. doi:10.1177/0165551506059220

Chen, W., Jiang, X., Li, X., Gao, J., Xu, X., & Ding, S. (2013). Wireless Sensor Network nodes correlation method in coal mine tunnel based on Bayesian decision. *Measurement*, *46-8*(8), 2335–2340. doi:10.1016/j.measurement.2013.04.018

Chen, Z. S., Yang, Y., & Hu, Z. (2012). A Technical Framework and Roadmap of Embedded Diagnostics and Prognostics for Complex Mechanical Systems in Prognostics and Health Management Systems. *IEEE Transactions on Reliability*, *61*(2), 314–322. doi:10.1109/TR.2012.2196171

Cheong, KFR. & Tsui, E. (2010a). The roles and values of personal knowledge management: an exploratory study. *VINE: The journal of information and knowledge management systems*, 40(2), 204–227.

Cheong, K. F. R., & Tsui, E. (2010b). Exploring the synergy between business process management and personal knowledge management. *Cutter IT Journal*, *23*(5), 28–33.

Chramcov, B. Baláte, Ja., Princ, M. (2010). Heat Demand Forecasting for Concrete District Heating System. International Journal of Mathematical Models and Methods in Applied Sciences, 4(4), 231-239.

Chua, A. Y. K., & Goh, D. H. (2008). Untying the knot of knowledge management measurement: A study of six public service agencies in Singapore. *Journal of Information Science*, *34*(3), 259–274. doi:10.1177/0165551507084139

Clarck, J. R. (1970). Fatigue Crack Growth Characteristics of Heavy Section. *ASTM A533 Grade B Class 1 Steel Weldments, ASME Paper N° 70-PVP 24*. American Society of Mechanical Engineers, New York.

Clark, J. R. (1974). How Fatigue Crack Initiation and Growth Properties Affect the Material Selection and Design Criteria. *Metals Engineering Quarterly*, Aug.

Claunch, J. W. (1996). *Set-up Time Reduction*. Chicago, USA: Irwin Professional Publishing.

Clintock, M. C. (1963). On the Plasticity of the Growth of Time Fatigue Cracks. In Fracture of Solids (p. 65). New York: Willey-Interscience.

Coble, J. B., & Hines, J. W. (2008, October 6-9). Prognostic algorithm categorization with PHM challenge application. *Paper presented at theInternational Conference on Prognostics and Health Management*, Denver, CO, USA.

Cohen, M. A., Agrawal, N., & Agrawal, V. (2006). Winning in the Aftermarket. *Harvard Business Review*, *84*(5), 129–138. PMID:16770900

Colino, J. & Martinez, J; Martínez-Carrasco, F. (2010). La gestión de la innovación en la industria. El caso de la región de Murcia. *Economía Industrial*, 377, 150-158.

Colino, J. & Riquelme, P. (2000). Estructura industrial y desarrollo tecnológico en la región de Murcia. *Economía Industrial*, 335/336, 171-183.

Construction Specification for Fixed Off-Shore Structures in the North Sea (1986). *The Engineering Equipment and Materials Users Association EEMUA*. Publication, *158*.

Contreras, J. (2014a). Gestión del mantenimiento. Una lista de indicadores o un sistema de indicadores de gestión. Retrieved from https://app.box.com/s/tnp0pw8kjo6s3o2vzek0

Contreras, J. (2014b). El Balanced Scorecard aplicado a la gestión del mantenimiento. Retrieved from https://app.box.com/s/syfy3mklpt8nca7iq2k9

Cooke-Davies, T. (2002). The "real" success factors on projects. *International Journal of Project Management*, *20*(3), 185–190. doi:10.1016/S0263-7863(01)00067-9

Cooper, R. J., Silva, H. F., & Trevisan, R. E. (2004). Influencia del precalentamiento en las propiedades de uniones soldadas de acero API 5L-X80 soldadas con alambre tubular autoprotegido. *Rev. Metal Madrid*, *40*(4), 280–270. doi:10.3989/revmetalm.2004.v40.i4.275

COSO. (2013). United States: Internal Control - Integrated Framework.

Crespo Márquez A., Gupta J. (2006). Contemporary Maintenance Management. Process, Framework and Supporting Pillars. *Omega: The international journal of management science*, 34(3), 325-338. DOI:.10.1016/j.omega.2004.11.003

Crespo Marquez, A. (2007). *The Maintenance Management Framework, Models and Methods for Complex Systems Maintenance*. London: Springer.

Crespo, A. (2007). *The Maintenance Management Framework: Models and Methods for Complex Systems Maintenance*. Springer Series in Reliability Engineering.

Crespo, A., Moreu de León, P., Sola, A., & Gómez, J. F. (2015). Criticality Analysis for Maintenance Purposes: A Study for Complex In-service Engineering Assets. *Quality and Reliability Engineering*, *02*. doi:10.1002/qre.1769

Crespo, A., & Sánchez, A. (2002). Models for maintenance optimization: A study for repairable system and finite time periods. *Reliability Engineering & System Safety*, *75*(3), 367–377. doi:10.1016/S0951-8320(01)00131-4

Cross, M. (1988). Engineering Maintenance Organisation Performance An Assessment of the Evidence from over 200 sites. *Management Research News*, *11*(1/2), 20–23. doi:10.1108/eb027938

Cuc, A. I. (2002). Vibration-Based Techniques for Damage Detection and Health Monitoring of Mechanical Systems. University of South Carolina.

Cumbane, J. (2007, December 5). Urban Environmental Management Needs And Priorities In Mozambique. *Proceedings of the SANORD Pre-Conference Workshop, Research and Knowledge for Development in Southern Africa*, Universidade Eduardo Mondlane, Department of Environmental and Geographical Sciences, Maputo.

Cumbe, F., Sharma, D., & Lucas, C. (2005). The Status of Clean Cooking Fuelsin Mozambique. Sydney: University of Technology. Retrieved from www.saee.ethz.ch

Custer, R. L., Scarcella, J. A., & Stewart, B. R. (1999). The modified Delphi technique: A rotational modification. *Journal of Vocational and Technical Education*, *15*(2), 1–10.

Cuvilas, C., Jirjis, R., & Lucas, C. (2010). Energy Situation in Mozambique: A Review. *Renewable & Sustainable Energy Reviews*, *14*(7), 2139–2146. doi:10.1016/j.rser.2010.02.002

da Costa, J., Navarro, A., Neves, J., & Martin, M. (2004). Household Wood And Charcoal Smoke Increases Risk Of Otitis Media In Childhood In Maputo. *International Journal of Epidemiology*, *33*(3), 573–578. doi:10.1093/ije/dyh071 PMID:15105407

Daniel, J., & Lauffenburger, J. P. (2014). Fusing navigation and vision information with the Transferable Belief Model: Application to an intelligent speed limit assistant. *Information Fusion*, *18*, 62–77. doi:10.1016/j.inffus.2013.05.013

Dasarathy, B. (1990). *Nearest neighbor: Pattern classification techniques (nn norms: Nn pattern classification techniques)*. IEEE Computer Society.

Dataware Technologies. (1998). *Seven Steps to Implementing Knowledge Management in Your Organization*. Manual de trabajo.

Davenport, H. (1997). Ten principles of knowledge management and four case studies. *Knowledge and Process Management*, *4*(3), 187–208. doi:10.1002/(SICI)1099-1441(199709)4:3<187::AID-KPM99>3.0.CO;2-A

De Groote, P. (1995). Maintenance performance analysis: A practical approach. *Journal of Quality in Maintenance Engineering*, *1*(2), 4–24. doi:10.1108/13552519510089556

de Leeuw, J., & Pries, F. (2014, June4). Deltaplan voor infrastructuur nodig: Verouderde infrastructuur in Nederland vereist gezamenlijke aanpak op verschillende niveaus. In *Het Financieele Dagblad* (p. 10). Opinie & Dialoog.

De Quevedo, E., De La Fuente, J., & Delgado, J. (2005). Reputación Corporativa y Creación de Valor. Marco Teórico De Una Relación Circular. *Investigaciones Europeas de Dirección y Economía de la Empresa*, *11*(2), 81–97.

Del Moral, A. (2007). *Gestión del Conocimiento*. Thompson Editores. España.

Dempster, A. (1967). Upper and lower probabilities induced by a multivalued mapping. *Annals of Mathematical Statistics*, *38*(2), 325–329. doi:10.1214/aoms/1177698950

Denoeux, T. (2008). Conjunctive and disjunctive combination of belief functions induced by non-distinct bodies of evidence. *Artificial Intelligence*, *172*(2-3), 234–264. doi:10.1016/j.artint.2007.05.008

Deshmukh, A., Ho Oh, E., & Hastak, M. (2011). Impact of flood damaged critical infrastructure on communities and industries. *Built Environment Project and Asset Management*, *1*(2), 156–175. doi:10.1108/20441241111180415

Dhouib, D. (2014). An extension of MACBETH method for a fuzzy environment to analyze alternatives in reverse logistics for automobile tire wastes. *Omega*, *42*(1), 25–32. doi:10.1016/j.omega.2013.02.003

Dibsdale, C. (2011). The Development of Maintenance and the Role of Integrated Vehicle Health Management. In I. K. Jennions (Ed.), *Integrated Vehicle Health Management: Perspectives on an Emerging Field* (pp. 9–16). Warrendale, PA, USA: SAE International.

Dibsdale, C. (2015). Holistic Prognostics. In R. Rajkumar & L. Redding (Eds.), *Through-life Engineering Services* (pp. 71–82). New York: Springer.

Dijoux, Y., Doyen, L., & Gaudoin, O. (2005) Simultaneous assessment of the efficiency of preventive and corrective maintenance for repairable systems in a competing risk framework. *Presented at 5th European safety and Reliability Conference,* Tri City (Gdynia-Sopot-Gdansk), Poland

District Heating Network Monitoring. Control and Optimization with CyberVille IoE Application Platform. (n. d.). Retrieved from http://www.iiconsortium.org/case-studies/Cyberlightning_Fortum_Case_Study.pdf

District Heating System Management Guide. (n. d.). Retrieved from http://projects.bre.co.uk/DHCAN/pdf/Management.pdf

Ditzel, B. (2005). *Desarrollo de un modelo de gestión del conocimiento para un departamento universitario. Tesis doctoral*. Campus Tecnológico de la Universidad de Navarra. Escuela Superior de Ingenieros de San Sebastián.

Dorfman, P. (2001). The accidental knowledge manager. *Knowledge Management*, *4*(2), 36–41.

Douglas, D. (2004). Grounded theory and the 'And' in entrepreneurship research. *Electronic Journal of Business Research Methods*, 2(2).

Doyen, L., & Gaudoin, O. (2004). Classes of imperfect repair models based on reduction of failure intensity or virtual age. *Reliability Engineering & System Safety*, *84*(1), 45–56. doi:10.1016/S0951-8320(03)00173-X

Doyen, L., & Gaudoin, O. (2011). Modelling and assessment of aging and efficiency of corrective and planned preventive maintenance. *IEEE Transactions on Reliability*, *60*(4), 759–769. doi:10.1109/TR.2011.2171115

Druzdel, M. J., & Flynn, R. R. (2002). Decision Support Systems. In A. Kent (Ed.), Encyclopedia of Library and Information Science, (2nd ed., Vol. 67). New York: Dekker, M.

Dubois, D., & Prade, H. (1988). Representation and combination of uncertainty with belief functions and possibility measures. *Computational Intelligence*, *4*(3), 244–264. doi:10.1111/j.1467-8640.1988.tb00279.x

Dubois, D., & Prade, H. (2009). Formal representation of uncertainty. In D. Bouyssou, D. Dubois, M. Pirlot, & H. Prade (Eds.), *Decision-Making Process*. UK and Wiley. doi:10.1002/9780470611876.ch3

Duenyas, I., Gupta, D., & Olsen, T. L. (1998). Control of a Single-Server Tandem Queueing System with Setups. *Operations Research*, *46*(2), 218–230. doi:10.1287/opre.46.2.218

Duenyas, L., & Van Oyen, M. (1996). Heuristic Scheduling of Parallel Heterogeneous Queues with Set-Ups. *Management Science*, *42*(6), 817–829. doi:10.1287/mnsc.42.6.814

Duffuaa, S. O. (2000). Mathematical models in maintenance planning and scheduling. In M. Ben-Daya, S. O. Duffuaa, & A. Raouf (Eds.), *Maintenance, Modelling and Optimization*. Boston, MA: Kluwer Academic Publishers. doi:10.1007/978-1-4615-4329-9_2

Duncan, G. L., & Gorsha, R. A. (1983). Project management: A major factor in project success. *IEEE Transactions on Power Apparatus and Systems*, *102*(11), 3701–3705. doi:10.1109/TPAS.1983.317755

Dunn-Cavelty, M., & Suter, M. (2009). Public–Private Partnerships are no silver bullet: An expanded governance model for Critical Infrastructure Protection. *International Journal of Critical Infrastructure Protection*, *2*(4), 179–187. doi:10.1016/j.ijcip.2009.08.006

Dvir, D., Raz, T., & Shenhar, A. J. (2003). An empirical analysis of the relationship between project planning and project success. *International Journal of Project Management*, *21*(2), 89–95. doi:10.1016/S0263-7863(02)00012-1

Edvinsson, L., & Malone, M. S. (1997). *Intellectual Capital: Realizing Your Company's True Value by Finding its Hidden Brainpower*. New York: Harper Business.

Edwards, I. (2011). *Overcoming challenges for the offshore wind industry and learning from the oil and gas industry*. Report from POWER cluster, natural power and the Interreg IVB North Sea Region Programme.

EFNMS. (2006). The requirements of Competencies and Responsibilities for an European Specialist in Maintenance Supervision.

Eguren Martí, M. L., & Castán Farrero, J. M. (2011). La cadena de valor bajo un enfoque sistémico y sus implicaciones en el análisis de la creación de valor en la empresa. *Proceedings of the XXI Congreso Nacional de la Asociación Científica de Economía y Dirección de la Empresa (ACEDE)*. Barcelona: ACEDE.

Eich, D. (2008). A Grounded Theory of High-Quality Leadership Programs: Perspectives From Student Leadership Development Programs in Higher Education. *Journal of Leadership & Organizational Studies*, *15*(2), 176–187. doi:10.1177/1548051808324099

Eisenhardt, K. (1989). Building theories from case studies research. *Academy of Management Review*, *14*(4), 532–550.

El Haddad, M. H., Dowling, N. E., Topper, T. H., & Smith, K. N. (1980). J Integral Application for Short Fatigue Cracks at Notches. *International Journal of Fracture*, *16*(1), 15–30. doi:10.1007/BF00042383

Elber, W. (1971). *The Significance of Crack Closure. Damage Tolerance Aircraft Structure. ASTM STP 486* (pp. 230–242). Philadelphia: American Society for Testting and Materials. doi:10.1520/STP26680S

Elices, M. (1992). Mecánica De La Fractura Aplicada A Sólidos Elásticos Bidimensionales. *Departamento de Ciencias de Materiales E.T. Superior de Ingenieros de Caminos de Madrid,* January.

Ellegard, A. (1996). Cooking Fuel Smoke and Respiratory Symptoms Among Women in Low-income Areas in Maputo. *Environmental Health Perspectives*, *104*(9), 980–985. doi:10.1289/ehp.96104980 PMID:8899378

Emans, B. (2002). *Interviewen: theorie, techniek en training* (4th ed.). Groningen: Stenfert Kroese.

Emblemsvag, J., & Tonning, L. (2003). Decision support in selecting maintenance organization. *Journal of Quality in Maintenance Engineering*, *9*(1), 1–24. doi:10.1108/13552510310466765

EN 13306:2001. (2001). Maintenance terminology. *European Standard. CEN. European Committee for Standardization*, Brussels.

EN 15221-1. (2006). European standard in facility management–Part 1: Terms and definitions.

Energy Management and Remote Energy Management . (n. d.). Retrieved from http://www.power-protection.co.uk/energy-management-and-remote-energy-management/

ENISA. (2011). Cooperative models for effective Public-Private Partnerships: Good Practice Guide. ISBN: 978-92-9204-054-3

Enoro Solutions . (n. d.). Retrieved from http://www.enoro.com/#!solutions/component_74511

Erkoyuncu, J. A., Khan, S., Hussain, S., & Roya, R. (2015). A framework to estimate the cost of No-Fault Found events. *International Journal of Production Economics,* 173, 207–222.

Eskom. (n.d.). Eskom load shedding. Retrieved September 30, 2015, from http://loadshedding.eskom.co.za/LoadShedding/Description

EU Commission. (2005). Green Paper on a European Programme for Critical Infrastructure Protection COM (2005) 576.

Eurobserver. (2014). *Wind energy barometer.*

Euroheat & Power Task Force Customer Installations . (2011). District Heating in Buildings. Retrieved from http://www.euroheat.org/Files/Filer/documents/Publications/District%20Heating%20in%20buildings_final.pdf

European Federation of National Maintenance Societies (2006). *The Requirements of Competencies and Responsibilities for a European Specialist in Maintenance Supervision.*

European Federation of National Maintenance Societies. (2007). *EN-15341 - Maintenance Key Performance Indicators.*

Eusgeld, I., Nan, C., & Dietz, S. (2011). "System-of-systems" approach for interdependent critical infrastructures. *Reliability Engineering & System Safety*, *96*(6), 679–686. doi:10.1016/j.ress.2010.12.010

Ewald's, H. L., & Wanhill, R. J. H. (1991). *Fracture Mechanics.* London: Edward Arnold.

EWEA. (2009). *Wind at Work: Wind energy and job creating in the EU.*

EWEA. (2011). *Wind in our Sails: the coming of Europe's offshore wind energy industry report.*

FAO. (2005). Special report: Crop and Food Supply Assessment Mission to Mozambique, Food and Agricultire Organization, Rome. Retrieved from www.fao.org

Farrar, C. R., & Lieven, N. A. J. (2007). Damage prognosis: the future of structural health monitoring. *Philosophical Transactions of the Royal Society A: Mathematical, Physical and Engineering Sciences, 365*(1851), 623-632. doi: 10.1098/rsta.2006.1927

Febles, J., & Oreja, J. (2008). Factores externos e internos determinantes de la orientación de la cultura estratégica de las empresas. *Investigaciones Europeas de Dirección y Economía de la Empresa, 14*(1), 13–32. doi:10.1016/S1135-2523(12)60009-4

Feng, M., Li, C., & McVay, S. (2009). Internal control and management guidance. *Journal of Accounting and Economics, 48*(2-3), 190–209. doi:10.1016/j.jacceco.2009.09.004

Fernandez, E., Avella, L., & Fernandez, M. (2006). Estrategia de Producción (2nd ed.). Madrid, España: Editorial McGraw Hill.

Fernández, J. C., & Canals, X. (2006). Quality standards for medical devices. In Proccedings of the XV Winter Course Standards in Telemedicine CATAI (pp. 133-140). Tenerife, Spain.

Ferrada, X., & Serpell, A. (2009). La Gestión del Conocimiento y la Industria de la Construcción. *Revista de la Construcción, 8*(1), 46–58.

Ferri, F. A. S., Rodrigues, L. R., Gomes, J. P. P., de Medeiros, I. P., Galvao, R. K. H., & Nascimento, C. L. (2013, April 15-18). Combining PHM information and system architecture to support aircraft maintenance planning. *Paper presented at the 2013 IEEE International Systems Conference (SysCon).*

Flammini, F., Marrone, S., Mazzocca, N., Nardone, R. & Vittorini, V. (2015). Using Bayesian Networks to evaluate the trustworthiness of '2 out of 3' decision fusion mechanisms in multi-sensor applications, *IFAC-PapersOnLine*, 48-21, 682-687.

Fleming, K. N. (2004). Markov models for evaluating risk-informed in-service inspection strategies for nuclear power plant piping systems. *Reliability Engineering & System Safety, 83*(1), 27–45. doi:10.1016/j.ress.2003.08.009

Flusser, V. (2010). *A Filosofia Do Design, A Forma Das Coisas*. Lisboa: Relógio D'água. [In Portuguese]

Flyvbjerg, B. (2011). Over budget, over time, over and over again: managing major projects. In Morris, P.W.G., Pinto, J.K. & Söderlund, J. (2011) The Oxford Handbook of Project Management (321–344). Oxford: Oxford University Press. doi:10.1093/oxfordhb/9780199563142.003.0014

Flyvbjerg, B. (2009). Survival of the unfittest: Why the worst infrastructure gets built and what we can do about it. *Oxford Review of Economic Policy, 25*(3), 344–367. doi:10.1093/oxrep/grp024

Flyvbjerg, B., Bruzelius, N., & Rothengatter, W. (2003). *Megaprojects and risk: an anatomy of ambition*. Cambridge: Cambridge University Press. doi:10.1017/CBO9781107050891

Forman, E., & Peniwati, K. (1998). Aggregating individual judgments and priorities with the Analytic Hierarchy Process. *European Journal of Operational Research, 108*(1), 165–169. doi:10.1016/S0377-2217(97)00244-0

Frangopol, D. M., Saydam, D., & Kim, S. (2012). Maintenance, management, life-cycle design and performance of structures and infrastructures: A brief review. *Structure and Infrastructure Engineering, 8*(1), 1–25. doi:10.1080/15732479.2011.628962

Frikha, A. & Moalla, H. (2014). Analytic hierarchy process for multi-sensor data fusion based on belief function theory. *European Journal of Operational Research*, 241-1, 133-147.

Frost. (1960). Notch Effects and the Critical Alternating Stress Required to Propagate a Crack in an Aluminium Alloy Subjected to Fatigue Loading. *Journal of the Mechanical Engineering Society, 2*(2), 109-119.

Fukuda, S. (2014). BIM for FM: A Case Support for Business Life Cycle. *PLM 2014. IFIP AICT*, *442*, 63–74.

Fund, M. I. (2004). *Public-Private Partnerships*. Fiscal Affairs Department.

Funtowicz, S., & Ravetz, J. (1993). Science For The Post-Normal Age. *Futures*, *25*(7), 739–755. doi:10.1016/0016-3287(93)90022-L

Gabrielaitiene, I., Kacianauskas, R., & Sunden, B. (2003). Thermo-hydraulic Finite Element Modelling of District Heating Network by the Uncoupled Approach. *Journal of Civil Engineering and Management*, *IX*(3), 153–162. doi:10.1080/13923730.2003.10531321

Gallaher, M. P., O'Connor, A. C., Dettbarn, J. L., & Gilday, L. T. (2004). Cost Analysis of Inadequate Interoperability in the U.S. Capital Facilities Industry. *National Institute of Standards and Technology*. Retrieved from http://fire.nist.gov/bfrlpubs/build04/PDF/b04022.pdf

Gane, M. (2007). Strategy. In M. Gane (Ed.), *Forest Strategy, Strategic Management and Sustainable Development for the Forest Sector* (pp. 266–392). Dordrecht: Springer.

García, M., & Leal, M. (2008). Evolución histórica del factor humano en las organizaciones: De recurso humano a capital intelectual. *Omnia*, *14*(3), 144–159.

Garg, A., & Deshmukh, S. G. (2006). Maintenance management: Literature review and directions. *Journal of Quality in Maintenance Engineering*, *12*(3), 205–238. doi:10.1108/13552510610685075

Garrido, C. (2007). La educación desde la teoría del capital humano y el otro. *Educere*, *11*(36), 73–80.

Geisler, E. (2007). A typology of knowledge management: Strategic groups and role behavior in organizations. *Journal of Knowledge Management*, *11*(1), 84–96. doi:10.1108/13673270710728259

Gelders, L., Mannaerts, P., & Maes, J. (1994). Manufacturing strategy, performance indicators and improvement programmes. *International Journal of Production Research*, *32*(4), 797–805. doi:10.1080/00207549408956971

Gerdes, G., Tiedemann, A. & Zeelenberg, S. (2005). *Case study: European offshore wind farms a survey for the analysis of the experiences and lessons learnt by developers of offshore wind farms - final report.* Report.

Ghosh, D., & Roy, S. (2009). A decision-making framework for process plant maintenance. *European Journal of Industrial Engineering*, *4*(1), 78–98. doi:10.1504/EJIE.2010.029571

Giezen, M. (2012). Keeping it simple? A case study into the advantage and disadvantages of reducing complexity in megaproject planning. *International Journal of Project Management*, *30*(7), 781–790. doi:10.1016/j.ijproman.2012.01.010

Gil, M.; Pérez, A. & López, G. (2008). La auditoría del conocimiento como etapa previa a la gestión del conocimiento en una institución educativa Mexicana. *Revista Ciencia Administrativa*, 2008(2), 17-27.

Gilabert, E., Fernandez, S., Arnaiz, A., & Konde, E. (2015) Simulation of predictive maintenance strategies for cost-effectiveness analysis, *Proceedings of the Institution of Mechanical Engineers, Part B: Journal of Engineering Manufacture*. doi:10.1177/0954405415578594

Glaser, B. G., & Strauss, A. L. (1967). *The discovery of grounded theory*. New York: Aldine deGruyter.

Goh, S. C. (2002). Managing effective knowledge transfer: An integrative framework and some practice implications. *Journal of Knowledge Management*, *6*(1), 23–30. doi:10.1108/13673270210417664

Goldemberg, J., & Johansson, T. (Eds.). (1995). Overview. In J. Goldemberg, & T. Johansson (Eds.), Energy As An Instrument for Socio-Economic Development (pp. 1-9). New York: United Nations Development Programme (UNDP).

Gómez de León, F. C., & Ruiz, J. J. (2006). Maintenance strategy based on a multicriterion classification of equipments. *Reliability Engineering & System Safety*, *91*(4), 444–451. doi:10.1016/j.ress.2005.03.001

Gómez, J. F., & Crespo, A. (2009). Framework for implementation of maintenance management in distribution network service providers. *Reliability Engineering & System Safety*, *94*(10), 1639–1649. doi:10.1016/j.ress.2009.04.003

González-Palma, R., Suárez-Bermejo, J. C., & Vicario, F. J., & Aurelio Muñoz. (2006). Procesos metalúrgicos de aceros microaleados de alta resistencia 450 EMZ tipo II en la zona afectada térmicamente debido al efecto multipasada en soldaduras por arco sumergido. *Rev. Metal. Madrid*, *42*, 256–269.

González, R., & García, A. (2010). Innovación abierta: Un modelo preliminar desde la gestión del conocimiento. *Intangible Capital*, *7*(1), 82–115. doi:10.3926/ic.2011.v7n1.p82-115

Goossens, A. J. M., & Basten, R. J. I. (2015). Exploring maintenance policy selection using the Analytic Hierarchy Process; An application for naval ships. *Reliability Engineering & System Safety*, *142*, 31–41. doi:10.1016/j.ress.2015.04.014

Gosselin, S. R., & Fleming, K. N. (1997). Evaluation of pipe failure potential via degradation mechanism assessment. *Proceedings of the 5th International Conference on Nuclear Engineering (INCONE)*. Nice, France.

Granderson, J., Piette, M. A., Ghatikar, G., & Price, P. (2009) Building Energy Information Systems: State of the Technology and User Case Studies. Retrieved from http://eis.lbl.gov/pubs/lbnl-2899e.pdf

Gray, P. H., & Meister, D. B. (2006). Knowledge sourcing methods. *Information & Management*, *43*(2), 142–156. doi:10.1016/j.im.2005.03.002

Greitzer, F. L., Hostick, C. J., Rhoads, R. E., & Keeney, M. (2001). Determining how to do prognostics, and then determining what to do with it. *Paper presented at theIEEE Systems Readiness Technology ConferenceAUTOTESTCON '01.*

Grubic, T., Redding, L., Baines, T., & Julien, D. (2011). The adoption and use of diagnostic and prognostic technology within UK-based manufacturers. *Proceedings of the Institution of Mechanical Engineers. Part B, Journal of Engineering Manufacture*, *225*(8), 1457–1470. doi:10.1177/0954405410395858

Grzenda, M. (2012). Consumer-oriented heat consumption prediction. *Control and Cybernetics*, *41*(1), 213–240.

Guerras, L., & Navas, J. (2015). *La dirección estratégica de la empresa. Teoría y aplicaciones*. Thomson Reuters.

Haddad, G., Sandborn, P. A., & Pecht, M. G. (2014). Using maintenance options to maximize the benefits of prognostics for wind farms. *Wind Energy (Chichester, England)*, *17*(5), 775–791. doi:10.1002/we.1610

Haffejee, M., & Brent, A. C. (2008). Evaluation of an integrated asset life-cycle management (ALCM) model and assessment of practices in the water utility sector. *Water S.A.*, *34*(2), 285–290.

Hall, M., Frank, E., Holmes, G., Pfahringer, B., Reutemann, P., & Witten, I. H. (2013). Weka 3: Data mining and open source machine learning software in java. Retrieved from http://www.cs.waikato.ac.nz/ml/weka/

Hallegatte, S., & Przyluski, V. (2010). The economics of natural disasters: concepts and methods. *World Bank Policy Research Working Paper Series 5507.*

Hamel, G., & Prahalad, C. (1989). Strategic intent. *Harvard Business Review*, 1989, 63–76. PMID:10303477

Handfield, R. B., & Melnyk, S. A. (1998). The scientific theory-building process: A premier using the case of TQM. *Journal of Operations Management*, *16*(4), 312–339. doi:10.1016/S0272-6963(98)00017-5

Hanlon, J., & Cunguara, B. (2010). *Poverty is not being reduced in Mozambique*. LSE Crisis States Research Centre.

Hansen, B., & Kautz, K. (2004). Knowledge zapping: A technique for identifying knowledge flows in software organisations. *EuroSPI.*, *2004*, 126–137.

Hansen, M. T., Nohria, N., & Tierney, T. (1999). What's your strategy for managing knowledge? *Harvard Business Review*, (March-April): 106–116. PMID:10387767

Hao, Yu., Hao, Yu., & Wilamowski, B. M. (2011). Levenberg–Marquardt Training. In Irwin J. David (Ed.), The Industrial Electronic Handbook: Intelligent Systems. USA: Taylor & Francis Group, CRC Press.

Hardy, M., & Bryman, A. (2004). *Handbook of data analysis*. London: Sage Publications. doi:10.4135/9781848608184

Harmon, R. L., & Peterson, L. D. (1990). *Reinventar la Fábrica*. Madrid, Spain: Ciencias de la Dirección.

Harrington, M. (2002, August 1). EMERGENCY: The critical condition of hospital power. Retrieved from http://ecmweb.com/cee-news-archive/emergency-critical-condition-hospital-power

Harris, F. J. (1978). On the use of windows for harmonic analysis with the discrete fourier transform. *Proceedings of the IEEE*, *66*(1), 51–83. doi:10.1109/PROC.1978.10837

Harrison, J. D. (1970). An Analysis of Data on non Propagating Fatigue Cracks on a Fracture Mechanics Basis, *British. Welding Journal*, *2*(3).

Hedblom, M. (2013). *A Cognition-Based Definition of Creativity and A Proposition for Approaching Creativity Artificially* [Master Thesis]. Institute of Computer and Information Science, Linköping University.

Hedlud, O. (1994). A model of knowledge management and the N-form corporation. *Strategic Management Journal*, *15*(Suppl. 2), 73–90. doi:10.1002/smj.4250151006

Heise, D., Strecker, S., & Frank, U. (2014). ControlML: A domain-specific modeling language in support of assessing internal controls and the internal control system. *The International Journal of Accounting*.

Henderson, J., & Venkatraman, N. (1999). Strategic Alignment: Leveraging Information Technology for Transforming Organisations. *IBM Systems Journal*, *38*(2/3), 472–484. doi:10.1147/SJ.1999.5387096

Heracleous, L. (1998). Strategic Thinking Or Strategic Planning? *Long Range Planning*, *31*(3), 481–487. doi:10.1016/S0024-6301(98)80015-0

Herr, K. (2013). *Quick Changeover Concepts Applied*. New York, USA: Productivity Press.

Hinton, G. E., Osindero, S., & Teh, Y. (2006). A fast learning algorithm for deep belief nets. *Neural Computation*, *18*(7), 1527–1554. doi:10.1162/neco.2006.18.7.1527 PMID:16764513

Hinton, G. E., & Salakhutdinov, R. R. (2006). Reducing the dimensionality of data with neural networks. *Science*, *313*(5786), 504–507. doi:10.1126/science.1127647 PMID:16873662

History.com Staff. (2009). Hurricane Katrina. Retrieved from http://www.history.com/topics/hurricane-katrina

Hlebnikov, A., Dementjeva, N., & Siirde, A. (2009). Optimization of Narva District Heating Network and Analysis of Competitiveness of Oil Shale CHP Building in Narva. *Oil Shale*, *26*(3), 269–282. doi:10.3176/oil.2009.3S.09

Hochreiter, S., Bengio, Y., Frasconi, P., Schmidhuber, J., & Elvezia, C. (2001). Gradient flow in recurrent nets: the difficulty of learning long-term dependencies. In S. C. Kremer & J. F. Kolen (Eds.), *A Field Guide to Dynamical Recurrent Neural Networks* (pp. 237–244). New York, NY: IEEE Press.

Hodge, G. A., & Greve, C. (2007). Public–private partnerships: An international performance review. *Public Administration Review*, *67*(3), 545–558. doi:10.1111/j.1540-6210.2007.00736.x

Holdeman, E. (2015, May 18). Critical Infrastructure Maintenance: It's Not Sexy But It's Necessary. *Emergency Management.com.* Retrieved from http://www.emergencymgmt.com/emergency-blogs/disaster-zone/Maintenanceofcriticalinfrastructure.html

Holmström, J., Ketokivi, M., & Hameri, A. (2009). Bridging practice and theory: A design science approach. *Decision Sciences*, *40*(1), 65–87. doi:10.1111/j.1540-5915.2008.00221.x

Holsapple, C., & Jones, K. (2004). Exploring primary activities in the knowledge chain. *Knowledge and Process Management*, *11*(3), 155–174. doi:10.1002/kpm.200

Hooff, B., & Huysman, M. (2009). Managing knowledge sharing: Emergent and engineering approaches. *Information & Management*, *46*(1), 1–8. doi:10.1016/j.im.2008.09.002

Hsu, C.-C., & Sandford, B. A. (2007). The Delphi Technique: Making Sense of Consensus. *Practical Assessment, Research & Evaluation*, *12*(10), 1–8.

Huang, E. G. T. (2011) Understanding the Requirements of the Energy Management System Certification. Retrieved from http://www.sgs.com/~/media/Global/Documents/White%20Papers/sgs-energy-management-whitepaper-en-11.ashx

Hubbard, D. (2007). *How to measure anything: finding the value of intangibles in business*. Hoboken, NJ: Wiley.

Huiskonen, J. (2001). Maintenance spare parts logistics: Special characteristics and strategic choices. *International Journal of Production Economics*, *71*(1-3), 125–133. doi:10.1016/S0925-5273(00)00112-2

Hwang, J., & Singh, M. R. (1998). Optimal Production Policies for Multi-Stage Systems with Setup Costs and Uncertain Capacities. *Management Science*, *44*(9), 1279–1998. doi:10.1287/mnsc.44.9.1279

IAM, & BSI. (2008). *Asset management - Publicly available standard.* IAM & BSI.

IAM. The Institute of Asset Management. (2013). What is asset management? Retrieved from http://theiam.org/what-asset-management

IEA. (2006). Energy For Cooking. In *Developing Countries, in IEA: World Energy Outlook 2006* (pp. 419–446). Paris: International Energy Agency, Organisation for Economic Cooperation and Development.

IFMA, International Facilities Management Association. (2013). What is FM? Retrieved from http://www.ifma.org/know-base/browse/what-is-fm

Ilangkumaran, M., & Kumanan, S. (2009). Selection of maintenance policy for textile industry using hybrid multi-criteria decision making approach. *Journal of Manufacturing Technology Management*, *20*(7), 1009–1022. doi:10.1108/17410380910984258

INE. (2007). *Third General Population and Housing Census- 2007.* Maputo: Instituto Nacional de Estatística.

Infogrid. (2011). InfoGrid Web Graph Database. Retrieved from http://infogrid.org

Infor EAM Energy Performance Management (n. d.). Retrieved from http://www.infor.com/product-summary/eam/energy-performance-management/

International Electrotechnical Commission. (2009). *IEC 60300-3-11 Dependability Management – Part 3:11 Application guide – Reliability centered maintenance.*

Ishikawa, K., & Ishikawa, K. (1982). *Guide to quality control* (Vol. 2). Asian Productivity Organization Tokyo.

Ishizaka, A., & Nemery, P. (2014). Assigning machines to incomparable maintenance strategies with ELECTRE-SORT. *Omega, 47*, 45–59. doi:10.1016/j.omega.2014.03.006

ISO 50001 : 2011 (en) Energy management systems – Requirements with guidance for use. (n. d.). Retrieved from https://www.iso.org/obp/ui/#iso:std:iso:50001:ed-1:v1:en

ISO 55000. (2014). Asset management — Overview, principles and terminology.

ISO 55001. (2014). Asset management — Management systems — Requirements.

ISO 55002. (2014). Asset management — Management systems — Guidelines on the application of ISO 55001.

ISO. (2013). ISO 55000 Asset Management - Overview, principles and terminology.

ISO. (2014). *ISO 55000 - Asset management*. Geneva, Switzerland: International Standards Organization.

ISO. (2014). ISO 55001 Asset management — Management systems — Requirements.

ISO. (2014). ISO 55002 Asset management — Management systems — Guidelines on the application of ISO 55001.

Ittner, C., & Oyon, D. (2014). *The Internal Organization of Enterprise Risk Management.*

Jaffe, E. (2015, February 12). America's Infrastructure Crisis Is Really a Maintenance Crisis. *CityLab.com.* Retrieved from http://www.citylab.com/cityfixer/2015/02/americas-infrastructure-crisis-is-really-a-maintenance-crisis/385452/

Jagger, P., & Shively, G. (2014). Land Use Change, Fuel Use And Respiratory Health In Uganda. *Energy Policy, 67*, 713–726. doi:10.1016/j.enpol.2013.11.068 PMID:24535892

James, I. J., Marshall, J., & Walls, L. (2002). Improving design for reliability with in-service data analysis. *Proceedings of theReliability and Maintainability Symposium '02* (pp. 417-422). IEEE. doi:10.1109/RAMS.2002.981677

James, I., Lumbard, D., Willis, I., & Goble, J. (2003). Investigating no fault found in the aerospace industry. Proceedings of the Reliability and Maintainability Symposium (pp. 441-446). doi:10.1109/RAMS.2003.1182029

Jamsek, M., Dobersek, D., Goricanec, D., Krope, J. (2010). Determination of Optimal District Heating Pipe Network Configuration. *WSEAS Transactions on Fluid Mechanics*, 3(5), 167-174

Jardine, A. (1999). Measuring maintenance performance: A holistic approach. *International Journal of Operations & Production Management, 19*(7), 691–715. doi:10.1108/01443579910271674

Jardine, A. K. S., Lin, D., & Banjevic, D. (2006). A review on machinery diagnostics and prognostics implementing condition-based maintenance. *Mechanical Systems and Signal Processing, 20*(7), 1483–1510. doi:10.1016/j.ymssp.2005.09.012

Jensen, F., & Nielsen, T. (2007). *Bayesian networks and decision graphs.* Springer. doi:10.1007/978-0-387-68282-2

Jetter, J., Zhao, Y., Smith, K., Khan, B., Yelverton, T., DeCarlo, P., & Hays, M. D. (2012). Pollutant Emissions And Energy Efficiency Under Controlled Conditions For Household Biomass Cookstoves And Implications For Metrics Useful In Setting International Test Standards. *Environmental Science & Technology, 46*(19), 10827–10834. doi:10.1021/es301693f PMID:22924525

Jianhui, L., Namburu, M., Pattipati, K., Qiao, L., Kawamoto, M., & Chigusa, S. (2003). Model-based prognostic techniques [maintenance applications]. *Paper presented at theIEEE Systems Readiness Technology ConferenceAUTOTESTCON '03*, Anaheim, CA, USA.

Jimenez-Gálvez, J. (2015, July 11). El riesgo mortal de volver a las carreteras. *El Pais.* Retrieved from http://politica.elpais.com/politica/2015/07/11/actualidad/1436571260_237661.html

Johansen, E., & Wilson, B. (2006). Investigating first planning in construction. *Construction Management and Economics, 24*(December), 1305–1314. doi:10.1080/01446190600863160

Johansson, C. (2010). Towards Intelligent District Heating. Retrieved from https://www.diva-portal.org/smash/get/diva2:833722/FULLTEXT01.pdf

Johnson, J., Karen, D., Boucher, K. C., & Robinson, J. (2001). *Collaborating on project success. Software Magazine.* February/March.

Jokipii, A. (2010). Determinants and consequences of internal control in firms: a contingency theory based analysis. *Journal of Management & Governance.*

Jones, J., Marshall, J., Aulak, G., & Newman, B. (2003). Development of an expert system for reliability task planning as part of the REMM methodology. *Proceedings of theReliability and Maintainability Symposium, 2003. Annual* (pp. 423-428). IEEE. doi:10.1109/RAMS.2003.1182026

Jones, J., & Hayes, J. (2001). Investigations of the occurrence of: No-faults-found in electronic systems. *IEEE Transactions on Reliability, 50*(3), 289–292. doi:10.1109/24.974126

Jones, R. (1995). *Risk-based management: A Reliability-Centered Approach* (1st ed.). Houston, Texas: Gulf Publishing Company.

Jongen, R. A. (2012). *Statistical lifetime management for energy network components.* Delft, The Netherlands: Technische Universiteit Delft.

Jongepier, A. G. (2007). *Ageing assets - Consume, prolong or replace. Leonardo Energy.* Arnhem, The Netherlands: KEMA Consulting.

Junginger, M., Faaij, A., & Turkenburg, W. C. (2004). Cost reduction prospects for offshore wind farms. *Wind Engineering, 28*(1), 97–118. doi:10.1260/0309524041210847

Jurow, S. (1996). Core competencies: Strategic thinking about the work we choose to do. *Journal of Academic Librarianship, 22*(4), 300–302. doi:10.1016/S0099-1333(96)90122-6

Kaplan, R. S., & Norton, D. P. (1992). The balanced scorecard – measures that drive performance. *Harvard Business Review, 70*(1), 71–79. PMID:10119714

Kaplan, R. S., & Norton, D. P. (1992). The Balanced Scorecard - measures that drive performance. *Harvard Business Review, 70*(1), 71–79. PMID:10119714

Kaplan, R. S., & Norton, D. P. (1992). The Balanced Scored – measures that drive performance. *Harvard Business Review, 70*(1), 71–79. PMID:10119714

Kaplan, R. S., & Norton, D. P. (1992). Using the Balanced Scorecard as a Strategic Management System. *Harvard Business Review, 74*(1), 75–85.

Karlsson, C. (2009). *Researching Operations Management.* New York: Routledge.

Kassem, M., Kelly, G., Dawood, N., Serginson, M., & Lockley, S. (2015). BIM in facilities management applications: A case study of a large university complex. *Built Environment Project and Asset Management*, *5*(3), 261–277. doi:10.1108/BEPAM-02-2014-0011

Katok, E., Serrander, T., & Wennstrom, M. (1998b). Setup Reduction through Process Improvement in Aluminum can Manufacturing. *Production and Inventory Management Journal, Fourth Quarter*, 77-82.

Katok, E., Lewis, H. S., & Harrison, T. P. (1998a). Lot Sizing in General Assembly Systems with Setup Costs, Setup Times, and Multiple Constrained Resources. *Management Science*, *44*(6), 859–877. doi:10.1287/mnsc.44.6.859

Kelly, A. (2006). *Strategic Maintenance Planning*. Oxford, UK: Butterwoth-Heinemann.

Kerkhof, R. M. d., Akkermans, H. A., & Noorderhaven, N. G. (2014). Knowledge lost in data: Organizational impediment to condition-based maintenance in the process industry. *Paper presented at the21st EurOMA Conference*, Palermo.

Kezner, H. (2009). *Project Management: a system approach to planning, scheduling, and controlling* (10th ed.). New York: John Wiley & Sons.

Khan, S., Phillips, P., Hockley, C., & Jennions, I. (2014). No Fault Found events in maintenance engineering Part 2: Root causes, technical developments and future research. *Reliability Engineering & System Safety*, *123*, 196–208. doi:10.1016/j.ress.2013.10.013

Khan, S., Phillips, P., Hockley, C., & Jennions, I. (2015). *No Fault Found: The Search for the Root Cause*. Published by SAE International. doi:10.4271/R-441

Khan, S., Phillips, P., Jennions, I., & Hockley, C. (2014). No Fault Found events in maintenance engineering Part 1: Current trends, implications and organizational practices. *Reliability Engineering & System Safety*, *123*, 183–195. doi:10.1016/j.ress.2013.11.003

Kijima, M. (1989). Some results for repairable systems with general repair. *Journal of Applied Probability*, *26*(1), 89–102. doi:10.2307/3214319

Kijima, M., Morimura, H., & Suzuki, Y. (1988). Periodical replacement problem without assuming minimal repair. *European Journal of Operational Research*, *37*(2), 194–203. doi:10.1016/0377-2217(88)90329-3

Kijima, M., & Sumita, U. (1986). A useful generalisation of renewal theory: Counting processes governed by non-negative Markovian increments. *Journal of Applied Probability*, *23*(1), 71–88. doi:10.2307/3214117

Kim, H., Breslin, J. G., Decker, S., Choi, J., & Kim, H. (2009). Personal knowledge management for knowledge workers using social semantic technologies. *International Journal Intelligent Information and Database Systems*, *3*(1), 28–43. doi:10.1504/IJIIDS.2009.023036

Kirkland, L. V. (2011, September). Why did we add LabVIEW applications to our ATLAS TPSs?Proceedings of*AUTOTESTCON* (pp. 266–271).

Kitagawa, & Takahashi, S. H. (1976). Applicability of Fracture Mechanics to Very Small Cracks or the Cracks in the Early Stages. *Proceedings of the Second International Conference on Mechanical Behaviour of Materials*, Metal Park, Ohio, USA. American Society for Metals.

Klir, G.J. (1995). Principles of uncertainty: What are they? Why do we need them? *Fuzzy Sets and Systems*, 74-1, 15-31.

Kloot, L. (1997). Organizational learning and management control systems: Responding to environmental change. *Management Accounting Research*, *8*(1), 47–73. doi:10.1006/mare.1996.0033

Knezevic, J. (1996). *Mantenimiento*. Madrid: Isdefe.

Kobbacy, K., & Jeon, J. (2001). The development of a hybrid intelligent maintenance optimisation system (HIMOS). *Journal of the Operational Research Society*, 2001, 762–778.

Koch, C. (2012). Contested overruns and performance of offshore wind power plants. *Construction Management and Economics*, *30*(August), 609–622. doi:10.1080/01446193.2012.687830

Kodali, R., & Chandra, S. (2001). Analytical Hierarchy Process for justification of total productive maintenance. *Production Planning and Control*, *12*(7), 695–705. doi:10.1080/09537280010024045

Kogut, B., & Zander, U. (1992). Knowledge of the firm, combinative capacities, and the replication of technology. *Organization Science*, *7*(3), 502–517.

Komass, T., Sniders, A., & Laizans, A. (2012) Statistical Analysis of Municipal Heat Supply.*Proceedings of 11th International Scientific Conference* (Vol. 11, pp. 1 – 8).

Komonen, K. (2002). A cost model of industrial maintenance for profitability analysis and benchmarking. *International Journal of Production Economics*, *79*(1), 15–31. doi:10.1016/S0925-5273(00)00187-0

Komonen, K., Kortelainen, H., & Räikkönen, M. (2012). Corporate asset management for industrial companies: An integrated business-driven approach. In T. van der Lei, P. Herder, & Y. Wijnia (Eds.), *Asset management: The state of the art in Europe from a life cycle perspective* (pp. 47–63). Dordrecht, The Netherlands: Springer Netherlands. doi:10.1007/978-94-007-2724-3_4

Kontu, K. (2013). Forecasting district heating consumption based on customer measurements. Report on the 36th Euroheat & Power Congress. Retrieved from http://www.euroheat.org/Files/Filer/Presentations/20130527_EHP_Congress/130527_1600_1730_Pl_6_Kaisa%20Kontu.pdf

Kotler, P., & Rath, A. (1984). Design: A Powerful But Neglected Strategic Tool. *The Journal of Business Strategy*, *5*(2), 16–21. doi:10.1108/eb039054

Kotter, J. P. (1995). Leading change: Why transformation efforts fail. *Harvard Business Review*, *73*(2), 59–67.

Kruke, B. I., & Olsen, O. E. (2012). Knowledge creation and reliable decision-making in complex emergencies. *Disasters*, *36*(2), 212–232. doi:10.1111/j.1467-7717.2011.01255.x PMID:21992233

Kujawski, D. (2013). Environment and Stress Interaction in Corrosion Fatigue and Stress Corrosion Cracking. *Metallurgical and Materials Transactions. A, Physical Metallurgy and Materials Science*, *44A*(3), 1340–1352. doi:10.1007/s11661-012-1507-8

Kujawski, D., & Sadananda, K. (2011). Effect of Crack-Tip Stresses on SCC Behavior. *Metallurgical and Materials Transactions. A, Physical Metallurgy and Materials Science*, *42A*(2), 377–382. doi:10.1007/s11661-010-0373-5

Kulkarni, U., Ravidran, S., & Freez, E. R. (2007). A knowledge management success model: Theoretical development and empirical validation. *Journal of Management Information Systems*, *23*(3), 309–347. doi:10.2753/MIS0742-1222230311

Kumar, U. (2006). Development and implementation of maintenance performance measurement system: *Issues and challenges. WCEAM Paper 127.*

Kusar, J, Berlec, T, Zefran, F, & Starbek, M (2.010). Reduction of Machine Setup Time. *Journal of Mechanical Engineering*, *56*(12), 833–845.

Labaka, L. (2013). *Resilience Framework for Critical Infrastructures* [Doctoral dissertation]. University of Navarra.

Labaka, L., Hernantes, J., Comes, T., & Sarriegi, J. M. (2014). Defining Policies to Improve Critical Infrastructure Resilience.*Proceedings of the 11th International ISCRAM Conference.*

Labib, W. A. (2004). Towards An Intelligent Holonic Maintenance System. *Maintenance Journal, 17*(2).

Labib, W. A., O'Connor, F. R., & Glyn, B. W. (1998). An effective maintenance system using the analytic hierarchy process. *Integrated Manufacturing Systems*, *9*(2), 87–98. doi:10.1108/09576069810202005

Laugé, A., Hernantes, J., Labaka, L., & Sarriegi, J. M. (2012). Análisis y clasificación de los impactos en situaciones de crisis. *Revista Internacional de Desastres Naturales, Accidentes e Infraestructura Civil*, 12(2).

Lawless, J. F. (1982). *Statistical Models and Methods for Lifetime Data*. New York: John Wiley and Sons.

Lechler, T. G., Edington, B. H., & Gao, T. (2012). Challenging classic project management: Turning project uncertainties into business opportunities. *Project Management Journal*, *43*(6), 59–69. doi:10.1002/pmj.21304

LeCun, Y. L., Bottou, L., Bengio, Y., & Haffner, P. (1998). Gradient-based learning applied to document recognition. *Proceedings of the IEEE*, *86*(11), 2278–2324. doi:10.1109/5.726791

Lee, J. (2003). E-manufacturing: Fundamental, tools, and transformation. *Robotics and Computer-integrated Manufacturing*, *19*(6), 501–507. doi:10.1016/S0736-5845(03)00060-7

Lee, J., Wu, F., Zhao, W., Ghaffari, M., Liao, L., & Siegel, D. (2014). Prognostics and health management design for rotary machinery systems—Reviews, methodology and applications. *Mechanical Systems and Signal Processing*, *42*(1), 314–334. doi:10.1016/j.ymssp.2013.06.004

Leisinger, K. M., Blasco, J., Trigo, J., Petitbò, A., Álvarez, J. M., & Argandoña, A. P. (2010). Hacia una nueva ética económica global. Innovación vs Status Quo. Barcelona, España: Editorial Huygens.

Leiva, R., Donoso, J. R., Muehlich, U., & Labbe, F. (2004). Un enfoque de dos parámetros para el análisis del efecto del cordón de soldadura sobre el constreñimiento. *Rev. Metal Madrid*, *40*, 280–287.

Lei, Y., & Zuo, M. J. (2009). Gear crack level identification based on weighted K nearest neighbor classification algorithm. *Mechanical Systems and Signal Processing*, *23*(5), 1535–1547. doi:10.1016/j.ymssp.2009.01.009

León, O. (1993). *Análisis de decisiones: técnicas y situaciones aplicables a directivos y profesionales*. Madrid: McGraw-Hill.

Lequeux, Q., & Ciavola, P. (2012). *Methods for estimating the costs of coastal hazards. Technical report*. CONHAZ.

Levis, A. H. (1984). Information Processing and Decision-Making Organizations: A Mathematical Description. *Large Scale Systems*, *7*, 151–163.

Levy, C., & Kaufmann, B. (2014). Charcoal for Food Livelihood Diversification In Two Peasant Communities In Mozambique, in Socio-Economic Insecurity. In K. Fakier, & E. Ehmke (Eds.), Emerging Economies: Building New Spaces (pp. 68-82). Routledge.

Lewis, J. (2010, December 25). Heathrow bosses knew an inch of snow would cripple airport. *Telegraph UK*. Retrieved from http://www.telegraph.co.uk/finance/newsbysector/transport/8225006/Heathrow-bosses-knew-an-inch-of-snow-would-cripple-airport.html

Li, C., Lim, J., & Wang, Q. (2007). Internal and external influences on IT control governance. *International Journal of Accounting Information*.

Li, B., Zhang, P.-l, Tian, H., Mi, S.-s, Liu, D.-s, & Ren, G.-q. (2011). A new feature extraction and selection scheme for hybrid fault diagnosis of gearbox. *Expert Systems with Applications*, *38*(8), 10000–10009. doi:10.1016/j.eswa.2011.02.008

Lim, S., Vos, T., Flaxman, A., Danaei, G., & Shibuya, K. et al.. (2012). A Comparative Risk Assessment Of Burden Of Disease And Injury Attributable To 67 Risk Factor Clusters In 21 Regions, 1990-2010: A Systematic Analysis For The Global Burden Of Disease Study 2010. *Lancet, 380(9859)*, 2224–2260. PMID:23245609

Lin, Z., Zheng, J., & Wang, S. (2007). Study on Relation of Internal Control and Enterprise Value: An Empirical Analysis from Shanghai and Shenzhen Stock Markets. *Journal of Finance and Economics*.

Lindner, B. M. (1965). Extremely Slow Crack Growth Bates Aluminium Alloy 7075 T6 [M.S Thesis]. Lehigh University, Bethlehem, PA.

Lindqvist, B. H. Langseth, H. & Stove, B. (2006) Modelling of dependence between critical failure and preventive maintenance: the repair alert model, *Journal of Statistical Planning and Inference (Special issue on competing risks)*, 136, 1701–1717

Lindqvist, B. H. (2006). On the statistical modeling and analysis of repairable systems. *Statistical Science*, *21*(4), 532–551. doi:10.1214/088342306000000448

Line, J. K., & Krishnan, G. (2008). Managing and predicting intermittent failures within long life electronics. *Proceedings of theAerospace Conference'08* (pp. 1-6). IEEE. doi:10.1109/AERO.2008.4526629

Lin, H. F. (2007). A stage model of knowledge management: An empirical investigation of process and effectiveness. *Journal of Information Science*, *33*(6), 643–659. doi:10.1177/0165551506076395

Liu, R., & Issa, R. R. A. (2013). Issues in BIM for facility management from industry practitioners' perspectives. *Computing in Civil Engineering*, 2013, 411-418.

Livi, L. & Sadeghian, A. (2015) Data granulation by the principles of uncertainty. *Pattern Recognition Letters*, 67-2, 113-121.

Loch, C.H. & Pich, M.T. (2002). Managing project uncertainty: from variation to chaos. *MIT Sloan Management Review*, Winter, 60-67.

Loch, C. H., DeMeyer, A., & Pich, M. T. (2006). *Managing the unknown: a new approach to managing high uncertainty and risk in projects*. New York: John Wiley & Sons. doi:10.1002/9780470172377

Lopez del Puerto, C., & Shane, J. S. (2014). Keys to success in megaproject management in Mexico and the United States: Case study. *Journal of Construction Engineering and Management*, *40*(4), 1–7.

Lorenz, J., Battiston, S., & Schweitzer, F. (2009). Systemic risk in a unifying framework for cascading processes on networks. *The European Physical Journal B*, *71*(4), 441–460. doi:10.1140/epjb/e2009-00347-4

Love, P. E. D., Matthews, J., Simpson, I., Hill, A., & Olatunji, O. A. (2014). A benefits realization management building information modelling framework for asset owners. *Automation in Construction*, *37*, 1–10. doi:10.1016/j.autcon.2013.09.007

Love, P. E. D., Simpson, I., Hill, A., & Standing, C. (2013). From justification to evaluation: Building information modelling for asset owners. *Automation in Construction*, *35*, 208–216. doi:10.1016/j.autcon.2013.05.008

Low, B. K. H. (1997). Managing business relationships and positions in industrial networks. *Industrial Marketing Management*, *26*(2), 189–202. doi:10.1016/S0019-8501(96)00091-0

Lu, D., & Qiao, W. (2014a). A GA-SVM hybrid classifier for multiclass fault identification of drivetrain gearboxes. *Proceedings of the IEEE Energy Conversion Congress and Exposition*, Pittsburgh, PA, USA (pp. 3894-3900). doi:10.1109/ECCE.2014.6953930

Lu, D., & Qiao, W. (2014b). Fault diagnosis for drivetrain gearboxes using PSO-optimized multiclass SVM classifier. *Proceedings of the IEEE PES General Meeting Conference & Exposition*, National Harbor, MD, USA. doi:10.1109/PESGM.2014.6939892

Magnanti, T. L., & Vachani, R. A. (1990). Strong Cutting-plane Algorithm for Production Scheduling with Changeover Costs. *Operations Research*, *38*(3), 456–473. doi:10.1287/opre.38.3.456

Mäkeläinen, E. (1998). Economic Value Added, EVA. Retrieved from http://www.evanomics.com

Malik, M. A. K. (1979) Reliable preventive maintenance scheduling. *AIIE Transactions,* 11, 221-228

Mamdani Fuzzy Models . (n. d.). Retrieved from http://www.bindichen.co.uk/post/AI/mamdani-fuzzy-model.html

Manjarres, M. (2013). Valor economico agregado, ebitda y flujo de caja ¿qué es el valor económico agregado? Retrieved from http://www.ebanque-pdf.com/fr_valor.html

Maraña, P., Hernantes, J., Labaka, L., & Sarriegi, J. M. (2015) CRISIS SCORECARD: How to reduce impacts increasing resilience and awareness.*Proceedings of The International Emergency Management Society Workshop (TIEMS).*

Márquez, A. C., Márquez, C. P., Fernández, J. F. G., Campos, M. L., & González-Prida Díaz, V. (2012). Life cycle cost analysis. In *Asset management: The state of the art in Europe from a life cycle perspective* (pp. 81–99). Dordrecht, The Netherlands: Springer Netherlands. doi:10.1007/978-94-007-2724-3_6

Marshall Institute. (1999) *Maintenance Effectiveness Survey*. Retrieved from http://www.marshallinstitute.com/inc/eng/consulting/areas/customized_reliability_solutions/body/measurvey.pdf

Martínez Marín, J. E., & Eguren Martí, M. L. (2013). Maritime Transport: A Theoretical Analysis Under a System's Approach. *Journal of Maritime Research.*

Martinez, I., & Ruiz, J. (2002). Los procesos de creación del conocimiento: El aprendizaje y la espiral de creación del conocimiento. *Proceedings of theXVI Congreso Nacional de AEDEM*, Alicante.

Martin, J. (2008). Personal knowledge management: the basis of corporate and institutional knowledge management. In J. Martin & K. Wright (Eds.), *Managing Knowledge: Case Studies in Innovation*. Alberta: Spotted Cow.

Martins, R. (2014). *The Wood Fuel Energy Systems Conceptual Design Metamodel, A Novel Approach To Participatory Design Contextualised To The Case Of Mozambique* [PhD Thesis]. Imperial College London.

Martorell, S., Sánchez, A., Muñoz, A., Pitarch, J. L., Serradell, V., & Roldan, J. (1999). The use of maintenance indicators to evaluate the effects of maintenance programs on NPP performance and safety. *Reliability Engineering & System Safety*, *65*(2), 85–94. doi:10.1016/S0951-8320(98)00088-X

Martorell, S., Villanueva, J. F., Carlos, S., Nebot, Y., Sanchez, A., Pitarch, J. L., & Serradell, V. (2005). RAMS+C informed decision-making with application to multiobjective optimization of technical specifications and maintenance using genetic algorithms. *Reliability Engineering & System Safety*, *87*(1), 65–75. doi:10.1016/j.ress.2004.04.009

Márvel, M., Rodríguez, D., & Núñez, M. A. (2011). La productividad desde una perspectiva humana: Dimensiones y factores. *Intangible Capital*, *7*(2), 549–584. doi:10.3926/ic.2011.v7n2.p549-584

Marzoli, A. (2007). *Relatório do inventário florestal nacional*. Maputo: Direcção Nacional de Terras e Florestas. Ministério da Agricultura.

Mather, D. (2005). *The Maintenance Scorecard – Creating Strategic Advantage*. New York: Industrial Press.

Mathur, A., Cavanaugh, K. F., Pattipati, K. R., Willett, P. K., & Galie, T. R. (2001). Reasoning and modeling systems in diagnosis and prognosis. *Paper presented at the the International Society for Optical Engineering.*

Matsuoka, S., Tanaka, H., Homma, N., & Murakami, Y. (2011). Influence of hydrogen and frequency on fatigue crack growth behavior of Cr-Mo steel. *International Journal of Fracture*, *168*(1), 101–112. doi:10.1007/s10704-010-9560-z

McArthur, J. J. (2015). A building information management (BIM) framework and supporting case study for existing building operations, maintenance and sustainability.*International Conference on Sustainable Design, Engineering and Construction. Procedia Engineering 00* doi:10.1016/j.proeng.2015.08.450

Mcintosh RI, Culley SJ, Mileham AR, &d Owen GW (2001). *Improving Changeover Performance*. Oxford, Great Britain: Butterworth-Heinemann.

Megavind. (2010), *Denmark - Supplier of Competitive Offshore Wind Solutions report.*

Methods for Crack Opening Displacement C.O.D. (Testing). (1979). Methods for Crack Opening Displacement C.O.D. (Testing) DO: 19, (1.972). Methods for Plane Strain Fracture Toughness K_{ic} Testing. DD3. (1971). *British Standard BS5762.*

Meyer, B. (2012). Unravelling the Mysteries of NFF. *Air Transport and Operations: Proceedings of the Third International Air Transport and Operations Symposium 2012* (p. 354). IOS Press.

Middelmann, M. H. (2007). *Natural hazards in Australia: identifying risk analysis requirements*. Geoscience Australia.

Miles, M., & Huberman, A. (1994). *Qualitative Data Analysis: A sourcebook of new methods*. California: Sage Publications.

Milne, A., & Leifer, L. (1999, August 24-26). The Ecology Of Innovation In Engineering Design. *Proceedings of the International Conference On Engineering Design ICED '99*, Munich.

Minto, B. (2008). *The pyramid principle - Logical writing, thinking and problem solving* (3rd ed.). Pearson Education Limited.

Mipro Company: Water and Energy Management Products and Services . (n. d.). Retrieved from http://www.mipro.fi/eng/businesslines/waterandenergymanagement/productsandservices

MIREME (2014): Estratégia De Conservação E Uso Sustentável Da Energia Da Biomassa Para O Período 2014-2025, (ECUSEB 2014-2015) Ministério da Energia (actual MIREME), Maputo [in Portuguese].

Mitchell, T. (1997). *Machine Learning*. McGrawHill International.

Mkrtchyan, L., Podofillini, L., & Dang, V. N. (2015). Bayesian belief networks for human reliability analysis: A review of applications and gaps. *Reliability Engineering & System Safety*, *139*, 1–16. doi:10.1016/j.ress.2015.02.006

Mobley, K. (1999). Root Cause Failure Analysis. Butterworth-Heinemann, Reed Elsevier Group.

Morris, R. (2009). *The Fundaments of Product Design*. AVA Publishing SA, Lousanne.

Moubray, J. (1991). *RCM II: Reliability-Centered Maintenance*. New York, USA: Industrial Press Inc.

Moubray, J. (1997). *Reliability centered maintenance* (2nd ed.). Oxford, UK: Butterworth-Heinemann.

Moubray, J. (1997). *Reliability-Centered Maintenance*. New York, NY: Industrial Press.

Moubray, J. (1997). *Reliability-centred Maintenance* (2nd ed.). Oxford: Butterworth-Heinemann.

Muller, A., Marquez, A. C., & Iung, B. (2008). On the concept of e-maintenance: Review and current research. *Reliability Engineering & System Safety*, *93*(8), 1165–1187. doi:10.1016/j.ress.2007.08.006

Munda, G. (2005). Multiple Criteria Decision analysis and sustainable development. In J. Figueira, S. Greco, & M. Ehrgott (Eds.), Multicriteria Decision analysis. State of the Art Surveys (pp. 953-988). Boston: Springer Verlag. doi:10.1007/0-387-23081-5_23

Mungaray, A., & Ramírez-Urquidy, M. (2007). *Human capital and productivity in microenterprises.* Universidad Autónoma de Baja California.

Muñoz, J. (1999). Sobre gestión del conocimiento, un intangible clave en la globalización. *Economía Industrial*, 330.

Muralidharan, R. (2004). A framework for designing strategy content controls. *International Journal of Productivity and Performance Management*, *53*(7), 590–601. doi:10.1108/17410400410561213

Murthy, D., Atrens, A., & Eccleston, J. (2002). Strategic maintenance management. *Journal of Quality in Maintenance Engineering*, *8*(4), 287–305. doi:10.1108/13552510210448504

Musee, N., Lorenzen, L., & Aldrich, C. (2006). Decision support for evaluating energy demand in vinification processes using fuzzy sets theory. *Journal of Energy in Southern Africa*, *17*(4), 4–18.

Naim, P., Wuillemin, P. H., Leray, P., Pourret, O., & Becker, A. (2011). *Réseaux bayésiens.* Algorithmes. Eyrolles.

Nakajima, S. (1988). *TPM: Introduction to total productive maintenance.* Cambridge, MA: Productivity Press Inc.

Nates, J. L. (2004). Combined external and internal hospital disaster: Impact and response in a Houston trauma center intensive care unit. *Critical Care Medicine*, *32*(3), 686–690. doi:10.1097/01.CCM.0000114995.14120.6D PMID:15090948

Neill, J. (2007). *Delphi study: research by iterative, consultative inquiry.* Retrieved from available at http://www.wilderdom.com/delphi.html

Nelson, W. (1969). Hazard Plotting For Incomplete Failure Data. *Journal of Quality Technology*, *1*, 27–52.

Nhantumbo, I., Dent, J., & Kowero, G. (2001). Goal Programming: Application In The Management Of The Miombo Woodland In Mozambique. *European Journal of Operational Research*, *133*(2), 310–322. doi:10.1016/S0377-2217(00)00300-3

Nicolaides, G. (2015, January 15). Eskom: Bad maintenance led to load shedding. Retrieved September 30, 2015.

Nix, O. (2011). Building Information Technology and Management. Retrieved from http://www.institutebe.com/InstituteBE/media/Library/Resources/Smart%20Grid_Smart%20Building/Issue_Brief---Building_Info_Tech_and_Mgmnt.pdf

NOAA. (2014). *Coastal ecosystem restoration: risk and uncertainty in environmental restoration programs*, Retrieved from http://coast.noaa.gov/archived/coastal/economics/riskinrest.htm?redirect=301ocm

Nonaka, I. (1991). The knowledge-creating company. *Harvard Business Review*, November-December, 96–104.

Nonaka, I. (1994). A dynamic theory of organizational knowledge creation. *Organizatioll Saence*, *5*(1), 14–37. doi:10.1287/orsc.5.1.14

Nonaka, I., & Konno, N. (1998). The concept of "Ba": Building a foundation for knowledge creation. *California Management Review*, *40*(3), 40–54. doi:10.2307/41165942

Nonaka, I., & Takeuchi, H. (1995). *The knowledge-Creating Company: How Japanese Companies Create the Dynamics of Innovation.* New York: Oxford University Press.

Nonaka, I., & Takeuchi, H. (1999). *La Organización Creadora de Conocimiento.* Oxford, México.

Norris, F. H., Stevens, S. P., Pfefferbaum, B., Wyche, K. F., & Pfefferbaum, R. L. (2008). Community resilience as a metaphor, theory, set of capacities, and strategy for disaster readiness. *American Journal of Community Psychology*, *41*(1-2), 127–150. doi:10.1007/s10464-007-9156-6 PMID:18157631

Norton, D., & Kaplan, R. (1996). *The Balanced Scorecard: Translating Strategy into Action*. Boston: Harvard Business School Press.

Nowlan, F. S., & Howard, F. H. (1978). *Reliability Centered Maintenance*. Springfield, VA. U.S. Department of Commerce, National Technical Information Services, Report Number AD/A066-579.

NVDO. (2015). *Wie zijn wij? – NVDO* [Who are we?]. Retrieved from http://www.nvdo.nl/wie-zijn-wij/

O'Connor, P., Neil, D., Liu, S.-C., Delbruck, T., & Pfeiffer, M. (2013). Real-time classification and sensor fusion with a spiking deep belief network. *Frontiers in Neuroscience*, *7*, 178. PMID:24115919

O'Dell, C., & Grayson, C. J. (1998). If only we knew what we know: Identification and transfer of internal best practices. *California Management Review*, *40*(3), 154–170. doi:10.2307/41165948

O'Keefe, R. M., Balci, O., & Smith, E. P. (1987). Validating expert system performance. *IEEE Expert*, *2*(Winter), 81–87. doi:10.1109/MEX.1987.5006538

O'Leary, D. E. (1987). Validation of expert systems—with application to auditing and accounting expert systems. *Decision Sciences*, *18*(3), 468–486. doi:10.1111/j.1540-5915.1987.tb01536.x

OCDE. (1996). The Knowledge-Based Economy.

OCDE. (2004). *Medición de la gestión de conocimientos en las empresas: primeros resultados*.

Olaz, A. (2003). *Análisis de los procesos de transformación en las cualificaciones profesionales en el sector financiero* [Thesis Doctoral]. Universidad de Murcia. Facultad de Economía y Empresa.

OPHI. (2011). *Country Briefing: Multidimensional Poverty Index At A Glance*. Oxford: Oxford Poverty and Human Development Initiative.

Ordoñez de Pablos, P. (1999). *Gestión del conocimiento y medición del capital intelectual en la empresa internacional*. Proyecto de Doctorado, Universidad de Oviedo.

Ordoñez de Pablos, P. (2001). *Capital intelectual, gestión del conocimiento y sistemas de gestión de recursos humanos: Influencia sobre los resultados organizativos* [Thesis Doctoral]. Universidad de Oviedo.

Ordoñez de Pablos, P., & Parreños, J. (2003). Aprendizaje organizativo en el contexto internacional: Implicaciones para la gestión del conocimiento. *Investigaciones Europeas de Dirección y Economía de la Empresa*, *9*(2), 205–216.

Ordoñez de Pablos, P., & Parreños, J. (2005). Aprendizaje organizativo y gestión del conocimiento: Un análisis dinámico del conocimiento en la empresa. *Investigaciones Europeas de Dirección y Economía de la Empresa*, *11*(1), 165–177.

Osterwalder, A. (2004). The Business Model Ontology- A Proposition In A Design Science Approach [Doctoral Thesis]. Lausanne University, Lausanne.

Osterwalder, A., & Pigneur, Y. (2010). *Business model generation: a handbook for visionaries, game changers, and challengers*. John Wiley & Sons.

Pace, S. (2004). A grounded theory of the flow experiences of Web users. *International Journal of Human-Computer Studies*, *60*(3), 327–363. doi:10.1016/j.ijhcs.2003.08.005

Palacios, D., & Garrigos, F. (2006). Propuesta de una escala de medida en la gestión del conocimiento en las industrias de biotecnología y telecomunicaciones. *Investigaciones Europeas de Dirección y Economía de la Empresa*, *12*(1), 207–224.

Palmer, R. D. (1999). *Maintenance Planning and Scheduling*. New York, NY: McGraw-Hill.

Paniagua, E. (2007). *La Gestión tecnológica del conocimiento*. Murcia: Universidad de Murcia, Servicio de Publicaciones.

Parfenenko, Yu., Shendryk, V., Nenja, V., & Vashchenko, S. (2014). Information System for Monitoring and Forecast of Building Heat Consumption. Communications in Computer and Information Science, 465 1-11

Parfenenko, Yu., Okopnyi, R., & Nenja, V. (2012). Creation of Control Instruments for Buildings' Heating. *The Journal of the National Technical University NTU KPI. Information Systems and Networks*, CCIS (Vol. 465, pp. 93–97).

Parida, A., & Kumar, U. (2006). Maintenance performance measurement (MPM): Issues and challenges. *Journal of Quality in Maintenance Engineering*, *12*(3), 239–251. doi:10.1108/13552510610685084

Paris, P. C. (1970). Testing for Very Slow Growth of Fatigue Cracks. *MTS Close Loop Magazine*, 2(5).

Paris, P. C., Bucci, R. J., Wessel, E. T., Clark, W. G., & Mager, T. R. (1972). *Extensive Study of Low Fatigue Crack Growth Rates in A533 and A508 Steels, ASTM STP513*. Philadelphia: American Society for Testing and Materials.

PARPA. (2001). *Action Plan For The Reduction Of Absolute Poverty 2001-2005, (PARPA), Strategy Document For The Reduction Of Poverty And Promotion Of Economic Growth*. Maputo: Mozambican Government. [in Portuguese]

Parra, C., & Omaña, C. (2001). Técnica de auditoría de la gestión del mantenimiento para el sector Refinación. *Proceedings of theVII Congreso de Ingeniería de Mantenimiento de Petróleos de Venezuela*, Caracas.

Parra, C., & Crespo, A. (2012). *Ingeniería de Mantenimiento y Fiabilidad Aplicada en la Gestión de Activos. Desarrollo y aplicación práctica de un Modelo de Gestión del Mantenimiento (MGM)*. Primera Edición. Editado por INGEMAN, Escuela Superior de Ingenieros Industriales, Universidad de Sevilla, España.

Parra, C., Crespo, A., Kristjanpoller, F., & Viveros, P. (2012). Stochastic model of reliability for use in the evaluation of the economic impact of a failure using life cycle cost analysis. Case studies on the rail freight and oil industries. *Journal of Risk and Reliability*, *226*(4), 392–405.

Parsanezhad, P., & Tarandi, V. (2013). Is the age of facility manager's papers boxes over? *Proceedings of theCIB World Building Congress*.

Partington, D. (2000). Building grounded theories of management action. *British Journal of Management*, *11*(2), 91–102. doi:10.1111/1467-8551.00153

PAS 1192-2:2013, Specification for information management for the capital/delivery phase of construction projects using building information modeling. (2013).

PAS 1192-3:2014, Specification for information management for the operational phase of assets using building information modeling. (2014).

Patel, S., Dale, B. G., & Shaw, P. (2001). Set-Up Time Reduction and Mistake Proofing Methods: An Examination in Precision Component Manufacturing. *The TQM Magazine*, *13*(3), 175–179. doi:10.1108/09544780110385528

Patton, M. Q. (2001). *Qualitative Research & Evaluation Methods* (3rd ed.). SAGE Publications, Inc.

Pauleen, D. (2009). Personal knowledge management: Putting the "person" back into the knowledge equation. *Online Information Review*, *33*(2), 221–224. doi:10.1108/14684520910951177

Pavez, A. S. (2000). *Modelo de implantación de Gestión del Conocimiento y Tecnologías de Información para la Generación de Ventajas Competitivas*. Valparaíso, Chile: Universidad Técnica Federico Santa María.

Pedraja, L. & Rodríguez, E. (2004). Efectos del estilo de liderazgo sobre la eficacia de las organizaciones públicas. *Revista facultad de ingeniería*, 12(2), 63-73.

Pedraja, L., Rodríguez, E., & Rodríguez, J. (2006). Liderazgo y decisiones estratégicas: Una perspectiva integradora. *Asociación Interciencia*, *31*(8), 577–582.

Peña, P. (2001). *To know or no to be. Conocimiento, el oro gris de las organizaciones*. Madrid: DINTEL.

Pérez de Miguel, A. (2006). Hacia el Índice de Conocimiento Roto: Introducción a los coeficientes de rotura de conocimiento. Documentos de trabajo “Nuevas tendencias en dirección de empresas”. DT 06/06.

Pérez, A. (2007). *Modelo para la Auditoría del Conocimiento Considerando los Procesos Clave de la Organización y Utilizando Tecnologías Basadas en Conocimientos. Tesis doctoral*. Universidad de Murcia.

Peris, F.; Loria, B. & Méndez, M. (2002). Creación de conocimiento y diseño de organizaciones: Equidad, confianza y objetivos compartidos como reto de la gestión del conocimiento. *Cuadernos de CC.EE. y EE.*, 43, 41-56.

Persson, F., & Saccani, N. (2009). Managing the after-sales logistic network–a simulation study. *Production Planning and Control*, *20*(2), 125–134. doi:10.1080/09537280802707530

Peters, R. A. (2006). *Maintenance Benchmarking and Best Practices* (1st ed.). New York, NY: McGraw-Hill.

Petrides, L. A., & Nodine, T. R. (2003). *Knowledge Management in Education: Defining the Landscape*. Half Moon Bay, California: March Institute for the Study of Knowledge Management in Education.

Peulic, A., Dragicevic, S., Jovanovic, Z., & Krneta, R. (2013). Flexible GPS/GPRS Based System for Parameters Monitoring in the District Heating System. *International Journal of Computers, Communications & Control*, *8*(1), 105–110. doi:10.15837/ijccc.2013.1.174

Pfister, J. (2009). *Managing Organizational Culture for Effective Internal Control: From Practice to Theory*. Berlin, Germany: Physica-Verlag. doi:10.1007/978-3-7908-2340-0

Pfurtscheller, C., Lochner, B., & Thieken, A. H. (2011). Costs of Alpine hazards. *WP8 report of the project ConHaz–Costs of Natural Hazards (D8. 1)*. Retrieved from http://conhaz. org

Pickthall, N. (2014). The Contribution of Maintenance Human Factors to no Fault Founds on Aircraft Systems Engineering. *Procedia CIRP*, *22*, 59–64. doi:10.1016/j.procir.2014.07.013

Pintelon, L. M., & Gelders, L. F. (1992). Maintenance management decision making. *European Journal of Operational Research*, *58*(3), 301–317. doi:10.1016/0377-2217(92)90062-E

Pisapia, J., Jelenc, L., & Mick, A. (2016). The Foundations Of Strategic Thinking: Effectual, Strategic, And Causal Reasoning. In *Neostrategic management* (pp. 45-55).

Plummer, M. (1999). To Fetch A Pail Of Water. *Natural History*, *108*(1), 56–65.

PM Network (2014). Standardised project management practices = success. *PMI Today,* April(2014), 8-9.

PMI. (2014). *What is project management.* Retrieved from http://www.pmi.org/About-Us/About-Us-What-is-Project-Management.aspx

Polanyi, M. (1958). *Personal Knowledge: Towards a Post-Critical Philosophy*. University of Chicago Press.

Polanyi, M. (1966). *The tacit dimension*. London: Routledge & Kegan Paul.

Porter, M., & M., K. (2011). The Big Idea. Creating Shared Value. How to reinvent capitalism and unleash a wave of innovation and growth. *Harvard Business Review*.

Porter, M. (1985). *Competitive Advantage: Creating and Sustaining Superior Performance*. Nueva York, USA: Free Press.

Porter, M., & Kramer, M. (2006). Strategy and Society: The link between competitive advantage and corporate social responsibility. *Harvard Business Review*. PMID:17183795

Prasad Mishra, R., Anand, D., & Kodali, R. (2006). Development of a framework for world-class maintenance systems. *Journal of Advanced Manufacturing Systems*, *5*(2), 141–165. doi:10.1142/S0219686706000807

Prawda, J. (1999). *Métodos y modelos de investigación de operaciones. Modelos estocásticos*. México: Limusa S.A.

Presidio Graduate School. (2006). The dictionary of sustainable management. Retrieved from http://www.sustainabilitydictionary.com/

Priemus, H., Flyvbjerg, B., & van Wee, B. (2008). *Decision-making on mega-projects: cost–benefit analysis, planning and innovation*. Cheltenham: Edward Elgar. doi:10.4337/9781848440173

Puddicombe, M. S. (2006). The limitations of planning: The importance of learning. *Journal of Construction Engineering and Management*, *132*(9), 949–955. doi:10.1061/(ASCE)0733-9364(2006)132:9(949)

Pudney, S. (2010). *Asset renewal decision modelling with application to the water utility industry* [Ph.D. Thesis]. Faculty of Built Environment and Engineering. Queensland University of Technology, Queensland, Australia.

Purnomo, H., Yasmi, Y., Prabhu, R., Hakim, S., & Jafar, A., Suprihatin. (2003). Collaborative Modelling to Support Forest Management: Qualitative Systems Analysis at Lumut Mountain, Indonesia, Small-scale Forest Economics. *Management and Policy*, *2*(2), 259–275.

Qi, H., Geneson, S., & Pecht, M. (2008). No-fault-found and intermittent failures in electronic products. *Microelectronics and Reliability*, *48*(5), 663–674. doi:10.1016/j.microrel.2008.02.003

Quintana Fundora, Y. (2006). *Gestión por el conocimiento en la carrera de Ingeniería Industrial. Administración de operaciones. Tesis en opción al título de Master en Ciencias*. Cuba: Facultad de Industrial Economía. Universidad de Matanzas.

Ramadhan, R. H., Hamad, I., Wahhab, A., & Duffuaa, S. O. (1999). The use of an analytical hierarchy process in pavement maintenance priority ranking. *Journal of Quality in Maintenance Engineering*, *5*(1), 25–39. doi:10.1108/13552519910257041

Ramani, S., & Krishnan Kutty, K. V. (1985). Management of multi-item, multi-group inventories with multiple criteria under service level constraints. *Engineering Costs and Production Economics*, *9*(1), 59–64. doi:10.1016/0167-188X(85)90010-2

Rani, N. A. A., Baharum, M. R., Akbar, A. R. N., & Nawawi, A. H. (2015). Perception of Maintenance Management Strategy on Healthcare Facilities. *Procedia: Social and Behavioral Sciences*, *170*(27), 272–281. doi:10.1016/j.sbspro.2015.01.037

Rao, B. K. N. (1996). *Handbook of Condition Monitoring*. Oxford, UK: Elsevier.

Rausand, M., & Høyland, A. (2004). *System reliability theory: models, statistical methods, and applications* (Vol. 396). John Wiley & Sons.

Reducing your ICT energy costs (n. d.). Retrieved from https://www.jisc.ac.uk/guides/reducing-your-ict-energy-costs

Reiman, M. I., & Wein, L. M. (1998). Dynamic Scheduling of a Two-Class Queue with Setups. *Operations Research*, *46*(4), 532–547. doi:10.1287/opre.46.4.532

Reisig, W. (2012). *Petri nets: an introduction* (Vol. 4). Springer Science & Business Media.

Ribstein, L. E. (2002). Market vs. Regulatory Responses to Corporate Fraud: A Critique of the Sarbanes-Oxley Act of 2002. *The Journal of Corporation Law*, *28*(1), 1–74.

Rigdon, S. E., & Basu, A. P. (2000). *Statistical Methods for the Reliability of Repairable Systems*. New York: John Wiley & Sons Inc.

Righter, R., & Shanthikumar, J. G. (1998). Multiclass Production Systems with Setup Times. *Operations Research*, *46*(3), S146–S153. doi:10.1287/opre.46.3.S146

Rinaldi, S. M. (2004, January). Modeling and simulating critical infrastructures and their interdependencies. *Proceedings of the 37th annual Hawaii international conference onSystem sciences* (pp. 8-pp). IEEE. doi:10.1109/HICSS.2004.1265180

Rivas, L., & Flores, B. (2007). La gestión de conocimiento en la industria automovilística. *Estudios Gerenciales*, *23*(102), 83–100. doi:10.1016/S0123-5923(07)70003-7

RMS Titanic. (2015, September 29). *Wikipedia*. Retrieved from https://en.wikipedia.org/wiki/RMS_Titanic

Robbins, S., & Judge, T. (2009). *Comportamiento Organizacional. Decimotercera edición*. México: Pearson Educación.

Roberts, K. H., Stout, S. K., & Halpern, J. J. (1994). Decision dynamics in two high reliability military organizations. *Management Science*, *40*(5), 614–624. doi:10.1287/mnsc.40.5.614

Rodrigues, T. C. (2014). The MACBETH Approach to Health Value Measurement: Building a Population Health Index in Group Processes. *Procedia Technology*, *16*, 1361–1366. doi:10.1016/j.protcy.2014.10.153

Rodríguez Méndez, M. (2003). *El Proceso de Cambio de Útiles. La Flexibilidad de una Fábrica*. Madrid, Spain: FC Editorial.

Rodríguez, D. (2006). Modelos para la creación y gestión del conocimiento: Una aproximación teórica. *Revista Educar*, *37*, 25–39.

Rojas, C. M., Montilva, C. J., & Barrios, A. J. (2009). *Revista Colombiana de Tecnologías Avanzadas*, *1*(13), 72-80.

Roke, S. (2009). *Harrier No Fault Found reduction* [MSc Dissertation in Engineering Business Management]. Faculty of Engineering and Computing Coventry University.

Romero, C. (1996). *Análisis de Decisiones Multicriterio*. Madrid: Isdefe.

Rosenau, P. V. (2000). The strengths and weaknesses of public-private policy partnerships. In *Public-private policy partnerships* (pp. 217-242).

Rothery, M. (2005). Critical infrastructure protection and the role of emergency services.

Rouse, P., & Chiu, T. (2009). Towards optimal life cycle management in a road maintenance setting using DEA. *European Journal of Operational Research*, *196*(2), 672–681. doi:10.1016/j.ejor.2008.02.041

Roweis, S. (n. d.). Levenberg-Marquardt Optimization. Retrieved from https://www.cs.nyu.edu/~roweis/notes/lm.pdf

Ruggles, R. (1998). The state of the notion: Knowledge management in practice. *California Management Review*, *40*(3), 80–89. doi:10.2307/41165944

Ruitenburg, R. J., Braaksma, A. J. J., & van Dongen, L. A. M. (2015). Towards a model for effective asset life cycle management control – A case study in rolling stock maintenance. *Proceedings of the 22nd EurOMA Conference* (pp. 1–10). Neuchatel, Switzerland: EurOMA.

Ruitenburg, R. J., & Braaksma, A. J. J. (in press). The multidisciplinary lifetime impact identification analysis to support strategic asset life cycle management. *CIRP Journal of Manufacturing Science and Technology.*

Ruitenburg, R. J., Braaksma, A. J. J., & van Dongen, L. A. M. (2014). A multidisciplinary, expert-based approach for the identification of lifetime impacts in asset life cycle management. *Procedia CIRP*, *22*, 204–212. doi:10.1016/j.procir.2014.07.007

Saaty, T. L. (1990). How to make a decision: The analytic hierarchy process. *European Journal of Operational Research*, *48*(1), 9–26. doi:10.1016/0377-2217(90)90057-I

Sagarna, I., Uribetxebarria, J., & Castellano, E. (2015, 19-20 November). Optimization of offshore wind parks maintenance service strategies using agent based simulation techniques.*Proceedings of the 4th International Conference on Business Servitization (ICBS)*, Madrid, Spain.

Salkeld, L. (2007). Is this the man who sank the Titanic by walking off with vital locker key? *Daily Mail*. Retrieved from http://www.dailymail.co.uk/news/article-478269/Is-man-sank-Titanic-walking-vital-locker-key.html

Sánchez, M., Chaminade, C., & Escobar, C. (1999). En busca de una teoría sobre medición y gestión de los intangibles en la empresa: Una aproximación metodológica. *Ekonomiaz*, 45, 188-213.

Sandborn, P. A., & Wilkinson, C. (2007). A maintenance planning and business case development model for the application of prognostics and health management (PHM) to electronic systems. *Microelectronics and Reliability*, *47*(12), 1889–1901. doi:10.1016/j.microrel.2007.02.016

Sapountzaki, K. (2007). Social resilience to environmental risks: A mechanism of vulnerability transfer? *Management of Environmental Quality: An International Journal*, *18*(3), 274–297. doi:10.1108/14777830710731743

Saravanan, N., & Ramachandran, K. (2009). Fault diagnosis of spur bevel gear box using discrete wavelet features and Decision Tree classification. *Expert Systems with Applications*, *36*(5), 9564–9573. doi:10.1016/j.eswa.2008.07.089

Saravanan, N., & Ramachandran, K. I. (2010). Incipient gear box fault diagnosis using discrete wavelet transform (DWT) for feature extraction and classification using artificial neural network (ANN). *Expert Systems with Applications*, *37*(6), 4168–4181. doi:10.1016/j.eswa.2009.11.006

Sarriegi, J. M., Sveen, F. O., Torres, J., & Gonzalez, J. J. (2008). Towards a research framework for critical infrastructure interdependencies. *International Journal of Emergency Management*, *5*(3), 235–249. doi:10.1504/IJEM.2008.025094

Savic, V., Wymeersch, H., & Zazo, S. (2013). Belief consensus algorithms for fast distributed target tracking in wireless sensor networks. *Signal Processing*, *95*, 149–160. doi:10.1016/j.sigpro.2013.09.005

Scandizzo, S. (2005). Risk mapping and key risk indicators in operational risk management. *Economic Notes*, *34*(2), 231–256. doi:10.1111/j.0391-5026.2005.00150.x

Schevers, H., Mitchell, J., Akhurst, P., Marchant, D., Bull, S., McDonald, K., & Linning, C. (2007). Towards digital facility modelling for Sydney opera house using IFC and semantic web technology. *ITcon*, *12*, 347–362.

Schmid, H. (2012). *How to use the FFT and Matlabs pwelch function for signal and noise simulations and measurements (Tech. Report)*. University of Applied Sciences of Northwestern Switzerland.

Schmidt, R. A., & Paris, P. C. (1973). Threshold for Fatigue Crack Propagation and Effect of Load Ratio and Frequency. In Progress in Flow Growth and Fracture (p. 79). Philadelphia: ASTM.

Schmidt, G., & Wilhelm, W. E. (2000). Strategic, tactical and operational decisions in multi-national logistics networks: A review and discussion of modelling issues. *International Journal of Production Research*, *38*(7), 1501–1523. doi:10.1080/002075400188690

Schuman, C. A., & Brent, A. C. (2005). Asset life cycle management: Towards improving physical asset performance in the process industry. *International Journal of Operations & Production Management*, *25*(5/6), 566–579. doi:10.1108/01443570510599728

Schweizer, C., Seifert, T., Nieweg, B., Von Hartrott, P., & Riedel, H. (2011). Mechanisms and modelling of fatigue crack growth under combined low and high cycle fatigue loading. *International International Journal of Fatigue*, *33*(2), 194–202. doi:10.1016/j.ijfatigue.2010.08.008

Scoones, I. (1998). *Sustainable Rural Livelihoods. A Framework For Analysis, IDS WORKING PAPER 72, Institute of Development studies*. Brighton: Sussex University.

Sedlacko, M., Martinuzzi, A., Røpke, I., Videira, N., & Antunes, P. (2014). Participatory Systems Mapping For Sustainable Consumption: Discussion Of A Method Promoting Systemic Insights. *Ecological Economics*, *106*, 33–43. doi:10.1016/j.ecolecon.2014.07.002

Seemann, P., & Cohen, D. (1997). The Geography of Knowledge: From Knowledge Maps to the Knowledge Atlas. *Knowledge and Process Management*, *4*(4), 247–260. doi:10.1002/(SICI)1099-1441(199712)4:4<247::AID-KPM107>3.0.CO;2-1

Sekine, K., & Arai, K. (1992). *Kaizen for Quick Changeover: Going Beyond SMED*. Cambridge, USA: Productivity Press.

Sen, A. (1997). Editorial: Human Capital And Human Capability. *World Development*, *25*(12), 1959–1961. doi:10.1016/S0305-750X(97)10014-6

Senouci, M. R. (2014). *On the use of the belief functions theory in the deployment and control of wireless sensor networks* [PhD thesis]. Paris-Est University.

Senouci, M. R., Mellouk, A., Oukhellou, L. & Aissani, A. (2012). Using the Belief Functions Theory to Deploy Static Wireless Sensor Networks. In *Belief Functions: Theory and Applications* (pp. 425-432). Springer Berlin Heidelberg.

SESCAM. (2007). *Valoración de la calidad de la atención sanitaria pública en hospitales*. Memory of activities of SESCAM.

Shafer, G. (1976). *A Mathematical Theory of Evidence*. Princeton University Press.

Shafiee, M. (2015). Maintenance strategy selection problem: An MCDM overview. *Journal of Quality in Maintenance Engineering*, *21*(4), 378–402. doi:10.1108/JQME-09-2013-0063

Shahin, A., Shirouyehzad, H., & Pourjavad, E. (2012). Optimum maintenance strategy: A case study in the mining industry. *International Journal of Services and Operations Management*, *12*(3), 368–386. doi:10.1504/IJSOM.2012.047626

Shang, C., Yang, F., Huang, D., & Lyu, W. (2014). Data-driven soft sensor development based on deep learning technique. *Journal of Process Control*, *24*(3), 223–233.

Sharma, R. K., Kumar, D., & Kumar, P. (2005). FLM to select suitable maintenance strategy in process industries using MISO model. *Journal of Quality in Maintenance Engineering*, *11*(4), 359–374. doi:10.1108/13552510510626981

Sheffi, Y., & Rice, J. B. Jr. (2005). A supply Chain View of the resilient Enterprise. *MIT. Sloan Management Review*, *47*(1).

Shendryk, V. V., & Vashchenko, S. M. (2011). The system of Data Collection, Allocation and Analysis. *Visnyk of the National University Lviv Polytechnic*, *715*, 343–353.

Shenoy, D., & Bhadury, B. (1998). *Maintenance resources management: Adapting MRP*. London: Taylor and Francis.

Shen, W., Hao, Q., Mak, H., Neelamkavil, J., Xie, H., Dickinson, J., & Xue, H. et al. (2010). Systems integration and collaboration in architecture, engineering, construction, and facilities management: A review. *Advanced Engineering Informatics*, *24*(2), 196–207. doi:10.1016/j.aei.2009.09.001

Sher, J. P., & Lee, C. V. (2004). Information technology as a facilitator for enhancing dynamic capabilities through knowledge management. *Information & Management*, *41*(8), 933–945. doi:10.1016/j.im.2003.06.004

Shingo, S. (1985). *A Revolution in Manufacturing: The Smed System*. Cambridge, USA: Productivity Press.

Sikorska, J., Hodkiewicz, M., & Ma, L. (2011). Prognostic modelling options for remaining useful life estimation by industry. *Mechanical Systems and Signal Processing*, *25*(5), 1803–1836. doi:10.1016/j.ymssp.2010.11.018

Silverio, R. C., Moura, T., & Guedes, P. (2003). Evaluation of cold crack susceptibility on HSLA steel welded joints. *Rev. Metal. Madrid*, *39*(2), 83–89. doi:10.3989/revmetalm.2003.v39.i2.318

Simmons, M. (1997). *COSO based auditing*. Internal Auditor.

Singh, K. (1997). The impact of technological complexity and interfirm cooperation on business survival. *Academy of Management Journal*, *40*(2), 339–367. doi:10.2307/256886

Singpurwalla, N. D. (1995). Survival in dynamic environments. *Statistical Science*, *10*(1), 86–103. doi:10.1214/ss/1177010132

Sitoe, A., Salomão, A., & Wertz-Kanounnikoff, S. (2012). The context of REDD+ in Mozambique- Drivers, agents and institutions. CIFOR Occasional Paper 79, Centre For International Forestry Research, Bogor.

Si, X. S., Wang, W., Hua, C. H., & Zhou, D. H. (2011). Remaining useful life estimation – A review on the statistical data driven approaches. *European Journal of Operational Research*, *213*(1), 1–14. doi:10.1016/j.ejor.2010.11.018

Skagestad, B., & Mildenstein, P. District Heating and Cooling Connection Handbook. Retrieved from http://www.districtenergy.org/assets/CDEA/Best-Practice/IEA-District-Heating-and-Cooling-Connection-Handbook.pdf

Smets, P. (1993). Belief functions: The disjunctive rule of combination and the Generalized Bayesian Theorem. *International Journal of Approximate Reasoning*, *9*(1), 1–35. doi:10.1016/0888-613X(93)90005-X

Smets, P. (2002). The application of the matrix calculus to belief functions. *International Journal of Approximate Reasoning*, *31*(1-2), 1–30. doi:10.1016/S0888-613X(02)00066-X

Smets, P. (2005). Decision making in the TBM: The necessity of the pignistic transformation. *International Journal of Approximate Reasoning*, *38*(2), 133–147. doi:10.1016/j.ijar.2004.05.003

Smets, P., & Kennes, R. (1994). The transferable belief model. *Artificial Intelligence*, *66*(2), 191–234. doi:10.1016/0004-3702(94)90026-4

Smith, K., Balakrishnan, K., Butler, C., Chafe, Z., Fairlie, I., . . . (2012). Energy And Health. In T. Johansson, N. Nakicenovic, A. Patwardhan, & L. Gomez-Echeverri (Eds.), Global Energy Assessment: Toward a Sustainable Future (pp. 255-324). Cambridge, Cambridge University Press/Laxenburg, Austria: Int. Inst. Appl. Syst. Anal. doi:10.1017/CBO9780511793677.010

Smith, A. M., & Hinchcliffe, G. R. (2003). *RCM--Gateway to world class maintenance*. Burlington, MA: Butterworth-Heinemann.

Smith, A. M., & Hinchcliffe, G. R. (2004). *RCM – Gateway to World Class Maintenance*. Burlington, MA, USA: Elsevier Butterworth-Heinemann.

Smit, K. (2014). *Maintenance engineering and management*. Delft, The Netherlands: DAP Delft Academic Press.

Society for Maintenance and Reliability Professionals. (2014, December 5). Retrieved from http://library.smrp.org/metrics/

Society of Automotive Engineers. (2009). *SAE JA1011 Evaluation Criteria for Reliability-Centered Maintenance (RCM) Processes*.

Society of Automotive Engineers. (2011). *A Guide to the Reliability-Centered Maintenance (RCM)*.

Sokovic, M., Pavletic, D., & Kern Pipan, K. (2010). Quality Improvement Methodologies – PDCA Cycle, RADAR Matrix, DMAIC and DFSS. *Journal of Achievements in Materials and Manufacturing Engineering*, *43* (1), 476-483.

Sols, A. (2000). Fiabilidad, Mantenibilidad, Efectividad, un enfoque sistémico, Comillas, Madrid.

Souchère, V., Millair, L., Echeverria, J., Bousquet, F., Le Page, C., & Etienne, M. (2010). Co-Constructing With Stakeholders A Role-Playing Game To Initiate Collective Management Of Erosive Runoff Risks At The Watershed Scale. *Environmental Modelling & Software*, *25*(11), 1359–1370. doi:10.1016/j.envsoft.2009.03.002

Spira, L. F., & Page, M. (2001). Risk management The reinvention of internal control and the changing role of internal audit. *Accounting, Auditing & Accountability Journal*, *16*(4), 640–661. doi:10.1108/09513570310492335

SPRI. (2015) *Business Development Basque Agency: Machinery and Equipment Sector*. Retrieved from http://www.spri.eus/en/invest-in-the-basque-country/significant_sectors/machinery_equipment_sector

Sree, P. C. R., & Kujawski, D. (2014). A Deviatoric Stress-Strain Function for Fatigue Life Analysis. *Advanced Materials Research*, *891-892*, 789–802. doi:10.4028/www.scientific.net/AMR.891-892.797

Standa-Gunda, W., Mutimukuru, T., Nyirenda, R., Prabhu, R., Haggith, M., & Vanclay, J. (2003). Participatory Modelling To Enhance Social Learning, Collective Action And Mobilization Among Users Of The Mafungautsi Forest, Zimbabwe. *Small-scale Forestry*, *2*(2), 313–326.

Standard method for estimating the average grain size of metals, ASTM: E 112-63. (n. d.).

Steptoe-Warren, G., Howat, D., & Hume, I. (2011). Strategic Thinking And Decision Making: Literature Review. *Journal of Strategy and Management*, *4*(3), 238–250. doi:10.1108/17554251111152261

Stewart, G. T., Kolluru, R., & Smith, M. (2009). Leveraging public-private partnerships to improve community resilience in times of disaster. *International Journal of Physical Distribution & Logistics Management*, *39*(5), 343–364. doi:10.1108/09600030910973724

Suárez, J. C., Molleda, F., González, R., & Jiménez, R. (1994). Correlation of modified crack tip opening distance with heat input to the heat affected zone of high-strength low-alloy steels. *Theoretical and Applied Fracture Mechanics*, *21*(1), 17–22. doi:10.1016/0167-8442(94)90004-3

Sugeno Fuzzy Models . (n. d.). Retrieved from http://www.bindichen.co.uk/post/AI/takagi-sugeno-fuzzy-model.html

Surucu1 B., & Sazak, H.S. (2013). Graphical Methods for Reliability Data. In *Wiley Encyclopedia of Operations Research and Management Science*. John Wiley & Sons, Inc.DOI:10.1002/9780470400531.eorms1062

Suzaki, K. (1991). Competitividad en la Fabricación en la Década de los 90. (Técnicas para la Mejora Continua). Madrid, Spain: TGP, S.A.

Swinford, S. (2010, December 27). Theresa Villiers: Government knew Heathrow was not prepared for snow. *Telegraph UK*. Retrieved from http://www.telegraph.co.uk/news/weather/8225790/Theresa-Villiers-government-knew-Heathrow-was-not-prepared-for-snow.html

Taghipour, S., Banjevic, D., & Jardine, A. K. S. (2011). Prioritization of medical equipment for maintenance decisions. *The Journal of the Operational Research Society*, *62*(9), 1666–1687. doi:10.1057/jors.2010.106

Taha, H. A. (2011). *Operations research. An introduction*. New York, NY: Prentice Hall.

Tamilselvan, P., Wang, Y., & Wang, P. (2012). Deep Belief Network based state classification for structural health diagnosis. *Proceedings of theIEEE Aerospace Conference*, Big Sky, Montana, USA (pp. 1-11). doi:10.1109/AERO.2012.6187366

Tamilselvan, P., & Wang, P. (2013). Failure diagnosis using deep belief learning based health state classification. *Reliability Engineering & System Safety*, *115*, 124–135. doi:10.1016/j.ress.2013.02.022

Tari, J., & Garcia, M. (2009). Dimensiones de la gestión del conocimiento y la gestión de la calidad: Una revisión de la literatura. *Investigaciones Europeas de Dirección y Economía de la Empresa*, *15*(3), 135–148. doi:10.1016/S1135-2523(12)60105-1

Taylor, P. (2015). Constructivism. In R. Gunstone (Ed.), Encyclopedia of Science Education (pp. 218-224). SpringerLink.

Teisman, G. R., & Klijn, E. H. (2002). 10 Public-private partnerships in the European Union. In *Public-Private Partnerships: theory and practice in international perspective* (p. 165).

Temperature Optimization saves millions for district heating. (n. d.). Retrieved from http://www.global-download.schneider-electric.com/mainRepository/EDMS_CORP7.nsf/69f5d72c7a0cf811c12573d800389503/a9c3fb166094c67b85257a78005c3f20/$FILE/TCC-Hjoerring-eng.pdf

Tersine, R. J. (1994). *Principles of Inventory and Materials Management* (4th ed.). Englewood Cliffs, NJ: Prentice-Hall.

The Concise Oxford Dictionary. (2012). Oxford: Oxford University Press.

The Council of the European Union. (2008). COUNCIL DIRECTIVE 2008/114/EC of 8 December 2008 on the identification and designation of European critical infrastructures and the assessment of the need to improve their protection. *European Union*.

The Guidance for district heating companies. (2012). The environmental benefits of district heating: using the new Ecoheat4cities label Guidance for district heating companies. Retrieved from http://www.bre.co.uk/filelibrary/rpts/ecoheat4cities/Ecoheat4Cities_WP4_Guidance_for_companies.pdf

The Heat Monitoring Solution of ENMAT Software. (n. d.). Retrieved from http://www.en-mat.com/solutions/heat-monitoring

The Official Website of ePortal: Energy Insight Software. (n. d.). Retrieved from http://eportal.eu/en/

The Official Website of Landis+Gyr Company. (n. d.). Retrieved from http://www.landisgyr.no/

The Official Website of Wattics Company. (n. d.). Retrieved from http://www.eltec-petrol.com/home/

Thi, C. H., & Swierczek, F. W. (2010). Critical success factors in project management: Implication from Vietnam. *Asia Pacific Business Review*, *16*(4), 567–589. doi:10.1080/13602380903322957

Tiddens, W. W., Braaksma, A. J. J., & Tinga, T. (2015). The Adoption of Prognostic Technologies in Maintenance Decision Making: A Multiple Case Study. *Procedia CIRP*, *38*, 171–176. doi:10.1016/j.procir.2015.08.028

Tiemessen, L., Lane, H. W., Crossan, M. M., & Yinkpen, A. C. (1997). Knowledge management in international joint ventures. In P.W. Beamish, & J.P. Killing (Eds.), Cooperative strategies. North American perspectives. The New Lexington Press.

Tinga, T. (2010). Application of physical failure models to enable usage and load based maintenance. *Reliability Engineering & System Safety*, *95*(10), 1061–1075. doi:10.1016/j.ress.2010.04.015

Tinga, T., & Janssen, R. (2013). The interplay between deployment and optimal maintenance intervals for complex multi-component systems. *Proceedings of the Institution of Mechanical Engineers, Part O: Journal of Risk and Reliability*, *227*(3), 227–240. doi:10.1177/1748006X13480743

Tirpak, T. M. (2005). Five steps to effective knowledge management. *Research Technology Management*, *48*(3), 15–16.

Tompkins, J. A. (1989). *Winning Manufacturing: The How-to Book of Successful Manufacturing*. Norcross, GA: Industrial Engineering and Management Press.

Tragett, C. (2012). *From romanticized Women to politicized Gender: How has the literature addressed Gender and Natural Resource Use?* [Master Thesis]. Imperial College London, London.

Tsang, A. H. C. (1998). A strategic approach to managing maintenance performance. *Journal of Quality in Maintenance Engineering*, *4*(2), 87–94. doi:10.1108/13552519810213581

Tse, P., & Atherton, D. (1999). Prediction of machine deterioration using vibration based fault trends and recurrent neural networks. *Journal of Vibration and Acoustics*, *121*(3), 355–362. doi:10.1115/1.2893988

Turell, J. (2013). 10 breakthrough technologies 2013: Deep learning. *MIT Technology Review*. Retrieved from **Error! Hyperlink reference not valid.**http://www.technologyreview.com/featuredstory/513696/deep-learning/

Turner, B. A. (1976). The organizational and interorganizational development of disasters. *Administrative Science Quarterly*, *21*(3), 378–397. doi:10.2307/2391850

U.S. Securities and Exchange Comission. (2013). The Laws That Govern the Securities Industry. Retrieved from http://www.sec.gov/about/laws.shtml#sox2002

UNDP. (2014). *Human Development Report 2014, Sustaining Human Progress: Reducing Vulnerabilities And Building Resilience*. New York: United Nations Development Program.

Ungar, L. (2007). Design for diagnosability guidelines. *IEEE Instrumentation & Measurement Magazine*, 2007, 24–32.

Vagliasindi, F. (1989). *Gestire la manutenzione. Perche e come*. Milan: Franco Angeli.

Valdimarsson, P. (2014). District Heat Distribution Networks. Retrieved from http://www.os.is/gogn/unu-gtp-sc/UNU-GTP-SC-18-27.pdf

van den Belt, M. (2004). *Mediated Modelling: A System Dynamics Approach To Environmental Consensus Building*. Washington: Island Press.

van Dongen, L. A. M. (2011). *Maintenance engineering: Maintaining links*. Enschede, The Netherlands: Faculty of Engineering Technology, University of Twente.

van Dongen, R. (2011). *Referentiemodel voor levensduurverlenging van technische assets* [Reference model for life extension of physical assets]. VITALE Mainnovation.

Van Horenbeek, A., Pintelon, L., & Muchiri, P. (2010). Maintenance optimization models and criteria. *Int J Syst Assur Eng Manag*, *1*(3), 189–200. doi:10.1007/s13198-011-0045-x van Kuijk, G. (2013), *P13-13 EUROS – Excellence in Uncertainty Reduction of Offshore wind Systems.* Retrieved from http://www.stw.nl/nl/content/p13-13-euros-%E2%80%93-excellence-uncertainty-reduction-offshore-wind- systems

Van Noortwijk, J. (2009). A survey of the application of gamma processes in maintenance. *Reliability Engineering & System Safety*, *94*(1), 2–21. doi:10.1016/j.ress.2007.03.019

Van Ostaeyen, J. (2014). *Analysis of the business potential of product-service systems for investment goods* [PhD thesis]. KU Leuven, Faculty of Engineering Science.

Vandermerwe, S., & Rada, J. (1988). Servitization of Business: Adding Value by Adding Services. *European Management Journal*, *6*(4), 314–324. doi:10.1016/0263-2373(88)90033-3

Vanneste, S. G., & van Wassenhove, L. N. (1995). An integrated and structured approach to improve maintenance. *European Journal of Operational Research*, *82*(2), 241–257. doi:10.1016/0377-2217(94)00261-A

Vatn, J. (1997). Maintenance optimisation from a decision theoretical point of view. *Reliability Engineering & System Safety*, *58*(2), 119–126. doi:10.1016/S0951-8320(97)00025-2

Vaz, K., Storey, P., Cipriano, A., Cuamba, B., D'Addario, P., Falcão, P., & Lehman, N. (2011). Green Investments. In *Mozambique*. Maputo: Banco de Moçambique, Agence Francaise de Development.

Veenman, A., & Besselink, L. (2010). *Asset owners & maintenance*. Breda, The Netherlands: Academic Press.

Veldman, J., Klingenberg, W., & Wortmann, H. (2011). Managing condition-based maintenance technology: A multiple case study in the process industry. *Journal of Quality in Maintenance Engineering*, *17*(1), 40–62. doi:10.1108/13552511111116240

Ventura, J. & Ordóñez de Pablos, P. (2007). Análisis de estrategias de conocimiento en la industria manufacturera española: Evidencia empíricas. *Tribuna de Economía*, 836, 141-161.

Ventura, J. (1996). *Análisis dinámico de la estrategia empresarial: Un ensayo interdisciplinar. Servicio de publicaciones*. Universidad de Oviedo.

Vera, D., Carabias, J., Jurado, F., & Ruiz-Reyes, N. (2010). A Honey Bee Foraging Approach For Optimal Location Of A Biomass Power Plant. *Applied Energy*, *87*(7), 2119–2127. doi:10.1016/j.apenergy.2010.01.015

Verma, T., & Pearl, J. (1991). A Theory of Inferred Causation. *Proceedings of theSecond International Conference on the Principles of Knowledge Representation and Reasoning*, Cambridge, Massachusetts, USA.

Verron, S. (2007). *Diagnostic et surveillance des processus complexes par réseaux bayésiens* [PhD thesis]. Institut des Sciences et Techniques de l'Ingénieur d'Angers.

Villeneuve, E., Beler, C., Peres, F., & Geneste, L. (2012). Hybridization of statistical and cognitive experience feedbacks to perform risk assessment. *Proceedings of theIEEE International Conference on Industrial Engineering and Engineering Management*, Hong-Kong, China.

Visser, J., & Pretorious, M. (2002). The development of a performance measurement system for maintenance. Retrieved from http://sajie.journals.ac.za

Volkel, M., & Abecker, A. (2008). Cost-benefit analysis for the design of personal knowledge management systems. Proceedings of the International Conference on Enterprise Information Systems ICEIS '08 (pp. 95–105).

Volkel, M., & Haller, H. (2009). Conceptual data structures for personal knowledge management. *Online Information Review*, *33*(2), 298–315. doi:10.1108/14684520910951221

Volk, R., Stengel, J., & Schultmann, F. (2014). Building Information Models (BIM) for existing buildings –literature review and future needs. *Automation in Construction*, *38*, 109–127. doi:10.1016/j.autcon.2013.10.023

Von Bertalanffy, L. (1989). Teoría General de los Sistemas (7th ed.). New York: George Braziller.

Waeyenbergh, G., & Pintelon, L. (2002). A framework for maintenance concept development. *International Journal of Production Economics*, *77*(3), 299–313. doi:10.1016/S0925-5273(01)00156-6

Wah, L. (1999). Behind the Buzz. *Management Review*, *88*(4), 16–19.

Wamukonya, N. (2007). Solar Home System Electrification As A Viable Technology Option For Africa's Development. *Energy Policy*, *35*(1), 6–14. doi:10.1016/j.enpol.2005.08.019

Wang, L., Chu, J., & Wu, J. (2007). Selection of optimum maintenance strategies based on a fuzzy analytic hierarchy process. *International Journal of Production Economics*, *107*(1), 151–163. doi:10.1016/j.ijpe.2006.08.005

Weber, C., Marechal, F., & Favrat, D. (2006). Design and Optimization of District Energy Systems. *Proceedings of the 10th International Symposium on District Heating and Cooling*, Lausanne, Switzerland. Retrieved from http://www.lsta.lt/files/events/16_weber.pdf

Weir, J., Crew, D., & Crew, J. (2013). Wetland Forest Culture: Indigenous Activity For Management Change In The Southern Riverina, New South Wales. *Australasian Journal of Environmental Management*, *20*(3), 193–207. doi:10.1080/14486563.2013.819303

Wernerfelt, B. (1984). A resource based view of the firm. *Strategic Management Journal*, *5*(2), 171–180. doi:10.1002/smj.4250050207

Wernstedt, F. (2003). Multi-Agent Systems for District Heating Management. Retrieved from http://www.bth.se/fou/forskinfo.nsf/0/d613f6536b050fa6c1256e2800391f91/$file/Lic.pdf

Wernstedt, F., Davidsson, P., & Johansson, C. (n. d.). Simulation of District Heating Systems for Evaluation of Real-time Control Strategies. Retrieved from http://cj.noda.se/papers/ESMC-WernstedtDavidssonJohansson.pdf

Wetzer, J. M., & Bouwman, S. (2007). *Managing risks in ageing networks: A case study. Leonardo Energy - Electricity briefing paper*. Arnhem, The Netherlands: KEMA Consulting.

Whittaker, B. (1999). What went wrong? Unsuccessful information technology projects. *Information Management & Computer Security*, *7*(1), 23–29. doi:10.1108/09685229910255160

WHO. (2002). *Addressing the Impact Of Household Energy And Indoor Air Pollution on the Health Of the Poor: Implications for Policy Action And Intervention Measures, World Health Organisation*. Geneva: WHO.

Wieland, J. (2005). Corporate governance, values management, and standards: A European perspective. *Business & Society*, *44*(1), 74–93. doi:10.1177/0007650305274852

Wiig, K. M. (1997, June). Integrating Intellectual Capital and Knowledge Management. *Long Range Planning*, *30*(3), 399–405. doi:10.1016/S0024-6301(97)90256-9

Williams, W. Asset Management Perspective (n. d.). Retrieved from http://bv.com/Home/news/solutions/Expert-Perspectives/realizing-value-with-iso-55001

Wilson, R. H. (1934). A Scientific Routine for Stock Control. *Harvard Business Review*, *13*, 116–128.

Wireman, T. (1998). *Developing Performance Indicators For Managing Maintenance*. New York, NY: Industrial Press.

Wireman, T. (2004). *Total productive maintenance*. Industrial Press.

Witten, I., Frank, E., & Hall, M. (2011). *Data mining. Practical Machine Learning Tools and Technques*. San Francisco: Morgan Kaufmann.

Wolbert-Haverkamp, M., Feil, J., & Mußhoff, O. (2014). The Value Chain Of Heat Production From Woody Biomass Under Market Competition And Different Intervention Systems: An Agent-Based Real Options Model (Discussion Paper 1407). Department of Agricultural Economics and Rural Development, Georg-August-University, Goettingen.

Wolf, M. (2008). Strategic management of medical equipment. Starting and operating a TSC. *Proceedings of the 20th Congress of the International Federation of Hospital Engineering*, Barcelona, Spain.

Wong, K. Y., & Aspinwall, E. (2004). Knowledge management implementation frameworks: A review. *Knowledge and Process Management*, *11*(2), 93–104. doi:10.1002/kpm.193

Woodhouse, J. (1993). *Managing industrial risk*. London: Chapman Hill.

Woodward, D. G. (1997). Life cycle costing theory. Information, acquisition and application. *International Journal of Project Management*, *15*(6), 335–344. doi:10.1016/S0263-7863(96)00089-0

World Bank. (2010). *Economics of Adaptation to Climate Change Mozambique*. Washington: The World Bank.

World Bank. (n. d.). World Bank World Data Base. Retrieved from http://data.worldbank.org

World Bank. Asian Development Bank, & Inter-American Development Bank. (2014). Public-Private Partnerships: Reference Guide, Version 2.0. Retrieved from https://openknowledge.worldbank.com/handle/10986/20118

WRI. (2010). Country Profiles- Mozambique, World Resources Institute. Retrieved from http://earthtrends.wri.org

Wright, K. (2005). Personal knowledge management: Supporting individual knowledge worker performance. *Knowledge Management Research and Practice*, *3*(3), 156–165. doi:10.1057/palgrave.kmrp.8500061

Wu, B., Saxena, A., Patrick, R., & Vachtsevanos, G. (2005). Vibration monitoring for fault diagnosis of helicopter planetary gears.*Proceedings of the 16th IFAC World Congress*.

Xiomara, P. (2009). La gestión del conocimiento y las Tics en el siglo XXI. *CONHISREMI, Revista Universitaria de Investigación y Diálogo Académico*, 5(1).

Xu, H., & Smets, P. (1994). Evidential reasoning with conditional belief functions.*Proceedings of the 10th Conference on Uncertainty in Artificial Intelligence* (pp. 598–605).

Yaghlane, B. B., Smets, P., & Mellouli, K. (2001). Belief function independence: I. The marginal case. *International Journal of Approximate Reasoning*, *29*(1), 47–70. doi:10.1016/S0888-613X(01)00055-X

Yaghlane, B. B., Smets, P., & Mellouli, K. (2004). Directed Evidential Networks with Conditional Belief Functions. Symbolic and Quantitative Approaches to Reasoning with Uncertainty, *LNCS* (Vol. *2711*, pp. 291–305). doi:10.1007/978-3-540-45062-7_24

Yang, J. (2006). La estrategia de gestión del conocimiento y su efecto en el crecimiento corporativo. *Economía Industrial*, 362, 123-133.

Yang, Y., & Shi, L. (2000). Integrating Environmental Impact Minimization Into Conceptual Chemical Process Design- A Process Systems Engineering Review. *Computers & Chemical Engineering*, *24*(2-7), 1409–1419. doi:10.1016/S0098-1354(00)00384-7

Yin, R. K. (1995). *Case study research: design and methods*. California: Sage Publications.

Yuan, G., Lan, Z., & Gan, M. (2013). Composite fault location of gearbox based on wavepack and BP neural network ensemble.*Proceedings of the 32nd Chinese Control Conference* (pp. 3795-3799).

Yukl, G. (2008). Liderazgo en las organizaciones (6th ed.). Madrid: Pearson Educación S.A.

Yusta, J. M., Correa, G. J., & Lacal-Arántegui, R. (2011). Methodologies and applications for critical infrastructure protection: State-of-the-art. *Energy Policy*, *39*(10), 6100–6119. doi:10.1016/j.enpol.2011.07.010

Zaal, T. M. E. (2013). *Profit-driven maintenance for physical assets* (2nd ed.). Maj Engineering Publishing.

Zahir, S. (1999). Geometry of decision making and the vector space formulation of the analytic hierarchy process. *European Journal of Operational Research*, *112*(2), 373–396. doi:10.1016/S0377-2217(98)00020-4

Zaim, S., Turkyílmaz, A., Acar, M. F., Al-Turki, U., & Demirel, O. F. (2012). Maintenance strategy selection using AHP and ANP algorithms: A case study. *Journal of Quality in Maintenance Engineering*, *18*(1), 16–29. doi:10.1108/13552511211226166

Zapata, L. (2001). *La Gestión del Conocimiento en Pequeñas Empresas de Tecnología de la Información: Una Investigación Exploratoria. Document de treball núm. 2001/8.* Universitat Autònoma de Barcelona. Facultat de Ciències Econòmiques i Empresarials.

Zeithaml, V. A., Parasuraman, A., & Berry, L. L. (1990). *Delivering quality service: Balancing customer perceptions and expectations*. Simon and Schuster.

Zerriffi, H., & Wilson, E. (2010). Leapfrogging Over Development? Promoting Rural Renewables For Climate Change Mitigation. *Energy Policy*, *38*(4), 1689–1700. doi:10.1016/j.enpol.2009.11.026

Zhang, X., Arayici, Y., Wu, S., Abbott, C., & Aouad, G. (2009). Integrating BIM and GIS for large scale (building) asset management: a critical review. *Proceedings of theTwelfth International Conference on Civil, Structural and Environmental Engineering Computing*. doi:10.4203/ccp.91.93

Zhou, Y., Kou, G. and Ergu, D. (2014). Analysing Operating Data to Measure the Maintenance Performance. *Journal of Quality and Reliability Engineering International.* DOI:10.1002/qre.1584

Zio, E., Broggi, M., Golea, L., & Pedroni, N. (2012). Failure and reliability predictions by infinite impulse response locally recurrent neural networks. *Chemical engineering transactions, 26.*

Zio, E. (2009). Reliability engineering: Old problems and new challenges. *Reliability Engineering & System Safety*, *94*(2), 125–141. doi:10.1016/j.ress.2008.06.002

Zio, E., & Compare, M. (2013). Evaluating maintenance policies by quantitative modelling and analysis. *Reliability Engineering and System Safety, Elsevier*, *109*, 53–65. doi:10.1016/j.ress.2012.08.002

Zwikael, O., & Globerson, S. (2006). Benchmarking of project planning and success in selected industries. *Benchmarking. International Journal (Toronto, Ont.)*, *13*(6), 688–700.

Zwikael, O., Pathak, R. D., Singh, G., & Ahmed, S. (2014). The moderating effect of risk on the relationship between planning and success. *International Journal of Project Management*, *32*(3), 435–441. doi:10.1016/j.ijproman.2013.07.002

Zwikael, O., & Sadeh, A. (2007). Planning effort as an effective risk management tool. *Journal of Operations Management*, *25*(4), 755–767. doi:10.1016/j.jom.2006.12.001

About the Contributors

María Carmen Carnero has a Doctorate in Industrial Engineering from the University of Castilla-La Mancha (2001) and is a lecturer at the Higher Technical School of Industrial Engineering at the University of Castilla-La Mancha. She has published two books, four chapters in science books, and research articles in Omega, Decision Support System, European Journal of Operational Research, Reliability Engineering and System Safety, European Journal of Industrial Engineering, Journal of Manufacturing Systems, The International Journal of Advanced Manufacturing Technology, Mechanical Systems and Signal Processing, Production Planning and Control, etc. She has 57 contributions to conferences. She was the lead researcher in three projects financed by the regional government, and has been a part of 17 other European, national, regional and local projects, and of seven contracts.

Vicente González -Prida is PhD (Summa Cum Laude) in Industrial Engineering by the University of Seville, and Executive MBA (First Class Honors) by the Chamber of Commerce. He has been honoured with the following awards and recognitions: - Extraordinary Prize of Doctorate by the University of Seville; - National Award for PhD Thesis on Dependability by the Spanish Association for Quality; - National Award for PhD Thesis on Maintenance by the Spanish Association for Maintenance; - Best Nomination from Spain for the Excellence Master Thesis Award bestowed by the EFNSM (European Federation of National Maintenance Societies). Dr. Gonzalez-Prida is member of the Club of Rome (Spanish Chapter) and has written multitude of articles for international conferences and publications. His main interest is related to industrial asset management, specifically the reliability, maintenance and aftersales organization. He currently works as Program Manager in the company General Dynamics - European Land Systems and shares his professional performance with the development of research projects in the Department of Industrial Organization and Management at the University of Seville.

* * *

Alfredo R. Aguilar Mechanical Engineer (2000) Universidad Central de Venezuela, Reliability Engineer (2015) Universidad Simon Bolivar. PDVSA Oil & Gas experience as Maintenance and Reliability Engineer. MWH Global Maintenance/Reliability Engineer (Third Set of Locks Project in the Panama Canal).

Olga Aleksenko is an Associate Professor in the Department of Computer Science at the Sumy State University where she has been a faculty member since 2005. She completed her Ph.D. and undergraduate studies at Sumy State University. Her research interests lie in the area of modelling of information systems and simulation of processes with a focus on probability problems.

David Almorza is Lecturer at the University of Cádiz. Statistics Department. Dr. in Mathematics. Born in Cádiz (Spain).

Luis Barberá Martínez is PhD (Summa Cum Laude) in industrial Engineering by the University of Seville, and Mining Engineer by UPC (Spain). He is a researcher at the School of Engineering of the University of Seville and author of more than fifty articles, national and main internationals. He has worked at different international Universities: Politecnico di Milano (Italy), CRAN (France), UTFSM (Chile), C-MORE (Canada), EPFL (Switzerland), University of Salford (UK) or FIR (Germany), among others. His line of research is industrial asset management, maintenance optimization and risk management. Currently, he is Spain Operations Manager at MAXAM Europe. He has three Master's degrees: Master of Industrial Organization and Management (2008/2009) (School of Engineering, University of Seville), Master of Economy (2008/2009) (University of Seville) and Master of Prevention Risk Work (2005/2006). He has been honoured with the following awards and recognitions: Extraordinary Prize of Doctorate by the University of Seville; three consecutive awards for Academic Engineering Performance in Spain by AMIC. (2004/2005), (2005/2006) and (2006/2007); honor diploma as a "Ten Competences Student by the University of Huelva and CEPSA Company (2007); graduated as number one of his class.

Cédrik Béler is associate professor at the National Engineering School of Tarbes of the National Polytechnic Institute of Toulouse (ENIT-INPT) since 2010. He received his PhD degree of the University Paul Sabatier (Toulouse) in 2008. He is currently member of the "Cognitive and Decisional Systems" of the Production Management Laboratory in Tarbes (LGP-SDC). His current research interests relate to knowledge engineering. More specifically, he works on knowledge representation and inference while considering uncertainty and risks.

Jan Braaksma is an assistant professor in the chair of Maintenance Engineering and director of the WCM Summer School. He is a member of the maintenance research centre TIME. He has worked for the University of Groningen (RuG) and the Dutch Defense Academy (NLDA). He holds a Master's degree in Business and ICT and a PhD degree in Economics and Business and conducts research on Maintenance Engineering, Asset information and life-cycle management.

Iñaki Bravo-Imaz Industrial Engineer, degreed from The College of Engineering of the University of Navarra (TECNUN). Having worked in the department of organization of Michelin, in Metalocaucho, at the department of new products, and in the research and development center CEIT studying issues related to material fatigue. Currently a member of the Intelligent Information Systems Unit in IK4-Tekniker involved in several statewide projects related to Health Monitoring of industrial equipment, as well as a PhD student in Novel Materials in the Faculty of Science and Technology of the University of the Basque Country. Nowadays working in the thesis entitled "Magnetoelastic sensor development for continuous monitoring of lubricant oils".

Javier Cárcel-Carrasco is an Industrial Engineer with a PhD in Industrial Engineering from the Universitat Politècnica de València (UPV), Spain. He holds a PhD in Economic and Business Sciences from UNED, Spain. He is currently Professor and Head of the area of Installation and Maintenance in the Department of Building Construction in the UPV. He is a member of the Institute of Energy Engineering of the UPV. His main research areas are renewable energy sources, energy efficiency and industrial maintenance engineering.

Eduardo Castellano is BSc. Mechanical and Electrical Engineering (Nottingham University), BSc. Industrial Organization Engineering (TECNUN), MSc. Complex Systems (UPV-EHU) and PhD in Management Systems (UPV-EHU). He also holds postgraduates in Logistics (ICIL) and Microeconomics of Competitiveness (Deusto-Harvard). Since 1997 he has been working at IK4-IKERLAN, first in the Production Systems Management area (1997-2000) and after that in the Strategic Innovation area (2001-2010) and in the Advance Manufacturing Technologies area (2011-nowadays). During 2003-2004 he worked as visiting researcher in the Complexity Research Group of the London School of Economics. His fields of expertise are Supply Networks, Product-Service Systems, Technology Innovation Management and Business Development. In each of these fields he has participated and leaded several national and European projects as well as SMEs and large enterprises application research projects; +20 new product-service and O&M logistics enterprise development actions, as well as the development of feasibility plans, investment projects analyses, and coached-mentored the business development of technology-based new businesses in the energy, capital goods, and facility management sectors. He has published numerous peer-reviewed papers/articles and is also Programme Committee Member (e.g. Industrial and Simulation Conference), Panel Group Member (e.g. McKinsey Quarterly Executive) and Journal Reviewer (e.g. International Journal of Information Management) in several conferences, journals and special interest groups related to the cited research domains.

Peter Eecen is the programme development manager Wind Energy at ECN. He is an expert in offshore wind power, wind turbine design and operation and maintenance aspects of offshore wind farms.

Santiago Fernandez received his B.Sc. in Physics from the University of Santiago de Compostela in 1996, and M.Sc. (equiv.) in 1997. In 2001 he received his Ph.D. degree for work in the area of speech perception and acoustic phonetics. From November 2001 till 2003, he held a part-time lecturer position in the Department of Applied Physics of the University of Santiago de Compostela. Between 2003-2005, he was a post-doctoral Marie Curie Fellow at University College London, where he studied the importance of the local segmental context in the recognition of phonemes in speech signals by humans and machines. Between 2005-2008 he was a post-doctoral researcher at the Dalle Molle Institute for Artificial Intelligence, Switzerland, focusing on artificial recurrent neural networks for sequence learning tasks and in particular in aspects of speech recognition. In 2008 he joined Tekniker, where he works on intelligent information systems. He has published more than 25 papers in conferences, and 7 journal articles.

Susana Ferrerio Del Río got her degree in Computing science at the Technical School in the University of the Basque Country (Spain) in 2005. In 2012 she received her PhD (cum laude) in the area of "Probabilistic Models for Artificial Intelligence and Data Mining". Since 2005 she has been working in Tekniker in national and European projects in the area of health assessment, diagnosis and prognosis applied to different industrial sectors such as manufacturing, aeronautics, chemistry, etc. She is an expert on several technologies related to data processing and analysis (such as artificial intelligence, data mining, machine learning, probabilistic models, etc) applied to diagnosis and prognosis for an intelligent and predictive maintenance support; and on the design and implementation of expert systems based on these techniques with friendly graphical user-interfaces (i.e. for the control of high precision laboratory equipment). Today, she is working specially on aviation projects aimed at the development of health monitoring systems, on electromechanical actuation systems of A/C and UAVs, to detect and predict system failures. She has published more than 10 papers in conferences, 5 articles in journals with impact index (ISSI) and some books collaboration.

Leo van Dongen (1954) has worked for the Netherlands Railways (NS, 100% state owned) for over 30 years and since 2003 has been a director of NedTrain, the full subsidiary company of NS which is responsible for maintaining rolling stock. Since 2010, he has been a part-time professor in Maintenance Engineering at the faculty of Engineering Technology, in the department of Design, Production and Management of the University of Twente. After his studies in mechanical engineering, he completed his doctoral research at Eindhoven University of Technology on energy efficiency of drive trains for electric vehicles. At DAF Trucks he was active in the development of diesel engines. His career within NS has concentrated mainly on technological functions. These include project manager of electric locomotives, secretary to the executive board, fleet manager at NS Reizigers, and now, within Ned Train, he is responsible for technical fleet management, maintenance systems, maintenance management and construction of new capital goods: workshops and other maintenance equipment. In addition to his work at NedTrain he promotes the academic development of the profession and encourages further research into maintenance processes: from design methodologies for capital goods to the development of associated maintenance concepts, not only for the initial investment, but also for the management during the entire life-cycle!

Andrés Gómez received his technical industrial engineering degree from the Polytechnic University of Madrid (Spain) and his industrial engineering degree from the University of Castilla-La Mancha. He has a Doctorate in Industrial Engineering from the University of Castilla-La Mancha (2012). He has worked in the Nuclear Power Stations of Ascó and Trillo and Repsol-YPF as maintenance engineer. He was Manager of the Maintenance Service of the Hospital Complex of Ciudad Real until 2004. Currently, he is Vice-director of the Technical Services of the University General Hospital of Ciudad Real (Spain). He has published a research article in Production Planning and Control and has participated in several conferences and research projects supported by Regional Administration.

Rafael González-Palma Department of Mechanic Engineering and Industrial Design, University of Cádiz (Spain). Honorary Professor of the University of Cádiz.

Antonio Jesús Guillén López is a PhD candidate and contracted researcher in Intelligent Maintenance System research group (SIM) of the University of Seville (USE), focusing his studies in Prognosis Health Management & Condition Based Maintenance applications and Assets Management. Industrial Engineering and Master in Industrial Organization & Business Management from the University of Seville (USE). From 2003 to 2004 working for the Elasticity and Strength of Materials Group (GERM) of the USE in aeronautical materials tests. From 2006 to 2009 member of the Department of Electronic Engineering of USE, working in numerous public-private international R&D project, developing new thermo-mechanical design applications for improving the performance and the life cycle of power electronics system. From 2009 to 2010 is Project Manager of "Solarkit" project, representing Andalucia's Government and USE in the international competition Solar Decathlon 2010. Is foundational partner of Win Inertia Tech., technological based company specializing in electronic R&D for Smart Grids and Energy Store fields, where he has carried out different roles, from R&D engineer to General Manager until September 2012. Nowadays is coordinator of the Spanish National Research Network of Asset Management.

Aitor Arnaiz Irigaray is head of the intelligent information systems (IIS) unit at IK4-Tekniker, where he works since 1991, having participated and led more than 15 international projects. He holds an MSc on 'Applied Artificial Intelligence' at Aberdeen University and a PhD at Sunderland Univ. with the thesis 'Development of Artificial Intelligence Technologies to support automation of condition monitoring tasks'. During last 5 years he has also been head of the diagnostics and prediction unit and responsible of the research line in maintenance and reliability at Tekniker. He has published more than 75 refereed papers in international journals and conferences, co-edited different books, and co-chaired several workshops and conferences, most of them in the field of condition monitoring, maintenance and reliability. He is also member of the steering board of ESREDA (European Safety, Reliability and Data Association), the editorial board of Risk and Reliability Journal (IMechE-PartO) and Comadem (Condition Monitoring and diagnostic engineering management) International Journal.

Samir Khan is a lecturer on aerospace engineering at Coventry University. He had gained his PhD in Control Theory from Loughborough University in 2010. He was the leading researcher on the No Fault Found research project between 2011-2015, at Cranfield University, collaborating with Rolls-Royce MRO, BAE Systems, UK MoD and Jaguar Land Rover. Dr Khan's current research work is focused on intelligent monitoring of intermittent failures and false alarms in electronic systems. He is a Chartered engineer and a member of the IEEE and IET.

Khairy A. H. Kobbacy has joined the College of Engineering at Taibah University, Saudia Arabia as the Scientific Chair of MM Bin Ladin for Operation and Maintenance Technology in April 2014. Previously he was a Professor of Management Science at Salford University, UK. He also lectured at Strathclyde University after gaining industrial experience in a major oil company. Khairy has long-standing interest in 'applied' operational research and his research interests are in maintenance modelling and optimisation, the development of intelligent management systems in operations and modelling and simulation in healthcare. His research has been funded by industry and the research councils and he published extensively in international journals. He is the co-author of Complex system maintenance handbook (Kobbacy and Murthy, Springer 2008). Prof Kobbacy was awarded the UK Operational Research President's medal in 1990, the Literati Club Award in 2001 and the Building and Environment Award in 2013 for best published papers. He was a Vice President of the Operational Research Society, UK (2002–2004).

Fredy Kristjanpoller is an Industrial Engineer and Master in Asset Management and Maintenance of Federico Santa Maria University (USM, Chile), and doctoral candidate in Industrial Engineering at the University of Seville. Researcher, academic and master program coordinator linked to asset management. He has important scientific papers on indexed journals and international proceedings congress on the following areas: Reliability engineering and Maintenance Strategies. He has developed consultant activities in the main Chilean companies.

Egoitz Konde received a degree in Industrial Engineering in 2006 at University of Navarra in Spain. Egoitz Konde worked in Spicer Ayra Cardan, a member company of Dana Holding Corporation world leader in the supply of products for automotive industry, implementing Lean Manufacturing projects in the maintenance area to improve the availability of machines and thereby optimize production. Nowadays, Konde works in the Maintenance and Reliability area of Tekniker on different project related to dependability, predictive strategies and new technologies to obtain maintenance optimization. Using different techniques related to failure modes and event analysis (FMEA, FTA), reliability calculation tools and statistical analysis. Konde worked as teacher in the Own Master in Industrial Maintenance of the Seville University and also in the Master of Industrial Maintenance of Mondragon Unibertsitatea, as expert in Reliability Engineering, and collaborated in the edition of the 7th Maintenance Book of the Spanish Maintenance Association, as well as in different articles and conferences related to dependability and maintenance.

Etienne Le Page is an engineering student in finance, logistics, project and asset management.

Leire Labaka, Industrial Engineer (2009, PhD 2013), is a professor of Operational Research, Modelling and Simulation and Accounting and Finance at TECNUN. Her research interests include complex system modelling, critical infrastructures protection and resilience. She has taken part in SEMPOC and ELITE European projects. She has published several papers in conference proceedings as well as in journals such as Journal of Homeland Security and Emergency Management, Reliability Engineering and System Safety, Journal of Technological Forecast and Social Change, and International Journal of Critical Infrastructures.

Gabriel Llort is Deputy Design Manager of Electromechanical Works and O&M of the Panama Canal Third Set of Locks.

Carlos López-Escobar is an engineer with working experience in large multinational organizations including Delphi automotive, Valeo automotive and Alcoa. He holds a consolidated international management experience with exposure to different businesses and processes, including automotive, aerospace, food packaging, primary metals, rolling products, engineered products and others. Most of his background is on Operational Excellence (OpEx) which includes strategy deployment, leading change, maintenance, production, operations, quality, process/quality management and other aspects related to operations. Dr. López Escobar has been exposed to demanding customers, including Toyota, VW group, PSA group, Nissan, the Coca-Cola company, Pepsi, and others. He is an experienced OpEx internal auditor and consultant. Dr. López-Escobar has lead Operational Excellence assessments mainly in Europe, Asia and Australia. He is Co-Author of articles published in technical magazines and a experienced

OpEx trainer. His quality management experience has consisted in roles as Alcoa CSI regional Quality manager for Europe and North Africa operations. In addition, he held positions as quality Director in Valeo automotive systems. Previous positions in Delphi automotive included maintenance management, process engineering and production supervisory roles.

Patricia Maraña is a Ph.D Student in the Management Engineering department of TECNUN, the Engineering School of the University of Navarra in San Sebastian, Spain. Currently, she is a PhD student interested in Public-Private Partnership, Critical Infrastructures, Crisis Management and Resilience. In September 2015 she received her Master Degree of Research in Applied Engineering at TECNUN. In 2014 she received her Degree in Industrial Management Engineering from TECNUN, University of Navarra.

María de Lourdes Eguren Martí PhD Business Administration (Universitat de Barcelona. Barcelona, España). Bs. Mathematics (Universidad Metropolitana. Caracas, Venezuela), System's Engineering (Universidad Metropolitana. Caracas, Venezuela). Master on EBusiness: Telecommunications and new business models (Universidad de Cantabria. Santander, España). Master of Research in Business Administration (Universitat de Barcelona. Barcelona, España). Professional experience of more than 11 years in risk management working for several multinationals from different sectors (Deloitte, Visteon, Novartis and Bunge), including financial controlling, project management and process reengineering. Currently in charge of the Internal Control coordination for Europe, Middle East and Africa in Bunge. Associate professor. Tecnocampus - Universitat Pompeu Fabra. Faculty of Social Sciences and Business.

Ricardo Martins is Mozambican by heart, and holds a PhD in Energy Systems Design from Imperial College in London, a Master in Environmental Technology from Tokyo Institute of Technology and a Degree in Chemical/Process Engineering from Instituto Superior Técnico in Lisbon, which complement the practice of 12 years of international experience working in Mozambique, Japan, Mongolia and Rwanda. Ricardo's work deals mostly with facilitating, assessing, managing, implementing and conceptually design bioenergy projects and strategies both in developed and developing countries. A systems and design thinker, Ricardo creates methodologies, tools and approaches to facilitate and catalyze relevant actors in designing and implementing creative, participated, socially fitted, technically robust, ecological and economically viable bioenergy systems, projects and strategies. Current work included the development of participatory modelling approaches for bioenergy assets and resource management.

José N. Contreras Márquez Aeronautical Engineer (IUPFAN-1979-Venezuela). MSc in Mechanical Engineering (USB-1983-Venezuela). Project Management Specialist (USB-1985-Venezuela). Currently, ASME certified instructor and INGEMAN instructor for Latin America in the following courses: 1) Maintenance Planning, Scheduling and Costs. 2) Maintenance Inventory Optimization Management. 3) Maintenance Project Management. 4) Integration of Maintenance Technical Indicators to Financial Strategies in Asset Management. Advisor in INGECON (www.confiabilidadoperacional.com).

Pedro Mayorga Indutrial Engineer since 1995. Pedro Mayorga is working as CTO of EnerOcean S.L. after having been working as Technology Area Manager at ITE in Valencia (Spain) and Manufacturing and Maintenance Engineer in Delphi Automotive Group (previously GM).

Carlos Parra Rewarded Titles: - Naval Engineering, Polytechnic Institute of the National Armed Forces (IUPFAN), 1986-1991, Caracas, Venezuela. - Master in Maintenance Engineering, Universidad de los Andes, School of Mechanical Engineering, Master Program in Maintenance Engineering, 1994 - 1996, Merida, Venezuela. - Reliability Specialist Engineering, PDVSA Convention - University Maryland - ASME, 2002 - 2003, United States. - Specialist in Industrial Organization Engineering, School of Industrial Engineering, University of Seville, 2004 - 2006, Seville, Spain. - Diploma of Advanced Studies, Industrial Engineering area Organization, PhD in Industrial Engineering, School of Industrial Engineering, University of Seville, 2006 - 2008, Seville, Spain. - Doctor (PhD.) In Industrial Organization Engineering, University of Sevilla, School of Industrial Engineering, Industrial Engineering Department Organization, 2004-2009, Sevilla, Spain. AWARDS / HONORS or Gran Mariscal de Ayacucho Scholarship to study: Master Maintenance Engineering at the University of the Andes, Merida, Venezuela 1994 or OAS Scholarship ALBAN ESI (Organization of American States - European Community - Association for the Development of Maintenance Engineering - School of Industrial Engineering at the University of Sevilla) to study: Doctorate in Industrial Organization Engineering, University of Seville, Spain, 2004. o Award for best technical work in the 1st. World Congress of Maintenance Engineering, Bahia / Brazil, September 2005. Presentation / Publication: "Optimizing Maintenance Management process in the Venezuelan oil industry from the use of Reliability Engineering Methodologies".

Manuel Rodríguez Méndez is an Industrial Engineer and Honours Degree in Chemistry with a PhD in Industrial Engineering from the Castilla La Mancha University, Spain. Currently, he is partner of ESeyPro S.L. a consult company to improve the industrial processes, and he is an external teacher in several masters in the Santiago de Compostela University.

Victor Nenja is an associated professor of Computer Science at Sumy State University since 1995. He has written over 100 articles and conference papers and presentations. She has also been a Member of the Academic Council on award of a scientific degree of Ph.D. on the specialty "Hydraulic and pneumatic machines" at Sumy State University for over 20 years. He is one of three founders of the specialty "Information technologies of Design". His research interest is focused on the field of mathematical and computer modeling of technical systems and facilities applied aspects of GIS technology in the regional administration.

Gerardo Pagalday is graduated in Industrial Engineering in 1988 at the ETSII in San Sebastian. In addition, he holds the APICS certificate (1996). Since 1990 he has worked at IK4-IKERLAN as project manager, first in the Production Systems department till 2001. From 2001 till 2014 worked in the Strategical Innovation department. Since then belonges to the Advanced Manufacturing Technologies Area. He has led and taken part in several R&D projects funding by the Spanish/Basque Governments related to Collaborative Business Processes, Scheduling Systems, Production Planning, Analysis and Simulation of Production Systems, Inventory management, Innovation Management, Technology Roadmaps and New Business Models. He has also been involved in different international projects as participant.

Yuliia Parfenenko is an assistant lecturer of the Department of Computer Science at the Sumy State University, Sumy, Ukraine. She received her Master's Degree of Information Technology of Design (Diploma with distinction) in Sumy State University in 2008. From November 2008 till November 2011 she was a PhD student at the Sumy State University. Her research direction is creation information technologies for energy management in district heating. Additional information you can find on the personal webpage: http://itp.elit.sumdu.edu.ua/index.php/home/staff/38-parfenenko-yuliya-viktorivna. She can be contacted at: yuliya.parfenenko@gmail.com, skype: yuliya_parfenenko.

François Pérès is a French engineer who received his PhD from the University of Bordeaux and entered the Ecole Centrale of Paris as assistant professor. He is now full professor within the Toulouse Midi-Pyrénées Federal University, working inside the Laboratoire Génie de Production as leader of the Uncertainty, Risk and Decision group. His research deals with risk management applied to the design and maintenance strategies of manufactured products. Lately some extensions of his work have concerned human factor-based risk assessment approach.

Richard Ruitenburg is a Ph.D. candidate in the chair of Maintenance Engineering at the University of Twente. He holds a Research Master's degree in Management, Economics & Consumer Studies from the Wageningen University. In December 2013, he has started his Ph.D. research project at Liander N.V., the largest network provider in the Netherlands. The project aims to develop new methods and tools to assist Liander in its strategic Asset Life Cycle Management. The project is carried out in active cooperation with NedTrain, the company responsible for the maintenance of the trains used by the Netherlands Railways. His main research interests lie in the strategic aspects of Asset Management. Currently, he is involved in a research on agility in Asset Management.

Jose Mari Sarriegi has published papers about knowledge management, crisis management and resilience. He has been coordinador of several European research projects. He has been lecturing about information systems and modeling and simulation at Tecnun, University of Navarra.

Samir M. Shariff is currently working at Taibah University as a faculty member and he is serving as the Director of the Strategic Research Unit and Head of the MM Binladin Scientific Chair on Operation And Maintenance. He has chaired the EE department between 2006 till 2012. Dr. Shariff has graduated with his BS and MS in Electrical Engineering from Washington University – St. Louis MO in 1994 and 1996 respectively. Dr. Shariff has received his Ph.D. in Electrical Engineering from Wichita State University in 2000 with an honorary award of Phi Kappa Phi (Academic). After that Dr. Shariff has started his practical life by working as a project manager at Saudi Binladin Group (SBG) for the operation and maintenance of the Makkah Harm projects. In 2002 Dr. Shariff has joined Saudi Aramco as a project engineer working with the PMT Department. In 2004 Dr. Shariff moved to the academic life and worked at Taibha University where he served at various academic, research as well as management posts.

Vira Shendryk started her academic career since Sept, 2002 at the Sumy State University, Ukraine. She was a visiting scholar at McMaster University, Canada in 2012 and was a visiting research fellow at the Faculty of Technology and Society, the Malmo University, Sweden in 2013 and at the Faculty of Computer Science and Information Technology, the Riga Technical University, Latvia in 2014. Her research interest is focused on the field of Information Systems and Decision Science particularly in decision making under uncertainty. She has written over 100 journal articles and conference papers and presentations. She has also been a Member of the Editorial Review Board of international multi-disciplinary quarterly journal «Information Technology and Economics», which focuses on the intersection of Information Technology and Economics and was Member of Programme Committees the International Conference on Information and Software Technologies (ICIST 2014-2016, Lithuania) and Advanced Information Systems and Technologies (AIST 2012-2016, Ukraine).

A. Syamsundar obtained his Ph.D. in Reliability Engineering in 2009 from the Indian Institute of Technology, Kharagpur and is presently working as Head of the Research and Development Centre of an integrated steel works, Rashtriya Ispat Nigam Limited's Visakhapatnam Steel Plant, Visakhapatnam. He has worked in various capacities in maintenance and operations of Steel Melting Shops and later as a Technical advisor to the Heads of Maintenance and Works and Director (Operations) of the integrated steel works. His research interests are in the fields of reliability and maintenance of repairable/industrial systems. He is the co-author of a book and has published a number of papers in international journals and conferences on these issues.

Wieger Tiddens is a PhD candidate at the University of Twente and the Netherlands Defence Academy. He holds a Master's degree in Technology Management obtained at the University of Groningen. He has professional experience in the application of maintenance concepts and analyses within industry. His research interests include the support of informed maintenance decision making and the application of advanced maintenance techniques within companies. His research is part of the Tools4LCM project, funded by the Netherlands Ministry of Defence and the National Aerospace Laboratory NLR. The research is also part of the MaSeLMA project (Integrated Maintenance and Service Logistics Concepts for Maritime Assets), funded by Dinalog (Dutch Institute for Advanced Logistics).

Tiedo Tinga is a full professor in dynamics based maintenance at the University of Twente since 2012 and associate professor Maintenance Technology at the Netherlands Defence Academy since 2007. He received his PhD degree in mechanics of materials from Eindhoven University in 2009. He is now leading a number of research projects on developing predictive maintenance concepts, mainly based on physics of failure models, but also following data-driven approaches.

Jone Uribetxebarria has a Matrise of computing by the University Paul Sabatier in Toulouse (1993) and holds a graduate diploma in the specialty of robotics (DESS in IRR) (1994). In 2000 she obtained a Master in Production Management at the University of Mondragon. In 1995 she began working in the Technology Research Center IK4-IKERLAN in the knowledge area of Robotics. Since 1999 she is working in the area of Design and Production Technologies. Her main research interests are related to the product definition and design (functional design, modular design and configuration) and supply network management and simulation in mass customization scenarios.

Jannes van der Wal is an MSc in Supply Chain and Technology Operations Management with a BSc in Business Administration. Currently he is working with a large Dutch dairy company which operates globally. There he applies the knowledge obtained from his Masters and research into practice where possible.

Jasper Veldman is assistant professor at the Operations department, University of Groningen (Faculty of Economics and Business). His research interests are in the area of offshore wind, asset management and supply chain management.

Éric Villeneuve is researcher at the Production Management Laboratory (LGP) of the National Engineering School of Tarbes of the National Polytechnic Institute of Toulouse (ENIT-INPT). After obtaining an engineering degree in 2009, he graduated with a PhD degree in the field of decision-support systems in 2012. His successive research works led him to focus on issues related to the knowledge management, experience feedback, risk management and uncertain knowledge processing.

Patxi Zubizarreta has a degree in Mechanical Engineering by the University of Mondragon (1986). In 1999 obtains a Master in Production Management at the University of Mondragon. From 1987 he has been working in the Technology Research Center IK4-Ikerlan in different areas related to Design and Production Technologies, participating in several R & D projects for the European Union and the Spanish and Basque Governments, as well as in transfer projects to companies. His main research interests are related to the product/process/service definition and modeling in mass customization scenarios.

Index

F

G

H

I

K

L

M

N

O

P

Q

R

S

T

U

V

www.ingramcontent.com/pod-product-compliance
Lightning Source LLC
LaVergne TN
LVHW080621300926
849162LV00005B/1222